Gastrointestinal
MICROBIOLOGY

Gastrointestinal MICROBIOLOGY

edited by

Arthur C. Ouwehand

Danisco Innovation
Kantvik, Finland

Elaine E. Vaughan

Unilever R&D, Vlaardingen and
Wageningen University, Wageningen,
The Netherlands

CRC Press
Taylor & Francis Group
Boca Raton London New York

CRC Press is an imprint of the
Taylor & Francis Group, an **informa** business

CRC Press
Taylor & Francis Group
6000 Broken Sound Parkway NW, Suite 300
Boca Raton, FL 33487-2742

First issued in paperback 2019

© 2010 by Taylor & Francis Group, LLC
CRC Press is an imprint of Taylor & Francis Group, an Informa business

No claim to original U.S. Government works

ISBN-13: 978-0-8247-2641-6 (hbk)
ISBN-13: 978-0-367-39074-7 (pbk)

A CIP record for this book is available from the British Library.

Library of Congress Cataloging-in-Publication Data available on application

Visit the Taylor & Francis Web site at
http://www.taylorandfrancis.com

and the CRC Press Web site at
http://www.crcpress.com

We dedicate this book to Willem M. de Vos and Seppo Salminen, who contribute significantly to insight in gut microbe-host interactions, and moreover, as our colleagues and mentors in this field.

Preface

The human gastrointestinal tract microorganisms, termed the "microbiota," have been investigated since the beginning of microbiological studies, when Antonie van Leeuwenhoek, the father of microbiology, investigated the microorganisms in his own stools. The human microbiota comprises trillions of microbes distributed in various niches throughout the intestinal tract and is one of the most complex microbial ecosystems on earth. The host and its microbiota have co-evolved together, and considering the staggering numbers and diversity, it is therefore not surprising that the microbiota exert a major influence on the host. The original term for the microbiota upon discovery was the "flora" or "microflora," literally translated as "small plants," which has a botanical connotation. These terms are still used widely today and internationally recognized. Nevertheless, it is considered more appropriate to use the term microbiota, i.e., "small life" taking into account that the human microbiota is comprised of bacteria, archaea, bacteriophage, a smaller number of yeasts, and some protozoa; hence, this term is mainly used throughout this book. With this book, we have made an attempt to cover all issues associated with the gastrointestinal microbiota, from health to disease and from sampling to identification. Although various books have addressed the intestinal microbiota, this has mainly been from the perspective of disease or nutrition, while the microbiota itself has rarely been the focus. This current book aims to fill this gap and provide the reader with a comprehensive overview of all aspects related to the gastrointestinal microbiota. There have been major scientific advances especially in human intestinal microbiology in the recent past, which are also covered by the contributions.

Early studies were limited to description of the culturable microbes, which as we now realize, made up only a minority of the gastrointestinal tract microbiota. Due to the development of molecular biological techniques over the last decade, microbes can now be detected and studied to a large extent, without the need for culturing. In the first chapter, Kaouther Ben Amor and Elaine Vaughan review the major achievements of recent times in determining the diversity of the microbiota using modern molecular techniques, based on 16S ribosomal RNA, as well as methods to evaluate their activity within the various niches. Research of the gastrointestinal tract microbiota, especially in the case of humans, is often restricted to fecal material. In fact, a range of other sampling techniques are available, which are presented by Angèle Kerckhoffs and colleagues, to access the small intestine, as well as noninvasive sampling methods that are routinely used in medical practice. This is an important issue since feces represent only the luminal material of the terminal colon and will provide insufficient information about other locations of the gut. Anne McCartney and Glenn Gibson describe the succession of the microbiota in infants, as well as the earlier culturing studies, and the methodology to characterize the microbiota down to subspecies level. It has long been recognized that the intestinal microbiota plays an important role in maintaining health in infants. Currently, much attention is also focused on the intestinal microbiota of the elderly, as is discussed in the chapter by Fang He. In western nations, the elderly are becoming a more numerous segment of the population, and it is becoming increasingly established that intestinal health has a major role in their quality of life.

While establishing the microbiota diversity and their activity (live versus dead) is a major challenge, it is essential to know and understand their effects on the host. The intestinal microbiota has a major influence on the development and maintenance of our immune system as described by Marie-Christiane Moreau. Because of their direct contact with the host, the activity and interaction of the microbiota with the intestinal mucosa may be more important than the activity of microbes in the lumen, as described by Wai Ling Chow and Yuan-Kun Lee. The human microbiota also play a major role in our nutrition. Barry Goldin reviews the myriad of metabolic possibilities of the human microbiota concerning the metabolism of food ingredients and drugs we consume, as well as host-derived substrates. Max Bingham focuses on the metabolism by the microbiota of polyphenols, which are considered to be key active constituents of fruits and vegetables and responsible for many of the health protective effects of diets rich in these foods. Today, functional genomics technologies are developing and will facilitate our ability to detect the microbes and determine the molecular mechanisms of their impact on the host. Through the sequencing of an ever-increasing number of microbiota genomes, and elegant molecular studies, a further understanding is being obtained into the molecular functioning of the host-microbiota interactions, a dynamic area that is discussed by Peter Bron, Willem de Vos, and Michiel Kleerebezem.

The gastrointestinal tract microbiota is receiving more attention than ever in particular in relation to disease. Fergus Shanahan, Barbara Sheil, and coworkers review the relationship between the intestinal microbiota and inflammatory bowel diseases, as well as give an overview of the probiotic clinical trials and the potential mechanisms of probiotics for ameliorating these intestinal diseases. Through its metabolism, the intestinal microbiota is thought to play an important role in both the etiology and prevention of colorectal cancer, as discussed by Patricia Heavey, Ian Rowland, and Joseph Rafter. In addition to diseases of the gastrointestinal tract, Pirkka Kirjavainen and Gregor Reid also discuss that diseases such as allergy are being recognized to have an "intestinal component," again mediated through the interaction between the microbiota and the intestinal immune system.

In order to gain a better understanding of the composition and functioning of the intestinal microbiota and how this can be influenced, intestinal models have been developed; this allows for a simplification of the complex intestinal ecosystem as presented by Harri Mäkivuokko and Päivi Nurminen. Experimental animals, as described by Anders Henriksson, have also been highly valuable for this purpose, especially with the availability of various knockout animal models for disease. Also the use of animals with a "human" microbiota provide valuable models to investigate the influence of substances on the microbiota and host physiology. The best animal models to show the importance of the intestinal microbiota are germ-free animals. Their physiological differences compared to conventional animals are striking and show clearer than any other model the role intestinal microbes play, as discussed by Elisabeth Norin and Tore Midtvedt. Because of its influence on the health and well being of the host, strategies have been devised to alter the composition and/or activity of the intestinal microbiota. Antibiotics have long been known to alter the composition of the intestinal microbiota, as discussed by Åsa Sullivan and Carl Erik Nord, which may lead to various side effects, depending on the activity spectrum of the antibiotic. Methods to improve the activity and composition of the intestinal microbiota include probiotics, microbes ingested orally that provide beneficial effects, and prebiotics substrates that are selectively metabolized by the beneficial native gastrointestinal tract microbes, as discussed in the chapters by Chandraprakash Khedkar and Arthur Ouwehand, and Ross Crittenden and Martin Playne, respectively.

The major part of the book deals with the microbiota of humans, and when animals are studied, it is often as a model for humans. Minna Rinkinen describes the microbiota of companion animals, an area that has received very little attention to date, although the well being of pets can contribute significantly to the well being of the owner. In the case of farm animals, discussed by Alojz Bomba and colleagues, there is an important economic drive where the role of the microbiota on performance is a major focus. This will only become more important from 2006 onward as antimicrobial growth promoters will be prohibited in the European Union.

Gastrointestinal Microbiology is a vibrant field of research that is benefiting from many interdisciplinary interactions between different research groups in the world, that are using, developing, and applying novel technologies. Exciting initiatives are emerging with high through put technologies such as sequence analysis of the human microbiome (collective genomes of the gut microbiota) and metabolomics applied to microbiota and nutritional research. There is occasionally some overlap in information scattered throughout the book that is valuable since the reader will get an appreciation for the different opinions and perspectives that reflect the current state of research findings in the literature for this subject. It remains a highly complex task to understand the mutual relationship between members of the microbial community in the gut and their interaction with the host.

Finally, we hope that all readers will share our excitement for this dynamic subject that impacts on all our lives.

Arthur C. Ouwehand
Elaine E. Vaughan

Acknowledgments

We are most grateful to the contributing authors who have been willing to share their knowledge and experience in their field of intestinal microbiology. They are all busy researchers and yet they committed themselves to writing these chapters. It has been a pleasure to cooperate with these experts for the production of this book. Together their excellent contributions provide the state-of-the-art research on the human intestinal microbiology as well as informative chapters about the animal microbiota for comparative purposes. Elaine Vaughan acknowledges the staff and colleagues in Unilever Research and Development, and in the Laboratory of Microbiology, especially the Molecular Ecology Group, Wageningen University, for inspiring discussions on intestinal microbiology and critical support in this field. She further acknowledges the enjoyable and stimulating collaborations with the Wageningen Center of Food Sciences. Arthur Ouwehand similarly acknowledges the support and inspiration from the colleagues at Danisco Innovation and the Functional Foods Forum, University of Turku. We thank the Egerton Group Ltd. for their excellent support during production of this book. Importantly, our families, especially our spouses (Dr. Patrick Wouters and Anna-Maija Ouwehand), who have learned to live with Gut Microbiology, we give our heartfelt thanks.

Contents

Contributors

Louis M. A. Akkermans   Department of Surgery, Utrecht University Medical Center, Utrecht, The Netherlands

Kaouther Ben Amor   Laboratory of Microbiology, Wageningen University, Wageningen, The Netherlands

Malik M. Anwar   Alimentary Pharmabiotic Centre, Departments of Medicine and Surgery, Microbiology, National Food Biotechnology Centre, National University of Ireland, Cork, Ireland

Max Bingham   Unilever Research and Development, Vlaardingen, The Netherlands

Alojz Bomba   Institute of Gnotobiology and Prevention of Diseases in Young, University of Veterinary Medicine, Kosice, Slovak Republic

Peter A. Bron   Wageningen Centre for Food Sciences and NIZO Food Research, BA Ede, Wageningen, The Netherlands

Wai Ling Chow   National University of Singapore, Department of Microbiology, Faculty of Medicine, Singapore

Ross Crittenden   The Preventative Health Flagship, Food Science Australia, Werribee, Victoria, Australia

Willem M. de Vos   Wageningen Centre for Food Sciences and Laboratory of Microbiology, Wageningen University, Wageningen, The Netherlands

Soňa Gancarčíková   Institute of Gnotobiology and Prevention of Diseases in Young, University of Veterinary Medicine, Kosice, Slovak Republic

Glenn R. Gibson   Food Microbial Sciences Unit, School of Food Biosciences, Whiteknights, University of Reading, Reading, U.K.

Barry R. Goldin   Department of Public Health and Family Medicine, Tufts University School of Medicine, Boston, Massachusetts, U.S.A.

Fang He   Technical Research Laboratory, Takanashi Milk Products Co., Ltd, Yokohama, Kanagawa, Japan

Patricia M. Heavey   School of Life Sciences, Kingston University, Kingston-upon-Thames, U.K.

Gerard P. van Berge Henegouwen   Department of Gastroenterology, Utrecht University Medical Center, Utrecht, The Netherlands

Anders Henriksson   DSM Food Specialties, Sydney, Australia

Zuzana Jonecová   Institute of Gnotobiology and Prevention of Diseases in Young, University of Veterinary Medicine, Kosice, Slovak Republic

Angèle P. M. Kerckhoffs   Department of Gastroenterology, Utrecht University Medical Center, Utrecht, The Netherlands

Chandraprakash D. Khedkar   Department of Dairy Microbiology and Biotechnology (Maharashtra Animal and Fishery Sciences University, Nagpur), College of Dairy Technology, Warud (Pusad), India

Pirkka V. Kirjavainen   Canadian Research and Development Center for Probiotics, The Lawson Health Research Institute, London, Ontario, Canada

Michiel Kleerebezem   Wageningen Centre for Food Sciences and NIZO Food Research, BA Ede, Wageningen, The Netherlands

Yuan-Kun Lee   National University of Singapore, Department of Microbiology, Faculty of Medicine, Singapore

Harri Mäkivuokko   Danisco Innovation, Kantvik, Finland

Jane McCarthy   Alimentary Pharmabiotic Centre, Departments of Medicine and Surgery, Microbiology, National Food Biotechnology Centre, National University of Ireland, Cork, Ireland

Anne L. McCartney   Food Microbial Sciences Unit, School of Food Biosciences, Whiteknights, University of Reading, Reading, U.K.

Tore Midtvedt   Microbiology and Tumor Biology Center, Karolinska Institutet, Stockholm, Sweden

Marie-Christiane Moreau   French Institute of Agronomical Research (INRA), Nancy, France

Radomíra Nemcová   Institute of Gnotobiology and Prevention of Diseases in Young, University of Veterinary Medicine, Kosice, Slovak Republic

Vincent B. Nieuwenhuijs   Department of Surgery, Utrecht University Medical Center, Utrecht, The Netherlands

Carl Erik Nord   Department of Laboratory Medicine, Karolinska University Hospital, Karolinska Institutet, Stockholm, Sweden

Elisabeth Norin   Microbiology and Tumor Biology Center, Karolinska Institutet, Stockholm, Sweden

Päivi Nurminen   Danisco Innovation, Kantvik, Finland

Liam O'Mahony   Alimentary Pharmabiotic Centre, Departments of Medicine and Surgery, Microbiology, National Food Biotechnology Centre, National University of Ireland, Cork, Ireland

Arthur C. Ouwehand   Danisco Innovation, Kantvik, and Functional Foods Forum, University of Turku, Turku, Finland

Martin J. Playne   Melbourne Biotechnology, Hampton, Victoria, Australia

Joseph J. Rafter   Department of Medical Nutrition, Novum, Huddinge University Hospital, Karolinska Institutet, Stockholm, Sweden

Gregor Reid   Canadian Research and Development Center for Probiotics, The Lawson Health Research Institute, London, Ontario, Canada

Minna Rinkinen   Department of Clinical Veterinary Sciences, Faculty of Veterinary Medicine, University of Helsinki, Helsinki, Finland

Ian R. Rowland   Northern Ireland Center for Food and Health, University of Ulster, Coleraine, Northern Ireland, U.K.

Melvin Samsom   Department of Gastroenterology, Utrecht University Medical Center, Utrecht, The Netherlands

Fergus Shanahan   Alimentary Pharmabiotic Centre, Departments of Medicine and Surgery, National University of Ireland, Cork, Ireland

Barbara Sheil   Alimentary Pharmabiotic Centre, Departments of Medicine and Surgery, Microbiology, National Food Biotechnology Centre, National University of Ireland, Cork, Ireland

Åsa Sullivan   Department of Laboratory Medicine, Karolinska University Hospital, Karolinska Institutet, Stockholm, Sweden

Elaine E. Vaughan   Unilever Research and Development, Vlaardingen, and Laboratory of Microbiology, Wageningen University, Wageningen, The Netherlands

Maarten R. Visser   Department of Microbiology, Utrecht University Medical Center, Utrecht, The Netherlands

1
Molecular Ecology of the Human Intestinal Microbiota

Kaouther Ben Amor
Laboratory of Microbiology, Wageningen University, Wageningen, The Netherlands

Elaine E. Vaughan
Unilever Research and Development, Vlaardingen, and Laboratory of Microbiology, Wageningen University, Wageningen, The Netherlands

INTRODUCTION

The human gastrointestinal (GI) tract is the home of a huge microbial assemblage, the microbiota, the vast extent of which is only now being revealed. The number of micro organisms within the intestine greatly exceeds human cells, resulting in one of the most diverse and dynamic microbial ecosystems. Relationships among the microbes, and between the microbiota and the host, have a profound influence on all concerned (1,2). The GI-tract offers various niches with nutrients, those ingested and generated by the host, and a relatively non-hostile environment to the microbes. The microbiota play essential roles in a wide variety of nutritional, developmental, and immunological processes and therefore significantly contribute to the well being of the host (3–6). During the last decade, specific bacterial isolates, termed "probiotics," have been extensively used in an attempt to modulate the intestinal microbiota to benefit the host. Today, there is persuasive evidence for probiotics in prevention or treatment of a number of intestinal disorders in humans, especially for reducing bouts of diarrhea and providing relief for lactose intolerant individuals (7,8). In order to rationally use probiotics, prebiotics or other functional foods as therapeutic agents, in-depth knowledge of the structure, dynamics, and function of the bacterial populations of the GI-tract microbiota is crucial.

Studying the microbial ecology in the intestine involves determining the abundance and diversity of the microorganisms present, their activity within this niche, and their interactions with each other and their host (symbiosis, commensalism and pathogenicity). Although the human intestinal microbiota have been extensively investigated by culture-based methods more than any other natural ecosystem (9–11), our knowledge about the culturable fraction of this community is limited. This is essentially due to the challenges of obtaining pure cultures of intestinal inhabitants, which are hindered by the largely anaerobic nature of this community, and the paucity of suitable enrichment strategies to simulate intestinal conditions. The advent of molecular techniques based on 16S ribosomal RNA (rRNA) gene analysis is now allowing a more complete assessment of this complex microbial ecosystem by unraveling the extent of the diversity, abundance and population dynamics of this community (12,13). These techniques have extended our view of those microorganisms that have proven difficult to culture and which play an important role in gut physiology. This huge intestinal microbial reservoir is estimated to contain 1000 bacterial species and as much as 10^{14} cells (1,14). Besides studying the diversity, it is essential to identify these microbes based upon their eco-physiological traits, i.e., those that are functionally active versus those that are effectively redundant and play little or no role at a particular time or at a given site of the intestinal tract. The latter requires the development of approaches that monitor the activity of these microorganisms at the single cell level in their natural habitat. This chapter initially reviews molecular techniques to study the diversity of the microbiota, and subsequently highlights newly developed molecular methods to study the eco-physiology of the GI-tract.

GI-TRACT MICROBIOTA AS IDENTIFIED BY 16S rRNA GENE ANALYSIS

The human GI-tract microbiota comprise bacteria, archaea and eukarya. It is by far the bacteria that dominate and reach the highest cell density documented for any microbial ecosystem (1). The comparative analysis of environmentally retrieved nucleic acid sequences, most notably of rRNA molecules and the genes encoding them, has become the standard over the last decade for cultivation-independent assessment of bacterial diversity in environmental samples (Fig. 1) (15,16). The 16S rRNA gene comprises highly variable to highly conserved regions, and the differences in sequence are used to distinguish bacteria at different levels from species to domain and determine phylogenetic relationships. rRNA gene fragments are today routinely retrieved without prior cultivation of the microbes by constructing 16S ribosomal DNA (rDNA) libraries. The procedure is based upon polymerase chain reaction (PCR)-mediated amplification of 16S rRNA genes or gene fragments, isolated from the environmental sample, followed by segregation of individual gene copies by cloning into *Escherichia coli*. In this way a library of community 16S rRNA genes is generated, the composition of which can be estimated by screening clones, full or partial sequence analysis, and comparing them with adequate appropriate reference sequences in databases to infer their phylogenetic affiliation. Large databases of 16S rRNA gene sequence information ($>200,000$ sequences) for described as well as uncultured microorganisms are available, which provide a high-resolution platform for the assignment of those new sequences obtained in 16S rDNA libraries. Databases harboring 16S rRNA sequences include the ARB software package (17), the Ribosomal Database project (http://rdp.cme.msu.edu/index.jsp) (18) and EMBL (www.embl-heidelberg.de/).

Sequencing of 16S rDNA clone libraries generated from various sites of the GI-tract including terminal ileum, colon, mucosa and feces have confirmed that relevant fractions of gut bacteria were derived from new, as yet undescribed bacterial phylotypes (19–23). Clearly the biases of culturing studies in the 1960s and 1970s such as incomplete knowledge of culture conditions and selectivity had prejudiced the outcome. The new molecular studies revealed that the vast majority of rDNA amplicons generated directly from fecal or biopsy samples of adults, originated from the phyla of the *Firmicutes* (including the large class of *Clostridia* and the lactic acid bacteria), *Bacteroidetes*, *Actinobacteria* (including *Atopobium* and *Bifidobacterium* spp.) and *Proteobacteria*

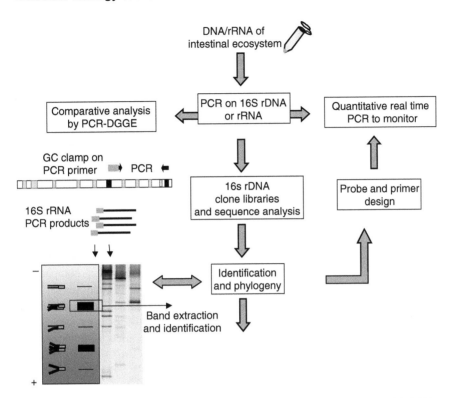

Figure 1 PCR-based approaches to monitor the GI-tract microbiota. The 16S rDNA or rRNA isolated from a GI-tract sample may be amplified by (reverse transcriptase-) PCR using primers that target all or some bacteria. The amplicons may be cloned and sequenced in order to identify the bacteria present in the sample. The 16S rRNA gene comprises highly variable to highly conserved regions, and the differences in sequence are used to determine phylogenetic relationships and distinguish bacteria at different levels from species to domain. The DGGE technique is based on 16S rRNA sequence-specific melting behavior of the PCR products, generated with primers one of which contains a 40-bp GC clamp. Statistical software enables the calculation of similarity indices and cluster analysis to compare the samples. The 16S rRNA sequences may also be used to design new primers specific for bacterial groups or species in order to quantify them in samples by real time PCR. *Abbreviations*: DGGE, denaturing gradient gel electrophoresis; DNA, deoxyribonucleic acid; GC, guanine cytosin; PCR, polymerase chain reaction; rDNA, ribosomal deoxyribonucleic acid; rRNA, ribosomal ribonucleic acid.

(including *Escherichia coli*). The large class of *Clostridia* comprises the *Clostridium coccoides-Eubacterium rectale* group, and the *Clostridium leptum* group consists of *Ruminococcus* species and *Faecalibacterium prausnitzii*. These analyses indicated that the adult intestinal microbiota constitutes a majority of low and high G+C content Gram-positive bacteria. The latter has been indirectly confirmed by analysis of the metagenome of bacterial viruses recovered from fecal samples that revealed predominantly viral sequences with similarity to genomes of bacteriophages specific for Gram-positive bacteria (24). In fact, this bacterial diversity at the division level relative to other microbial ecosystems is quite low, mainly deriving from the divisions Firmicutes and the Cytophaga-Flavobacterium-Bacteroides (9,19).

Interestingly, molecular inventories based on 16S rDNA clone libraries of microbial communities in inflammatory bowel disease (IBD) patients differed from healthy subjects (25). In several Crohn's disease (CD) patients numerous clones were isolated belonging to phylogenetic groups that are commonly not dominant in adult fecal microbiota of healthy persons, while *Bacteroides vulgatus* was the only molecular species shared by all patients, and *E. coli* clones were also detected unlike in healthy persons (25). In another study, 16S rDNA libraries generated from mucosa-associated microbiota of patients with IBD revealed a reduction in diversity due to a loss of normal anaerobic bacteria, especially those belonging to the *Bacteroides*, *Eubacterium* and *Lactobacillus* species. Most of the sequenced clones retrieved (70%) were assigned to known intestinal bacteria, but a significant number of the cloned sequences were affiliated to normal residents of the oral mucosa such as *Streptococcus* species (26). It was suggested that alteration of the microbiota in mucosal inflammation reflects a metabolic imbalance of the complex microbial ecosystem with severe consequences for the mucosal barrier rather than disrupted defense to single microorganisms (26).

Even though sequencing of cloned 16S rDNA amplicons provides relevant information about the identity of uncultured bacteria, the data are not quantitative. Moreover, PCR and cloning steps are not without bias (27): a recent comparative analysis of clone libraries from a fecal sample pointed out that the number of PCR cycles may affect the diversity of the amplified 16S rDNAs and thus should be minimized (28). More rapid culture-independent options to the cloning procedures include exploring of the complex microbial populations using a variety of fingerprinting methods. See Table 1 for an overview of some current methods used to investigate the intestinal microbiota.

Table 1   Potential and Limitations of Various Methods for Investigating the Diversity of the Human Intestinal Microbiota

Method	Application	Comments
Culturing	Isolation of pure cultures, enumeration	Not representative for microbiota; insufficient selective media; time consuming
16S rRNA gene libraries and sequencing	Identification and phylogeny	Large scale cloning is laborious; primer bias can be an issue
Dot-blot hybridization	Detection, quantification and activity	Gives information about activity of microbiota; of rRNA; comprehensive set of probes published
FISH	Single cell detection and enumeration	High throughput with image analysis software and flow cytometry; requires probe design; comprehensive set of probes published
PCR-DGGE/TGGE	Rapid profiling of total microbiota	Detection of specific groups possible; semi-quantitative identification by band extraction and sequencing
T-RFLP	Rapid profiling of total microbiota	Identification by cloning and sequencing; bank of T-RF under construction
Quantitative real time PCR	Detection and quantification	Requires probe/primers design; very high throughput

Abbreviations: DGGE, denaturing gradient gel electrophoresis; FISH, fluorescent in situ hybridization; PCR, polymerase chain reaction; rRNA, ribosomal ribonucleic acid; TGGE, temperature gradient gel electrophoresis; T-RF, terminal restriction fragment; T-RFLP, terminal restriction fragment length polymorphism.

FINGERPRINTING REVEALS CHARACTERISTICS OF THE MICROBIOTA

PCR-Denaturing Gradient Gel Electrophoresis

The most commonly applied fingerprinting methods used to study the GI-tract microbiota are denaturing and temperature gradient gel electrophoresis (DGGE and TGGE, respectively) of PCR-amplified genes coding for 16S rRNA (Fig. 1) (12,23). Other techniques such as terminal restriction fragment length polymorphism (T-RFLP) and single strand conformation polymorphism (SSCP) analysis are being applied but less frequently (26,29). The common principle of these methods is based on the separation of PCR-amplified segments of 16S rRNA genes of the same length, but with different sequence to visualize the diversity within the PCR amplicons by a banding pattern. One of the PCR primers has a 40-bp GC clamp to hold the DNA strands of the PCR product or amplicon together. With DGGE/TGGE, separation is based on the decreased electrophoretic mobility of partially melted double-stranded DNA molecules in polyacrylamide gels containing a linear gradient of DNA denaturants (a mixture of formamide and urea) or a linear temperature gradient, respectively. As a result mixed amplified PCR products will form a banding pattern after staining that reflects the different melting behaviors of the various sequences (30,31). Subsequent identification of specific bacterial groups or species present in the sample can be achieved either by cloning and sequencing of the excised bands or by hybridization of the profile using phylogenetic probes (30). Furthermore, complementation of the fingerprinting results with statistical analysis provides additional information of the observed diversity by highlighting some putative correlation between different sets of variables (32).

Since its application to study the intestinal microbiota, PCR-DGGE/-TGGE fingerprinting has advanced our knowledge of the intestinal microbiota by unraveling the complexity of this ecosystem and providing insight in the establishment and succession of the bacterial community within the host (23,33). The succession of the microbiota in the feces of infants over the first year of life has been visualized using DGGE profiles of the total microbial community, which showed the relatively simple and unstable infant fecal ecosystem (31). In healthy adults, the predominant fecal microbiota was shown to be complex, host-specific and remarkably stable in time (23,34,35). DGGE profiles for monozygotic twins were significantly more similar than for unrelated individuals, while marital partners showed less similar profiles than twins, indicating the influence of genotype over dietary or environmental factors (35). DGGE profiles also revealed that the predominant bacterial species associated with the colonic mucosa are uniformly distributed along the colon, but significantly different from the predominant fecal community (36,37).

Under certain environmental circumstances and/or in genetically susceptible individuals, there is clear evidence that the GI-tract microbiota may play a role in the pathogenesis and etiology of a number of inflammatory diseases such as ulcerative colitis (UC), and CD (30,38,39). Using DGGE, TGGE and SSCP fingerprinting analyses, it was demonstrated that fecal and mucosal-associated microbiota of patients with UC and CD is altered, less complex, and also unstable over time as compared to matched healthy people (26,40,41). In subjects with irritable bowel syndrome (IBS), higher temporal instability was also seen in comparison to healthy persons, but this was likely influenced by antibiotics used during the study (42).

Group-Specific PCR-DGGE

Bands originating from lactobacilli in fecal samples could not be detected on the DGGE profiles since they represent less than 1% of the community, which is approximately the detection limit of this method (43,44). The dominant fecal microbiota of adults as assessed by DGGE was not significantly altered following consumption of certain probiotic strains (34,43). Although DGGE or TGGE were initially developed for total ecosystem communities, the sensitivity of the method for detecting specific groups that are present in lower numbers in the GI-tract such as bifidobacteria and especially lactobacilli has been considerably enhanced by using group- or genus-specific primers (34,45–47). Consequently, it was possible to monitor the effect of the administration of prebiotics and/or probiotics on the composition of indigenous bifidobacterial species, and to track the probiotic strain itself (46). In the latter case, DGGE profiles showed that the simultaneous administration of the prebiotic and probiotic (synbiotic approach) did not improve the colonization of the probiotic strain in the gut of the tested individuals. In another study, the DGGE profiles generated from fecal samples of healthy individuals fed a probiotic strain *Lactobacillus paracasei* F19, allowed the tracing of the probiotic and supported its presence as autochthonous within the intestinal community of a number of individuals (45). A nested PCR-DGGE approach has been developed to determine the diversity of sulfate-reducing bacteria (SRB) in complex microbial communities (48). SRB have been implicated in the pathogenesis of IBD, and consequently are an interesting population to investigate.

Recently an approach combining GC fractionation with DGGE (GC-DGGE) effectively reduced the complexity of the community DNA mixture being analyzed such that the total diversity within each fraction could be more effectively assessed (49). Thus, initially the total DNA of the complex community was fractionated using buoyant density gradient centrifugation based on the % G+C content, using bisbenzimidazole which preferentially binds to A+T rich regions (50). This fractionation based on G+C content effectively reduced the complexity of the community DNA mixture being analyzed and the total diversity within each fraction could be more effectively assessed by the subsequent DGGE.

Terminal-Restriction Fragment Length Polymorphism

Another community fingerprinting technique which is gaining in popularity is T-RFLP (51). The basis is a PCR reaction for the 16S rRNA gene in the complex community followed by restriction enzyme digestion that generates the terminal restriction fragments (T-RFs). The latter are separated by electrophoresis or by using a capillary electrophoresis sequencer, which is more high throughput and reproducible (52), to produce a fingerprint. The technique has been used in several studies, including characterizing the human fecal bifidobacteria, as well as the tracking of probiotic *Lactobacillus* strains, and monitoring antibiotic-induced alterations in intestinal samples (53,54). Further improvements in this technique include the application of new primer-enzyme combinations for specifically bacterial populations in human feces (29). Furthermore, a novel phylogenetic assignment database for the specific T-RFLP analysis of human fecal microbiota (PAD-HCM) has been designed, which enables a high-level prediction of the terminal-restriction fragments at the species level (55). This will facilitate the use of this technique in studies on the microbiota.

While the application of 16S rDNA-based fingerprinting methods are particularly well suited for examining time series and population dynamics, a more quantitative approach is useful to complement our knowledge about the composition and structure of this complex intestinal ecosystem.

16S rRNA-TARGETED PROBES QUANTIFY THE GI-TRACT MICROBIOTA

Hybridization with rRNA-targeted oligonucleotide probes has become the method of choice for the direct cultivation-independent identification of individual bacterial cells in natural samples. During the last decade, this technique has extended our view of bacterial assemblages and the population dynamics of complex microbial communities (15,56,57). The most commonly used biomarker for hybridization techniques, whether dot-blot or fluorescent in situ hybridization (FISH), is the 16S rRNA molecule because of its genetic stability, domain structure with conserved and variable regions, and high copy number. Highly conserved stretches may thus be used to design domain-specific probes such as EUB338/EUBII /EUBIII which collectively target most of the bacteria, whereas specific probes for each taxonomic level, between bacterial and archaeal, down to genus-specific and species-specific, can be designed according to the highly variable regions of the 16S rRNA (15,58–60). The increasing availability of 16S rRNA sequences has contributed significantly to the development of the hybridization methods and their application in different microbial ecosystems. Unquestionably, the success of the implementation of 16S rRNA hybridization strategies depends on different factors, among them rational design and validation of newly designed rRNA-targeted probes.

Probe Design and Validation

There is an online resource for oligonucleotide probes, called probeBase (142), which contains published FISH rRNA-targeted probes as well as recommended conditions of use, and many probes for dominant or interesting microbiota groups are described here (61). When designing new probes, one must consider specificity, sensitivity and accessibility to the target sequence. Nucleic acid probes can be designed to specifically target taxonomic groups at different levels of specificity (from species to domain) by virtue of variable evolutionary conservation of the rRNA molecules. The probes are typically 15–25 nucleotides in length. Appropriate software such as the ARB software package (17) and availability of large databases (http://rdp.cme.msu.edu/html/) are useful tools for rapid probe design and *in silico* specificity profiling. Additional experimental evaluation of the probes with target and non-target microorganisms is necessary to ensure the specificity and the sensitivity of the newly designed probe. It is important to notice that the validation of a newly designed probe requires different procedures for the dot blot (62) and FISH format (60). Moreover, the hybridization and washing conditions (temperature, salt concentration and detergent) are also crucial for obtaining a detectable probe signal (63). The accessibility of the probe to its target site is another factor to be considered when designing new probes. The accessibility of probe target sites on the 16S and 23S rRNA of *Escherichia coli* has been mapped systematically by flow cytometry (FCM) and FISH, and it was shown that probe-conferred signal intensities vary greatly among different targets sites (64,65). More recently, it was demonstrated that accessibility patterns of 16S rRNA's are more similar for phylogenetically related organisms; these findings may be the first description of consensus probe accessibility maps for prokaryotes (66).

Hybridization Techniques

Nucleic acid probing of complex communities comprises two major techniques: dot blot hybridization and FISH. In the dot blot format, total DNA or RNA is extracted from the sample and is immobilized on a membrane together with a series of RNA from reference strains. Subsequently, the membrane is hybridized with a radioactively labeled probe and

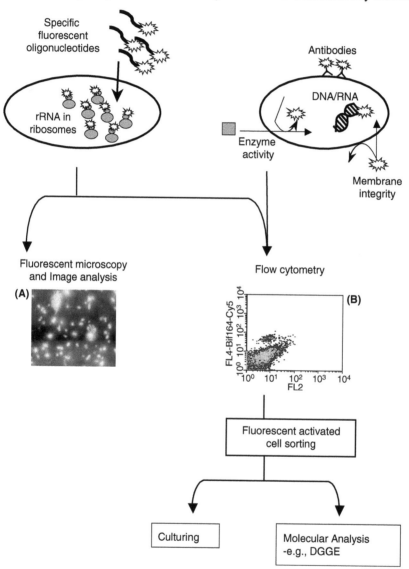

Figure 2 FISH involves whole cell hybridization with fluorescent oligonucleotide probes targeted against specific bacterial groups and species (*left-hand scheme*). The fluorescent probe hybridized cells may be visualized and/or counted using fluorescent microscopy and image analysis. The right-hand scheme illustrates how the viability of the cells may be assessed using functional probes that can also be visualized by fluorescent microscopy (**A**). FISH-labeled or functional probe-labeled cells may also be detected and enumerated using the flow cytometer (FCM). (**B**) shows a dot blot of fecal cells that were hybridized with a *Bifidobacterium*-specific probe. Following FCM the cells can be sorted according to the functional properties based on the probe stains, and subjected to further analysis. *Abbreviation*: DGGE, denaturing gradient gel electrophoresis.

after a stringent washing step the amount of target rRNA is quantified. The membrane can be rehybridized with a general bacterial probe, and the amount of population-specific rRNA detected with the specific probe is expressed as a fraction of the total bacterial RNA. Quantification of the absolute and relative (as compared to total rRNA) amounts of a specific rRNA reflects the abundance of the target population. Consequently this technique does not represent a direct measure of cell number since cellular rRNA content varies with the current environmental conditions and the physiological activity of the cells at the time of sampling (67). Dot-blot hybridization has been successfully used to quantify rRNA from human fecal and cecal samples (68,69). It was found that strict anaerobic bacterial populations represented by the *Bacteroides*, *Clostridium leptum* and *Clostridium coccoides* groups were significantly lower in the cecum (right colon) than in the feces, while the *Lactobacillus* group was significantly higher in the feces than in the cecum (68).

In contrast to dot-blot hybridization, FISH is applied to morphologically intact cells and thus provides a quantitative measure of the target organism without the limitation of culture-dependent methods (Fig. 2) (15,70). Following fixation, bacteria from any given sample can be hybridized with an appropriate probe or set of probes. The fixation allows permeabilization of the cell membrane and thus facilitates the accessibility of the fluorescent probes to the target sequence. For some Gram-positive bacteria, especially lactobacilli, additional pre-treatments including the use of cell wall lytic enzymes e.g., lysozyme, mutanolysin, protease K or a mixture is needed (71–73). Prior to hybridization, the cells can be either immobilized on gelatine-treated glass slides or simply kept in suspension when analyzed by FCM. The oligonucleotide probe is labeled covalently at the $5'$ end with a fluorescent dye, such as fluorescein iso(thio)cyanate, while any necessary competitor probes are unlabeled. The stringency, i.e., conditions of hybridization that increase the specificity of binding between the probe and its target sequence, can be adjusted by varying either the hybridization temperature or formamide concentration. Under highly stringent conditions oligonucleotide probes can discriminate closely related target sites. Post-hybridization stringency can be achieved by lowering the salt concentration in the washing buffer in order to remove unbound probe and avoid unspecific binding.

Quantification of FISH Signals

Over the past years, significant methodological improvements of the probe fluorescent-conferred signal have been reported. These include the use of brighter fluorochromes including Cy3 and Cy5 (74,75), and unlabeled helper oligonucleotide probes (76) that bind adjacent to and increase the accessibility of the selected target site. Horseradish peroxidase labeled probes and tyramide signal amplification (also termed CARD-FISH) can be used to significantly enhance the signal intensity of hybridized cells (77). However, the latter requires effective permeabilization for the large enzyme-probe complex to enter the cell with the risk of damaging and lysing fixed cells. A further possibility is the use of peptide nucleic acid (PNA) probes which can confer very bright signals to the cell (78,79). However, currently PNA probes are rather expensive and previously published oligonucleotide probes cannot be simply translated into PNA probes.

Epifluorescence microscopy is the standard method by which fluorescent-stained cells are enumerated; however, the method is time consuming and subjective (56,57). This technique has been improved by development of automated image acquisition and analysis software allowing accurate microscopic enumeration of fecal bacterial cells (73). Alternatively, FCM offers a potential platform for high-resolution, high throughput identification and enumeration of microorganisms using fluorescent rRNA-targeted oligonucleotides with the possibility of cell sorting (40,80–84).

An FCM method for direct detection of the anaerobic bacteria in human feces was first described over a decade ago (85). A membrane-impermeable nucleic acid dye propidium iodide (PI) was used in combination with the intrinsic scatter parameters of the cells to discriminate fecal cells from large particles. Coupling FCM results and image analysis, the authors showed that most of the particles detected with a large forward scatter value corresponded to aggregates most likely representing mucus fragments and undigested dietary compounds. They confirmed by means of cell sorting that the PI-stained cells (fecal cells) corresponded to a 2-D surface area of $<1.5\ \mu m^2$ while the unstained particles (aggregates) were around $5.0\ \mu m^2$ (85). The work highlighted the potential of FCM to study anaerobic fecal bacteria without culturing. Despite this valuable work and to quote from Shapiro "the subject matter may stink, but the method is superb" (86), the application of FCM to study the intestinal microbiota is still in development.

FISH-FCM was applied to detect and accurately quantify both fecal and mucosa-associated bacteria, and statistical analysis showed a high correlation between the FCM counts and microscopic counts (Fig. 2) (37,44,84). Using FCM, several thousands of cells can be counted accurately in a few seconds. Following the hybridization step, fecal cells are stained with a nucleic acid dye, for example PI, SYTO BC, and TOTO-1, to detect the total cells and subsequently spiked with standard beads of known size and concentration. The beads are thus used as an internal standard to calibrate the measured volume and to determine the absolute count of the probe-detected cells (40,87). In addition to the determination of the absolute cell counts, the fluorescence intensity signal can also be quantified using fluorescent beads with known fluorescent intensities (86). This is of major importance for determining optimal hybridization conditions for newly designed probes (37,82,88). FCM is becoming a popular method for high-resolution, high throughput identification of microorganisms using fluorescent rRNA-targeted oligonucleotides.

Application of FISH to Study the GI-Tract Ecosystem

During the last five years, hybridization studies with rRNA-targeted probes have provided significant knowledge about the composition and structure of the gut microbiota. A large panel of oligonucleotide probes specific for various genera predominant in the GI tract have been designed and validated (Table 2), and have been used intensively in these studies.

The uniqueness and complexity of the human gut microbiota revealed by finger-printing techniques were supported by results of analysis using nucleic-acid probe-based methods. These studies revealed that the majority of fecal bacteria belong to the *Clostridium coccoides-Eubacterium rectale* group and the *Clostridium leptum* group ($\sim$20–30% each), Bacteroides ($\sim$10%), *Atopobium* and bifidobacteria groups in that order of abundance (81,89,91,96,97). The *Clostridium coccoides-Eubacterium rectale* probe (Erec482) (Table 2) covers *Eubacterium hallii*, *Lachnospira* and *Ruminococcus* members, while the *Clostridium leptum* group comprises members of *Ruminococcus* species and *Faecalibacterium prausnitzii* (89,98). In particular members of *C. coccoides-E. rectale*, *C. leptum*, and the *Bacteroides* groups constituted more than half of the fecal microbiota. *Atopobium* and bifidobacteria groups comprised typically 4–5% each. The *Lactobacillus-Enterococcus* group, Enterobacteriaceae, *Phascolarctobacterium* and relatives, and *Veillonella* were less dominant (0.1 to a few percent) (90,91). However, differences in the occurrence of these bacterial groups have been reported by different research groups. These deviations may be due to the different methods or probes used, but it is also likely that the observed variance is due to the differences in the genetic background, lifestyle,

Table 2 FISH Probes Used to Study the Gastrointestinal Microbiota

Probe	Probe sequence (5″–3″)	Target organism	% Formamide	Reference
Eub338	GCTGCCTCCCGTAGGAGT	Most bacteria	0–80	(58)
EubII	GCAGCCACCCGTAGGTGT	*Planctomycetes*	0–60	(60)
EubIII	GCTGCCACCCGTAGGTGT	*Verrucomicrobia*	0–60	(60)
Bac303	CCAATGTGGGGGACCTT	*Bacteroides/ Prevotella*	0	(59)
Bdis656	CCGCCTGCCTCAAACATA	*Bacteroides distasonis*	0	(89)
Bfra602	GAGCCGCAAACTTTCACAA	*Bacteroides fragilis*	30	(89)
Bvulg1017	AGATGCCTTGCGGCT-TACGGC	*Bacteroides vulgatus*	30	(82)
Bfrag998	GTTTCCACATCATTCCACTG	*Bacteroides fragilis*	30	(83)
Bdist1025	CGCAAACGGCTATTGGTAG	*Bacteroides distasonis*	30	(68)
Erec482	GCTTCTTAGTCAR^{a}GTACCG	*Clostridium coccoides* group	0	(89)
Clep866	GGTGGATWACTTATTGTG	*Clostridium leptum* group	30	(90)
Rfla729	AAAGCCCAGTAAGCCGCC	*Ruminococcus flavefaciens*	20	(91)
Rbro730	TAAAGCCCAGY^{a}AGGCCGC	*Ruminococcus bromii*		(91)
Rcal733	CAGTAAAGGCCCAG-TAAGCC	*Ruminococcus callidus*	30	(90)
Elgc01	GGGACGTTGTTTCTGAGT	*Clostridium leptum* subgroup	0	(89)
Fprau645	CCTCTGCACTACTCAA-GAAAA	*Faecalibacterium prausnitzii*	15	(92)
Bif164	CATCCGGCATTACCACCC	Bifidobacteria	0	(93)
Ato291	GGTCGGTCTCTCAACCC	*Atopobium* group	0	(94)
Veil223	AGACGCAATCCCCTCCTT	*Veillonella*	0	(91)
Ecyl387	CGCGGCATTGCTGCTTCA	*Eubacterium cylindroides*	20	(91)
Cvir1414	GGGTGTTCCCGRCTCTCA	*Clostridium viride*	30	(90)
Edes635	AGACCARCAGTTTGAAA	*Eubacterium desmolans*	30	(90)
Lach571	GCCACCTACACTCCCTTT	*Lachnospira* group	40	(91)
Ehal1469	CCAGTTACCGGCTCCACC	*Eubacterium halii* group	20	(91)
Phasco741	TCAGCGTCAGACACAGTC	*Phascolarctobacte rium* group	0	(91)
Enter1432	CTTTTGCAACCCACT	Enteric group	30	(68,69)
Strc498	GTTAGCCGTCCCTTTCTGG	*Lactococcus lactis ssp. lactis*	30	(89,90)
Lab158	GGTATTAGCAY^{a}CTGT TTCCA	*Lactobacillus/ Enterococcus*	0	(95)
Urobe63	AATAAAGTAATTCCCGTTCG	Uncultured *Ruminococcus obeum*-like bacteria	20	(84)
Urobeb	AAARAARTATTTCCCGTTCG			
Non338	ACATCCTACGGGAGGC	Negative control		(83)

^{a} N, R, W, and Y are the International Union of Pure and Applied Chemistry codes for ambiguous bases.

and diet in the human populations studied. Two large studies, where an extensive array of oligonucleotide probes that targeted the major bacterial groups in the GI-tract of northern European adults was used, showed that 62–75% of the fecal bacteria could be detected and identified. The remainder ($\approx$ 30%) could either belong to members of the Archaea, Eukarya or most likely to yet unknown bacteria (90,91). These types of studies provide a valuable basis in order to eventually determine factors that change the microbiota such as lifestyle, diet or illness. Interestingly, FISH-FCM analysis of fecal microbiota of patients with UC revealed substantial temporal variations in the major bacterial groups studied (i.e., *Bacteroides*, *C. coccoides-E. rectale*, *Atopobium*, bifidobacteria and lactobacilli), which was further was supported by PCR-DGGE profiles (40).

NEW MOLECULAR DIVERSITY APPROACHES

Real Time PCR

Real time quantitative PCR (qPCR) of the 16S rRNA gene is being developed the last few years for the detection and quantification of human intestinal microbiota, which has the advantages of being high throughput and measuring from 1 to up to 10^8 CFU (99). Both SYBR Green I and TaqMan chemistries have been used to target *Bacteroides fragilis*, *Bifidobacterium* species, *E. coli*, *L. acidophilus* and *Ruminococcus productus*, and the method was demonstrated to be easier and faster than dot-blot hybridization methodology (100). Real-time qPCR ($5'$ nuclease PCR assay) has been used to study the microbiota that adhere to the colonic mucosa (101). The primer-probe combinations were applied to DNA for the detection of *E. coli* and *Bacteroides vulgatus* from pure cultures and colonic biopsy specimens. The assay was very sensitive detecting as little as 1 and 9 CFU of *E. coli* and *B. vulgatus*, respectively. Many of the qPCR assays being developed target the lactobacilli and *Bifidobacterium* species that may be incorporated in functional foods (102,103). Besides real time PCR of the 16S rRNA gene, the option to use the transaldolase gene of *Bifidobacterium* species has also been investigated and appeared to be superior to the former in quantifying bifidobacterial populations in infants (104). The qRT-PCR assays have been used for various applications such as comparison of healthy persons versus patients suffering from IBS (105), and in patients with active IBD (26). Recently, a TaqMan real-time PCR-based method for the quantification of 20 dominant bacterial species and groups of the microbiota was developed (106). This method involved a pair of conserved primers, as well as universal and specific quantification probes, for species, group or genus in question, in a single reaction, and allowed relative and absolute quantification of bacteria in human biopsy and fecal samples. Further developments in real-time qPCR will facilitate our insight into the dynamics of the microbiota.

Diagnostic DNA Microarrays

The development of DNA oligonucleotide microarrays offer a fast, high throughput option for detection and estimation of the diversity of microbes in a complex ecosystem (107). Alternative terms for the microarrays are phylochips, microbial diagnostic microarrays and identification arrays. Their principle is based on the dot-blot hybridization described above. Typically microarrays contain hundreds of oligonucleotide probes, usually based on the 16S rRNA gene, specific for different strains or species or genera of microorganisms that are detected in a single assay. Total DNA or RNA is isolated from the sample, fragmented, and amplified by PCR with the simultaneous incorporation of labeled nucleotides, or directly chemically labeled. The labeled fragments are hybridized to the probes immobilized on a surface, and following washing hybridized fragments are detected by a fluorescence scanner. There are many different forms of arrays to which the probes can be attached including macroarrays, and glass microarrays that are low to medium density, and very high density Affymetric microarrays ($> 10^4$ probes typically 25 mer per chip) (108). Three-dimensional form microarrays such as the Pamgene system and gel-pads allow the option for quantitative detection (109). Studies are underway to apply microarray technology to the human intestinal microbiota (16). A macroarray membrane-based method with 60 40-mer oligonucleotide probes specific for the dominant microbiota demonstrated the feasibility of arrays for detection (110). The high throughput potential of arrays will undoubtedly encourage further efforts in this area in the coming years.

ASSESSMENT OF MICROBIOTA VITALITY AND METABOLIC ACTIVITY

The aforementioned molecular techniques have greatly contributed to our fundamental understanding of the biodiversity, establishment, succession and structure of the intestinal microbiota; yet little is known about the in situ association between the microbial diversity and the metabolic activity of a phylogenetic affiliated group. A further challenge is to determine the physiological activity of the detected cells. This includes those cells that are naturally present within the ecosystem as well as the ingested members from fermented or functional foods. Moreover, the use of specific food-grade lactic acid bacteria as vectors for therapeutic delivery of molecules with targeted activity in the host is being investigated (111,112). These bacteria appear capable of surviving and of being physiologically active at the mucosal surfaces in animal models. Biological containment systems are being developed for these genetically modified lactic acid bacteria to limit their activity to the host and allow their use in human healthcare (113).

In Situ Activity

Quantitative hybridization with fluorescent rRNA probes (as in FISH) is a useful indicator of activity as there is a correlation between the growth rate, which is coupled to efficient protein synthesis, and the number of ribosomes. The FISH technique has been used to estimate growth rates of *Escherichia coli* cells colonizing the intestinal tract of mice (114). In situ activity of pure cultures of the human commensal *Lactobacillus plantarum* strain has been measured by correlating the rRNA, as determined by fluorescence intensity, with the cell growth rate (72). However, at very high cell densities, a typical property of *L. plantarum* at late stages of growth, changes in the cell envelope appeared to prevent effective entry of the probe into the cells. Permeabilization issues may confound application of this technique to certain microbes in complex environments like the intestine. Furthermore, recent data suggest that cellular ribosome content is not always an indicator of physiological activity. Apparently some bacterial cells might be highly active but possess a low ribosome content (115), while other bacterial types possess high RNA even after extended starvation periods (116).

During the last years several innovative methods have been developed to resolve the linkage between taxonomic identity, activity and function in microbial communities. One of these techniques involves microautoradiography (MAR), which when combined with FISH (MAR-FISH), determines the uptake of specific radiochemicals by individual cells (117,118). MAR-FISH allows monitoring of the radio-labeled substrate uptake patterns of the probe-identified organisms under different environmental conditions (117,119). This method has been applied with high throughput DNA microarray analysis to study the complex activated sludge ecosystem (120).

Linking Taxomony to Function

Another recently developed molecular technique coupled with substrate labeling is stable isotope probing (SIP) (121,122). In SIP, either lipid biomarkers (123), DNA (121) or RNA (124) are extracted from microbial communities incubated with ^{13}C-labeled substrates. If cells grow on the added compounds, their pool of macromolecules will be isotopically enriched (heavy) compared to those of inactive organisms. For DNA- or RNA-SIP, identification of the metabolically active organisms (heavy) is achieved by separation of community DNA/RNA according to their buoyant density by means of equilibrium density-gradient centrifugation, followed by PCR-amplification of 16S rRNA genes in the isotopically heavy DNA/RNA pool, cloning and sequencing. The use of RNA was proposed as a more responsive biomarker as its turnover is much higher than that of DNA (124). Phospholipid fatty acids are also used as biomarker for ^{13}C enrichments, but their resolution for diversity analysis is less powerful than for sequence analysis.

Reporters to Monitor Gene Expression

Molecular reporter systems may also be used to monitor activity of specific genes of a microbe of the complex intestinal ecosystem. Generally this involves fusing the reporter gene to the promoter of the bacterial gene of interest, such as stress- and starvation-induced genes and other growth physiology-related genes. It is noteworthy that this approach involves a genetically modified microbe, and consequently, its application is limited to animal studies. The adaptation of ingested lactic acid bacteria has received particular attention in terms of how they adapt their metabolism in order to survive and colonize within the gastrointestinal niches.

The fusion of bacterial promoters from *Lactococcus lactis* with genes of the reporter protein luciferase (*luxA-luxB* genes of *Vibrio harveyi*) was developed to investigate gene expression of this food-grade bacteria in the mouse intestinal tract (125). *L. lactis* strains marked with reporter genes for luciferase and the green fluorescent protein (GFP; from *Aequorea victoria*) were studied for their metabolic activity and survival by assessment of lysis, respectively, which revealed differential expression depending on the intestinal conditions and mode of administration (126). Following consumption by rats and analysis of the strains in the different regions of the intestinal tract, the lactococci were demonstrated to survive gastric transit quite well but the majority lost activity and underwent lysis in the duodenum. The luciferase gene reporter system has also been applied to a probiotic *Lactobacillus casei* strain that is added to fermented dairy products. The luciferase-harboring *L. casei* derivative was consumed in milk by mice harboring human microbiota. Luciferase activity was undetectable in the stomach to jejunum, but detected when the cells reached the ileum, and the activity remained at a maximum level in the cecum, confirming reinitiation of protein synthesis in the ileal and cecal compartments (127,128).

Several variants of the GFP have been developed such as GFPs with alternative emission wavelengths, or with reduced stability to monitor shifts in gene expression (129,130).

Flow Cytometry-Based Approaches

FCM in combination with a variety of fluorescent physiological probes and cell sorting analysis is invaluable for measuring viability of cells in environmental samples (80,87,131,132). Ability to grow in medium is the current standard to assess viability, but it is recognized that some cells enter a non-culturable state although still exhibit metabolic activity. The criteria by which viability is evaluated by the FCM include membrane permeability or integrity, enzyme activity, and/or maintenance of a membrane-potential (Fig. 2). One of the most widely used dyes for assessment of viability is carboxy-fluorescein diacetate, a non-fluorescent precursor that diffuses across the cell membrane, but is retained only by viable cells with intact membranes which convert it into a membrane-impermeant fluorescent dye by non-specific esterases of active cells. Another probe is PI, a nucleic acid dye, which is excluded by viable cells with intact membranes, but enters cells with damaged membranes and binds to their DNA or RNA. Simultaneous staining of fecal *Bifidobacterium* species with these two probes was used to assess their viability during bile salt stress (133). Subsequent detection with the FCM and cell sorting revealed three populations representing viable, injured and dead cells, whereby a significant portion (40%) of the injured cells could be cultured. This approach highlights the importance of multi-parametric FCM as a powerful technique to monitor physiological heterogeneity within stressed populations at the single cell level.

FCM also allows monitoring of bacterial heterogeneity at the single cell level and provides a mean to sort sub-populations of interest for further molecular analysis (15). Recently, the viability of fecal microbiota in fecal samples was assessed by combining a viability assay with flow sorting, and subsequent analysis by PCR-DGGE and identification by cloning and 16S rRNA sequencing (80). The fecal cells of four adults were initially discriminated with physiological probes PI and SYTO BC into viable, injured and dead cells. This revealed that only approximately half of the microbial community in fecal samples is viable, while the remainder was injured or dead (about a quarter each of the total community). This is in agreement with a previous analysis of proportions of dead bacteria in 10 persons which ranged from 17% to 34%, as assessed by PI only (134). The 16S rRNA analysis indicated which bacterial groups comprised live, dead or injured populations, for example many butyrate-producers were in the live fractions, while many clones from Bacteroides were found in the dead fractions (80). Specific PCR-DGGE and 16S rRNA analysis of the bifidobacterial and lactobacilli populations showed sequences with low similarity to the characterized species suggested the potential of as yet uncultured novel species in humans (80,135). This interesting combination of technologies provided ecological information on the in situ diversity and activity of the fecal microbes.

PERSPECTIVES

This chapter has highlighted the extraordinary advances in the molecular technologies that have substantially contributed to our knowledge and understanding of the human intestinal microbiota. The application of these molecular tools has greatly facilitated our analysis of the composition of the human microbiota. A picture of the "typical" microbiota for at least the northern European population of infants and adults is emerging, as are differences in individuals with intestinal diseases. The diversity is far greater than previously predicted from the initial culturing studies in the 1960s. Consequently, further technological improvements to perform the techniques at higher throughput, and for measurement of more subtle changes in the diversity of the microbiota due to, for example, specific dietary components, require further development. Microarray technology is amenable to both these requirements, and currently DNA microarrays are being constructed for the human microbiota using 16S rRNA sequences of microbiota (136); [Mirjana Rajilic and Willem M. de Vos, personal communication]. FCM with its unique capacity for quantitative and high throughput analysis is resulting in the development of an alternative type of array using beads with oligonucleotide probes on the surface that can be applied in hybridization assays in suspension (137–139).

The substantial impact of this highly diverse microbiota on the health of the human host is now well recognized, such as processing of undigested food, contributing to the host defense and regulating fat storage amongst others (6,140,141). It is a particular challenge to develop methods that allow monitoring of microorganisms according to their eco-physiological traits in situ. The application of cytometric protocols using fluorescent probes in combination with molecular techniques opens the potential for examining key microbial processes and community function in complex microbial ecosystems. Further efforts to determine the molecular foundations of the host-microbiota interactions will require multi-disciplinary approaches. The rewards of this research in terms of promoting host health via our microbiota and diet can be substantial, as well as novel approaches for treating intestinal diseases and infections caused by pathogens.

REFERENCES

1. Bäckhed F, Ley RE, Sonnenburg JL, Peterson DA, Gordon JI. Host-bacterial mutualism in the human intestine. Science 2005; 5717:1915–1920.
2. Rawls JF, Samuel BS, Gordon JI. Gnotobiotic zebrafish reveal evolutionarily conserved responses to the gut microbiota. Proc Natl Acad Sci USA 2004; 101:4596–4601.
3. Falk PG, Hooper LV, Midtvedt T, Gordon JI. Creating and maintaining the gastrointestinal ecosystem: what we know and need to know from gnotobiology. Microbiol Mol Biol Rev 1998; 62:1157–1170.
4. Stappenbeck TS, Hooper LV, Gordon JI. Developmental regulation of intestinal vasculogenesis by indigenous microbes via Paneth cells. Proc Natl Acad Sci USA 2002; 99:15451–15455.
5. Rakoff-Nahoum S, Paglino J, Eslami-Varzaneh F, Edberg S, Medzhitov R. Recognition of commensal microflora by toll-like receptors is required for intestinal homeostasis. Cell 2004; 118:229–241.
6. Hooper LV, Gordon JI. Commensal host-bacterial relationships in the gut. Science 2001; 292:1115–1118.
7. Ouwehand AC, Salminen S, Isolauri E. Probiotics: an overview of beneficial effects. Antonie Van Leeuwenhoek 2002; 82:279–289.
8. Rachmilewitz D, Katakura K, Karmeli F, et al. Toll-like receptor 9 signaling mediates the anti-inflammatory effects of probiotics in murine experimental colitis. Gastroenterology 2004; 126:520–528.
9. Finegold SM, Sutter VL, Mathisen GE. Normal indigenous flora. In: Hentges DJ, ed. Human Intestinal Microflora in Health and Disease. New York: Academic Press, 1983:3–31.
10. Holdeman LV, Good IJ, Moore WE. Human fecal flora: variation in bacterial composition within individuals and a possible effect of emotional stress. Appl Environ Microbiol 1976; 31:359–375.
11. Moore WE, Holdeman LV. Human fecal flora: the normal flora of, 20 Japanese-Hawaiians. Appl Microbiol 1974; 27:961–979.
12. Vaughan EE, Schut F, Heilig HGHJ, Zoetendal EG, de Vos WM, Akkermans ADL. A molecular view of the intestinal ecosystem. Curr Issues Intest Microbiol 2000; 1:1–12.

13. Zoetendal EG, Cheng B, Koike S, Mackie RI. Molecular microbial ecology of the gastrointestinal tract: from phylogeny to function. Curr Issues Intest Microbiol 2004; 5:31–47.

14. Whitfield J. Features-science and health: microbial soup of life is sieved for treasure. Financial Times, 16 April 2004 http://news.ft.com.

15. Amann R, Ludwig W, Schleifer K. Phylogenetic identification and in situ detection of individual microbial cells without cultivation. Microbiol Rev 1995; 59:143–169.

16. Namsolleck P, Thiel R, Lawson P, et al. Molecular methods for the analysis of gut microbiota. Microb Ecol Health Dis 2004; 16:71–85.

17. Ludwig W, Strunk O, Westram R, et al. ARB: a software environment for sequence data. Nucleic Acids Res 2004; 32:1363–1371.

18. Cole JR, Chai B, Marsh TL, et al. The Ribosomal database project (RDP-II): previewing a new autoaligner that allows regular updates and the new prokaryotic taxonomy. Nucleic Acids Res 2003; 31:442–443.

19. Eckburg PB, Bik EM, Bernstein CN, et al. Diversity of the human intestinal microbial flora. Sci Express 2005; 14:1–4.

20. Hayashi H, Sakamoto Y, Benno M. Phylogenetic analysis of the human gut microbiota using 16S rDNA clone libraries and strictly anaerobic culture-based methods. Microbiol Immunol 2002; 46:535–548.

21. Hold GL, Pryde SE, Russell VJ, Furrie E, Flint HJ. Assessment of microbial diversity in human colonic samples by 16S rDNA sequence analysis. FEMS Microbiol Ecol 2002; 39:33–39.

22. Suau A, Bonnet R, Sutren M, et al. Direct analysis of genes encoding 16S rRNA from complex communities reveals many novel molecular species within the human gut. Appl Environ Microbiol 1999; 65:4799–4807.

23. Zoetendal EG, Akkermans ADL, De Vos WM. Temperature gradient gel electrophoresis analysis of 16S rRNA from human fecal samples reveals stable and host-specific communities of active bacteria. Appl Environ Microbiol 1998; 64:3854–3859.

24. Mangin I, Bonnet R, Seksik P, et al. Molecular inventory of fecal microflora in patients with Crohn's disease. FEMS Micro Ecol 2004; 50:25–36.

25. Breitbart M, Hewson I, Felts B, et al. Metagenomic analyses of an uncultured viral community from human feces. J Bacteriol 2003; 185:6220–6223.

26. Ott SJ, Musfeldt M, Wenderoth DF, et al. Reduction in diversity of the colonic mucosa associated bacterial microflora in patients with active inflammatory bowel disease. Gut 2004; 53:685–693.

27. Von Wintzingrode F, Göbel UB, Stackebrandt E. Determination of microbial diversity in environmental samples: pitfalls of PCR-based rRNA analysis. FEMS Microbiol Rev 1997; 21:213–229.

28. Bonnet R, Suau A, Dore J, Gibson GR, Collins MD. Differences in rDNA libraries of fecal bacteria derived from 10- and 25-cycle PCRs. Int J Syst Evol Microbiol 2002; 52:757–763.

29. Nagashima K, Hisada T, Sato M, Mochizuki J. Application of new primer-enzyme combinations to terminal restriction fragment length polymorphism profiling of bacterial populations in human feces. Appl Environ Microbiol 2003; 69:1251–1262.

30. Muyzer G, de Waal E, Uitterlinden A. Profiling of complex microbial populations by denaturing gradient gel electrophoresis analysis of polymerase chain reaction-amplified genes coding for 16S rRNA. Appl Environ Microbiol 1993; 59:695–700.

31. Rosenbaum V, Riesner D. Temperature-gradient gel electrophoresis. Thermodynamic analysis of nucleic acids and proteins in purified form and in cellular extracts. Biophys Chem 1987; 26:235–246.

32. Formin N, Hamelin S, Tarnawski S, et al. Statistical analysis of denaturing gel electrophoresis (DGE) fingerprinting patterns. Environ Microbiol 2002; 4:634–643.

33. Favier CF, Vaughan EE, De Vos WM, Akkermans ADL. Molecular monitoring of succession of bacterial communities in human neonates. Appl Environ Microbiol 2002; 68:219–226.

34. Tannock GW, Munro K, Harmsen HJM, Welling GW, Smart J, Gopal PK. Analysis of the fecal microflora of human subjects consuming a probiotic product containing *Lactobacillus rhamnosus* DR20. Appl Environ Microbiol 2000; 66:2578–2588.

35. Zoetendal EG, Akkermans ADL, Akkermans-van Vliet WM, de Visser JAGM, de Vos WM. The host genotype affects the bacterial community in the human gastrointestinal tract. Microbiol Ecol Health Dis 2001; 13:129–134.

36. Nielsen D, Moller P, Rosenfeldt V, Paerregaard A, Michaelsen K, Jakobsen M. Case study of the distribution of mucosa-associated *Bifidobacterium* species, and *Lactobacillus* species, and other lactic acid bacteria in the human colon. Appl Environ Microbiol 2003; 69:7545–7548.

37. Zoetendal EG, von Wright A, Vilpponen-Salmela T, Ben-Amor K, Akkermans ADL, de Vos WM. Mucosa-associated bacteria in the human gastrointestinal tract are uniformly distributed along the colon and differ from the community recovered from feces. Appl Environ Microbiol 2002; 68:3401–3407.

38. Campieri M, Gionchetti P. Bacteria as the cause of ulcerative colitis. Gut 2001; 48:132–135.

39. Shanahan F. Probiotics: a perspective on problems and pitfalls. Scand J Gastroenterol Suppl 2003;(237):34–46.

40. Ben-Amor K, Heikamp-de jong I, Verhaegh S, de Vos WM, Vaughan EE. Population dynamics and diversity of fecal microbiota of patients with ulcerative colitis participating in a probiotic trial. In: Ben-Amor K, eds. Microbial ecophysiology of the human intestinal tract: A flow cytometric approach. Netherland: PhD Thesis Wageningen University, 2004:83–121.

41. Seksik P, Rigottier-Gois L, Gramet G, et al. Alterations of the dominant fecal bacterial groups in patients with Crohn's disease of the colon. Gut 2003; 52:237–242.

42. Matto J, Maunuksela L, Kajander K, et al. Composition and temporal stability of gastrointestinal microbiota in irritable bowel syndrome-a longitudinal study in IBS and control subjects. FEMS Immunol Med Microbiol 2005; 43:213–222.

43. Vaughan EE, Heilig HGHJ, Zoetendal EG, et al. Molecular approaches to study probiotic bacteria. Trends Food Sci Technol 1999; 10:400–404.

44. Zoetendal EG, Ben Amor K, Akkermans AD, Abee T, de Vos W. DNA isolation protocols affect the detection limit of PCR approaches of bacteria in samples from the human gastrointestinal tract. Syst Appl Microbiol 2001; 24:405–410.

45. Heilig HGHJ, Zoetendal EG, Vaughan EE, Marteau P, Akkermans ADL, de Vos WM. Molecular diversity of *Lactobacillus* spp. and other lactic acid bacteria in the human intestine as determined by specific amplification of, 16S ribosomal DNA. Appl Environ Microbiol 2002; 68:114–123.

46. Satokari RM, Vaughan EE, Akkermans AD, Saarela M, De Vos WM. Polymerase chain reaction and denaturing gradient gel electrophoresis monitoring of fecal *Bifidobacterium* populations in a prebiotic and probiotic feeding trial. Syst Appl Microbiol 2001; 24:227–231.

47. Walter J, Hertel C, Tannock GW, Lis CM, Munro K, Hammes WP. Detection of *Lactobacillus*, *Pediococcus*, *Leuconostoc*, and *Weissella* species in human feces by using group-specific PCR primers and denaturing gradient gel electrophoresis. Appl Environ Microbiol 2001; 67:2578–2585.

48. Dar SA, Kuenen JG, Muyzer G. Nested PCR-denaturing gradient gel electrophoresis approach to determine the diversity of sulfate-reducing bacteria in complex microbial communities. Appl Environ Microbiol 2005; 71:2325–2330.

49. Holben WE, Feris KP, Kettunen A, Apajalahti JH. GC fractionation enhances microbial community diversity assessment and detection of minority populations of bacteria by denaturing gradient gel electrophoresis. Appl Environ Microbiol 2004; 70:2263–2270.

50. Holben WE, Harris D. DNA-based monitoring of total bacterial community structure in environmental samples. Mol Ecol 1995; 4:627–6331.

51. Kitts CL. Terminal restriction fragment patterns: a toll for comparing microbial communities and assessing community dynamics. Curr Issues Intest Microbiol 2001; 2:17–25.

52. Osborn AM, Moore ER, Timmis KN. An evaluation of terminal-restriction fragment length polymorphism (T-RFLP) analysis for the study of microbial community structure and dynamics. Environ Microbiol 2000; 2:39–50.

53. Jernberg C, Sullivan A, Edlund C, Jansson JK. Monitoring of antibiotic-induced alterations in the human intestinal microflora and detection of probiotic strains by use of terminal restriction fragment length polymorphism. Appl Environ Microbiol 2005; 71:501–506.

54. Sakamoto M, Hayashi H, Benno Y. Terminal restriction fragment length polymorphism analysis for human fecal microbiota and its application for analysis of complex bifidobacterial communities. Microbiol Immunol 2003; 47:133–142.

55. Matsumoto M, Sakamoto M, Hayashi H, Benno Y. Novel phylogenetic assignment database for terminal-restriction fragment length polymorphism analysis of human colonic microbiota. J Microbiol Methods 2005; 61:305–319.

56. Lipski A, Friedrich U, Altendorf K. Application of rRNA-targeted oligonucleotide probes in biotechnology. Appl Microbiol Biotechnol 2001; 56:40–57.

57. Moter A, Gobel UB. Fluorescence in situ hybridization (FISH) for direct visualization of microorganisms. J Microbiol Methods 2000; 41:85–112.

58. Amann RI, Binder B, Olson R, Chisholm SW, Devereux R, Stahl D. Combination of 16S rRNA-targeted oligonucleotide probes with flow cytometry for analyzing mixed microbial populations. Appl Environ Microbiol 1990; 56:1919–1925.

59. Manz W, Amann R, Ludwig W, Vancanneyt M, Schleifer K. Application of a suite of 16S rRNA-specific oligonucleotide probes designed to investigate bacteria of the phylum cytophaga-flavobacter-bacteroides in the natural environment. Microbiology 1996; 142:1097–1106.

60. Daims H, Bruhl A, Amann R, Schleifer KH, Wagner M. The domain-specific probe EUB338 is insufficient for the detection of all Bacteria: development and evaluation of a more comprehensive probe set. Syst Appl Microbiol 1999; 22:434–444.

61. Loy A, Horn M, Wagner M. Probebase: an online resource for rRNA-targeted oligonucleotide probes. Nucleic Acids Res 2003; 31:514–516.

62. de los Reyes FL, Ritter W, Raskin L. Group-specific small-subunit rRNA hybridization probes to characterize filamentous foaming in activated sludge systems. Appl Environ Microbiol 1997; 63:1107–1117.

63. Stahl DA, Amann R. Development and application of nucleic acid probes. In: Stackerbrandt E, Goodfellow M, eds. Nucleic Acid Techniques in Bacterial Systematics. Chichester: John Wiley & Sons Ltd, 1991:205–248.

64. Fuchs BM, Syutsubo K, Ludwig W, Amann R. In situ accessibility of *Escherichia coli*, 23S rRNA to fluorescently labeled oligonucleotide probes. Appl Environ Microbiol 2001; 67:961–968.

65. Fuchs BM, Wallner G, Beisker W, Schwippl I, Ludwig W, Amann R. Flow cytometric analysis of the in situ accessibility of *Escherichia coli* 16S rRNA for fluorescently labeled oligonucleotide probes. Appl Environ Microbiol 1998; 64:4973–4982.

66. Behrens S, Fuchs BM, Mueller F, Amann R. Is the in situ accessibility of the 16S rRNA of *Escherichia coli* for Cy3-labeled oligonucleotide probes predicted by a three-dimensional structure model of the 30S ribosomal subunit? Appl Environ Microbiol 2003; 69:4935–4941.

67. Molin S, Givskov M. Application of molecular tools for in situ monitoring of bacterial growth activity. Environ Microbiol 1999; 1:383–391.

68. Marteau P, Pochart P, Dore J, Bera-Maillet C, Bernalier A, Corthier G. Comparative study of bacterial groups within the human cecal and fecal microbiota. Appl Environ Microbiol 2001; 67:4939–4942.

69. Sghir A, Gramet G, Suau A, Rochet V, Pochart P, Dore J. Quantification of bacterial groups within human fecal flora by oligonucleotide probe hybridization. Appl Environ Microbiol 2000; 66:2263–2266.

70. Amann R, Fuchs BM, Behrens S. The identification of microorganisms by fluorescence in situ hybridization. Curr Opin Biotechnol 2001; 12:231–236.

71. Beimfohr C, Krause A, Amann R, Ludwig W, Schleifer KH. In situ identification of lactococci, enterococci and streptococci. Syst Appl Microbiol 1993; 16:450–456.

72. de Vries MC, Vaughan EE, Kleerebezem M, De Vos WM. Optimising single cell activity assessment of Lactobacillus plantarum by fluorescent in situ hybridization as affected by growth. J Microbiol Methods 2004; 59:109–115.

73. Jansen GJ, Wildeboer-Veloo ACM, Tonk RHJ, Franks AH, Welling GW. Development and validation of an automated, microscopy-based method for enumeration of groups of intestinal bacteria. J Microbiol Methods 1999; 37:215–221.

74. Glockner FO, Amann R, Alfreider A, et al. An in situ hybridization protocol for detection and identification of planktonic bacteria. Syst Appl Microbiol 1996; 19:403–406.

75. Southwick PL, Ernst LA, Tauriello EW, et al. Cyanine dye labeling reagents—carboxymethylindocyanine succinimidyl esters. Cytometry 1990; 11:418–430.

76. Fuchs BM, Glockner FO, Wulf J, Amann R. Unlabeled helper oligonucleotides increase the in situ accessibility to 16S rRNA of fluorescently labeled oligonucleotide probes. Appl Environ Microbiol 2000; 66:3603–3607.

77. Pernthaler A, Pernthaler J, Amann R. Fluorescence in situ hybridization and catalyzed reporter deposition for the identification of marine bacteria. Appl Environ Microbiol 2002; 68:3094–3101.

78. Oliveira K, Brecher SM, Durbin A, et al. Direct identification of *Staphylococcus aureus* from positive blood culture bottles. J Clin Microbiol 2003; 41:889–891.

79. Perry-O'Keefe H, Rigby S, Oliveira K, et al. Identification of indicator microorganisms using a standardized PNA FISH method. J Microbiol Methods 2001; 47:281–292.

80. Ben-Amor K, Heilig HG, Smidt H, Vaughan EE, Abee T, de Vos WM. Genetic diversity of viable, injured and dead fecal bacteria assessed by fluorescence activated cell sorting and, 16S rRNA gene analysis. Appl Environ Microbiol 2005; 71:4679–4689.

81. Rigottier-Gois L, Le Bourhis A-G, Gramet G, Rochet V, Dore J. Fluorescent hybridization combined with flow cytometry and hybridization of total RNA to analyse the composition of microbial communities in human faeces using 16S rRNA probes. FEMS Microbiol Ecol 2003; 43:237–245.

82. Rigottier-Gois L, Rochet V, Garrec N, Suau A, Dore J. Enumeration of Bacteroides species in human faeces by fluorescent in situ hybridization combined with flow cytometry using 16S rRNA probes. Syst Appl Microbiol 2003; 26:110–118.

83. Wallner G, Amann R, Beisker W. Optimizing fluorescent in situ hybridization with rRNA-targeted oligonucleotide probes for flow cytometric identification of microorganisms. Cytometry 1993; 14:136–143.

84. Zoetendal EG, Ben-Amor K, Harmsen HJM, Schut F, Akkermans ADL, de Vos WM. Quantification of uncultured *Ruminococcus obeum*-like bacteria in human fecal samples by fluorescent in situ hybridization and flow cytometry using 16S rRNA-targeted probes. Appl Environ Microbiol 2002; 68:4225–4232.

85. van der Waaij LA, Mesander G, Limburg PC, Wan der Waaij D. Direct flow cytometry of anaerobic bacteria in human feces. Cytometry 1994; 16:270–279.

86. Shapiro HM. Practical Flow Cytometry. New York: Wiley-Liss Inc., 1995.

87. Ben-Amor K, Abee T, De Vos WM. Flow cytometric analysis of microorganisms. In: Ben-Amor K, eds. Microbial Eco-physiology of the Human Intestinal Tract: A flow cytometric approach. Netherland: PhD thesis Wageningen University, 2004:21–47.

88. Derrien M, Ben-Amor K, Vaughan EE, de Vos WM. Validation of 16S rRNA probe specific for the novel intestinal mucin-degrader *Akkermansia muciniphila*. In: Ahonen R, Saarela M, Mattila-Sandholm T, eds.The FOOD, GI-tract Functionality and Human Heatlh Cluster. PROEUHEALTH. Vol. VTT Biotechnolgy. Sitges, Spain, 2004. http://www.vtt.fi/inf/pdf/.

89. Franks AH, Harmsen HJM, Raangs GC, Jansen GJ, Schut F, Welling GW. Variations of bacterial populations in human feces measured by fluorescent in situ hybridization with group-specific 16S rRNA-targeted oligonucleotide Probes. Appl Environ Microbiol 1998; 64:3336–3345.

90. Lay C, Rigottier-Gois L, Holmstrum K, et al. Colonic microbiota signatures across five northern European countries. Appl Environ Microbiol 2005; 717:4153–4155.

91. Harmsen HJM, Raangs GC, He T, Degener JE, Welling GW. Extensive set of 16S rRNA-based probes for detection of bacteria in human feces. Appl Environ Microbiol 2002; 68:2982–2990.

92. Suau A, Rochet V, Sghir A, et al. *Fusobacterium prausnitzii* and related species represent a dominant group within the human fecal flora. Syst Appl Microbiol 2001; 24:139–145.

93. Langendijk P, Schut F, Jansen G, et al. Quantitative fluorescence in situ hybridization of *Bifidobacterium* spp. with genus-specific 16S rRNA-targeted probes and its application in fecal samples. Appl Environ Microbiol 1995; 61:3069–3075.

94. Harmsen HJM, Wildeboer-Veloo ACM, Grijpstra J, Knol J, Degener JE, Welling GW. Development of 16S rRNA-based probes for the *Coriobacterium* group and the *Atopobium* cluster and their application for enumeration of *Coriobacteriaceae* in human feces from volunteers of different age groups. Appl Environ Microbiol 2000; 66:4523–4527.

95. Harmsen JHM, Elfferich P, Schut F, Welling GW. A 16S rRNA-targeted probe for detection of lactobacilli and enterococci in fecal samples by fluorescent in situ hybridization. Micro Ecol Health Dis 1999; 11:3–12.

96. Lay C, Sutren M, Rochet V, Saunier K, Doré J, Rigottier-Gois L. Design and validation of 16S rRNA probes to enumerate members of the *Clostridium leptum* subgroup in human fecal microbiota. Environ Microbiol 2005; 7:933–946.

97. Rochet V, Rigottier-Gois L, Rabot S, Doré J. Validation of fluorescent in situ hybridization combined with flow cytometry for assessing interindividual variation in the composition of human fecal microflora during long-term storage of samples. J Microbiol Methods 2004; 59:263–270.

98. Collins MD, Lawson PA, Willems A, et al. The phylogeny of the genus *Clostridium*: proposal of five new genera and eleven new species combinations. Int J Syst Bacteriol 1994; 144:812–826.

99. Nadkarni MA, Martin FE, Jacques NA, Hunter N. Determination of bacterial load by real-time PCR using a broad-range (universal) probe and primers set. Microbiology 2002; 148:257–266.

100. Malinen E, Kassinen A, Rinttila T, Palva A. Comparison of real-time PCR with SYBR Green I or 5′-nuclease assays and dot-blot hybridization with rDNA-targeted oligonucleotide probes in quantification of selected fecal bacteria. Microbiology 2003; 149:269–277.

101. Huijsdens XW, Linskens RK, Mak M, Meuwissen SG, Vandenbroucke-Grauls CM, Savelkoul PH. Quantification of bacteria adherent to gastrointestinal mucosa by real-time PCR. J Clin Microbiol 2002; 40:4423–4427.

102. Bartosch S, Fite A, Macfarlane GT, McMurdo ME. Characterization of bacterial communities in feces from healthy elderly volunteers, and hospitalized elderly patients by using real-time PCR and effects of antibiotic treatment on the fecal microbiota. Appl Enivr Microbiol 2004; 70:3575–3581.

103. Matsuki T, Watanabe K, Fujimoto J, Takada T, Tanaka R. Use of 16S rRNA gene-targeted group-specific primers for real-time PCR analysis of predominant bacteria in human feces. Appl Environ Microbiol 2004; 71:7220–7228.

104. Requena T, Burton J, Matsuki T, et al. Identification, detection, and enumeration of human Bifidobacterium species by PCR targeting the transaldolase gene. Appl Environ Microbiol 2002; 68:2420–2427.

105. Malinen E, Rinttila T, Kajander K, et al. Analysis of the fecal microbiota of irritable bowel syndrome patients and healthy controls with real-time PCR. Am J Gastroenterol 2005; 100:373–382.

106. Ott SJ, Musfeldt M, Ullmann U, Hampe J, Schreiber S. Quantification of intestinal bacterial populations by real-time PCR with a universal primer set and minor groove binder probes: a global approach to the enteric flora. J Clin Microbiol 2004; 42:2566–2572.

107. Bodrossy L, Sessitsch A. Oligonucleotide microarrays in microbial diagnostics. Curr Opin Microbiol 2004; 7:245–254.

108. Wilson KH, Wilson WJ, Radosevich JL, et al. High-density microarray of small-subunit ribosomal DNA probes. Appl Environ Microbiol 2002; 68:2535–2541.

109. Guschin DY, Mobarry BK, Proudnikov D, Stahl DA, Rittmann BE, Mirzabekov AD. Oligonucleotide microchips as genosensors for determinative and environmental studies in microbiology. Appl Environ Microbiol 1997; 63:2397–2402.

110. Wang RF, Kim SJ, Robertson LH, Cerniglia CE. Development of a membrane-array method for the detection of human intestinal bacteria in fecal samples. Mol Cell Probes 2002; 16:341–350.

111. Geoffroy MC, Guyard C, Quatannens B, Pavan S, Lange M, Mercenier A. Use of green fluorescent protein to tag lactic acid bacterium strains under development as live vaccine vectors. Appl Environ Microbiol 2000; 66:383–391.

112. Hanniffy S, Wiedermann U, Repa A, et al. Potential and opportunities for use of recombinant lactic acid bacteria in human health. Adv Appl Microbiol 2004; 56:1–64.

113. Steidler L. Live genetically modified bacteria as drug delivery tools: at the doorstep of a new pharmacology? Expert Opin Biol Ther 2004; 4:439–441.

114. Rang CU, Licht TR, Midtvedt T, et al. Estimation of growth rates of *Escherichia coli* BJ4 in streptomycin-treated and previously germfree mice by in situ rRNA hybridization. Clin Diagn Lab Immunol 1999; 6:434–436.

115. Pernthaler A, Pernthaler J, Schattenhofer M, Amann R. Identification of DNA-synthesizing bacterial cells in coastal North Sea plankton. Appl Environ Microbiol 2002; 68:5728–5736.

116. Morgenroth E, Obermayer A, Arnold E, Bruhl A, Wagner M, Wilderer PA. Effect of long-term idle periods on the performance of sequencing batch reactors. Water Sci Technol 2000; 41:105–113.

117. Lee N, Nielsen PH, Andreasen KH, et al. Combination of fluorescent in situ hybridization and microautoradiography-a new tool for structure-function analyses in microbial ecology. Appl Environ Microbiol 1999; 65:1289–1297.

118. Nielsen JL, Christensen D, Kloppenborg M, Nielsen PH. Quantification of cell-specific substrate uptake by probe-defined bacteria under in situ conditions by microautoradiography and fluorescence in situ hybridization. Environ Microbiol 2003; 5:202–211.

119. Daims H, Nielsen JL, Nielsen PH, Schleifer KH, Wagner M. In situ characterization of Nitrospira-like nitrite-oxidizing bacteria active in wastewater treatment plants. Appl Environ Microbiol 2001; 67:5273–5284.

120. Adamczyk J, Hesselsoe M, Iversen N, et al. The isotope array, a new tool that employs substrate-mediated labeling of rRNA for determination of microbial community structure and function. Appl Environ Microbiol 2003; 69:6875–6887.

121. Radajewski S, Ineson P, Parekh NR, Murrell JC. Stable-isotope probing as a tool in microbial ecology. Nature 2000; 403:646–649.

122. Radajewski S, McDonald IR, Murrell JC. Stable-isotope probing of nucleic acids: a window to the function of uncultured microorganisms. Curr Opin Biotechnol 2003; 14:296–302.

123. Boschker HTS, Middelburg JJ. Stable isotope and biomarkers in microbial ecology. FEMS Microbiol Ecol 2002; 40:85–95.

124. Manefield M, Whiteley AS, Griffiths RI, Bailey MJ. RNA stable isotope probing, a novel means of linking microbial community function to phylogeny. Appl Environ Microbiol 2002; 68:5367–5373.

125. Corthier G, Delorme C, Ehrlich SD, Renault P. Use of luciferase genes as biosensors to study bacterial physiology in the digestive tract. Appl Environ Microbiol 1998; 64:2721–2722.

126. Drouault S, Corthier G, Ehrlich SD, Renault P. Survival, physiology, and lysis of *Lactococcus lactis* in the digestive tract. Appl Environ Microbiol 1999; 65:4881–4886.

127. Oozeer R, Mater DD, Goupil-Feuillerat N, Corthier G. Initiation of protein synthesis by a labeled derivative of the *Lactobacillus casei* DN-114 001 strain during transit from the stomach to the cecum in mice harboring human microbiota. Appl Environ Microbiol 2004; 70:6992–6997.

128. Oozeer R, Furet JP, Goupil-Feuillerat N, Anba J, Mengaud J, Corthier G. Differential activities of four *Lactobacillus casei* promoters during bacterial transit through the gastrointestinal tracts of human-microbiota-associated mice. Appl Environ Microbiol 2005; 71:1356–1363.

129. Cormack BP, Valdivia RH, Falkow S. FACS-optimized mutants of the green fluorescent (GFP). Gene 1996; 173:33–38.

130. Andersen JB, Sternberg C, Poulsen LK, Bjorn SP, Givskov M, Molin S. New unstable variants of green fluorescent protein for studies of transient gene expression in bacteria. Appl Environ Microbiol 1998; 64:2240–2246.

131. Bernard L, Courties C, Duperray C, Schafer H, Muyzer G, Lebaron P. A new approach to determine the genetic diversity of viable and active bacteria in aquatic ecosystems. Cytometry 2001; 43:314–321.

132. Whiteley AS, Griffiths RI, Bailey MJ. Analysis of the microbial functional diversity within water-stressed soil communities by flow cytometric analysis and CTC+cell sorting. J Microbiol Methods 2003; 54:257–267.

133. Ben-Amor K, Breeuwer P, Verbaarschot P, et al. Multiparametric flow cytometry and cell sorting for the assessment of viable, injured, and dead *Bifidobacterium* cells during bile salt stress. Appl Environ Microbiol 2002; 68:5209–5216.

134. Apajalahti JHA, Kettunen A, Nurminen PH, Jatila H, Holben WE. Selective plating underestimates abundance and shows differential recovery of bifidobacterial species from human feces. Appl Environ Microbiol 2003; 69:5731–5735.

135. Vaughan EE, Heilig H, Ben Amor K, De Vos WM. Diversity, vitality and activities of intestinal lactic acid bacteria and bifidobacteria assessed by molecular approaches. FEMS Microbiol Rev 2005; 29:477–490.

136. Palmer C, Bik EM, Eckburg MB, Eisen MB, Relman DA, Brown PO. Microarray-based characterisation of the human colonic flora: a direct comparison to 16S rDNA sequencing—21C abstract. p37. In Beneficial Microbes ASM conferences, April 2005, Nevada U.S.A.

137. Kohara Y, Noda H, Okano K, Kambara H. DNA probes on beads arrayed in a capillary, 'Bead-array,' exhibited high hybridization performance. Nucl Acids Res 2002; 30:e87.

138. Nolan JP, Sklar LA. Suspension array technology: evolution of the flat-array paradigm. Trends Biotechnol 2002; 20:9–12.

139. Spiro A, Lowe M. Quantitation of DNA sequences in environmental PCR products by a multiplexed, bead-based method. Appl Envir Microbiol 2002; 68:1010–1013.

140. Hooper LV, Midtvedt T, Gordon JI. How host-microbial interactions shape the nutrient environment of the mammalian intestine. Ann Rev Nutr 2002; 22:283–307.

141. Backhed F, Ding H, Wang T, et al. The gut microbiota as an environmental factor that regulates fat storage. Proc Natl Acad Sci USA 2004; 101:15718–15723.

142. www.microbial-ecology.net/probebase/.

2

Sampling Microbiota in the Human Gastrointestinal Tract

Angèle P. M. Kerckhoffs, Melvin Samsom, and Gerard P. van Berge Henegouwen
Department of Gastroenterology, Utrecht University Medical Center, Utrecht, The Netherlands

Louis M. A. Akkermans and Vincent B. Nieuwenhuijs
Department of Surgery, Utrecht University Medical Center, Utrecht, The Netherlands

Maarten R. Visser
Department of Microbiology, Utrecht University Medical Center, Utrecht, The Netherlands

INTRODUCTION

General Introduction

Antonie van Leeuwenhoek (1632–1723) was the first to describe numerous micro-organisms from the gastrointestinal tract, which he described as "animalcules," having designed the first glass lenses for the microscope that were powerful enough to observe bacteria. His curiosity brought him to investigate samples taken from his own mouth and other people who never brushed their teeth, and he compared these findings with people who brushed their teeth daily and used large amounts of alcohol. He even investigated his own fecal samples in a period of diarrhea, compared these findings with fecal samples of animals, and reported these observations to the Royal Society in London (1).

We now know that the mucosal surface of the human gastrointestinal tract is about 300 m^2 and is colonized by 10^{13}–10^{14} bacteria consisting of hundreds of different species. The prevalence of bacteria in different parts of the gastrointestinal tract depends on pH, peristalsis, oxidation-reduction potential within the tissue, bacterial adhesion, bacterial cooperation, mucin secretion containing immunoglobulins (Ig), nutrient availability, diet, and bacterial antagonism. The composition of the Gram-negative, Gram-positive, aerobic, and anaerobic microbiota has been extensively studied by culturing methods, and shown to change at the various sites of the gastrointestinal tract (Fig. 1).

The stomach and proximal small bowel normally contain relatively small numbers of bacteria because of peristalsis, and the antimicrobial effects of gastric acidity. An intact ileocecal valve is likely to be an important barrier to backflow of colonic bacteria into the

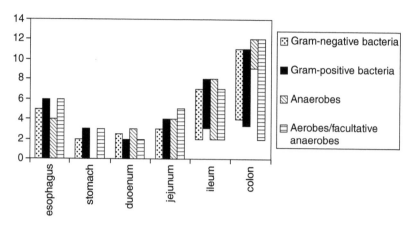

Figure 1 Numbers (^{10}log) of gram-negative bacteria, gram-positive bacteria, anaerobes and aerobes and facultative/anaerobes per gram of intestinal material in the human intestinal tract. *Source*: From Refs. 2–8.

ileum. The intestinal microbiota play a prominent role in gastrointestinal physiology and pathology. A bacterial population is essential for the development of the gastrointestinal mucosal immune system, for the maintenance of a normal physiological environment, and for providing essential nutrients (9). Culturing techniques suggested that dietary changes had a negligible effect on the intestinal microbiota composition (2,10). More recently molecular techniques indicated that diet can alter the microbiota composition, but the predominant groups are generally not substantially altered (11,12). In contrast, antibiotics can dramatically alter the composition of the intestinal microbiota.

Physiology of Microbiota Host Interaction in Humans

Normal gastrointestinal tract microbiota is essential for the physiology of its host. The microbiota in the gastrointestinal tract have important effects on nutrient processing, immune function, and a broad range of other host activities some of which are briefly described below (13). Pasteur (1822–1895) suggested that the intestinal microbiota might play an essential role in the digestion of food. We now know that bacteria harbor unique metabolic capabilities which enable otherwise poorly utilizable nutrients to be metabolized (14). The intestinal microbiota possess enzymes that can convert endogenous substrates, and dietary components, such as fibers, to provide short-chain fatty acids, and other essential nutrients, which are absorbed by the host (10). This interaction of host and bacteria, when one or both members derive specific benefits from metabolic capabilities, is defined as mutualism. Bacteria also produce a number of vitamins that the host can utilize, especially those of the B-complex (15).

The microbiota affords resistance to colonization by potential pathogens that cannot compete with entrenched residents of the microbial community for nutrients (13). Autochthonous or native microorganisms colonize specific intestinal habitats, whereas allochthonous or transient bacteria can only colonize particular habitats under abnormal conditions. The normal microbiota prevent colonization of allochthonous species or potential pathogens by releasing metabolic waste products as well as bacteriocins, and colicins which have antibacterial activity. A pathogenic relationship results in damage to the host. Most pathogens are allochthonous microorganisms. However, some pathogens can be autochthonous to the ecosystem, and live in harmony with the host unless the system is disturbed. Antibiotic therapy can drastically reduce the normal microbiota, and the host may then be overrun by introduced pathogens or by overgrowth of commensal microbial members normally present in small numbers. One notable example is following treatment with clindamycin, overgrowth by *Clostridium difficile* that survives the antibiotic treatment can give rise to pseudomembranous colitis (10,16).

Microbial factors are known to influence host postnatal development. Commensals acquired during the early postnatal life are essential for the development of tolerance, not only to themselves but also to other luminal antigens. Development of B- and T-cell responses depend on the microbiota. The natural antibodies that arise in response to the antigens of the normal gut microbiota are of great importance in immunity to a number of pathogenic species. Somatic hypermutation of Ig genes in intestinal lymphoid follicles plays a key role in regulating the composition of the microbial community (14).

The microbiota participate in bile acid metabolism. In the colon, bacterial enzymes convert cholic acid and chenodeoxycholic acid into the secondary bile acids deoxycholic acid and lithocholic acid, respectively, which in general are poorly reabsorbed; most of these are then eliminated in the stool. In patients with small bowel bacterial overgrowth (SBBO), bile acids are deconjugated and metabolized more proximally in the small bowel, and removed from further participation in the normal enterohepatic circulation, resulting in bile acid malabsorption and steatorrhea. Steatorrhea is defined as excessive loss of fat in the stool, i.e., greater than 7 g or 9% of intake for 24 hours (3).

The effects of having a normal intestinal microbiota has been determined by comparing the characteristics of germ-free and conventionally reared animals. In the small bowel of germ-free animals there are dramatic reductions in leukocytic infiltration of the lamina propria, and both the size and number of Peyer's patches. Moreover, the intraluminal pH is more alkaline, and the reduction potential more positive. Colonization of the intestinal tract of germ-free animals with even a single strain of bacteria is followed by the rapid development of physiologic inflammation of the mucosa resembling that of conventional animals. The migrating motor complex (MMC) is a cyclic pattern of motility that occurs during fasting, and is an important mechanism in controlling bacterial overgrowth in the upper small bowel. Gut transit is slow in the absence of the intestinal microbiota. The effect of selected microbial species in germ-free rats on small intestinal myoelectric activity is promotion or suppression of the initiation and migration of the MMC depending on the species involved. Anaerobes, which have a fermentative metabolism, emerge as important promoters of regular spike burst activity in the small intestine. Introduction of the fermentative species *Clostridium tabificum*, *Lactobacillus acidophilus*, and *Bifidobacterium bifidum* into the gastrointestinal tract of germ-free rats significantly reduces the MMC period, and accelerates small intestinal transit. In contrast introduction of bacteria with respiratory potential such as *Micrococcus luteus* and *Escherichia coli* in the germ-free rats prolongs the MMC period. Intestinal microbiota accelerate transit through the small intestine in the fasting state compared to the unchanged intestinal myoelectric response to food. Overall, the promoting influence of the conventional intestinal microbiota on MMC reflects the net effect of bacterial species with partly opposite effects (17–19).

In conclusion, the bacterial microbiota has a range of specific functions including intestinal transit, absorption of nutrients, and in the modulation of the immune system of the gastrointestinal tract. The introduction of pathogen bacteria can disturb the normal physiological functions of the gastrointestinal tract to a great extent. A number of functional tests for the detection of intestinal pathogenic bacteria have been developed, and are described below.

Importance of Sampling the Gastrointestinal Tract

The current knowledge of the human intestinal microbiota is mostly based on culture techniques but also more recently on molecular biology techniques that are applied to feces and gastrointestinal fluids or biopsies. Sampling of the gastrointestinal tract is clinically necessary for the diagnosis of *Helicobacter pylori*, and the etiology of diarrhea. The gastrointestinal tract is also sampled for research questions on SBBO or for the investigation of host-bacterial relationships in the gut. There are various methods of obtaining material to study the microbiota. Research or diagnosis of bacteria anywhere in the gastrointestinal tract can be performed using invasive or noninvasive methods. The various methods of investigating microbiota in the gastrointestinal tract will be specified for different compartments of the gastrointestinal tract, and the advantages and disadvantages of the sampling methodologies will be described below.

ESOPHAGUS: MICROBIOTA AND SAMPLING TECHNIQUES

Normal Microbiota

The mouth and the oropharynx predominantly harbor Gram-positive organisms (20). The most numerous species comprise the streptococci, *Neisseria*, and *Veillonella*, but *Fusobacteria*, *Bacteroides*, lactobacilli, staphylococci, yeasts, and *Enterobacteria* are also present in smaller amounts (4). The esophagus is covered with a stratified squamous epithelium layer, which is a mechanical barrier coated with saliva and mucus, that has high peristalsis and Ig containing mucus secretion, all of which contribute to prevention of infection. Because of the lack of absolute anatomic or known physiological barriers, bacteria can be introduced into the esophagus by the swallowing of food, by resident oral microbiota or by reflux from a colonized stomach (21). The esophagus, with its large mucosal surface located just downstream of the bacterial species-rich oropharynx, provides a potential environment for bacterial colonization, but so far limited research has been performed. A recent molecular analysis of the distal esophagus indicated members of 6 phyla, of which *Streptococcus* (39%), *Prevotella* (17%), and *Veillonella* (14%) were the most prevalent, and also demonstrated that most esophageal bacteria are similar or identical to residents of the upstream oral microbiota (21). Quantitative cultivation-based studies indicated that aerobic organisms were present in all, and obligate anaerobes in 80% of the subjects investigated. No differences in frequencies of isolation or composition of the microbiota were found between different subjects (5,22).

Disease-Causing Microbiota

A pathogen is a microorganism which by direct contact with or infection of another organism causes disease in that organism. Thus a microbe which produces a toxin that causes disease in the absence of the microbe itself would not be regarded a pathogen. Members of the commensal microbiota may become pathogenic and cause disease if the host defense mechanisms are compromised, or if they are introduced into normally sterile body sites. The esophagus of individuals with deficient immune systems (HIV or post-transplantation patients) may become infected with *Candida albicans*, cytomegalovirus, herpes simplex virus, *Histoplasma capsulatum, Mycobacterium avium*, and *Cryptosporidium*. These microorganisms are usually not seen in immunocompetent persons. With the exception of *Mycobacterium* species, bacterial etiologies for inflammation involving the distal esophagus have not been explored (23). Mycobacterial involvement of the esophagus is rare (incidence 0.14%) in both immunocompromised and immunocompetent hosts with advanced pulmonary tuberculosis (23).

Luminal Washes

Luminal washes to sample esophageal bacteria give poor yields. The washes may contain a few transient bacteria of oropharyngeal origin, or even no microbes at all, or an average of 16 colony forming units per ml (CFU/ml) with no common species found (24,25). Either intestinal contents are passed through the alimentary canal with high peristalsis, and prevent bacteria from residing in the esophagus, or the bacteria present in the washes are not culturable. Another possibility is that the bacteria are very closely associated with the esophageal mucosa, and cannot be removed by simple washes. This technique is not commonly used for research questions, and is clinically irrelevant.

Biopsy

Esophageal mucosal biopsy specimens from the distal esophagus can be obtained during upper endoscopy. The endoscope passes orally into the esophagus, and the biopsy forceps can be shielded from the oral microbiota. The forceps consists of a pair of sharpened cups. Forceps with a central spike make it easier to take specimens from lesions which have to be approached tangentially (such as in the esophagus). The maximum diameter of the cups is limited by the size of the operating channel. The length of the cups is limited by the radius of curvature through which they must pass in the instrument tip (26). Patients are instructed not to eat or drink for at least 4–6 hours before endoscopy (small sips of water are permissible for comfort) (27). The channel of the endoscope can also harbor bacteria if secretions have inadvertently been suctioned while advancing the endoscope. Oropharyngeal and gastric bacteria can contaminate the biopsy. Chlorhexidine or acidified sodium chlorite mouth rinse has been used to decontaminate the oropharynx. To compare biopsy samples of two individuals or to compare the reproducibility in one subject the biopsies have to be taken at the same level (28).

STOMACH: MICROBIOTA AND SAMPLING TECHNIQUES

Normal Microbiota

The human stomach is lined with columnar secreting epithelium. Normally most of the bacteria in the stomach are killed because of the low pH levels, and the typical numbers detected are less than 10^3 CFU/ml (2,6,26). Lactic acid bacteria are commonly isolated from the human gastric acid contents, especially when good anaerobic techniques are used. Candida and some other yeast species are also detected. Bacteria isolated from gastric contents are considered transient members. These bacteria have been passed down from habitats above the stomach or have been present in ingested materials (29). The normal resident microbiota of the stomach consists mainly of Gram-positive aerobic bacteria, such as streptococci, staphylococci, and lactobacilli (2,6,26,30). The microbiota isolated from gastric contents are presented in Table 1. In healthy fasting patients large numbers of *Enterococcus*, *Pseudomonas*, *Streptococcus*, *Staphylococcus*, and *Rothia* (*Stomatococcus*) may be isolated in culture when acidity is physiologically reduced, as occurs at night, and during phase I (motor quiescence) of the MMC (32–34).

Table 1 Microorganisms Isolated from the Stomach by Culturing

Microbial type
Lactobacilli
Streptococci
Bifidobacteria
Clostridia
Veillonella
Coliforms
Peptostreptococcus, Bacteroides
Staphylococcus, Actinobacillus
Candida albicans
Torulopsis
Unidentified yeasts
Neisseria
Micrococcus

Note: The most prevalent bacterial types are italicized.
Source: From Refs. 2, 15, 22, 31.

Disease-Causing Microbiota

Bacteria closely associated, and attached to the epithelium like *Helicobacter pylori*, may be sampled from gastric contents with difficulty (29). *H. pylori* is a Gram-negative bacterium that resides below the mucous layer next to the gastric epithelium. *H. pylori* is rarely found before age 10 but increases to 10% in those between 18 and 30 years of age, and to 50% in those older than age 60 (35). In developing nations the majority of children are infected before age 10, and adult prevalence peaks at more than 80% before age 50. Thus *H. pylori* infection ranges depend on age and socioeconomic differences (36). *H. pylori* produces urease, an enzyme that breaks down urea into ammonium and bicarbonate. Ammonium provides an alkaline environment, which helps the bacterium protect itself from gastric acid injury. Most infected subjects do not have symptoms of *H. pylori* infection. However, *H. pylori* may induce acute gastritis with symptoms such as epigastric pain, bloating, nausea and vomiting, and/or chronic gastritis. Furthermore, it may also be associated with ulcer disease and gastric carcinomas.

Other gastric bacteria besides Helicobacter species only become apparent in patients with reduced acidity (achlorhydria). Achlorhydria may occur in elderly persons (37). Colonization of the gastric lumen may occur in patients on anti-secretory medication meant to reduce gastric acid secretion. Many subjects regularly use these anti-secretory drugs. Acid suppression may allow bacteria to survive in the stomach which results in gastric bacterial overgrowth with the degree of overgrowth depending upon the elevation of the pH (20). Infectious gastritis is more rarely caused by *Mycobacterium tuberculosis*, *Mycobacterium avium, Actinomyces israellii*, and *Treponema pallidum* (3).

Biopsy

To investigate the gastric microbiota, tissue is generally obtained by an endoscopic biopsy. Slightly less invasive methods are available to obtain a specimen such as the use of a small bowel biopsy tube or capsule, or biopsy forceps that can be passed through a modified nasogastric tube positioned either in the gastric body or antrum. A biopsy is clinically unnecessary to diagnose *H. pylori* via microbiological methods unless one wishes to isolate the organism for antibiotic susceptibility testing. Recommendations to maximize the diagnostic yield of endoscopic biopsies include the use of large-cup biopsy forceps, obtaining at least two samples from the lesser curvature and the greater curvature (the prepyloric antrum and the body), and proper mounting and preparation of the samples. Special stains (H&E, Giemsa, and Warthin-Starry staining) are often used to help detect the presence of *H. pylori* (38).

The rapid urease test (by agar gel slide tests) involves placing a biopsy specimen from the antrum of the stomach on a test medium that contains urea (39). The biopsy specimens for the rapid urease test have to be removed from the sterilized biopsy forceps with a sterile toothpick, and have to be placed immediately into a tube. The urea is hydrolyzed by urease enzymes of *H. pylori*, and the ammonium formed increases the pH. A phenol indicator that changes the color from yellow at pH 6.8 to magenta at pH 8.4 can detect the pH alteration. The color change read off 1 hour after and 24 hours after the introduction of the gastric biopsy is an indication for the presence of *H. pylori*. Recommendations to maximize the rapidity and sensitivity of rapid urease tests are to warm the slide, and to use two regular or one jumbo biopsy specimen(s) (40). Increasing the number of biopsies to more than two biopsies from the antrum may increase the sensitivity, given that this probably increases the *H. pylori* load, and therefore the amount of urease. However, this will prolong the endoscopy time and add to the discomfort of the patient. The agar gel test may take up to 24 hours to turn positive, particularly in the presence of a low bacterial density. Recent use of antibiotics, bismuth, or proton pump inhibitors may render rapid urease tests falsely negative. Compared with histology as the gold standard in the diagnosis of *H. pylori* infection, the sensitivity of the rapid urease test is 70–99%, and the specificity is 92–100% in untreated patients (40). Mucosal biopsies can be fixed in neutral buffered formaldehyde, and if the rapid urease test is negative the biopsy can sent in the next day for histologic assessment. The presence or absence of *H. pylori* can be established by examining three sets of tissue levels within 12 consecutive sections. On microscopic examination of the tissue obtained by biopsy, the bacteria may be seen lining the surface epithelium. The sensitivity for histologic examination is 70–90%. Giemsa staining is required for *H. pylori* diagnosis. Culture for *H. pylori* is insensitive. Biopsies should be plated within 2 hours (or transported in a special medium) on nonselective media enriched with blood or serum, and incubated in a moist and microaerobic atmosphere. The identity of any colonies grown can be confirmed using Gram's stain and biochemical tests.

Aspiration

In order to sample gastric fluid a Shiner tube may be used. This is a polyvinyl tube with a stainless steel sampling capsule at the end with which the specimens are obtained by suction. This tube can be sterilized in the autoclave or by boiling (6). Sampling the luminal content of the stomach may lead to underestimation of the size or even misinterpretation of the composition of gastric microbial communities (29). Estimates per unit weight of material of the population levels of microbes attached to an epithelium surface made from samples of the mucosa itself have been found to be higher than estimates made from the luminal content in the region (29). This technique is not clinically relevant, and is hardly ever used in research models.

Urea Breath Test

The urea breath test is a noninvasive test that detects radio-labeled carbon dioxide excreted in the breath of persons with *H. pylori* infection; orally administered urea is hydrolyzed to carbon dioxide and ammonium in the presence of the enzyme urease, which is present in *H. pylori*. In non-infected subjects, urea leaves the stomach unchanged, unless there is urease activity from bacteria in the oral cavity or in situations of gastric bacterial overgrowth. The urea breath test is a highly sensitive (93.3%) and specific (98.1%) method (41). The two breath tests available are the ^{14}C urea (radioactive), and ^{13}C urea (stable isotope) breath tests. The ^{13}C urea breath test avoids radioactivity, and is the test of choice for children and pregnant women. The major limitation is the need for a gas isotope mass spectrometer to analyze the breath samples and calculate the ratio of ^{12}C to ^{13}C. A 4-hour fast is generally recommended before the urea breath test, and a test meal is given before the solution of labeled urea. This test meal delays gastric emptying, and increases contact time with the bacterial urease. It is relatively inexpensive compared to the "gold standard" of endoscopy with biopsy, and histological examination described above. The urea breath test avoids sampling errors that can occur with random biopsy of the antrum. False positive results can occur if gastric bacterial overgrowth with urease-producing bacteria other than *H. pylori* are present. False positive results can also occur if the measurements are taken too soon after the urea ingestion because the action of the oral microbiota on the urea may be measured. False negative results can be obtained if the patients were recently treated with antibiotics, bismuth preparations or acid suppression therapy, because the test is dependent on the numbers of *H. pylori* (42). Performance of the urea breath test has been associated with several disadvantages especially in infants, toddlers or handicapped children because one needs active collaboration. False positive results in infants affect the accuracy of the test, but correction for the carbon dioxide production of the tested individual will improve the specificity (43,44).

Other tests that do not require a mucosal biopsy include serologic tests and stool antigen tests. Chronic *H. pylori* infection elicits a circulating IgG antibody response that can be quantitatively measured by enzyme-linked immunosorbent assay (ELISA tests). The ELISA is based on a specific anti-*H. pylori* immune response, and this serologic test is as sensitive (95.6%) and specific (92.6%) as biopsy-based methods (41). The presence of IgG does not indicate an active infection. IgG antibody titers may decrease over time (6–12 months) in patients who have been successfully treated. ELISA or immuno-chromatographic methods can be performed on the fecal samples to detect *H. pylori* antigen. The limit of sensitivity of the test is 10^5 *H. pylori* cells per g of feces (45). Sensitivities and specificities of 88–97% and 76–100% have been reported (41,44–47). The stool antigen test is not used for follow-up evaluation of the *H. pylori* eradication as it gives false positive results. In conclusion, the noninvasive tests are sufficiently accurate for the diagnosis of *H. pylori* infection.

SMALL INTESTINE: MICROBIOTA AND SAMPLING TECHNIQUES
Normal Microbiota

The small intestine comprises the proximal, mid, and distal areas, which are designated the duodenum, jejunum, and ileum. The velocity of the intraluminal content of the small intestine decreases from the duodenum to the ileum. The microbes isolated from the small intestine include those descending from habitats above the small intestine such as the mouth, and ingested food. The microbes pass through the intestine with the chyme, and in the fasting state by the MMC. The MMC interdigestive motility prevents colonic microbiota from entering the proximal small intestine which would cause SBBO. The microbial species isolated from the small intestine are listed in Table 2. The density of microbiota increases towards the distal small intestine. The upper two thirds of the small intestine (duodenum and jejunum) contain only low numbers of roughly the same microorganisms, which range from 10^3 to 10^5 bacteria/ml (2). Culturing studies indicated that acid- and aero-tolerant Gram-positive species such as lactobacilli and streptococci dominate in the proximal part, while distally anaerobic, and more Gram-negative bacteria increasingly dominate. Whipple's disease is a rare multisystemic bacterial infection caused by *Tropheryma whipplei. T. whipplei* could not be cultured from the small intestine for decades, and was diagnosed by histopathology. Nowadays *T. whipplei* can be detected using polymerase chain reaction (PCR) or ribosomal RNA techniques on duodenal biopsies or fecal samples (48). The rich microbiota of the initial section of the large intestine (cecum) find their way through the ileocecal valve back into the ileum. The microbiota of the ileum begins to resemble that of the colon with around 10^7 to 10^8 bacteria/ml of the intestinal contents. With decreased intraluminal transit, decreased acidity, and lower oxidation-reduction potentials, the ileum maintains a more diverse and numerous microbial community (29). Factors that compromise the oxidation-reduction potential within the tissues are obstruction and stasis, tissue anoxia, trauma to tissues, vascular insufficiency, and foreign bodies (49). Decreased oxidation-reduction potential specifically predisposes to infection with anaerobes (50).

Disease-Causing Microbiota

Pathogenic bacteria of the small intestine, which cause severe diarrhea, are enterotoxic *Escherichia coli* (ETEC) and *Vibrio cholerae. V. cholerae* is diagnosed when it is present in fecal material. ETEC produces enterotoxins that cause intestinal secretion and diarrhea, and is a common cause of traveler's diarrhea. In SBBO, the proximal small intestine is populated by a substantially higher number of microorganisms than usual. These are frequently anaerobic bacteria that are normally not present in large numbers in the duodenum and the proximal jejunum. A total count of microorganisms exceeding 10^5 colony forming units/ml in a duodenal or jejunal aspirate is generally accepted as SBBO (51). Some gastroenterologists

Table 2 Microorganisms Isolated from the Small Intestine by Culturing

Microbial types	Most prevalent microbes in duodenum and proximal jejunum	Most prevalent microbes in distal jejunum and ileum
Lactobacilli	Lactobacilli	
Streptococci	Streptococci	
Bifidobacteria		Bifidobacteria
Clostridia		Clostridia
Coliforms		
Bacteroides		Bacteroides
Veillonellae	*Veillonellae*	
Gram positive nonsporing anaerobes		
Staphylococci	*Staphylococci*	
Actinobacilli	*Actinobacilli*	
Yeasts	*Yeasts*	
	Candida albicans	
	Haemophilus	
		Fusobacterium

Note: The most prevalent bacterial types are italicized.
Source: From Refs. 2, 15, 22.

also accept a concentration of colonic microorganisms above 10^3 CFU/ml as positive for SBBO. A profound suppression of gastric acid may facilitate the colonization of the upper small intestine (20). To diagnose SBBO, the quantitative culture of a small intestine is used, and considered to be the gold standard. Fluid aspirated from the descending part of the duodenum may be cultured in order to detect bacterial overgrowth in diffuse small bowel disorders.

Biopsy

To obtain biopsy samples from the small intestine upper endoscopy has to be performed. Upper endoscopy is performed after an overnight fast of at least 10 hours. An endoscope has a length of approximately 1 meter, and has a biopsy channel. During endoscopy the esophagus, stomach, and duodenal wall can be systematically inspected. To allow a good view air insufflation is required; the patient may complain of bloating during the endoscopy. When the endoscope reaches the site of interest, the biopsy from the small intestinal mucosa is rapidly taken by standard biopsy forceps. Figure 2 shows the size (in centimeters) of the tip of an endoscope, and a biopsy forceps. The distal part of the jejunum and the ileum cannot be reached using a standard endoscope, and therefore is not sampled. Endoscopic biopsies are an adequate substitute for jejunal suction biopsies. The advantage over capsule biopsy is that the site of interest can be inspected before the biopsy is taken (52–54). Adequacy of mucosal biopsies is a function of size and numbers of biopsies obtained (54). Alligator-type forceps obtain larger specimen pieces than oval-shaped forceps (55). Forceps with a needle, or the multibite forceps, allow more biopsies to be taken per passage, and improve the quality of tissue obtained (55). Biopsy forceps without a needle can be used to obtain two samples per passage through the endoscope that are quantitatively as good as when only one sample is collected. This approach can save time, and causes no significant damage to the biopsy specimens. Because air insufflation may distort the intraluminal anaerobic environment, nitrogen could be used as a substitute if the intention is to culture anaerobic bacteria. There is also the risk of contamination with microbiota from more proximal habitats that were passed along via the endoscope.

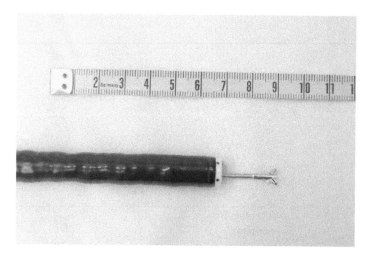

Figure 2 Tip of a standard endoscope and biopsy forceps with needle (tape measure in centimeters).

The biopsies have to be taken at a certain distance from the endoscope to prevent sampling contaminated parts of the intestine.

Intestinal biopsies taken from living persons may not yield satisfactory results because the biopsies are only a minimal part of the total intestinal wall (56). The number of persons sampled must be large to generate reliable results. The best source of information on microbiota in the small intestine so far has been achieved with sampling from autopsy studies of accident victims. As slow cooling of the gastrointestinal tract can cause alterations in bacterial localization the samples have to be taken immediately after death (57), and the number of individuals sampled must still be quite large.

Full Thickness Biopsy

Full thickness biopsy is a peroperative or laparoscopic biopsy (muscularis-containing biopsy) used to diagnose motility disturbances. One incision is situated below the umbilicus, and one in the left fossa. The bowel loop is identified laparoscopically, and will then be exteriorized through the incision below the umbilicus. The full thickness biopsy of at least 10×10 mm will then be taken with a surgical knife. The bowel loop is closed with absorbable sutures, and repositioned into the abdomen (56). Drawbacks of biopsies taken at surgery are the manipulation of the patients' diet (fasting), and the bowel preparation or preoperative treatment with antibiotics (29,58). Biopsies taken at surgery have the advantage of larger sample size than endoscopic biopsies, and various analyses may be applied such as molecular typing of bacteria in intestinal tissue of Crohn's patients (59).

Mucosal Brushings

Mucosal brushings may be used to sample bacteria from the intestinal mucosa. The cytology brush, protected by a sheath, is passed through the instrument channel of the endoscope. After the endoscope is placed at the location of interest, the brush is advanced from its sleeve within sight of the mucosal surface, and rubbed and rolled across the surface. Thereafter, the brush is pulled back into the sleeve. Normally, cytology brushes are only covered with a plastic sleeve to protect the specimen during withdrawal. This sleeve, however, does not protect against contamination; the use of suction of saliva and gastric fluid during endoscopy contaminates the suction channel of the endoscope, and the subsequent passage of the brush without a sheath through the suction channel causes loss of sterility of the brush (27). These brushes cannot be used for sampling bacteria in the lumen of the gastrointestinal tract. Avoidance of any suction during endoscopy is extremely difficult. To obtain small bowel samples without contamination one could utilize a catheter with a specimen brush plugged with sterile Vaseline. Brushes cannot be protected from contact with air, so it is not useful for the isolation of anaerobes for culture. To determine the concentration of bacteria obtained by the brush present per milliliter, one has to standardize the loading capacity of the brush used. Brushing is a highly reproducible technique (92%) (60).

Peroperative Needle Aspiration

Peroperative needle aspiration is useful for relatively inaccessible locations within the intestinal tract. The technique is only applicable for patients with an underlying disease who will undergo laparotomy or laparoscopy. The microbiota may be influenced by pre-operative fasting, antibiotic prophylaxis, and anesthesia. Until 1959 the peroperative needling technique was regularly performed at operation (61,62) but is currently no longer performed routinely.

The advantages of this technique are asepsis and the lack of contamination from other regions of the gastrointestinal tract.

Self-Opening Capsule

The Crosby capsule, first applied in 1957, was used to obtain biopsies from the small intestine before the introduction of the endoscope. This self-opening capsule is a metallic capsule of 19 to 11 mm with a round opening of 4 mm (53). A long tiny tube is attached to the capsule, and this is muscle loaded through an endoscope which is passed into the second part of the duodenum. Intestinal mucosa is sucked into the tube by suction and excised. Every part of the stomach and the small intestine can be reached (63). Sizes of the biopsies are 5–8 mm, with stomach biopsies usually being smaller. Failure of obtaining biopsies is 6%. The mucous membrane is very mobile with respect to the muscular layer so only mucosa is sucked into the capsule, and the risk of perforation is very small. Muscularis propria is never cut. The risk of bleeding (0.14%) and intestinal perforation is very small (64).

Capsules that can be opened electronically are also available. They have the disadvantage of a long interval between sample collection and culturing. During this interval, bacteria inside the capsule can replicate, and influence growth of other bacteria in the capsule. It is a very imprecise method. The advantage of this technique is that, like the Crosby capsule, every part of the small intestine can be examined. The disadvantage of the suction biopsy capsule used to provide specimens from the proximal jejunum is the need for radiological screening for the location of the capsule. This makes it unsuitable for repeated use in young children, and women who are or might be pregnant. There may be some discomfort when the procedure is prolonged. The technique fails in up to 10% of the cases. To overcome the problem of determining the sampling location with the capsule biopsy, it is better to take specimens with endoscopic forceps. Capsule biopsies are not common in current clinical gastroenterology practice (52).

Aspirate

Small bowel aspiration for quantitative and qualitative culture specimens is still regarded as the gold standard for diagnosis of SBBO. The sample should be properly harvested with respect to sterile technique and accurate location. The exact composition of the microbiota is not important for the diagnosis of SBBO if one uses the definition that more than 10^5 colony forming units/ml small intestinal fluid represents SBBO, but it is of use when antibiotic therapy is being considered. It should be realized that cultures of randomly harvested samples can produce false-negative results if the sample is not taken from the actual site of bacterial overgrowth.

Culturing is not necessary if one uses gas chromatographic detection and analysis of volatile fatty acids in the aspirates. The volatile fatty acids are produced by the metabolism of microorganisms such as *Bacteroides* and *Clostridia*. This is essentially a rapid test for the presence of anaerobic bacteria. When gas chromatography of volatile fatty acids is compared with cultures of jejunal aspirates, it shows a sensitivity of 56% and specificity of 100% (51). When the tests for volatile fatty acids in jejunal aspirates are positive, this always indicates the presence of bacterial overgrowth. This procedure avoids the more complicated, time-consuming, and expensive bacteriological analysis of jejunal samples (51,65,66). The numbers of bacteria per milliliter of intestinal fluid taken at two different levels of the proximal jejunum show highly significant correlations ($rs = 0.90$, $p < 0.001$;

thus one does not have to obtain the aspirate from the exact same location in the proximal jejunum (51).

Aspirate can be acquired by intestinal intubation with sterile or nonsterile tubes, the capsule method, direct needle aspiration of the gut contents, peroral intubation, and by the string test as described below.

Intubation with Sterile or Nonsterile Tubes

This endoscopic method for collection of proximal gastrointestinal fluid for culture is simple and can be performed during routine endoscopy. When the endoscope reaches the descending part of the duodenum, the polyethylene tube will pass through the biopsy channel into the intestinal lumen. Intestinal intubation seems to be the most suitable and reliable method for studying small intestinal microbiota, because of the short sample collection time and minimal disturbance of physiological conditions. Care must be taken to prevent contamination with upper respiratory tract microbiota during the passage of the tube, and to maintain oxygen-free conditions for anaerobic culturing. A closed polyethylene tube filled with water through the suction channel of the endoscope is therefore recommended, as it is not necessary to keep the suction channel sterile. The water has to have been boiled for sterilization and the removal of dissolved oxygen. The distal end is closed with a plug of agar. Because the innertube remains sterile even after the passage through the nonsterile suction channel of the endoscope, the use of an overtube eliminates the possibility of contamination. The proximal end can be attached to a double way stopcock connected to a syringe containing boiled water. In the duodenum the agar plug can be expelled from the tube by injection of the water in the syringe. After several minutes the expelled water has gone through and the duodenal contents can be aspirated into the tube, after which the tube is removed from the endoscope. Precision of the sample site and proven absence of contamination are the main advantages. Since fresh aspirate is known to tolerate oxygen fairly well for an exposure time of at least 8 hours, it is a good method for obtaining aerobic and anaerobic samples (60,62,67).

Highly significant correlations (rs=0.84, $p<0.001$) were found between the numbers of bacteria/ml of jejunal aspirate obtained from the closed and open tubes, confirming that the intubation method is highly reproducible (51). The use of suction during endoscopy contaminates the suction channel of the endoscope. The first milliliter of aspirate can be discarded to avoid this, although this is very difficult in the duodenum, where at best only a few milliliters of aspirate will be found (67). Using an open tube for collection of small bowel fluid can theoretically lead to contamination, but according to reported studies this does not seem to be the case (68,69).

Duodenal String Test (Enterotest)

The duodenal string test capsule is a cheap and simple device used for sampling the contents of the upper gastrointestinal tract. It has been used for the diagnosis of typhoid fever, whereby sampling duodenal contents by a "string" test yields a positive culture in 70% of patients (70). The weighted gelatin capsule contains a silicone rubber bag and a 140 cm highly absorbent nylon string. After a 10-hour fast the device is administered. The first 10 cm of the nylon line is pulled out from the capsule by the protruding loop. The capsule is then swallowed with water while the loop is held outside the mouth. The loop is then taped to the face to secure the line. After approximately 3.5 hours the thread has moved into the duodenum. The volume of the duodenal fluid absorbed by the distal end of the thread is calculated by subtracting the dry weight of the segment. The distal end is squeezed out between sterile gloved fingers in order to collect the intestinal contents.

Its major applications in pediatrics are the diagnosis of enteric parasitic infestations, and the diagnosis of *Salmonella* infection, *Giardia lamblia*, and assessment of duodenal bile salts in the diagnosis of neonatal cholestasis in duodenal contents. A drawback of the Enterotest is that when the string is pulled out of the gastrointestinal tract, the intestinal contents adhering to it are exposed first to the sterilizing effect of gastric acid, and afterwards to contamination with microbiota present in the esophagus and pharynx. The Enterotest is not useful for the isolation of anaerobes because samples cannot be protected from contact with air. The clinical value of the string test compared with a sterile endoscopic method for sampling small bowel secretions is limited by poor sensitivity, specificity, and positive predictive value. Thus the string test is not an adequate substitute for oro-duodenal intubation for the detection of SBBO (60,71).

Peroral Intubation

Peroral intubation and aspiration of luminal contents can be achieved using Miller-Abbott or Levin tubes. These tubes were modified to suit the special needs for culture studies. The headpiece of a Miller-Abbott tube comprises a capsule, which may be opened and closed by hydraulic pressure. The capsule has an advantage of large size (44.5×12 mm), but it has been proven possible for bacteria to gain access into the closed capsule in vitro. A Levin tube is clinically used as a gastroduodenal feeding tube with a length of 125 centimeters. A long radio-opaque tube is used, marked for accurate placement, either single- or double-lumened, with or without balloons, and perforated by one or more holes at its distal end. These perforations were either left free or were protected by means of a collodion membrane, a thin rubber sheath, or by plugs, which could be either dissolved or dispelled by positive pressure at the moment of taking samples for culture. Contamination of the tubes depends on the degree of contamination of the surrounding fluid, the exposure time, and the static environment. The small intestine contains only a very small quantity of fluid in contrast to gastric juice, which may be aspirated in large quantities. A disadvantage of peroral intubation is the lack of certainty that the specimen obtained from the desired level of the intestine has not been contaminated by bacteria from a higher position during its passage.

Noninvasive Methods

Because small intestinal intubation for quantitative culture is inconvenient, expensive, and not widely available, a variety of surrogate tests for bacterial overgrowth in the small intestine have been devised based on the metabolic actions of enteric bacteria rather than on increases in the number of bacteria. Several indirect methods have been developed to overcome the problem of location-dependence of aspirates for culturing. A comparison between the small intestinal noninvasive tests versus invasive methods with culture of material obtained for diagnosis of SBBO is presented in Table 3. Most of these indirect tests lack sensitivity for reliable detection of SBBO. The main reason for this is the great variability of the microbiota and its metabolic profile. The tests are based on a specific bacterial metabolic activity. Thus, if this particular activity is not present in the microbiota of a SBBO patient, the test will yield a false-negative result. For this reason urinary excretion tests (e.g., indican excretion, D-xylose, conjugated para-aminobenzoic acid), and analysis of intestinal aspirates for bacterial metabolic products (e.g., deconjugated bile acids in serum) lack the required reliability for detection of SBBO, and have become obsolete (71–75). These tests will not be described further.

Table 3 Small Intestinal Noninvasive Tests Compared to Jejunal Culture (Gold Standard)

Test	Sensitivity (%)	Specificity (%)	Simplicity
14C-D-xylose BT	42–100	85–100	Excellent
Lactulose H_2 BT	68	44	Excellent
Glucose H_2 BT	62–93	78–83	Excellent
13C and 14C- glycocholate BT	20–70	76–90	

Abbreviations: BT, breath test; H_2, hydrogen.
Source: From Refs. 42, 51, 78, 80, 84, 101–105.

To diagnose bacterial overgrowth, various breath tests may be used including the ^{14}C-glycocholate, ^{14}C-D-xylose, lactulose-H_2, and glucose-H_2 tests. The rationale for the breath test is the production of volatile metabolites i.e., carbon dioxide (CO_2), hydrogen (H_2) or methane (CH_4), by intraluminal bacteria from the administered substrates, which can be measured in the exhaled air. The most successful and popular methods analyze either expired isotope-labeled CO_2 after timed oral administration of [14]C- or ^{13}C-enriched substrates, or breath hydrogen following feeding of a non-labeled fermentable carbohydrate substrate.

The [14]C- and ^{13}C-breath tests measure the pulmonal excretion of labeled CO_2 produced by the fermentation of labeled substrates, using either a radioactive or a stable isotope. The increasing availability of methods for analyzing stable isotopes has raised interest in replacing the radioactive ^{14}C by non-radioactive [13]C. The use of radioactive isotopes is not recommended for study of children or women who are or might be pregnant. [13]CO_2 can be measured by mass spectrometry. Because of concerns about diagnostic accuracy, costs of the substrates and equipment, and limited availability, these tests have not gained widespread acceptance.

The first breath test to diagnose SBBO was the hydrogen breath test described by Levitt in 1969 (76). Hydrogen is a constituent of human breath derived exclusively from bacterial fermentation reactions in the intestinal lumen. Detection of hydrogen in expired breath is considered a measure of the metabolic activity of the hydrogen-producing bacteria. Bacteria produce hydrogen from carbohydrate substrates, and human tissue does not generate hydrogen. The colon is considered to be the only place in the human body where hydrogen is produced, because of the high amount of hydrogen-producing bacteria. In cases of SBBO, hydrogen is also produced in the small intestine. Part of the produced hydrogen is reabsorbed from the intestine into the blood, and is exhaled. Measurement of breath hydrogen could circumvent the administration of a radioactive isotope in testing for bacterial overgrowth. This test assumes the presence of a hydrogen-producing microbiota, but in 15–20% of humans the microbiota of the subject does not meet this condition. Hydrogen breath analysis is therefore not sufficiently reliable as a diagnostic tool in SBBO.

[14]C-Glycocholate Breath Test

[14]C-glycocholate breath test or bile acid test is based on the bile salt deconjugating capacity of bacteria in the proximal small bowel. Conjugated bile acids are excreted through the bile in the duodenum, and they are reabsorbed in the terminal ileum. Conjugated bile acids are in the enterohepatic circulation. Physiologically, less than 5% of the conjugated bile acids reach the colon. After excretion in the duodenum, bile acids stimulate micellization of dietary lipids. After oral administration of glycocholic bile

acid (a normal component of bile) this is normally reabsorbed in the terminal ileum. In cases of SBBO some bacteria split off glycine on the amide bond of cholylglycine. Glycine is absorbed, and fermented in the liver to CO_2, H_2O, and ammonia (NH_4); the CO_2 produced is exhaled. When using ^{14}C glycocholate, the [14]CO_2 in the exhaled air can be measured.

The sensitivity is too low (20–70%) to allow SBBO to be demonstrated without additional intestinal culturing. A rise in labeled CO_2 does not differentiate bile salt wastage from bacterial overgrowth. This is a disadvantage given that a significant number of SBBO patients may have had ileal resection. Ruling out bile salt malabsorption as an explanation for a positive breath test can be done with stool collection (42,77).

The false negative rate for the ^{14}C-glycocholate breath test is 30–40%. There are three reasons for false negative outcomes. Firstly, one needs anaerobic organisms to deconjugate bile salts. Secondly, not all cases of bacterial overgrowth involve bile salt deconjugation. Lastly, the fatty meal (usually a polymeric supplement) given with the cholylglycine may, in theory, affect the ratio of labeled and unlabeled carbon dioxide absorbed, diluting the labeled carbon dioxide with that produced from the metabolism of the meal. False positive results are possible in case of ileal pathology, ileal resection, and increased intestinal transit. In those cases bile acids are deconjugated by the (anaerobic) colonic microbiota. The disadvantage of using radioactivity in ^{14}C-substrate breath tests can be overcome by using the stable ^{13}C-isotope, which is measured by mass spectrometry in breath samples. However, the use of ^{13}C-isotope does not improve the sensitivity.

[14]C-D-xylose Breath Test

The ^{14}C-D-xylose breath test was considered to be the only breath test for the detection of bacterial overgrowth with high sensitivity (95–100%) and 100% specificity, but these promising results have not been sustained (42). Compared with cultures of the duodenal aspirates, the sensitivity and specificity are 60% and 40%, respectively (78).

This test is based on the assumption that the overgrown aerobic Gram-negative microbiota ferment D-xylose. The [14]CO_2 produced, and unmetabolized xylose are absorbed by the proximal small bowel, which thus avoids confusion of results caused by metabolism of substrate by colonic bacteria. Subjects must fast at least 8 hours before the test, and no smoking or exercise is permitted for 12 hours before the breath test. Following a 1 g oral dose of ^{14}C-D-xylose in water, elevated [14]CO_2 levels are detected in the breath within 60 minutes in 85% of patients with SBBO.

False negative rates for the ^{14}C-D-xylose breath test are 35–78%. False negative results cannot be entirely attributed to the absence of D-xylose fermentation of the microbiota (overgrown bacteria in 81.8% of SBBO patients are capable of D-xylose fermentation); body weight is correlated to endogenous CO_2 production, and should therefore also be taken into account (79). Disturbed gastric emptying and small intestinal motility can also contribute to a false-negative result of the ^{14}C-D-xylose breath test because of delayed delivery of the labeled substrate to the metabolizing microbiota. Refinement of the ^{14}C-D-xylose breath test to include a transit marker for intestinal motility increases its specificity. With the transit marker one can determine whether the site of metabolism is in the small intestine or the colon (80).

Lactulose Hydrogen Breath Test

Lactulose is an easily fermented disaccharide, and is used for the detection of bacterial overgrowth, and for determination of the orocecal transit time. The lactulose hydrogen breath test is a simple, inexpensive, and noninvasive technique to diagnose SBBO. The lactulose breath test is performed after 12 hours fasting previous to the test. Hydrogen breath samples are taken at baseline, and subsequently every 10–30 minutes after the test meal that contains 10–12 g of lactulose. The hydrogen breath samples are analyzed gas chromatographically (81). Baseline samples average 7.1 ± 5 parts per million (ppm) of H_2 and 0–7 ppm for CH_4 (82). Values of the baseline sample over 20 ppm H_2 are suspect for bacterial overgrowth. Values between 10 and 20 suggest incomplete fasting before the test or ingestion of slowly digested foods the day before the test, the colon being the source of the elevated levels (82). Slowly digested foods like beans, bread, pasta, and fiber must not be consumed the night before the test because these foods produce prolonged hydrogen excretion (82). The patient is not allowed to eat during the complete test. Antibiotics and laxatives must be avoided for weeks prior to breath hydrogen testing. Cigarette smoking, sleeping, and exercise must be avoided at least a half hour before and during the test because these may induce hyperventilation (42). Chlorhexidine mouthwash must be used before the test to eliminate oral bacteria, which might otherwise contribute to an early hydrogen peak after the substrate is given. Lactulose, which reaches the colon, shows peaks usually more than 20 ppm above baseline after 2–3 hours of testing. Lactulose is not absorbed in the small intestine so every patient should have a colonic peak, assuming the colonic microbiota has not been altered. Peaks associated with SBBO occur within 1 hour, and are less prominent. Some laboratories measure H_2 and CH_4 simultaneously whereas others test CH_4 selectively after flat lactulose tests (42). Figure 3 shows lactulose breath test results in a patient with small bowel bacterial overgrowth.

The lactulose hydrogen breath test is positive for small intestinal bacterial overgrowth if there is an increase in breath hydrogen of > 10 parts per million above basal that occurs at least 15 minutes before the cecal peak. Strict interpretative criteria, such as requiring two consecutive breath hydrogen values more than 10 ppm above the baseline reading, and recording a clear distinction of the small bowel peak from the subsequent colonic peak (double peak criterion), are recommended. Application of the double peak criterion alone for interpretation of the lactulose hydrogen breath test is inadequately sensitive, even with scintigraphy, to diagnose bacterial overgrowth. Twenty-seven percent of normal subjects have no peak due to organic acid reduction or dilution from voluminous diarrhea (42).

The disadvantage of this test is that it is not always easy to distinguish breath hydrogen arising from small bowel colonization from that resulting from cecal fermentation in patients with an exceptionally rapid orocecal transit time. A comparison with the jejunal culture sensitivity of 68% and specificity of 44% has been described (51). A sensitivity of 16% for SBBO has been described (83).

Despite the attractive aspects of ease of performance and avoidance of a radioactive tracer, breath hydrogen tests are not sufficiently sensitive or specific to justify their substitution for the ^{14}C-D-xylose breath test for noninvasive detection of intestinal bacterial overgrowth.

Glucose Hydrogen Breath Test

Glucose hydrogen breath tests can also be used to detect SBBO. Glucose is completely absorbed before reaching the colon even in patients with previous gastric surgery, who have faster than normal transit. Patients receive a solution containing 50–80 g of glucose dissolved in 250 ml water after fasting for 12 hours. Breath hydrogen concentrations are analyzed with an H_2 monitor after direct expiration through a Y-piece that prevents air from mixing with the exhaled hydrogen (84). Hydrogen concentration is determined every 10–15 minutes for two hours. Results of the hydrogen breath test are considered

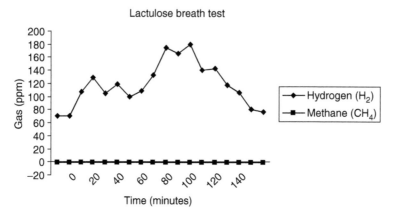

Figure 3 Production of hydrogen (H_2) and methane (CH_4) in a patient with bacterial overgrowth of the small bowel (SBBO). Fasting H_2 and CH_4 production at–10 and 0 minutes; 10 grams of lactulose was administered at 0 minutes.

positive when the hydrogen concentration increases by 14–20 ppm (85). Smoking and exercise are not allowed during the test, and the day previous to the test (86). The hydrogen breath test shows stable intra-individual results in healthy people. However, in patients with high values there is a large day-to-day variation (87). The coefficient of variation is 5–10% (84,88). Sensitivity of 93% and specificity of 78% have been described (85). The glucose hydrogen breath test has a sensitivity of 62% and a specificity of 83% compared with jejunal culture (51). Poor sensitivity due to rapid absorption of glucose substrate in the proximal small bowel, which inhibits hydrogen generation, can be explained by a washout effect of concomitant diarrhea, loss of bacterial microbiota because of recent antibiotic therapy, or an acidic bowel lumen.

LARGE INTESTINE: MICROBIOTA AND SAMPLING TECHNIQUES

Normal Microbiota

The large intestine including the cecum, colon, and the rectum harbors over 500 species of bacteria, mainly obligate anaerobes (99.9%) with 10^{11}–10^{12} CFU/g (2,10). Microorganisms isolated from large intestine and fecal samples are listed in Table 4. *Bacteroides, Bifidobacteria, Eubacteria, Clostridia*, and *Enterobacteriaceae* can predominantly be found in the colon. Novel molecular methods are aiding better understanding of the microbiota, which is challenging to culture due to the anaerobic nature of most of the microbiota, and insufficient knowledge of the culturing conditions (90,91). Knowledge about the mucosa-associated bacterial communities in different parts of the colon is limited as most attention has been focused on bacteria present in feces. Enormous microbial populations can develop in the lumen of the large bowel, and especially in that of the cecum because these areas have a relative stagnation in the flowing stream (up to 60 hours) and very low oxidation-reduction potentials. The transit time of the lumenal content exceeds the doubling times of bacteria. Whether the microbiota is transient or truly autochthonous to habitats in the region remains a main concern. Bacteria in food are known to pass into human feces at high population levels. Bacteria from habitats above the large bowel pass down into the lumen of that region. The population levels of transients probably do not contribute significantly to

Table 4 Microbiota Isolated from the Large Intestine and Feces by Culturing

Microbial types in large intestine	Microbial types in feces
Lactobacilli	Lactobacilli
Streptococci	Streptococci
Bifidobacteria	Bifidobacteria
Clostridia	Clostridia
Propionibacterium	*Propionibacterium*
Eubacterium	*Eubacterium*
Bacteroides	*Bacteroides*
Fusobacterium	*Fusobacterium*
Veillonella	*Veillonella*
Staphylococcus	*Staphylococcus*
Coliforms	Coliforms
Bacillus sp	*Bacillus* sp
Yeasts	Yeasts
Spiral shaped microbes	Spiral shaped bacteria
Actinobacillus	*Peptococcus*
Enterobacteriaceae	*Ruminococcus*
Enterococci	*Coprococcus*
	Acidaminococcus, Succinivibrio, Butyrivibrio, Megasphaera, Gemminger
	Catenabacterium
	Peptostreptococcus

Note: The most prevalent bacterial types are italicized.
Source: From Refs. 2, 10, 15, 22, 89.

the level in the region. Bacteria in the colon are important in processing maldigested carbohydrates (92).

Disease-Causing Microbiota

Yersinia enterocolitica, Salmonella, Shigella, Campylobacter, Clostridium difficile, enterohemorragic *Escherichia coli* (EHEC), and enteropathogenic *Escherichia coli* (EPEC) are the most common pathogenic bacteria in the colon that cause diarrhea. Diarrhea can also occur after oral antibiotic treatment. Poorly absorbed antibiotics change the normal composition of the microbiota in the colon (93). Suppression of the normal microbiota may lead to reduced colonization resistance with subsequent overgrowth of resistant microbiota, yeasts, and *Clostridium difficile*. This organism produces a protein toxin which causes necrosis and ulceration of the colonic mucosa, called antibiotic-associated hemorrhagic colitis.

Biopsy

A standard colonoscope has a length of 1.30 to 1.60 m, so that the colon and the distal ileum can be evaluated. Long colonoscopes (165–180 cm) are able to reach the cecum even in overly long and tortuous colons (27). Biopsy specimens can be collected with a flexible colonoscope and flexible biopsy forceps. Patients are given a laxative solution to drink the day before the examination. The object of full preparation is to cleanse the entire colon of fecal material, especially the proximal parts, to allow a clear view (27). So it is very likely that the bacteria in the biopsy sample are mucosa associated as the luminal bacteria will have been washed away (94). Typically biopsy samples contain 10^5–10^6 bacteria, and the predominant mucosa-associated bacterial community is host specific and uniformly distributed along the colon but differs significantly from the fecal community (95). Biopsy samples are very small in size, and therefore more easily exposed to oxygen during sampling; therefore, the number of viable strict anaerobes might be reduced easily. Relatively high levels of facultative anaerobes are reported to be present in intestinal biopsy samples. To minimize contamination during sampling, the colonoscope jaws will have to be washed in tap water after each biopsy is performed.

Pyxigraphy

Pyxigraphy is a technique which makes use of a capsule that can be swallowed, and by which contents of the gastrointestinal tract can be sampled under remote control. Pyxigraphy is a simple and safe sampling method that allows the microbial population of the proximal colon to be studied (96).

Fecal Samples

Feces are a complex microbial habitat, with many niches occupied by bacteria. It is estimated that bacteria account for about 30% of the fecal mass, and 40–55% of fecal solids. All of the bacteria in feces are exposed to the influences of dehydrating and concentrating mechanisms of the colon and rectum, and intense biochemical activity of the organisms living in the material. When the samples consist of only feces, the composition and localization of communities anywhere in the tract cannot be revealed. *Bacteroides* accounts for nearly 20% of the species that can be cultivated from feces (10). The *Bacteroides* and *Prevotella* group (gram-negative anaerobes), and *Eubacterium rectale* and *Clostridium coccoides* species (gram-positive anaerobes) are predominantly present in the fecal samples (90,92). The predominant bacterial community in feces is stable in time, host specific, affected by ageing, and not significantly altered after consumption of probiotic strains (97).

Fecal samples have to be collected in sterile bags, and kept at low temperature ($-80°C$ to $+4°C$) before processing (88). Stool specimens or rectal swabs can be used for the diagnosis of cholera. Dipsticks in rectal swabs are used for the rapid diagnosis of cholera caused by *Vibrio cholerae*. Dipstick analysis uses colloidal gold particles, and is based on a one-step immunochomatography principle. The sensitivity and the specificity of the dipsticks is greater than 92% and 91% respectively. This rapid test (diagnosis within 10 minutes) requires minimal technical skills (98,99).

Most knowledge of the gastrointestinal microbiota stems from colon or feces bacteriology. A major limitation in studying the proximal human colonic microbiota is the lack of suitable sampling methods. Studies in which only feces are sampled can never reveal the composition and localization of epithelial and cryptal communities anywhere in the tract. Such studies reveal little about the composition of lumenal communities in any area except perhaps the large bowel (29).

Low fecal pH is caused by ingestion of poorly absorbed carbohydrates or carbohydrate malabsorption in the small intestine, and consequently, the bacteria in the colon ferment the carbohydrate. Fecal pH of less than six is highly suggestive of carbohydrate malabsorption. A breath hydrogen test with lactose can confirm carbohydrate malabsorption. In this test a fasting patient is given 25 g of lactose dissolved in water, and exhaled breath is assayed for hydrogen content at baseline, and at intervals

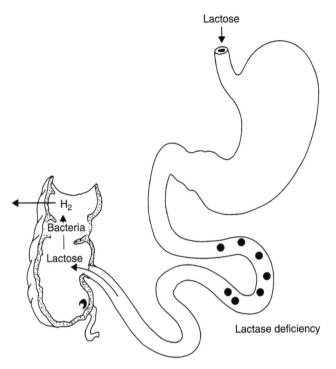

Figure 4 Principle of the hydrogen breath test with lactose to determine carbohydrate malabsorption in the small intestine.

for several hours as described in Figure 4. As explained above, because hydrogen is not a normal product of human metabolism, any increase in breath hydrogen concentration represents bacterial fermentation, and indicates that unabsorbed lactose has reached the colon.

CONCLUSION

The different methods of investigating the intestinal microbiota in humans all have their advantages, and their drawbacks as described above. If one desires information about the gastrointestinal tract one should also weigh the benefits of the (research) question, and their financial consequences. Sampling of the gastrointestinal tract in humans is far more difficult than in animal models. The sampled area is relatively small in comparison with the total area. In animal models the animal can be sacrificed so that the complete intestinal tract can be sampled and investigated. Unfortunately, individuals who are killed in accidents are the best source of complete information about microbiota in the gastrointestinal tract (29).

In general, the patient prefers the noninvasive method. Noninvasive methods are of particular importance for very young pediatric patients, pregnant women, and the elderly, as well as for research purposes. The difficulties of sampling the entire gastrointestinal tract are reduced by the noninvasive tests. However, noninvasive methods are often less sensitive and less specific. Invasive methods, such as endoscopy, are extremely unpleasant but are highly sensitive and specific, and have the advantage of sampling at the accurate location. The conditions that have to be satisfied in obtaining an uncontaminated specimen from anywhere in the gastrointestinal tract have to include: (1) strict asepsis of method, which necessitates that the instrument used must be suitable for sterilization by heat or gas; (2) prevention of contamination of the internal channels in which the culture specimen is to be lodged, until the site of sampling is reached, and protection against further contamination on withdrawal of the instrument; and (3) verification of the location from which cultures have been obtained.

As the development of molecular biology techniques increases the current sampling techniques can be revised. The condition of anaerobic sampling is becoming less important. Possible improvement of the current sampling methods only seems possible in small details. Nanotechnology is one of the promising techniques for possible improvement of sampling and analysis of bacteria in the human gastrointestinal tract.

REFERENCES

1. Leeuwenhoek AV, Dobell C. Antony van Leeuwenhoek and His "Little Animals": Being Some Account of the Father of Protozoology and Bacteriology and His Multifarious Discoveries in these Disciplines. London: Staples Press, 1932.
2. Simon GL, Gorbach SL. The human intestinal microbiota. Dig Dis Sci 1986; 31:147S–162S.
3. Feldman M, Friedman LS, Sleisenger MH, Fortran JS. Sleisenger & Fortran's Gastrointestinal and Liver Disease: Pathophysiology, Diagnosis, Management. 7th ed. Philadelphia: W.B. Saunders, 2002.
4. Galatola G, Grosso M, Barlotta A, et al. Diagnosis of bacterial contamination of the small intestine using the 1 g [14C] xylose breath test in various gastrointestinal diseases. Minerva Gastroenterol Dietol 1991; 37:169–175.
5. Mannell A, Plant M, Frolich J. The microbiota of the oesophagus. Ann R Coll Surg Engl 1983; 65:152–154.
6. Drasar BS, Shiner M, McLeod GM. Studies on the intestinal flora. I. The bacterial flora of the gastrointestinal tract in healthy and achlorhydric persons. Gastroenterology 1969; 56:71–79.
7. King TS, Elia M, Hunter JO. Abnormal colonic fermentation in irritable bowel syndrome. Lancet 1998; 352:1187–1189.
8. Mallory A, Savage D, Kern F, Jr., Smith JG. Patterns of bile acids and microbiota in the human small intestine. II. Microflora. Gastroenterology 1973; 64:34–42.
9. Cunningham-Rundles S, Ahrn S, Abuav-Nussbaum R, Dnistrian A. Development of immunocompetence: role of micronutrients and microorganisms. Nutr Rev 2002; 60:S68–S72.
10. Carman RJ, Van Tassell RL, Wilkins TD. The normal intestinal microbiota: ecology, variability and stability. Vet Hum Toxicol 1993; 35:11–14.
11. Blaut M, Collins MD, Welling GW, Dore J, van Loo J, de Vos W. Molecular biological methods for studying the gut microbiota: the EU human gut flora project. Br J Nutr 2002; 87:S203–S211.
12. Blaut M. Relationship of prebiotics and food to intestinal microbiota. Eur J Nutr 2002; 41:I11–I16.
13. Hooper LV, Gordon JI. Commensal host-bacterial relationships in the gut. Science 2001; 292:1115–1118.
14. Gordon JI, Stappenbeck TS, Hooper LV, et al. Response from Jeffrey I. Gordon. Commensal bacteria make a difference. Trends Microbiol 2003; 11:150–151.
15. Zinsser H. Zinsser Microbiology. 20th ed. Norwalk, CT: Appleton & Lange, 1992.
16. Hopkins MJ, Macfarlane GT. Changes in predominant bacterial populations in human feces with age and with *Clostridium difficile* infection. J Med Microbiol 2002; 51:448–454.
17. Husebye E, Hellstrom PM, Sundler F, Chen J, Midtvedt T. Influence of microbial species on small intestinal myoelectric activity and transit in germ-free rats. Am J Physiol Gastrointest Liver Physiol 2001; 280:G368–G380.

18. Husebye E, Hellstrom PM, Midtvedt T. Intestinal microbiota stimulates myoelectric activity of rat small intestine by promoting cyclic initiation and aboral propagation of migrating myoelectric complex. Dig Dis Sci 1994; 39:946–956.

19. Nieuwenhuijs VB, Verheem A, Duijvenbode-Beumer H, et al. The role of interdigestive small bowel motility in the regulation of gut microbiota, bacterial overgrowth, and bacterial translocation in rats. Ann Surg 1998; 228:188–193.

20. Williams C. Occurrence and significance of gastric colonization during acid-inhibitory therapy. Best Pract Res Clin Gastroenterol 2001; 15:511–521.

21. Pei Z, Bini EJ, Yang L, Zhou M, Francois F, Blaser MJ. Bacterial biota in the human distal esophagus. Proc Natl Acad Sci USA 2004; 101:4250–4255.

22. Sjostedt S. The upper gastrointestinal microbiota in relation to gastric diseases and gastric surgery. Acta Chir Scand Suppl 1989; 551:1–57.

23. Jain SK, Jain S, Jain M, Yaduvanshi A. Esophageal tuberculosis: is it so rare? Report of 12 cases and review of the literature Am J Gastroenterol 2002; 97:287–291.

24. Gagliardi D, Makihara S, Corsi PR, et al. Microbial flora of the normal esophagus. Dis Esophagus 1998; 11:248–250.

25. Pajecki D, Zilberstein B, dos Santos MA, et al. Megaesophagus microbiota: a qualitative and quantitative analysis. J Gastrointest Surg 2002; 6:723–729.

26. Franklin MA, Skoryna SC. Studies of the natural gastric flora: I Bacterial flora of fasting human subjects. Can Med Assoc J 1966; 95:1349–1355.

27. Cotton Peter B, Williams Christopher B. 4th ed. Practical Gastrointestinal Endoscopy. London: Blackwell Scientific Publications, 1996.

28. Osias GL, Bromer MQ, Thomas RM, et al. Esophageal bacteria and Barrett's esophagus: a preliminary report. Dig Dis Sci 2004; 49:228–236.

29. Savage DC. Microbial ecology of the gastrointestinal tract. Annu Rev Microbiol 1977; 31:107–133.

30. Simon GL, Gorbach SL. Intestinal flora in health and disease. Gastroenterology 1984; 86:174–193.

31. Franklin MA, Skoryna SC. Studies on natural gastric flora: survival of bacteria in fasting human subjects. Can Med Assoc J 1971; 105:380–386.

32. Monstein HJ, Tiveljung A, Kraft CH, Borch K, Jonasson J. Profiling of bacterial flora in gastric biopsies from patients with *Helicobacter pylori*-associated gastritis and histologically normal control individuals by temperature gradient gel electrophoresis and 16S rDNA sequence analysis. J Med Microbiol 2000; 49:817–822.

33. Verhagen MA, Roelofs JM, Edelbroek MA, Smout AJ, Akkermans LM. The effect of cisapride on duodenal acid exposure in the proximal duodenum in healthy subjects. Aliment Pharmacol Ther 1999; 13:621–630.

34. Pasricha PJ. Effect of sleep on gastroesophageal physiology and airway protective mechanisms. Am J Med 2003; 115:114S–118S.

35. Pounder RE, Ng D. The prevalence of *Helicobacter pylori* infection in different countries. Aliment Pharmacol Ther 1995; 9:33–39.

36. Graham DY, Adam E, Reddy GT, et al. Seroepidemiology of *Helicobacter pylori* infection in India. Comparison of developing and developed countries. Dig Dis Sci 1991; 36:1084–1088.

37. Guerre J, Vedel G, Gaudric M, Paul G, Cornuau J. Bacterial flora in gastric juice taken at endoscopy in 93 normal subjects. Pathol Biol (Paris) 1986; 34:57–60.

38. Wong BC, Wong WM, Wang WH, et al. An evaluation of invasive and non-invasive tests for the diagnosis of *Helicobacter pylori* infection in Chinese. Aliment Pharmacol Ther 2001; 15:505–511.

39. Cotton, Williams. Practical gastrointestinal endoscopy. 4th ed. London: Blackwell Scientific Publications, 1996.

40. Lim LL, Ho KY, Ho B, Salto-Tellez M. Effect of biopsies on sensitivity and specificity of ultra-rapid urease test for detection of *Helicobacter pylori* infection: A prospective evaluation. World J Gastroenterol 2004; 10:1907–1910.

41. Monteiro L, de Mascarel A, Sarrasqueta AM, et al. Diagnosis of *Helicobacter pylori* infection: noninvasive methods compared to invasive methods and evaluation of two new tests. Am J Gastroenterol 2001; 96:353–358.
42. Romagnuolo J, Schiller D, Bailey RJ. Using breath tests wisely in a gastroenterology practice: an evidence-based review of indications and pitfalls in interpretation. Am J Gastroenterol 2002; 97:1113–1126.
43. Koletzko S, Feydt-Schmidt A. Infants differ from teenagers: use of non-invasive tests for detection of *Helicobacter pylori* infection in children. Eur J Gastroenterol Hepatol 2001; 13:1047–1052.
44. Makristathis A, Pasching E, Schutze K, Wimmer M, Rotter ML, Hirschl AM. Detection of *Helicobacter pylori* in stool specimens by PCR and antigen enzyme immunoassay. J Clin Microbiol 1998; 36:2772–2774.
45. MacKay WG, Williams CL, McMillan M, Ndip RN, Shepherd AJ, Weaver LT. Evaluation of protocol using gene capture and PCR for detection of *Helicobacter pylori* DNA in feces. J Clin Microbiol 2003; 41:4589–4593.
46. Basset C, Holton J, Ricci C, et al. Review article: diagnosis and treatment of *Helicobacter*: a 2002 updated review. Aliment Pharmacol Ther 2003; 17:89–97.
47. Vakil N, Rhew D, Soll A, Ofman JJ. The cost-effectiveness of diagnostic testing strategies for *Helicobacter pylori*. Am J Gastroenterol 2000; 95:1691–1698.
48. Maibach RC, Dutly F, Altwegg M. Detection of *Tropheryma whipplei* DNA in feces by PCR using a target capture method. J Clin Microbiol 2002; 40:2466–2471.
49. Edmiston CE, Jr., Krepel CJ, Seabrook GR, Jochimsen WG. Anaerobic infections in the surgical patient: microbial etiology and therapy. Clin Infect Dis 2002; 35:S112–S118.
50. Finegold SM. Host factors predisposing to anaerobic infections. FEMS Immunol Med Microbiol 1993; 6:159–163.
51. Corazza GR, Menozzi MG, Strocchi A, et al. The diagnosis of small bowel bacterial overgrowth. Reliability of jejunal culture and inadequacy of breath hydrogen testing. Gastroenterology 1990; 98:302–309.
52. Mee AS, Burke M, Vallon AG, Newman J, Cotton PB. Small bowel biopsy for malabsorption: comparison of the diagnostic adequacy of endoscopic forceps and capsule biopsy specimens. Br Med J (Clin Res Ed) 1985; 291:769–772.
53. Crosby WH, Kugler HW. Intraluminal biopsy of the small intestine; the intestinal biopsy capsule. Am J Dig Dis 1957; 2:236–241.
54. Flick AL, Quionton WE, Rubin CE. A peroral hydraulic biopsy tube for multiple sampling at any level of the gastro-intestinal tract. Gastroenterology 1961; 40:120–126.
55. Chu KM, Yuen ST, Wong WM, et al. A prospective comparison of performance of biopsy forceps used in single passage with multiple bites during upper endoscopy. Endoscopy 2003; 35:338–342.
56. Tornblom H, Lindberg G, Nyberg B, Veress B. Full-thickness biopsy of the jejunum reveals inflammation and enteric neuropathy in irritable bowel syndrome. Gastroenterology 2002; 123:1972–1979.
57. Davis CP. Postmortem alterations of bacterial localization. Scan Electron Microsc 1980; 523-6:542.
58. Bengmark S, Jeppsson B. Gastrointestinal surface protection and mucosa reconditioning. JPEN J Parenter Enteral Nutr 1995; 19:410–415.
59. Tiveljung A, Soderholm JD, Olaison G, Jonasson J, Monstein HJ. Presence of eubacteria in biopsies from Crohn's disease inflammatory lesions as determined by 16S rRNA gene-based PCR. J Med Microbiol 1999; 48:263–268.
60. Leon-Barua R, Gilman RH, Rodriguez C, et al. Comparison of three methods to obtain upper small bowel contents for culture. Am J Gastroenterol 1993; 88:925–928.
61. Shiner M, Waters TE, Gray JD. Culture studies of the gastrointestinal tract with a newly devised capsule. Results of tests in vitro and in vivo. Gastroenterology 1963; 45:625–632.
62. Tally FP, Stewart PR, Sutter VL, Rosenblatt JE. Oxygen tolerance of fresh clinical anaerobic bacteria. J Clin Microbiol 1975; 1:161–164.

63. Nakshabendi IM, McKee R, Downie S, Russell RI, Rennie MJ. Rates of small intestinal mucosal protein synthesis in human jejunum and ileum. Am J Physiol 1999; 277:E1028–E1031.
64. Sheehy TW. Intestinal biopsy. Lancet 1964; 41:959–962.
65. Mayhew JW, Gorbach SL. Internal standards for gas chromatographic analysis of metabolic end products from anaerobic bacteria. Appl Environ Microbiol 1977; 33:1002–1003.
66. Gorbach SL, Mayhew JW, Bartlett JG, Thadepalli H, Onderdonk AB. Rapid diagnosis of anaerobic infections by direct gas-liquid chromatography of clinical speciments. J Clin Invest 1976; 57:478–484.
67. Bardhan PK, Gyr K, Beglinger C, Vogtlin J, Frey R, Vischer W. Diagnosis of bacterial overgrowth after culturing proximal small-bowel aspirate obtained during routine upper gastrointestinal endoscopy. Scand J Gastroenterol 1992; 27:253–256.
68. Tabaqchali S. The pathophysiological role of small intestinal bacterial flora. Scand J Gastroenterol Suppl 1970; 6:139–163.
69. Norby RS, Haagen NO, Justesen T, Jacobsen IE, Lave J, Magid E. Comparison of an open and a closed tube system for collection of jejunal juice. Scand J Gastroenterol 1983; 18:353–357.
70. Muraca M, Vilei MT, Miconi L, Petrin P, Antoniutti M, Pedrazzoli S. A simple method for the determination of lipid composition of human bile. J Lipid Res 1991; 32:371–374.
71. Riordan SM, McIver CJ, Duncombe VM, Bolin TD. An appraisal of a 'string test' for the detection of small bowel bacterial overgrowth. J Trop Med Hyg 1995; 98:117–120.
72. Bardhan PK, Feger A, Kogon M, et al. Urinary choloyl-PABA excretion in diagnosing small intestinal bacterial overgrowth: evaluation of a new noninvasive method. Dig Dis Sci 2000; 45:474–479.
73. Mayer PJ, Beeken WL. The role of urinary indican as a predictor of bacterial colonization in the human jejunum. Am J Dig Dis 1975; 20:1003–1009.
74. Krawitt EL, Beeken WL. Limitations of the usefulness of the D-xylose absorption test. Am J Clin Pathol 1975; 63:261–263.
75. Einarsson K, Bergstrom M, Eklof R, Nord CE, Bjorkhem I. Comparison of the proportion of unconjugated to total serum cholic acid and the [^{14}C]-xylose breath test in patients with suspected small intestinal bacterial overgrowth. Scand J Clin Lab Invest 1992; 52:425–430.
76. Levitt MD. Production and excretion of hydrogen gas in man. N Engl J Med 1969; 281:122–127.
77. Hepner GW. Increased sensitivity of the cholylglycine breath test for detecting ileal dysfunction. Gastroenterology 1975; 68:8–16.
78. Valdovinos MA, Camilleri M, Thomforde GM, Frie C. Reduced accuracy of ^{14}C-D-xylose breath test for detecting bacterial overgrowth in gastrointestinal motility disorders. Scand J Gastroenterol 1993; 28:963–968.
79. Riordan SM, McIver CJ, Duncombe VM, Bolin TD, Thomas MC. Factors influencing the 1-g ^{14}C-D-xylose breath test for bacterial overgrowth. Am J Gastroenterol 1995; 90:1455–1460.
80. Lewis SJ, Young G, Mann M, Franco S, O'Keefe SJ. Improvement in specificity of [^{14}C]D-xylose breath test for bacterial overgrowth. Dig Dis Sci 1997; 42:1587–1592.
81. Metz G, Gassull MA, Drasar BS, Jenkins DJ, Blendis LM. Breath-hydrogen test for small-intestinal bacterial colonisation. Lancet 1976; 1:668–669.
82. Perman JA, Modler S, Barr RG, Rosenthal P. Fasting breath hydrogen concentration: normal values and clinical application. Gastroenterology 1984; 87:1358–1363.
83. Riordan SM, McIver CJ, Walker BM, Duncombe VM, Bolin TD, Thomas MC. The lactulose breath hydrogen test and small intestinal bacterial overgrowth. Am J Gastroenterol 1996; 91:1795–1803.
84. Brummer RJ, Armbrecht U, Bosaeus I, Dotevall G, Stockbruegger RW. The hydrogen (H_2) breath test. Sampling methods and the influence of dietary fibre on fasting level. Scand J Gastroenterol 1985; 20:1007–1013.
85. Kerlin P, Wong L. Breath hydrogen testing in bacterial overgrowth of the small intestine. Gastroenterology 1988; 95:982–988.

86. Thompson DG, O'Brien JD, Hardie JM. Influence of the oropharyngeal microbiota on the measurement of exhaled breath hydrogen. Gastroenterology 1986; 91:853–860.
87. Riordan SM, McIver CJ, Bolin TD, Duncombe VM. Fasting breath hydrogen concentrations in gastric and small-intestinal bacterial overgrowth. Scand J Gastroenterol 1995; 30:252–257.
88. Rumessen JJ, Kokholm G, Gudmand-Hoyer E. Methodological aspects of breath hydrogen (H_2) analysis. Evaluation of a H_2 monitor and interpretation of the breath H_2 test. Scand J Clin Lab Invest 1987; 47:555–560.
89. Moore WE, Holdeman LV. Special problems associated with the isolation and identification of intestinal bacteria in fecal flora studies. Am J Clin Nutr 1974; 27:1450–1455.
90. Mai V, Morris JG, Jr. Colonic bacterial flora: changing understandings in the molecular age. J Nutr 2004; 134:459–464.
91. Tlaskalova-Hogenova H, Stepankova R, Hudcovic T, et al. Commensal bacteria (normal microbiota), mucosal immunity and chronic inflammatory and autoimmune diseases. Immunol Lett 2004; 93:97–108.
92. Zhong Y, Priebe MG, Vonk RJ, et al. The role of colonic microbiota in lactose intolerance. Dig Dis Sci 2004; 49:78–83.
93. Edlund C, Nord CE. Effect on the human normal microbiota of oral antibiotics for treatment of urinary tract infections. J Antimicrob Chemother 2000; 46:41–48 discussion 63–65: 41–48.
94. Huijsdens XW, Linskens RK, Mak M, Meuwissen SG, Vandenbroucke-Grauls CM, Savelkoul PH. Quantification of bacteria adherent to gastrointestinal mucosa by real-time PCR. J Clin Microbiol 2002; 40:4423–4427.
95. Zoetendal EG, von Wright A, Vilpponen-Salmela T, Ben Amor K, Akkermans AD, de Vos WM. Mucosa-associated bacteria in the human gastrointestinal tract are uniformly distributed along the colon and differ from the community recovered from feces. Appl Environ Microbiol 2002; 68:3401–3407.
96. Pochart P, Lemann F, Flourie B, Pellier P, Goderel I, Rambaud JC. Pyxigraphic sampling to enumerate methanogens and anaerobes in the right colon of healthy humans. Gastroenterology 1993; 105:1281–1285.
97. Zoetendal EG, Cheng B, Koike S, Mackie RI. Molecular microbial ecology of the gastrointestinal tract: from phylogeny to function. Curr Issues Intest Microbiol 2004; 5:31–47.
98. Bhuiyan NA, Qadri F, Faruque AS, et al. Use of dipsticks for rapid diagnosis of cholera caused by *Vibrio cholerae* O1 and O139 from rectal swabs. J Clin Microbiol 2003; 41:3939–3941.
99. Nato F, Boutonnier A, Rajerison M, et al. One-step immunochromatographic dipstick tests for rapid detection of *Vibrio cholerae* O1 and O139 in stool samples. Clin Diagn Lab Immunol 2003; 10:476–478.
100. Sayler GS, Simpson ML, Cox CD. Emerging foundations: nano-engineering and bio-microelectronics for environmental biotechnology. Curr Opin Microbiol 2004; 7:267–273.
101. Stotzer PO, Kilander AF. Comparison of the 1-gram (^{14}C)-D-xylose breath test and the 50-gram hydrogen glucose breath test for diagnosis of small intestinal bacterial overgrowth. Digestion 2000; 61:165–171.
102. King CE, Toskes PP, Guilarte TR, Lorenz E, Welkos SL. Comparison of the one-gram d-[^{14}C]xylose breath test to the [^{14}C]bile acid breath test in patients with small-intestine bacterial overgrowth. Dig Dis Sci 1980; 25:53–58.
103. Sherr HP, Sasaki Y, Newman A, Banwell JG, Wagner HN, Jr., Hendrix TR. Detection of bacterial deconjugation of bile salts by a convenient breath-analysis technic. N Engl J Med 1971; 285:656–661.
104. Metz G, Drasar BS, Gassull MA, Jenkins DJA, Blendis LM. Breath-hydrogen test for small-intestinal bacterial overgrowth. Lancet 1976; 1:668–669.
105. Rumessen JJ, Gudmand-Höyer E, Bachmann E, Justesen T. Diagnosis of bacterial overgrowth of the small intestine. Comparison of the ^{14}C-D-xylose breath test and jejunal cultures in 60 patients. Scan J Gastroenterol 1985; 20:1267–1275.

3

The Normal Microbiota of the Human Gastrointestinal Tract: History of Analysis, Succession, and Dietary Influences

Anne L. McCartney and Glenn R. Gibson
Food Microbial Sciences Unit, School of Food Biosciences, Whiteknights, University of Reading, Reading, U.K.

INTRODUCTION

The human body is a wonderland for the microbial world, with harsh uninhabitable lands in some regions and lush fertile metropolis in others. The normal microbiota of humans is an extensive and diverse microbial community, which is composed primarily of bacteria from numerous phylogenetic clusters (1–5). The largest proportion of the human microbiota is found in the gastrointestinal (GI) tract, or more specifically the colon. Other regions of the body harboring indigenous bacterial populations include the skin, oral cavity, upper respiratory tract, and urogenital tract (3). This chapter aims to discuss the normal microbiota of the human GI tract and our current understanding of its composition and role in human health. Discussion of the interactions between the gut microbiota and the host will also abridge the impact of extrinsic factors, such as diet and environment.

The GI tract of humans can be divided into three anatomical regions, namely, the stomach, small intestine (comprising duodenum, jejunum, and ileum) and large intestine or colon. Distinctive physicochemical environments are found within the different regions and the microbial populations harbored reflect this, both quantitatively and qualitatively (3,6,7). Thus, the normal microbiota of the human GI tract is often subdivided into three distinct bacterial communities: that of the upper GI tract, the ileum and the colon.

The rapid transit time and acidic conditions of the stomach restrict the levels of microbial colonization of this region (6,8). Gastric juices and small-intestine secretions (bile and pancreatic fluids) amplify the hostile nature of the upper GI tract to microbial establishment. However, some aciduric Gram-positive bacteria (lactobacilli and streptococci) can be detected in this region ($\sim 10^2$–10^4 bacterial cells per milliliter of contents). In addition, some micro-organisms, such as *Helicobacter pylori* (the possible etiological agent in peptic ulcers and Type B gastritis), are able to survive, evade or combat the harsh conditions of the stomach (9–12). *Helicobacter* spp. use their flagellae to avoid peristaltic movement and burrow into the mucosal lining of the stomach, where they are partially protected from the acidic conditions by producing NH_3 from urea to neutralize the acid (11,12).

The flow of digesta (intestinal motility) is somewhat slower in the ileum, compared with the upper GI tract, and conditions are thus more favorable for microbial colonization. Available data indicate increasing bacterial population levels (10^6–10^8 bacterial cells per milliliter of contents) and a higher diversity of micro-organisms, with the presence of Gram-negative facultatively anaerobic bacteria (such as members of the family *Enterobacteriaceae*) and obligate anaerobes (including *Bacteroides*, *Veillonella*, *Fusobacterium* and *Clostridium* species) in conjunction with lactobacilli and entero-cocci (1,3,6).

The typical GI transit time is between 55 and 70 hours (13,14). Taken together with a more neutral pH and relative abundance of nutrients (including non-digestible carbohydrates and food components which have escaped digestion in the upper GI tract, sloughed off epithelial cells and microbial cell debris), this region of the human GI tract is an oasis for microbial growth, attaining levels of 10^{10}–10^{12} bacterial cells per gram of contents (3,6,8). The composition of the colonic microbiota is extremely complex, generally estimated to comprise greater than 500 bacterial species, although it is thought that 30–40 predominate. The majority of members of the colonic microbiota are obligate anaerobic genera, including *Bacteroides*, *Bifidobacterium*, *Clostridium*, *Enterococcus*, *Eubacterium*, *Fusobacterium*, *Peptococcus*, *Peptostreptococcus* and *Ruminococcus* (2,3,15). Our understanding of the composition of the normal colonic microbiota has largely resulted from studies of the fecal microbiota. Questions regarding the accuracy of fecal samples to represent the colonic microbiota have been initially addressed by bacteriological analysis of the intestinal contents of sudden-death victims (14,16). This work demonstrated that cultivation studies of the fecal microbiota accurately reflected the culturable component of the distal colon. However, with recent advances in molecular technology (and indeed in cultivation assays), as well as sampling methods (including medical advances affording biopsy samples), analysis of the microbiota in different regions of the GI tract is now feasible as discussed below and in previous chapters. Future studies will, no doubt, begin to unravel the impact of impairment or disease on the mucosal microbiota, as well as the interaction between the luminal microbiota, the mucosally associated microbiota and the host.

ROLE OF THE GASTROINTESTINAL MICROBIOTA IN HUMANS

Traditionally, the colon has been considered to largely be the human sewage system which, as well as storing and removing waste material from the GI tract, was capable of recycling water (i.e., absorption). However, we now recognize that the GI tract is one of the most metabolically and immunologically active organs of the human body. Indeed, the primary function of the microbiota is generally considered to be salvage of energy via fermentation of carbohydrates, such as indigestible dietary residues (plant cell walls, non-digestible fibers and oligosaccharides), mucin side-chains and sloughed-off epithelial cells (5,6,8,13,17). It has been estimated that between 20 and 60 grams of carbohydrate are available in the colon of healthy human adults per day, as well as 5–20 grams of protein. In addition to salvaging energy, principally through production of short-chain fatty acids (SCFAs) and their subsequent absorption and use by the host, microbial fermentation produces gases (principally hydrogen, carbon dioxide and methane) and increases biomass. These all impact upon gut physiology. Components of the gut microbiota also synthesize certain B and K vitamins, metabolize xenobiotics, contribute to amino acid homeostasis, may impact drug efficacy and are an integral part of the host defense (both through host-microbe and microbe-microbe interactions; including colonization resistance) (6,17,18). Recent observations, using molecular techniques and germ-free/gnotobiotic animals, have also identified that intestinal bacteria can influence gene expression of epithelial cells (5,19). Taken together, the activity of the microbiota, or certain components thereof, may be more important to the homeostasis of the ecosystem than specific numerics. Although the combination of all these factors, as well as host and environmental factors, will ultimately determine the equilibrium of the colon.

Three main SCFAs are produced by microbial fermentation in the human colon: acetate, butyrate, and propionate (the approximate molar ratio for which is 70:10:20—although diet and microbiota composition influence the exact ratio) (5). SCFAs supply energy to cells (acetate, muscle; butyrate, colonocytes; propionate, liver), affect colonic metabolism, control epithelial cell proliferation and differentiation, and impact upon bowel motility and circulation (including water absorption and the hepatic regulation of lipids and sugars) (5,8,13).

Uptake and utilization of acetate is the primary method of the host salvaging energy from non-digestible dietary carbohydrates. Acetate may also play a role in lipogenesis by adipocytes and, together with propionate, may be involved in modulation of glucose metabolism (via the glycaemic index). Butyrate is estimated to provide between 40 and 70% of the required energy of the colonic mucosa (5,6). In vitro studies have demonstrated inhibition of proliferation of neoplastic cell lines by butyrate, suggesting a possible beneficial role of butyrate against the progression of colorectal carcinoma. Such work has also shown that butyrate stimulates cell differentiation, promoting reversion to non-neoplastic phenotypes.

In addition to carbohydrate fermentation, bacterial metabolism of amino acids may generate branched-chain fatty acids (such as isobutyrate, isovalerate, and 2-methyl butyrate), whilst microbial degradation of peptides and proteins forms potentially toxic compounds (including ammonia, amines, phenols, and indoles) (8,17).

The colonic microbiota impacts upon amino acid homeostasis, with 1–20% of circulating plasma lysine being derived from the activity of gut bacteria (18). In addition, microbial hydrolysis of urea to ammonia by the gut microbiota is important in the recycling of nitrogen in the intestine.

The protective effect of the gut microbiota against pathogenic microorganisms falls under two umbrellas: 1, colonization resistance and, 2, stimulation of immune function. In the healthy state, the resident microbiota effectively inhibits the establishment and/or overgrowth of harmful bacteria. A number of mechanisms appear to be responsible, including competition for adhesion sites, competition for nutrients, production of environmental conditions restrictive to pathogenic growth (pH, redox potential), production of anti-microbial compounds (either toxic metabolites or bacteriocins) and/or generation of signals which interact with gene expression of exogenous organisms (3,8,13). In addition, certain members of the intestinal microbiota are known to stimulate immune function (both locally and systemically) (17,20,21). Interactions between the mucosal barrier, the indigenous microbiota and the gut-associated lymphoid tissue (GALT) are paramount to the host defense against pathogenic invasion and infection. This three-component system is integral to the equilibrium of the GI tract ecosystem and defines the balance between oral tolerance and mounting an immune response.

Bacterial-host cell communications can also impact upon expression by host cells. One example of this is the ability of *Bact. thetaiotaomicron* to influence fucosylated glycoconjugate production by intestinal cells in relation to the availability of fucose (a substrate for the organism) (5,19). In this manner, the bacteria can essentially order nutrients from the epithelial cells as necessary. Such microbial induced signals may also act in cell-cell communications between different bacterial species and play an important role in homeostasis of their environmental niche.

ACQUISITION OF THE GUT MICROBIOTA

Acquisition of the normal microbiota is a biological succession which commences during or immediately following birth (depending on the mode of delivery). During natural birth, the neonate is exposed to the maternal microbiota, both vaginal and fecal (22–24). However, colonization is delayed in infants born via Caesarian section and the major source of inoculation is thought to be from the environment (including nosocomially from within the maternity ward) (23). Caesarean section delivery has been correlated with an increased clostridial component in the infant microbiota. Indeed, recent studies have demonstrated that higher clostridial counts in children delivered by Caesarean section relative to children delivered vaginally persist even after 7 years of age (25).

During the initial phase of acquisition, facultative anaerobes predominate (enterobacteria and streptococci) and effectively reduce the redox potential of the gut environment enabling colonization by obligate anaerobes (including bacteroides, bifidobacteria, clostridia, and eubacteria). Factors such as diet and host genetics play important roles in the development of the microbiota (with some bacterial populations eliminated and others maintained) (3,24). The classical studies by Tissier almost a century ago first highlighted the significant difference of the fecal microbiota harbored by breast-fed and formula-fed infants. Indeed, Tissier described three phases of microbial acquisition in infants: 1, initial hours of life when the fecal bacterial content was nil; 2, beginning between the tenth and twentieth hour of life, comprising a heterogeneous microbiota; 3, after passage of maternal milk through the intestinal tract, the microbiota being predominated by bifidobacteria (an obligately anaerobic Gram-positive bacillus which often exhibits bifurcating morphology, formerly named *Bacillus bifidus* by Tissier) (3,26,27). A fourth phase, following introduction of solid foods (weaning), was later described and is characterized by modulation of the breast-fed microbiota towards an adult-type microbiota (climax community) harboring a more complex and diverse bacterial community (13,28,29). It is worth noting that Tissier also speculated that subdominant populations (including facultative anaerobes) were harbored during phase three of acquisition and that complete bacteriological examination was necessary to determine this. No doubt some such populations are then re-established as predominant members within the heterogeneous climax community through the introduction of complex carbohydrates into the diet.

Bottle-fed infants did not demonstrate the same succession of micro-organisms as seen in their breast-fed counterparts. Indeed, Tissier observed that formula-fed infants maintained a heterogeneous fecal microbiota beyond day 4. Much work has been compiled over the last 30 years comparing the fecal microbiota of exclusively milk-fed infants. Until recently, such studies were performed using traditional cultivation techniques. A range of data has accumulated and while notable differences may still be observed between breast-fed and formula-fed infants, they are not as startling as those shown by Tissier. In general, the bifidobacterial microbiota, both carriage (percentage of infants harboring bifidobacteria) and population level, of exclusively milk-fed infants was not significantly different (30–33). However, levels of other organisms, notably *Bacteroides*, clostridia and enterobacteria, were significantly higher in formula-fed infants. Thus, breast-fed infants harbored a bifidobacterially predominant fecal microbiota, whereas formula-fed infants harbored a larger bacterial load comprising greater heterogeneity with higher levels of *Bacteroides*, enterobacteria and clostridia. Studies investigating the fecal microbiota of infants fed different formulae (for example, following fortification with iron and/or oligosaccharides) have shown that the constituents of the infant formulae impact upon the microbial composition (24,34). Recent studies employing molecular biological methods have further clarified the situation, demonstrating an initial diverse microbiota during the first 4–6 days of life (phase 2) followed by establishment of a bifidobacterially predominant microbiota in breast-fed infants (phase 3) which is not as obvious in formula-fed infants. Namely, bifidobacteria formed 60–91% of the bacterial composition of breast-fed infants ($n=6$) and between 28 and 75% of the total microbial load of formula-fed infants ($n=6$) after day six (35). Inter-individual differences were noted in both feeding groups, with respect to the relative proportions of the bacterial groups studied. Molecular characterization studies of the predominant isolates from concurrent cultivation work further highlighted the distinction between the microbiota harbored by infants [both between feeding groups and inter-individually (35)].

COMPOSITION OF THE ADULT FECAL MICROBIOTA ASSESSED BY CULTURING

Much of the early information on the composition of the human colonic microbiota was elucidated using traditional cultivation techniques. The majority of such work was driven by the quest to determine the relationship between diet and colonic cancer (16,36–38). Epidemiological studies had identified that risk of colon cancer correlated with dietary habit, with higher colorectal cancer incidence in populations consuming a high-fat, low-fiber diet. In 1969, Aries and coworkers (39) postulated that this correlation between diet and cancer should be reflected in the composition of the colonic microbiota. Thus, interest in the effect of diet on the GI microbiota began in earnest. The majority of these early studies compared the fecal microbiota of individuals from different populations which had significantly different incidences of colon cancer. For example, Aries and coworkers (39) compared the fecal microbiota of English subjects (relatively high incidence) to that of Ugandans (low incidence). Significantly higher numbers of *Bacteroides* and bifidobacteria were enumerated from English individuals (Table 1), whilst enterococci, lactobacilli, streptococci, and yeasts were present at higher numbers in the fecal microbiota of Ugandan subjects. Subsequent studies compared the microbial compositions of multiple populations with either a high or a low incidence of colon cancer (38). Again, higher yields of bacteroides were seen for the high-risk populations (Table 1). However, an even more striking observation was the higher anaerobe-to-aerobe ratio in fecal samples from the high-incidence populations. Moore and colleagues (40) similarly showed higher levels of *Bacteroides* and bifidobacteria in subjects from high-risk populations (North Americans), when compared to low-risk populations (Africans). However, these observations were not consistent for a second low-risk population (Japanese) for whom the greatest percentage of isolates was *Bacteroides* (Table 2). More detailed characterization of these isolates identified that *Bacteroides vulgatus*, *Bacteroides distasonis* and *Peptostreptococcus productus* (reclassified as *Ruminococcus productus*) were the more predominant members of the fecal microbiota of high-risk populations (40). In addition, a notably higher percentage of isolates in the low-risk populations belonged to the species *Bacteroides fragilis*, *Eubacterium aerofaciens* (reclassified as *Collinsella aerofaciens*) and *Escherichia coli* (Table 2). Such detailed analyses of the microbial community have highlighted the

Table 1 Investigations of the Bacterial Composition of Fecal Samples Collected from Individuals from Countries with High or Low Incidence of Colon Cancer

	No. of samples	Bacteroides	Bifidobacteria	Clostridia	Veillonella	Lactobacilli	Yeasts	Enterobacteria	Streptococci	Enterococci	Total anaerobes	Total aerobes	Anaerobe/aerobe Log$_{10}$ ratio	Anaerobe/aerobe ratio[a]
English[b]	40	9.7	9.9	4.4	4.4	6.0	1.3	7.5	7.0	5.7	ND	ND	ND	ND
Ugandan[b]	48	8.2	9.3	4.0	5.3	7.2	3.1	8.0	7.8	7.0	ND	ND	ND	ND
English[c]	68	9.8	9.8	5.8	4.2	6.5	ND	7.9	7.1	5.8	10.1	8.0	2.1	125.9
Scottish[c]	23	9.8	9.9	5.7	3.8	7.7	ND	7.6	6.8	5.3	10.2	7.7	2.5	316.2
White American[c]	22	9.7	10.1	5.5	3.4	6.5	ND	7.4	7.0	5.9	10.2	7.5	2.7	501.2
Black American[c]	12	9.8	9.9	5.0	5.2	8.0	ND	7.3	7.1	5.0	10.2	7.5	2.7	501.2
Ugandan[c]	48	8.2	9.3	5.2	5.3	7.2	ND	8.0	7.8	7.0	9.3	8.2	1.1	12.6
Japanese[c]	17	9.4	9.7	5.6	4.7	7.4	ND	9.3	8.5	8.1	9.9	9.4	0.5	3.2
Indian[c]	51	9.2	9.6	5.8	5.8	7.6	ND	7.9	7.9	7.3	9.7	8.2	1.5	31.6

Mean log$_{10}$ counts per gram of feces (wet weight).
[a] Anaerobe/aerobe ratio using real numbers.
[b] *Source*: From Ref. 39.
[c] *Source*: From Ref. 38.
Abbreviation: ND, not determined.

Table 2 Incidence of Bacterial Populations in Fecal Samples of Individuals from Countries with High or Low Risk of Colon Cancer

Bacterial population	High incidence	Low incidence	
	North American and polyp patients $(n = 40–160)^a$	Japanese $(n = 10)$	Africans $(n = 4)$
Bacteroides spp.	29.2^b	34.4	23.1
Bacteroides vulgatus	12.5	7.7	2.6
Bacteroides distasonis	4.0	1.7	0.9
Bacteroides thetaiotaomicron/ uniformis group	5.2	7.0	1.7
Bacteroides fragilis	2.3	3.2	8.0
Bifidobacterium spp.	7.7	7.8	1.8
Bifidobacterium adolescentis	4.3	6.1	1.2
Peptostreptococcus productus I	3.0	2.1	1.3
Peptostreptococcus productus II	5.7	2.2	1.9
Eubacterium aerofaciens II	0.8	2.7	9.2
Escherichia coli	0.5	1.0	4.6
Fusobacterium prausnitzii	5.6	3.4	3.5

[a] Incidence of colon cancer per 100,000.
[b] Percentage of isolates.
Source: From Ref. 40.

importance of investigating population dynamics and not merely population levels. For more information on the influence of the intestinal microbiota and diet on the risk for colon cancer, see the chapter by Rafter and Rowland in this book.

At this time (mid-1970s), researchers became concerned with the inherent variation between the different populations and the possible impact this may have on interpretation of the data (e.g., geographical, and genetic differences between the study groups). Subsequent investigations concentrated on comparing dietary changes within cultural populations. Initial work included comparison of two generations of Japanese living in Los Angeles, one maintaining the traditional Japanese (low-risk) diet and the other having adopted a high-risk Western diet (41). Interestingly, no statistically significant differences were seen in the predominant genera of the fecal microbiota of the two groups. In addition, though significant differences in the prevalence of certain species were observed between the dietary groups, the average age of the two groups was also significantly different (Table 3). So commenced the era of longitudinal studies, using individual subjects as their own controls. One of the first such studies investigated the fecal microbiota of three North Americans over several months and different dietary regimens (42). Greater inter-individual variation (between different subjects) in species composition was seen than intra-individual variation (between multiple samples from the same subject). Drasar and coworkers (36) monitored volunteers' fecal habits and composition over a six-week period (3 weeks on a conventional diet, followed by 3 weeks on a high-fiber diet). The only significant changes corresponded to stool weight and transit times. Hentges and colleagues (43) followed 10 subjects during baseline (1 month on a typical American diet; control), a meatless diet (1 month), a high-beef diet (1 month) and control diet again (1 month). Three stool samples were collected from each subject during the fourth week of each dietary period. *Bacteroides* spp. counts were significantly higher during the high-beef diet than the meatless diet ($P < 0.01$). Similar statistically significant observations were, however, seen

Table 3 Summary of the Statistically Significant Differences Between Japanese Subjects Consuming Different Diets

	Japanese diet	Western diet	P value[a]
Age (years)	60.30	41.3	0.013[b]
Streptococcus faecalis var *faecalis*	9.83[c]	8.46	0.038[b]
Other facultative or aerobic organisms	7.20	4.75	<0.01
Eubacterium contortum	9.58	ND	0.033
Eubacterieum lntum	10.20	10.07	0.015
Bifidobacterium infantis other	ND	10.29	0.009
Peptostreptococcus sp. 1	10.53	0	0.033
Peptostreptococcus sp. 1-25	8.29	4.64	0.001[b]

[a] Based on contingency table analysis (Fisher's exact probability statistic).
[b] Confirmed by Student's *t*-test.
[c] Mean $\log_{10}$ counts per gram feces (dry weight).
Abbreviation: ND, none detected.
Source: From Ref. 37.

between the *Bacteroides* spp. counts of the two control diet periods. Perhaps a better study strategy would have incorporated a control diet between the cross-over from meatless to high-beef diets. Indeed, this work demonstrated that short-term cross-over design dietary investigations may hinder identification of the effects of the different diets on the microbiota. Studies incorporating either prolonged diet regimens (allowing the microbiota to stabilize) or interspersed with control diet (enabling a return to baseline) may better demonstrate the microbial impact of each diet.

Another important aspect in studies monitoring the microbial composition over time and between subjects is the analytical method employed. For example, in the study by Hentges and colleagues (43), the data were essentially averaged twice (first by subject, then by dietary period). Such analysis is flawed due to the inter-individual variation which negates significance observed intra-individually. Indeed, this was discussed by Cummings (44), who concluded that overall changes in a group would be obscured due to inter-individual variations using such analytical methods.

Overall, data from early cultivation studies have indicated that the major bacterial populations harbored by individuals within a given society (i.e., Japanese, British, American) are reasonably stable to species level (3,45). Intra-individual, as well as inter-individual (even within a given society), subspecies variation has been documented in a number of studies following stability of the *E. coli* biotypes in humans (46–50). As will be discussed below, molecular fingerprinting techniques have demonstrated the complexity and dynamics of the bifidobacterial and lactobacilli populations of healthy New Zealanders (51,52). Such studies have highlighted the complex nature of the bacterial community residing in the distal regions of the human GI tract, with variation observed both in stability and in composition.

COMPOSITION OF THE ADULT FECAL MICROBIOTA ASSESSED BY MOLECULAR TECHNIQUES

With the advent of molecular-based techniques, bacterial characterization has become much more accurate, since it no longer relies upon phenotypic traits (which often vary due to the elastic nature of bacterial growth). In addition, more direct comparisons can be made between laboratories and across different studies. Initial work employed molecular methods to identify and/or discriminate different bacterial isolates from cultivation studies. One such study demonstrated that the majority of bacterial isolates from six healthy humans belonged to either the *Bact. fragilis* group or the *Clostridium coccoides* group (53). *Bifidobacterium*, the *Clostridium leptum* subgroup, *Collinsella* and *Prevotella* were also shown to be common phylogenetic lineages represented in healthy humans. Recent developments in molecular biology afford not only accurate and reproducible identification techniques for microbial isolates, but also strategies for direct community analysis at a number of genetic levels. Improved understanding of microbial taxonomy has generated a wealth of probing and polymerase chain reaction (PCR)-based strategies for quantification and/or qualification studies. Community profiling assays, including denaturing gradient gel electrophoresis (DGGE) and sequencing of clonal libraries from GI samples, have revolutionized our knowledge of the microbial composition of the GI tract.

The development and application of PCR-based methods and probing strategies, which have circumvented cultivation, highlighted the "tip-of-the-iceberg" scenario that our knowledge of the GI tract microbiota amounted to. The coverage that cultivation studies afforded has been calculated to be as low as 10%, although others suggest it may be as high as 40–58% (15,54–56). Modern cultivation media and incubation conditions enable greater diversity, and therefore coverage, to be recognized. However, many components of the human gut microbiota remain elusive to cultivation in vitro. Molecular strategies also have their limitations, including detection limits and inherent biasing. As such, the overall objective of the study generally determines which assay is most appropriate. In the case of investigations to elucidate the diversity and dynamics of the human gut microbiota, a polyphasic approach is best, allowing thorough analysis at multiple taxonomic levels.

Microbiota Assessed by Clone Libraries and Community Profiling Techniques

Two PCR-based profiling strategies have been used to obtain an overall profile of complex bacterial communities—clone libraries and PCR-DGGE [or alternatively PCR-TGGE (temperature gradient gel electrophoresis)]. Both utilize universal PCR primers to amplify the 16S rRNA genes from total DNA isolated from samples.

Suau and colleagues (15) prepared a detailed phylogenetic inventory of the fecal microbiota of a healthy 40-year-old male subject using PCR-cloning. A total of 520 clones were obtained from two transformations of the same ligation product from the 10-cycle PCR amplification (120 from the first and 400 from the second). The 282 clones that were sequenced were classified as belonging to 82 molecular species, 20 of which corresponded to bacteria previously cultivated from human stool samples (i.e., 24% corresponded to sequences available in public databases). Three major monophyletic groups contained 270 (95.7%) of the 282 clones; the *Clos. coccoides* group (125 clones), the *Bacteroides* group (88 clones) and the *Clos. leptum* group (57 clones). The remainder of the clones were distributed among a variety of phylogenetic clusters; two belonged to recognized molecular species (*Streptococcus salivarius* and *Streptococcus parasanguinis*), whilst the remainder were potentially novel molecular species. Most interesting was a lack of bifidobacterial sequences amongst the clones analyzed (even though rRNA dot-blot hybridizations indicated the carriage of bifidobacteria). Two possibilities could explain this: (1) lack of amplification of bifidobacterial rRNA genes, due to DNA extraction protocol, denaturation conditions during PCR, or amplification efficiency; and (2) coverage of the biodiversity provided by the 282 clones was insufficient (coverage was calculated to be 85%; thus, the probability that the 283rd clone was a different molecular species from the 82 already observed was 15%). An investigation of the 25-cycle PCR clone library was performed in parallel to this work, using the same subject (57). Comparison of the 10- and 25-cycle approaches demonstrated that PCR cycle number influences the diversity of the resulting phylogenetic profile. The clonal library obtained from the 25-cycle PCR was less diversified than that from the 10-cycle PCR. However, differences in diversity were seen between the two methods. That is, molecular species or operational taxonomic units (OTUs) were present in the 25-cycle PCR clone library that were not represented in the 10-cycle PCR clone library.

Previous work by Wilson and Blitchington (58) demonstrated somewhat similar results, with 25 of 50 clones (50%) classified as *Clos. leptum* subgroup, 34% as *Bacteroides* group and 10% as *Clos. coccoides* group. The disparity in the clostridial representation of the different clone libraries most probably reflects either inter-individual variations or disparity of the protocols. However, bifidobacteria were again absent from the clone library. In addition, *Eubacterium rectale* was not covered in the clone library in this earlier study, although *Eub. rectale* isolates were cultured from the same sample (58). These data highlight the difficulty to approach full coverage of the complex microbiota and further demonstrate that a polyphasic approach is pertinent. However, such work has enabled identification of previously unknown components of the fecal microbiota, and the sequence data can be used to develop new probing strategies to accurately quantify such bacteria.

Work carried out as part of the European Union (EU) human gut microbiota project using PCR clone libraries demonstrated that microbial diversity increased with age (57). In addition, the percentage of OTUs corresponding to known molecular species was highest in infants and lowest in the elderly subjects. Thus, not only was the microbial diversity greater in the elderly subjects, but also 92% of OTUs were undescribed (potentially novel) species.

The alternative to sequencing and subsequent phylogenetic analysis of clone libraries is to employ TGGE or DGGE to separate the 16S rRNA gene clones. Such techniques essentially provide a fingerprint representation of the numerically dominant members of the microbial community and allow rapid profiling of the microbial diversity of different samples (59). In addition, the TGGE/DGGE patterns can be used to selectively identify 16S rRNA amplicons of interest for characterization (which is achieved by sequencing and phylogenetic analysis). Recent years have seen an explosion in the development and application of TGGE and DGGE in human gut microbiology (Table 4). Zoetendal and coworkers (56) demonstrated the use of TGGE for monitoring the bacterial composition of human fecal samples. They compared the PCR-TGGE profiles of 16 healthy adults and identified host-specific patterns reflecting inter-individual variation in the predominant microbiota of stool samples. Some bands were seen in samples from multiple subjects, suggesting that certain members of the predominant human fecal microbiota were common across the volunteers (56). In addition, the study encompassed longer-term surveillance of the microbial community of two subjects. The PCR-TGGE profiles of each individual did not differ greatly with time, demonstrating that the predominant bacterial species were relatively stable. Phylogenetic analysis of the predominant bacteria was performed via cloning and sequencing. PCR-TGGE of each clone enabled mobility comparisons and showed 45 of the 78 clones had similar mobility to one of the 15 prominent bands of the fecal PCR-TGGE profile. This work demonstrated that the majority of predominant bacterial species represented in the fingerprint did not

Table 4 Application of TGGE/DGGE in Human Gut Microbiology

Reference	Target population	Subjects	Investigation	Overall results
Zoetendal et al. (56)	Total community	16 Finnish and Dutch adults (7 males, 9 females)	Inter-individual variation; stability over 6 months was monitored for 2 subjects	Differences in TGGE banding profiles demonstrated each individual harbored a unique microbiota (inter-individual variation), although some common bands were seen indicating some dominant bacteria were present in all samples; TGGE profiles were highly consistent over time for the same subject (intra-individual stability)
Satokari et al. (60)	Bifidobacteria	6 Finnish adults (3 males, 3 females)	Stability of bifidobacterial component over 4 weeks	Multiple bifidobacterial biotypes were seen in 5 of the 6 subjects; inter-individual variation; bifidobacterial PCR-DGGE profiles were generally stable (minor pattern changes seen for one subject; low bifidobacterial levels resulted in no PCR product for other subject
Walter et al. (61)	Lactic acid bacteria (LAB)	4 NZ adults (2 males, 2 females)	Development and validation of group-specific PCR primers for human studies	Lac1–Lac2GC PCR-DGGE enabled detection of *Lactobacillus* species present at levels $> 10^5$ CFU/g of feces (wet weight); inter-individual variation; intra-individual variation over 6 months
		2 NZ adults on probiotic trial	Monitor changes in LAB during *Lactobacillus* feeding trial	PCR-DGGE amplicon profile of probiotic *Lactobacillus* was only seen during feeding period of trial; dominant *Lactobacillus* stable for subject 2; intra-individual variation for subject 4
Zoetendal et al. (62)	Total community	50 adults of varying relatedness +4 different primates	Impact of genetic relatedness on composition of the fecal microbiota	Positive linear relationship between host genetic relatedness and similarity of PCR-DGGE profiles; no significant difference between similarity indices of unrelated persons grouped by either gender or living arrangements; significantly higher similarity between unrelated humans than compared to other primates; no relationship between similarity indices and age difference of siblings (range: 21–56 years)

(Continued)

Table 4 Application of TGGE/DGGE in Human Gut Microbiology (*Continued*)

Reference	Target population	Subjects	Investigation	Overall results
Favier et al. (63)	Total community	2 infants (1 breast-fed, 1 mixed-fed)	Feasibility of DGGE to monitor bacterial succession in neonates	Initial DGGE profiles were simple; bifidobacteria seen after 3 days; more complex DGGE profiles were seen when breast-feeding was supplemented (mixed-feeding and weaning); bacterial succession was seen for both infants
Heilig et al. (64)	*Lactobacillus* group including *Leuconostoc, Pediococcus* spp.	12 adults	*Lactobacillus* diversity and stability (0, 6 and 20 months)	Total community PCR-DGGE: inter-individual variation and intra-individual stability
				Lactobacillus spp. PCR-DGGE: inter-individual variation and variable intra-individual stability (stable over time for some subjects and more dynamic in others)
		1 baby boy	*Lactobacillus* diversity monitored over time (one day old to 5 months)	No PCR product seen prior to day 55 (indicating *Lactobacillus* were below the detection limit); two prominent amplicons (*Lb. rhamnosus* and *Lb. casei*) persisted throughout (day 55 to day 147), bacterial succession corresponding to dietary change (solid foods introduced at 3 months of age)
		4 infants (10–18 months old)	*Lactobacillus* diversity during *Lb. paracasei* feeding trial	Inter-individual variation of *Lactobacillus* community, with variable intra-individual stability; *Lb. paracasei* F19 present in profile during administration period, also present in one infant receiving placebo (both baseline and 2 week sample)
Zoetendal et al. (65)	Total community and lactobacillus group	10 adult patients (5 males, 5 females)	Mucosally associated bacteria in different regions of colon and fecal microbiota	Total community PCR-DGGE: no PCR product for ascending and transverse colon biopsy samples of one individual; large inter-individual variation of fecal and biopsy profiles; profiles of biopsy samples from same subject highly similar; fecal microbiota and mucosal microbiota profiles were significantly different
				Lactobacillus spp. PCR-DGGE: fecal and biopsy samples similar in 6/10 subjects (with single prominent amplicon); minor differences in profiles of different biopsy samples seen in 3/10 subjects

Abbreviations: DGGE, denaturing gradient gel electrophoresis; PCR, polymerase chain reaction; TGGE, temperature gradient gel electrophoresis.

correspond to known species. However, the 15 prominent bands were identified as belonging within different *Clostridium* clusters. In addition, the common biotypes found in virtually all subjects' TGGE patterns were identified as *Ruminococcus obeum*, *Eubacterium hallii* and *Fusobacterium prausnitzii* (reclassified as *Faecalibacterium prausnitzii*) (56).

More recent studies by this group have shown a positive linear relationship between host genetic relatedness and the similarity index of PCR-DGGE profiles (Table 4) (62). Higher similarity was seen between profiles obtained for monozygotic twins living apart than that seen for married couples. In addition, similarity was highest between monozygotic twin individuals than between pairs of twins. No correlation was shown between similarity index and gender or living arrangements of unrelated individuals, suggesting these factors did not significantly impact upon the bacterial composition. Inclusion of samples collected from four different primates (chimpanzee, gorilla, macaque, and orangutan) and subsequent analysis demonstrated that PCR-DGGE profiles of unrelated humans showed significantly greater similarity than that between humans and other primates. This work has indicated that host genotype factors have an important impact upon the bacterial composition of the gut microbiota (62).

A number of studies have also evaluated the application of PCR-DGGE to monitor the composition and dynamics of particular components of the human gut microbiota (60,61,64). To date, such research has concentrated on the lactic acid bacteria (LAB), as well as bifidobacteria. Each of these studies has displayed evidence of the ability to use PCR-DGGE for group- or genus-specific investigations. Overall, these studies demonstrated inter-individual variation within specific bacterial populations (Table 4). Differences were seen regarding the dynamics of different bacterial groups over time: fluctuations were seen in the LAB of two New Zealand adults over 6 months (61); the bifidobacterial population of five Finnish adults remained relatively stable over 4 weeks (60); *Lactobacillus* spp. PCR-DGGE of several healthy adults displayed varying stability over 20 months (stable for certain individuals and more dynamic for others) (64). The study by Heilig and colleagues (64) also monitored the lactobacilli diversity in one baby boy, from birth to 5 months of age. No *Lactobacillus* spp. PCR product was obtained for the first 55 days (suggesting this population was either absent or below the detection limit). Subsequently, two prominent amplicons were seen to persist throughout the study period (64). These were identified as belonging to the species *Lactobacillus rhamnosus* and *Lactobacillus casei*. In addition, this work displayed bacterial succession of the lactobacilli corresponding to the introduction of solid foods ($\sim$3 months of age), from which time a third prominent amplicon was observed (*Lactobacillus salivarius*). Two of these studies further investigated the usefulness of this technique in probiotic feeding trials (61,64). Both groups demonstrated the ability to identify probiotic-specific amplicons within the group-specific bacterial profiles.

Favier and colleagues (63) performed a pilot study with two infants investigating the feasibility of DGGE profiling to monitor bacterial succession during the first 10–12 months of life. One infant was exclusively breast-fed prior to weaning, whilst the other was breast-fed for a fortnight and then mixed-fed (both formula- and breast-milk) until weaning. The results demonstrated a simple fecal microbiota initially, which progressively diversified with time. Bifidobacterial amplicons were predominant in the fecal microbiota of both infants during the first 6 months. Alterations in diet, such as the supplementation of breast-feeding with formula-milk and introduction of solid foods (weaning), was associated with changes in the bacterial profiles. The shift in bacterial profiles seen following weaning, was more pronounced in the exclusively breast-fed infant (compared to the mixed-fed infant)—although this may be a reflection of the relative simplicity of the pre-weaning profile of this infant (compared to the more complex pre-weaning profile of the mixed-fed infant, comprised of multiple dominant amplicons) (63).

PCR-DGGE has also been used to compare the microbial component of biopsy samples taken from different regions of the colon, both with each other and the fecal microbiota (65). Inter-individual variation was shown for both fecal and biopsy samples. Interestingly, the biopsy samples taken from three distinct regions of the colon (ascending, transverse, and descending colon) of the same individual provided extremely similar DGGE profiles (total community). Significant differences were evident in the total community PCR-DGGE of fecal and biopsy samples. This is by no means alarming, as one can readily appreciate the distinction of the two ecological niches (i.e., the luminal microbiota and mucosally associated community), and the numbers within the species are likely to differ and result in different profiles. However, *Lactobacillus* spp. PCR-DGGE patterns from fecal and biopsy samples were very similar in 6/10 subjects. Minor differences were seen in the *Lactobacillus* spp. PCR-DGGE profiles of the different biopsy samples from three of the 10 individuals. Overall, no differences were noted in the mucosally associated lactobacilli of different individuals based on host health (i.e., healthy versus diseased tissues).

In summary, molecular methods enabling community analysis of the human fecal microbiota have demonstrated that a large proportion of the predominant microbial component are novel or unknown species—which have not yet been cultivated. Inter-individual variation and intra-individual stability are consistent features of studies of the prominent members of the total community. However, investigations of specific bacterial groups or genera indicate varying levels of stability, with fluctuations seen in some cases. Host genetic factors appear to play an important role in the microbial composition of healthy human adults, though it is as yet undetermined what impact bacterial acquisition and succession during childhood plays.

Directed PCR Analysis

In addition to PCR-cloning and PCR-DGGE profiling techniques, PCR strategies have been employed in gut microbiology for many years to investigate the presence/absence or activity of bacterial groups, genera, and even species. Such methods were initially developed for identification purposes but have subsequently been utilized for detection, essentially allowing qualitative analysis of the microbial component of samples. Modern developments in PCR technology now afford quantitative PCR assays (e.g., real-time PCR), though the major application of such methods to date has been clinical diagnostics.

Wang and coworkers (66) developed 12 species-specific PCR primer sets to monitor the predominant gut microbiota of humans (*Bact. distasonis, Bacteroides thetaiotaomicron, Bact. vulgatus, Bifidobacterium adolescentis, Bifidobacterium longum, Clostridium clostridioforme, E. coli, Eubacterium biforme, Eubacterium limosum, Fuso. prausnitzii, Lactobacillus acidophilus* and *Pep. productus*). During validation of the species-specific PCR assays, the sensitivity of each primer set was examined with DNA extracts from pure cultures. Interestingly, such work demonstrated that PCR sensitivities varied markedly. Following validation of the PCR assays, Wang and coworkers (66) examined the presence of the bacterial species in fecal samples from humans (seven adults and two infants), two BALB/c mice, two Fischer rats, two cats, one dog, one rhesus monkey and one rabbit. High titers of *Clos. clostridioforme, Fuso. prausnitzii* and *Pep. productus* were detected in all samples examined. High titers of *Bact. thetaiotaomicron, Bact. vulgatus* and *Eub. limosum* were also detected in all adult human samples, whereas the *Bacteroides* spp. specific assays gave either weak or no

PCR products for infant samples. Bifidobacterial levels were higher in human infants compared to adults and other animals.

Similar research by Matsuki and colleagues (53) developed four group-specific primer sets to monitor the predominant bacteria in human feces. These 16S rRNA gene-targeted primer sets included group-specific primers for the *Bact. fragilis* group and the *Clos. coccoides* group, and genus-specific primers for *Bifidobacterium* and *Prevotella*. DNA extracts were prepared from fecal samples collected from six healthy adults (five males and one female) and used for the group-specific PCR detection assays. The *Bact. fragilis* group, *Bifidobacterium* and *Clos. coccoides* group were detected in all six subjects, whilst PCR detected *Prevotella* in only two of the six subjects (53).

PCR techniques have also been developed for identification and detection of bacterial isolates or components at species level. One bacterial group that has enjoyed particular interest in this regard is bifidobacteria (67–69). Investigation of the distribution of the nine bifidobacterial species known to be harbored by humans was performed by Matsuki and coworkers (68), using fecal samples from 48 healthy adults and 27 breast-fed infants. No *Bifidobacterium gallicum* amplification products were obtained from any sample. In addition, no *Bifidobacterium infantis* products were seen from the adult samples. The bifidobacterial species that were most consistently detected in adult samples were *Bifidobacterium catenulatum* (44/48), *Bif. longum* (31/48), *Bif. adolescentis* (29/48) and *Bifidobacterium bifidum* (18/48). Overall, 29 of the 48 adult samples contained three or four different bifidobacterial species, with 17 of the remaining 18 samples comprising less than three species. The majority of breast-fed infants harbored *Bifidobacterium breve* (19/27), with a smaller proportion of samples containing *Bif. infantis* (11/27) and *Bif. longum* (10/27; six of which were positive for *Bif. infantis*). Interestingly, three breast-fed infant samples were negative with all nine bifidobacterial species-specific primers. In general, breast-fed infant fecal samples were positive for three or less bifidobacterial species (23/27).

Germond and coworkers (67) designed and validated species-specific primers for human bifidobacterial species and then developed PCR primer mixtures which enabled detection of multiple species concurrently (i.e., in a single reaction). PCR mix one comprised species-specific primers for seven bifidobacterial species: namely, *Bif. adolescentis*, *Bif. angulatum*, *Bif. bifidum*, *Bif. breve*, *Bif. catenulatum/pseudocatenulatum*, *Bif. infantis* and *Bif. longum*. Application of this PCR primer mixture with fecal DNA from two healthy human adults demonstrated both subjects harbored *Bif. longum* and *Bif. adolescentis*, whilst a weak PCR amplification product was also seen for *Bif. angulatum* for one subject. Confirmation assays performed with the individual species-specific primer sets indicated that *Bif. bifidum* was under-represented during concurrent PCR analysis as amplification was positive for both subjects when single species PCR was used but negative using PCR mix one.

Requena and colleagues (69) investigated the use of the transaldolase gene in identification and detection of nine bifidobacterial species (*Bif. adolescentis*, *Bif. angulatum*, *Bif. bifidum*, *Bif. breve*, *Bif. catenulatum*, *Bif. infantis*, *Bifidobacterium lactis*, *Bif. longum* and *Bif. pseudocatenulatum*). These workers examined its application for both PCR-DGGE and real-time PCR. Seven of the nine bifidobacterial species could be differentiated by transaldolase gene PCR-DGGE; *Bif. angulatum* and *Bif. catenulatum* displayed the same mobility characteristics. Examination of the bifidobacterial species diversity in fecal samples using this method showed 6/10 healthy adults contained two amplicons, one being *Bif. adolescentis*. In four of the six profiles the second amplicon was *Bif. longum*, the fifth profile also contained an unidentified amplicon and the sixth profile contained two *Bif. adolescentis* amplicons. One sample gave no PCR-DGGE product, two of the remaining three samples contained *Bif. longum,* the final sample contained *Bif. bifidum.* This strategy was also employed to assess the fecal bifidobacterial diversity of 10 babies. One sample gave no PCR product, 8/10 contained *Bif. bifidum* (one of which harbored a second unidentified amplicon) and the final sample comprised *Bif. infantis, Bif. longum* and an unidentified amplicon.

Comparison of bifidobacterial enumerations obtained from plate counts and bifidobacterial-specific real-time PCR with either transaldolase gene primers or 16S rRNA primers has been performed (69). Good correlation was seen between all three enumeration methods when healthy adult samples were used ($n=7$). Correlation of bifidobacterial levels in infant samples ($n=10$) was better between cultivation work and 16S rRNA gene real-time PCR than between cultivation work and transaldolase gene real-time PCR. Under-representation of the *Bif. bifidum* component of samples during transaldolase gene real-time PCR was largely responsible for this discrepancy.

Probing Strategies

As well as affording design of PCR primers for specific bacterial populations, the improved 16S rRNA gene sequence information has greatly enhanced the development of probing strategies for gut micro-organisms. Two probing strategies have generally been employed, namely, dot-blot hybridization and fluorescent in situ hybridization (FISH). The nature of the 16S rRNA gene of bacteria also enables development of oligonucleotide probes targeting different taxonomic levels, i.e., domain level (Bact 338), group level (e.g., Chis 150), genus level (e.g., Bif 164) or species level (e.g., Bdis 656) (57,70). The last 5 years have seen enormous development and application of these strategies in gut microbiology (71–76).

A longitudinal study was performed with nine healthy human volunteers (five males, four females) monitoring the fecal microbiota using FISH (72). The results demonstrated that 90–100% of $4'$, 6-diamidino-2-phenylindol dihydrochloride (DAPI)-stained cells were hybridized by the bacterial probe (Bact 338), and that the *Clos. coccoides/Eub. rectale* group (Erect 482) and *Bacteroides* group (Bfra 602 and Bdis 656) represented almost 50% of the total bacteria of healthy humans. In addition, the Low G+C #2 group (Lowgc2P) comprised 12% of the total bacteria, and *Bifidobacterium* (Bif 164) 3%. Initial data indicated that the *Clostridium lituseburense* group (Clit 135), the *Clostridium histolyticum* group (Chis 150), and the *Streptococcus/Lactococcus* group (Strc 493) all formed less than 1% of the total bacteria and so were not included in the longitudinal study. In general, the fecal microbiota of individuals was shown to fluctuate during the 8-month study. Interestingly, the greatest variation was seen in the bifidobacterial component of the microbiota. A more recent study from the same laboratory group employed a set of 15 probes to investigate the microbial composition of 11 healthy volunteers (73). Again, the *Bacteroides* group (27.7%) and the *Clos. coccoides/Eub. rectale* group (22.7%) were seen to be the numerically predominant bacterial components. In addition, three other predominant groups were identified: *Atopobium* group (11.9%), *Eubacterium* low G+C #2/*Fuso. prausnitzii* group (10.8%), and *Ruminococcus* and relatives (10.3%). *Bifidobacterium* (4.8%), *Eub. hallii* and relatives (3.8%), *Lachnospira* and relatives (3.6%), and *Eubacterium cylindroides* and relatives (1.8%) were also dominant members of the microbiota. However, *Enterobacteriaceae,* the *Lactobacillus/Enterococcus* group, *Phascolarctobacterium* and relatives, and *Veillonella* were all subdominant (each forming 1%). Taken together, this afforded 90.5% coverage of the total bacteria hybridized with the Bact 338 probe. (N.B.: *Eub. hallii* and relatives, and *Lachnospira* and relatives are subsets of the *Clos. coccoides*/*Eub. rectale* group, so were not included in summation). However, a large proportion of the DAPI-stained cells ($\sim$40%) were not accounted for by the Bact 338 probe. The question arises as to whether these cells are non-viable or metabolically inactive (low rRNA), impermeable, or represent novel bacterial groups whose 16S rRNA differs within the "conserved" region the Bact 338 probe targets or in the secondary structure surrounding it.

Other research groups have developed and validated additional oligonucleotide probes suitable for FISH, for potentially important members of the human GI tract microbiota (75,76). *Ruminococcus obeum*-like bacteria have been frequently identified in ribosomal clonal libraries of human fecal samples and the development of probing strategies was thus considered pertinent (76). Following validation, the Urobe 63 probe was used to examine the *Rum. obeum* group in three healthy Dutch males (three samples were collected from each subject over one month). FISH enumeration was performed both by epifluorescent microscopy and by flow cytometry (which require different handling and thus different protocols). The two methods gave comparable results, demonstrating that *Rum. obeum*-like bacteria comprise $\sim$2.5% of the total bacterial count (Bact 338). A further six individuals (two males, four females) provided stool samples and the results were consistent in all subjects. In addition, counts of the *Clos. coccoides*/*Eub. rectale* group were made which indicated that the *Rum. obeum* group accounted for $\sim$16% of this group (76). Similarly, the *Fuso. prausnitzii* cluster has been shown in numerous molecular analyses to be part of the dominant microbiota of healthy humans. As such, Suau, and coworkers (75) developed an oligonucleotide probe for this cluster which was applicable both for FISH and dot-blot hybridizations. Overall, 16.5% (range 5–28%) of the DAPI-stained cells hybridized with the Fprau 645 probe ($n = 10$ healthy adults). Samples from a further 10 healthy individuals were used for dot-blot analysis with the same probe and showed the *Fuso. prausnitzii* cluster accounted for 5.3% of the total bacterial 16S rRNA (range 1.5–9.5%). Unfortunately, these data are not comparable as different samples were used for each assay. In addition, the two assays provide distinctive enumeration: FISH provides counts of the number of cells in the sample (which can be represented as a percentage of total bacterial (Bact 338) cells or total cells (DAPI), whereas dot-blot provides an index of the percentage of total 16S rRNA the specific population forms. The index obtained by dot-blot is further complicated as it is not only proportional to the number of cells in the sample, but also the number of copies of the rRNA gene in each cell and the activity of the cells.

Dot-blot analyses of the healthy human fecal microbiota using an array of probes have, once again, highlighted the inter-individual variation (71,74). Both studies employed six oligonucleotide probes to monitor the predominant bacterial groups. The work by Sghir and colleagues (74) ($n = 27$ healthy adults; 13 males, 14 females) was consistent with earlier work which showed that the *Bacteroides* group (including *Bacteroides*, *Prevotella* and *Porphyromonas*; 37%), the *Clos. leptum* subgroup (16%) and the *Clos. coccoides*/*Eub. rectale* group (14%) were predominant, accounting for 67% of the total rRNA. *Bifidobacterium* and the enteric group each made up less than 1% of the total rRNA, whilst the low-G+C Gram-positive group (including *Lactobacillus*, *Streptococcus* and *Enterococcus*) represented 1% (74). Marteau and coworkers (71) similarly demonstrated the predominant fecal rRNA ($n = 8$ healthy adults; four males, four females) corresponded to the *Clos. coccoides*/*Eub. rectale* group (23%), the *Clos. leptum* subgroup (13%) and the *Bacteroides* group (8%) using the same probes as Sghir and colleagues (74). Although, using different probes, this later study indicated higher bifidobacterial and *Lactobacillus*/*Enterococcus* rRNA indices, 3% and 7%, respectively. Interestingly, Marteau and coworkers (71) compared the fecal rRNA indices of these bacterial groups and *E. coli* species with cecal rRNA indices. Overall, the indices for the *Bacteroides* group and the *Clos. leptum* subgroup were significantly higher in fecal samples than cecal samples, and the *Lactobacillus/Enterococcus* fecal rRNA index was significantly lower than that of the cecum. The *Clos. coccoides/Eub. rectale* rRNA index was higher in fecal samples than cecal samples, but the inter-individual variation meant that this was not statistically significant. Concurrent cultivation analysis monitoring total anaerobes, facultative anaerobes, bifidobacteria and *Bacteroides* demonstrated significantly higher levels of total anaerobes, bifidobacteria, and bacteroides populations in fecal samples compared to cecal samples (71).

Most recent developments in probing strategies include membrane-array and/or microarray methodologies (77,78). The results of these assays were in agreement with previous studies, demonstrating inter-individual variation in the fecal microbiota of different healthy human subjects. The predominant microbiota of healthy humans determined by the membrane-array technique (employing 60 oligonucleotide probes targeting 20 bacterial species) included *Bacteroides* species, *Clos. clostridioforme*, *Clos. leptum*, *Fuso. prausnitzii*, *Pep. productus*, *Ruminococcus* species, *Bifidobacterium* species and *E. coli* (78). In addition, analysis of the fecal microbiota of an individual suffering long-term diarrhea demonstrated a loss of a number of bacterial species common to the normal microbiota of healthy subjects. These results were replicated in a microarray study using the same probe array (77), where the probes were printed on aldehyde slides rather than applied to membranes.

Overall, probing and PCR-based strategies have been shown to afford good coverage of the predominant microbiota of the GI tract. This situation is likely to improve with continued development of specific primer sets and/or oligonucleotide probes, particularly in the light of increased diversity as elucidated by community analysis work. Indeed, such community profiling studies provide excellent direction for the development of novel probes and primers.

INVESTIGATIONS AT THE SUBSPECIES LEVEL

One aspect of gut microbiology that is not amenable to current probing or PCR-based methodologies is subspecies differentiation (i.e., investigations of the microbial complexity and dynamics below the phylogenetic level of species). A number of studies have, however, demonstrated the importance of such research (48,51,52). One study monitored the composition of the bifidobacterial and lactobacilli populations of two healthy humans over a 12-month period (52). Overall, the bifidobacterial levels of both individuals were relatively stable throughout the study period [$\sim 10^{10}$/g of feces (wet weight)]. Lactobacilli levels were relatively constant in subject one ($\sim 10^9$/g) but fluctuated considerably in samples from subject two (10^6–10^9 per gram). *Bacteroides* and enterobacterial levels were also examined during the study; the former remained stable for both individuals and the latter displayed marked variability (especially in subject two, for whom numbers were below the detection limit in weeks 17 and 20). Genetic fingerprinting techniques were used to differentiate the predominant bacterial isolates of the bifidobacterial and lactobacillus populations. Interestingly, two distinct bifidobacterial profiles were seen, with one individual harboring a simple, stable bifidobacterial population (five distinct strains of bifidobacteria were detected during 12 months, one of which was numerically predominant throughout the study). In contrast, the second subject harbored a more complex and dynamic bifidobacterial population (36 distinct strains were seen over the

12 months, with between four and nine strains at any given time). The *Lactobacillus* populations of both subjects were simple and stable. None of the lactic acid bacterial strains isolated during the study were common to both individuals, further accentuating the inter-individual variations of these two microbial ecosystems (52). Subsequent work from the same laboratory extended these investigations to a further eight healthy humans (four males, four females) (51). Two separate samples were collected and processed for each individual. The results of this secondary study confirmed the findings of the former, with bifidobacterial levels remaining relatively stable and *Lactobacillus* numbers varying greatly. Again, each individual harbored unique bifidobacterial and *Lactobacillus* strains (at least in regard to the numerically predominant microbiota). Intriguingly, half of the subjects were shown to harbor a complex bifidobacterial microbiota (five or more predominant strains).

McBurney and colleagues (48) examined the perturbation of the enterobacterial populations of the initial long-term study by McCartney and colleauges (52). Similar to the bifidobacterial populations of these two individuals, the enterobacterial population of subject one was relatively simple and stable (predominated by a single strain) whilst subject two harbored a diverse and dynamic enterobacterial biota (27 distinct strains were identified over 12 months). As mentioned previously, enterobacterial levels were below detection for subject two during weeks 17 and 20. This individual undertook a 7-day course of amoxicillin for a respiratory infection during weeks 21 and 22, after which time the enterobacterial population re-emerged. Most interesting, though, was the antibiotic-resistance profiles of this bacterial group before and after treatment. Strains isolated prior to antibiotic administration were susceptible to a wide range of antibiotics tested, whereas strains isolated following treatment were resistant to a number of antibiotics. Thirteen weeks after amoxicillin administration, multiple drug-resistant enterobacterial strains were still present. In the following 2 months, strains resistant to ampicillin were still harbored, and only after 25 weeks post-treatment did the predominant enterobacterial microbiota return to a simple, stable, susceptible composition.

Taken together, the above-mentioned work clearly demonstrates the value of investigations at the subspecies level, as such studies afford more detailed analysis of the diversity and dynamics of the gut microbiota. Furthermore, such strategies allow the detection of microbial perturbations which are often not evident at the bacterial group, genus or species levels (79).

CONCLUSION

The normal microbiota of the human GI tract is a complex microbial community whose composition is defined by a number of factors (including host genomics, diet, age, bacterial succession, immune function and health status). In general, the predominant bacterial groups are relatively stable in healthy human adults. However, inter-individual variations are evident, reflecting the unique equilibrium of each person's GI ecosystem. In addition, examination of the microbial populations in more detail (i.e., investigations at the subspecies level) further demonstrates the complexity and dynamics of this bacterial community, and most probably reflects its adaptive nature. Interactions between the host and the gut microbiota have led some researchers to acknowledge that the human intestine is, indeed, *"intelligent"—based on Alfred Binet's definition of intelligence: "intelligence is the range of processes involved in adapting to the environment"* (13).

REFERENCES

1. Simon GL, Gorbach SL. Intestinal flora in health and disease. Gastroenterology 1984; 86:174–193.
2. Macfarlane GT, Cummings JH. The colonic flora, fermentation, and large bowel digestive function. In: Phillips SF, Pemberton JH, Shorter RG, eds. The Large Intestine: Physiology, Pathophysiology, and Disease. New York: Raven Press Ltd, 1991:51–92.
3. Tannock GW. An Introduction to Microbes Inhabiting the Human Body. London: Chapman & Hall, 1995.
4. Mitsuoka T. Intestinal flora and human health. Asia Pacific J Clin Nutr 1996; 5:2–9.
5. Hooper LV, Midtvedt T, Gordon JI. How host-microbial interactions shape the nutrient environment of the mammalian intestine. Annu Rev Nutr 2002; 22:283–307.
6. Holzapfel WH, Haberer P, Snel J, Schillinger U, Huis in't Veld JHJ. Overview of gut flora and probiotics. Int J Food Microbiol 1998; 41:85–101.
7. Rabiu BA, Gibson GR. Carbohydrates: a limit on bacterial diversity within the colon. Biol Rev 2002; 77:443–453.
8. Guarner F, Malagelada J-R. Gut flora in health and disease. Lancet 2003; 361:512–519.
9. Blaser MJ, Kirschner D. Dynamics of *Helicobacter pylori* colonization in relation to the host response. Proc Natl Acad Sci USA 1999; 96:8359–8364.
10. Coconnier M-H, Lievin V, Hemery E, Servin AL. Antagonistic activity against *Helicobacter* infection in vitro and in vivo by the human *Lactobacillus acidophilus* strain LB. Appl Environ Microbiol 1998; 64:4573–4580.
11. Gibson GR, McCartney AL. Modification of the gut flora by dietary means. Biochem Soc Trans 1998; 26:222–228.
12. Marshall BJ. *Helicobacter pylori*. Am J Gastroenterol 1994; 89:S116–S128.
13. Bourlioux P, Koletzko B, Guarner F, Braesco V. The intestine and its microflora are partners for the protection of the host: report on the Danone Symposium "The Intelligent Intestine," held in Paris, June 14, 2002. Am J Clin Nutr 2003; 78:675–683.
14. Macfarlane GT, Macfarlane S, Gibson GR. Validation of three-stage continuous culture system for investigating the effect of retention time on the ecology and metabolism of bacteria in the human colon. Microb Ecol 1998; 35:180–187.
15. Suau A, Bonnet R, Sutren M, et al. Direct analysis of genes encoding 16S rRNA from complex communities reveals many novel molecular species within the human gut. Appl Environ Microbiol 1999; 65:4799–4807.
16. Moore WEC, Holdeman LV. Discussion of current bacteriological investigations of the relationships between intestinal flora, diet, and colon cancer. Cancer Res 1975; 35:3418–3420.
17. Salminen S, Bouley C, Boutron-Ruault M-C, et al. Functional food science and gastrointestinal physiology and function. Br J Nutr 1998; 80:S147–S171.
18. Metges CC. Contribution of microbial amino acids to amino acid homeostasis of the host. J Nutr 2000; 130:1857S–1864S.
19. Gordon JI, Hooper LV, McNevin MS, Wong M, Bry L. Epithelial cell growth and differentiation. III. Promoting diversity in the intestine: conversations between the microflora, epithelium, and diffuse GALT. Am J Physiol 1997; 273:G565–G570.
20. Borruel N, Casellas F, Antolín M, et al. Effects of nonpathogenic bacteria on cytokine secretion by human intestinal mucosa. Am J Gastroenterol 2003; 98:865–870.
21. Perdigón G, Maldonado Galdeano C, Valdez JC, Medici M. Interaction of lactic acid bacteria with the gut immune system. Eur J Clin Nutr 2002; 56:S21–S26.
22. Tannock GW. The acquisition of the normal microflora of the gastrointestinal tract. In: Gibson SAW, ed. Human Health: The Contribution of Microorganisms. London: Springer-Verlag, 1994:1–16.
23. Bezirtzoglou E. The intestinal microflora during the first weeks of life. Anaerobe 1997; 3:173–177.

24. Mountzouris KC, McCartney AL, Gibson GR. Intestinal microflora of human infants and current trends for its nutritional modulation. Br J Nutr 2002; 87:405–420.

25. Salminen S, Gibson GR, McCartney AL, Isolauri E. Influence of mode of delivery on gut microbiota composition in seven year old children. Gut 2004; 53:1388–1389.

26. Biavati B, Sgorbati B, Scardovi V. The Genus *Bifidobacterium*. In: Balows A, Trüper HG, Dworkin M, Harder W, Schleifer KH, eds. The Prokaryotes. A Handbook on the Biology of Bacteria: Ecophysiology, Isolation, Identification, Applications. 2nd ed., Vol. 1. New York: Springer-Verlag, 1992:816–833.

27. Scardovi V. Genus *Bifidobacterium* Orla-Jensen 1924, 472[AL]. In: Sneath PHA, Mair NS, Sharpe ME, Holt JG, eds. In: Bergey's Manual of Systematic Bacteriology, Vol. 2. Baltimore: Williams and Wilkins, 1986:1418–1434.

28. Edwards CA, Parrett AM. Intestinal flora during the first months of life: new perspectives. Br J Nutr 2002; 88:S11–S18.

29. Salminen S, Bouley C, Boutron Ruault M-C, et al. Functional food science and gastrointestinal physiology and function. Br J Nutr 1998; 80:S147–S171.

30. Kleessen B, Bunke H, Tovar K, Noack J, Sawatzki G. Influence of two infant formulas and human milk on the development of the faecal flora in newborn infants. Acta Paediatrica 1995; 84:1347–1356.

31. Mitsuoka T, Kaneuchi C. Ecology of the bifidobacteria. Am J Clin Nutr 1977; 30:1799–1810.

32. Stark PL, Lee A. The microbial ecology of the large bowel of breast-fed and formula-fed infants during the first year of life. J Med Microbiol 1982; 15:189–203.

33. Yoshioka H, Fujita K, Sakata H, Murono K, Iseki K. Development of the normal intestinal flora and its clinical significance in infants and children. Bifidobacteria Microflora 1991; 2:33–39.

34. Boehm G, Lidestri M, Casetta P, et al. Supplementation of a bovine milk formula with an oligosaccharide mixture increases counts of faecal bifidobacteria in preterm infants. Arch Dis Child 2002; 86:178–181.

35. Harmsen HJM, Wildeboer-Veloc ACM, Raangs GC, et al. Analysis of intestinal flora development in breast-fed and formula-fed infants by using molecular identification and detection methods. J Pediatr Gastroenterol Nutr 2000; 30:61–67.

36. Drasar BS, Jenkins DJA. Bacteria, diet, and large bowel cancer. Am J Clin Nutr 1976; 29:1410–1416.

37. Finegold SM, Sutter VL. Fecal flora in different populations, with special reference to diet. Am J Clin Nutr 1978; 31:S116–S122.

38. Hill MJ, Drasar BS, Aries V, Crowther JS, Hawksworth G, Williams REO. Bacteria and aetiology of cancer of large bowel. Lancet 1971; 1:95–100.

39. Aries V, Crowther JS, Drasar BS, Hill MJ, Williams REO. Bacteria and the aetiology of cancer of the large bowel. Gut 1969; 10:334–335.

40. Moore WEC, Cato EP, Holdeman LV. Some current concepts in intestinal bacteriology. Am J Clin Nutr 1978; 31:S33–S42.

41. Finegold SM, Attebery HR, Sutter VL. Effect of diet on human fecal flora: comparison of Japanese and American diets. Am J Clin Nutr 1974; 27:1456–1469.

42. Holdeman LV, Good IJ, Moore WEC. Human fecal flora: variation in bacterial composition within individuals and a possible effect of emotional stress. Appl Environ Microbiol 1976; 31:359–375.

43. Hentges DJ, Maier BR, Burton GC, Flynn MA, Tsutakawa RK. Effect of a high-beef diet on the fecal flora of humans. Cancer Res 1977; 37:568–571.

44. Cummings JH. Dietary fibre and the intestinal microflora. In: Hallgren B, ed. Nutrition and the Intestinal Flora. Stockholm: Almqvist and Wiksell International, 1983:77–86.

45. Bornside GH. Stability of human fecal flora. Am J Clin Nutr 1978; 31:S141–S144.

46. Hartley CL, Clements HM, Linton KB. *Escherichia coli* in the faecal flora of man. J Appl Bacteriol 1977; 43:261–269.

47. Mason TG, Richardson G. A review. *Escherichia coli* and the human gut: some ecological considerations. J Appl Bacteriol 1981; 51:1–16.

48. McBurney WT, McCartney AL, Apun K, McConnell MA, Tannock GW. Perturbation of the enterobacterial microflora detected by molecular analysis. Microb Ecol Health Dis 1999; 11:175–179.

49. Shooter RA, Bettelheim KA, Lennox-King SMJ, O'Farrell S. *Escherichia coli* serotypes in the faeces of healthy adults over a period of several months. J Hyg 1977; 78:95–98.

50. Wallick H, Stuart CA. Antigenic relationships of *Escherichia coli* isolated from one individual. J Bacteriol 1943; 45:121–126.

51. Kimura K, McCartney AL, McConnell MA, Tannock GW. Analysis of fecal populations of bifidobacteria and lactobacilli and investigation of the immunological responses of their human hosts to the predominant strains. Appl Environ Microbiol 1997; 63:3394–3398.

52. McCartney AL, Wenzhi W, Tannock GW. Molecular analysis of the composition of the bifidobacterial and lactobacillus microflora of humans. Appl Environ Microbiol 1996; 62:4608–4613.

53. Matsuki T, Watanabe K, Fujimoto J, et al. Development of 16S rRNA-gene-targeted group-specific primers for the detection and identification of predominant bacteria in human feces. Appl Environ Microbiol 2002; 68:5445–5451.

54. Tannock GW. Analysis of the intestinal microflora using molecular methods. Eur J Clin Nutr 2002; 56:S44–S49.

55. Vaughan EE, Schut F, Heilig HGHJ, Zoetendal EG, de Vos WM, Akkermans ADL. A molecular view of the intestinal microflora. Curr Issues Intest Microbiol 2000; 1:1–12.

56. Zoetendal EG, Akkermans ADL, de Vos WM. Temperature gradient gel electrophoresis analysis of 16S rRNA from human fecal samples reveals stable and host-specific communities of active bacteria. Appl Environ Microbiol 1998; 64:3854–3859.

57. Blaut M, Collins MD, Welling GW, Doré J, van Loo J, de Vos W. Molecular biological methods for studying the gut microflora: the EU human gut flora project. Br J Nutr 2002; 87:S203–S211.

58. Wilson KH, Blitchington RB. Human colonic flora studied by ribosomal DNA sequence analysis. Appl Environ Microbiol 1996; 62:2273–2278.

59. Konstantinov SR, Fitzsimons N, Vaughan EE, Akkermans ADL. From composition to functionality of the intestinal microbial communities. In: Tannock GW, ed. Probiotics and Prebiotics. Where are we Going?. Norfolk, U.K.: Caister Academic Press, 2002:59–84.

60. Satokari RM, Vaughan EE, Akkermans ADL, Saarela M, de Vos WM. Bifidobacterial diversity in human feces detected by genus-specific PCR and denaturing gradient gel electrophoresis. Appl Environ Microbiol 2001; 67:504–513.

61. Walter J, Hertel C, Tannock GW, Lis CM, Munro K, Hammes WP. Detection of *Lactobacillus*, *Pediococcus*, *Leuconostoc*, and *Weissella* species in human feces by using group-specific PCR primers and denaturing gradient gel electrophoresis. Appl Environ Microbiol 2001; 67:2578–2585.

62. Zoetendal EG, Akkermans ADL, Akkermans, van Vliet WM, de Visser JAGM, de Vos WM. The host genotype affects the bacterial community in the human gastrointestinal tract. Microb Ecol Health Dis 2001; 13:129–134.

63. Favier CF, Vaughan EE, de Vos WM, Akkermans ADL. Molecular monitoring of succession of bacterial communities in human neonates. Appl Environ Microbiol 2002; 68:219–226.

64. Heilig HGHJ, Zoetendal EG, Vaughan EE, Marteau P, Akkermans ADL, de Vos WM. Molecular diversity of *Lactobacillus* spp. and other lactic acid bacteria in the human intestine as determined by specific amplification of 16S ribosomal DNA. Appl Environ Microbiol 2002; 68:114–123.

65. Zoetendal EG, von Wright A, Vilpponen-Salmela T, Ben-Amor K, Akkermans ADL, de Vos WM. Mucosa-associated bacteria in the human gastrointestinal tract are uniformly distributed along the colon and differ from the community recovered from faeces. Appl Environ Microbiol 2002; 68:3401–3407.

66. Wang R-F, Cao W-W, Cerniglia CE. PCR detection and quantitation of predominant anaerobic bacteria in human and animal fecal samples. Appl Environ Microbiol 1996; 62:1242–1247.

67. Germond J-E, Mamin O, Mollet B. Species specific identification of nine human *Bifidobacterium* spp. in feces. Syst Appl Microbiol 2002; 25:536–543.
68. Matsuki T, Watanabe K, Tanaka R, Fukuda M, Oyaizu H. Distribution of bifidobacterial species in human intestinal microflora examined 16S rRNA-gene-targeted species-specific primers. Appl Environ Microbiol 1999; 65:4506–4512.
69. Requena T, Burton J, Matsuki T, et al. Identification, detection, and enumeration of human *Bifidobacterium* species by PCR targeting the transaldolase gene. Appl Environ Microbiol 2002; 68:2420–2427.
70. Harmsen HJM, Welling GW. In: Tannock GW, ed. Probiotics and Prebiotics. Where are we going?. Norfolk, U.K.: Caister Academic Press, 2002:41–58.
71. Marteau P, Pochart P, Doré J, Béra-Maillet C, Bernalier A, Corthier G. Comparative study of bacterial groups within the human cecal and fecal microflora. Appl Environ Microbiol 2001; 67:4939–4942.
72. Franks AH, Harmsen HJM, Raangs GC, Jansen GJ, Schut F, Welling GW. Variations of bacterial populations in human feces measured by fluorescence in situ hybridization with group-specific 16S rRNA-targeted oligonucleotide probes. Appl Environ Microbiol 1998; 64:3336–3345.
73. Harmsen HJM, Raangs GC, He T, Degener JE, Welling GW. Extensive set of 16S rRNA-based probes for detection of bacteria in human feces. Appl Environ Microbiol 2002; 68:2982–2990.
74. Sghir A, Gramet G, Suau A, Rochet V, Pochart P, Dore J. Quantification of bacterial groups within human fecal flora by oligonucleotide probe hybridization. Appl Environ Microbiol 2000; 66:2263–2266.
75. Suau A, Rochet V, Sghir A, et al. *Fusobacterium prausnitzii* and related species represent a dominant group within the human fecal flora. Syst Appl Microbiol 2001; 24:139–145.
76. Zoetendal EG, Ben-Amor K, Harmsen HJM, Schut F, Akkermans ADL, de Vos WM. Quantification of uncultured *Ruminococcus obeum*-like bacteria in human fecal samples by fluorescent in situ hybridization and flow cytometry using 16S rRNA-targeted probes. Appl Environ Microbiol 2002; 68:4225–4232.
77. Wang R-F, Beggs ML, Robertson LH, Cerniglia CE. Design and evaluation of oligonucleotide-microarray method for the detection of human intestinal bacteria in fecal samples. FEMS Microbiol Lett 2002; 213:175–182.
78. Wang R-F, Kim S-J, Robertson LH, Cerniglia CE. Development of a membrane-assay method for the detection of intestinal bacteria in fecal samples. Mol Cell Probes 2002; 16:341–350.
79. Tannock GW. Molecular assessment of intestinal microflora. Am J Clin Nutr 2001; 73:410S–414S.

4

The Intestinal Microbiota of the Elderly

Fang He
Technical Research Laboratory, Takanashi Milk Products Co., Ltd, Yokohama, Kanagawa, Japan

INTRODUCTION

With the significant progress in medical science and health care, the average life expectancy has increased by nearly three decades over the last century (1). The old (>65 years) and the "oldest" (>85 years) age groups are the fastest growing subpopulation in the world, especially in industrialized societies referred to as "aged societies." World Health Organization (WHO) figures indicate there are currently about 580 million people in the world aged 60 or older, and this figure is expected to rise to over a billion within the next 20 years (2).

It has been well known that many physiological functions, such as immunity and gut function, in humans usually decline progressively with age after approximately the 30th birthday (1). The elderly are an increased-risk population with high rates of morbidity and mortality due to their susceptibility to degenerative and infectious diseases. A major consequence of people living longer is an increased incidence in health problems. In fact, industrial societies are now suffering from a sharp increase in medical costs to the age-related infectious and autoimmune diseases, malignancies, allergies, and digestive problems. Therefore, effective measures to redress the age-related decline (or imbalance) in physiological function should be much sought.

The intestinal microbiota mediates many crucial events towards the protection or alteration of health. This chapter summarizes the current knowledge and findings about the intestinal microbiota in the elderly, although a limited but growing body of literature on this subject is available.

COLONIZATION AND SUCCESSION OF HUMAN INTESTINAL MICROBIOTA WITH AGE

The gastrointestinal tract (GI tract) serves as one of the biggest interfaces between the body and the external environment (3). This GI tract is a highly specialized organ system that allows us to consume food in discrete meals as well as a very diverse array of foodstuffs to meet our nutrient needs. The organs of the GI tract include the oral cavity, esophagus, stomach, small and large intestine; in addition, the pancreas and liver secrete into the small intestine. The system is connected to the vascular, lymphatic, and nervous systems to facilitate regulation of the digestive response, delivery of absorbed compounds to organs of the body, and regulation of the food intake.

One of the characteristic aspects of the GI tract is the presence of numerous endogenous microbes colonizing the surface of the GI tract throughout the life of the host. It consists of a complex community inside the host, known as the intestinal microbiota. In healthy adults, the intestinal microbial cells have been estimated to outnumber the host's somatic and germ cells by a ratio of 10:1 (4). The development of this microbiota is initiated during the birth process. The fetus exists in a sterile environment until birth. After being born, the infant is progressively colonized by bacteria from the mother's vagina and feces and from the environment. As long as nutrients and space are not limited, the commensals with high division rates predominate, e.g., enterobacteria (*Escherichia coli*) and *Enterococcus* appear. The succession of microbes in an infant's intestinal tract also depends on the feeding mode. The fecal microbiota of breast-fed babies has been found to be relatively simple, usually exclusively dominated by *Bifidobacterium* (5). However, recent comparative studies showed that bifidobacteria were the predominate fecal bacteria in both group of infants (6,7). In bottle-fed infants, the count and frequencies of occurrence of *Bacteroides*, *Enterobacteriaceae* and streptococci were significantly higher than those in the breast-fed infants (6,7). After weaning, when solid food is consumed, the stools of infants begin to shift to an adult-like microbiota: bifidobacteria decrease remarkably and constitute only 5% to 15% of total microbes. The number of *Bacteroidacecea*, eubacteria, *Peptococcaceae* and usually clostridia outnumber bifidobacteria, while aerobic bacteria such as *E. coli* and streptococci, which have been regarded as the predominant species are always detected, but account for less than 1% of the total bacterial count. Lactobacilli, *Megasphaerae* and *Veillonellae* are often found in adult feces, but the counts are usually less than 10^7 per gram of feces. By the end of the secondary year of life, the microbiota becomes more stable and resembles that of an adult (see also the chapter by McCartney and Gibson). As the microbial population increases nutrients become scarce and the intestinal niches become occupied with more specialized species with an advanced symbiotic relationship between the host and microbiota. Once the climax microbiota has become established, the major bacterial groups in the intestine of an adult usually remain relatively constant over time.

The habitats of the intestinal microbiota vary in different parts of the human GI tract (8). In healthy persons, acid stomach contents usually contain few microbes. Immediately after a meal, counts of around 10^5 bacteria per milliliter of gastric juice can be recorded: bacteria including streptococci, enterobacteriaceae, *Bacteroides* and bifidobacteria derived from the oral cavity and the meal. The microbiota of the small intestine is relatively simple and no large numbers of organisms are found. Total counts are generally 10^4 or less per milliliter, except for the distal ileum, where the total counts are usually about 10^6/ml. In the duodenum and jejunum, streptococci, lactobacilli and *Veillonellae* are mainly found. Towards the ileum, *E. coli* and anaerobic bacteria increase in number. In the caecum, the composition suddenly changes and is similar to that found in feces, and the concentration may reach 10^{11} per gram of content.

As more than 400 species have been estimated to reside in the colon of healthy adults, which may attain population levels nearly as high as 10^{12} /g in the colon and may make up almost half the content by weight (8,9). This bacterial community is dominated by strict anaerobes, and contains less facultative anaerobes with a rate of anaerobes and aerobes as 1000:1. In accordance with the metabolic activity, the major bacteria present in the intestinal microbiota of the healthy adult can be divided roughly into three groups (10).

Group one is lactic acid–producing bacteria including *Bifidobacterium, Lactobacillus* and *Streptococcus* (including *Enterococcus*), which may possess a symbiotic relationship with the host. Group two includes putrefactive bacteria such as *Clostridium prefringens, Clostridium* spp. *Bacteroides*, Peptococcaceae, *Veillonella, E. coli, Staphylococcus* and *Pseudomonas aeruginosa*. Others are like *Eubacterium, Ruminococcus, Megasphaera, Mitsuokello, C. butyricum* and *Candida*, group three. Normally, near-stability exists in these habitats and each person has an individually fixed microbiota as far as qualitative composition is concerned.

The intestinal microbiota play an important role in normal bowel function and maintenance of host health, through the formation of short chain fatty acids, modulation of immune responses, and development of colonization resistance (8,10). These functions of the intestinal microbiota are the consequence of the activities of the numerous intestinal bacteria as a whole community with a well-organized structure built on the balance among the various bacterial members. Therefore, the functions of the intestinal microbiota are very sensitive to factors that can alter the structure of the intestinal microbiota qualitatively and quantitatively such as aging, physiological state, disease, medication, diet, and stresses.

Age-Altered Aspects of the Intestinal Microbiota

Normal aging is associated with significant changes in the function of most organs and tissues, such as decreased taste thresholds, hypochlorhydria due to atrophic gastritis, and decreased liver blood flow and size (11). The GI tract is no exception, and there is increased evidence of impaired gastrointestinal function with aging (3,11–13). In the GI tract of the elderly, the age-related changes include decreased acid secretion by the gastric mucosa, and greater permeability of mucosal membranes which have been linked to increase in circulating antibodies to components of the intestinal microbiota in elderly subjects. Therefore, certain microbes which can take advantage of new ecological niches are assumed to become predominant inhabitants, leading to a dramatic shift in the composition of the gut microbiota upon age.

Although the knowledge about the age-related alteration of the human intestinal microbiota is still limited, the structure of the intestinal microbiota in the healthy elderly has been suggested to be different from that of the healthy adults. This phenomenon is considered to be a result of aging, but it may also accelerate senescence.

As early as in the 1960s, the scientific attention has been focused to characterize the intestinal microbiota of the elderly. In several works conduced in the different geographic regions, reduced presence of bifidobacteria was often observed in the fecal microbiota of the elderly compared to that of the healthy adults, as well as more putrefactive bacteria *Enterobacteriaceae, Streptococcus, Staphylococcus, Proteus* and *C. perfringens* (14–17).

Mitsuoka and his colleagues (18,19) analyzed the composition of the intestinal microbiota in the various stages of life and observed an age-dependent change in the composition of the fecal microbiota. Bifidobacteria were less present in the fecal microbiota of senile (65- to 85-year-old) persons than in those of younger adults, while more lactobacilli and clostridia were found in the fecal samples of the elderly.

Mitsuoka and coworkers (20) compared the fecal flora of the elderly (61–95 years old) with healthy adults (31.8±6.6 years old) using optimized culture procedures for members of the anaerobic microbiota. Total bacterial count, *Bidifobacterium Veillonella, Eubacterium* were decreased significantly, whereas *C. perfringens* and *Lactobacillus* were increased significantly. Furthermore, the frequencies of occurrence in *Bifidobacterium,*

Micrococcaceae was decreased, while those of *C. perfringens*, other *Clostridium* sp., and yeasts were increased significantly in the elderly compared with that of the healthy adults.

Using the same method, Benno and Mitsuoka (21) did not find significant differences in *Bacteroidaceae, Eubacterium, Peptostreptococcus* and *Megasphaera* between the healthy elderly and healthy adults. However, *Bifidobacterium (Bif. adolescentis* and *Bif. longum), Enterococcus* were less in the elderly compared with the healthy adults, while lecithinase-negative *Clostridium* and *C. paraputrificum* were increased in the elderly.

Recently, non–culture-dependent molecular methods have been used to investigate the intestinal microbiota (22). The advent of these molecular methods, which do not rely on our ability to culture bacteria prior to quantification, allow additional information to be gained on the gut microbiota as a whole. Another method that allows ecological analysis without the need to culture the organism is that of community cellular fatty acid (CFA) analysis. Numerous environmental factors affect bacterial fatty acid synthesis, but certain signature fatty acids have been used to indicate the presence of specific groups of organisms in soil and marine environments, and have also been used to study community structure in human fecal samples.

Direct polymerase chain reaction analysis was performed on elderly persons' fecal samples (22). Over 280 clones were generated and characterized by sequence analyses, providing a molecular taxonomic inventory. Phylogenetic analysis showed that the microbiota of the elderly was more diversified than that of younger adults. The proportion of unknown molecular species was very high among the clones derived from fecal samples of elderly persons. It is evident from this study that the fecal microbiota of the elderly person is very complex. The microbial diversity of the intestinal microbiota appears to be increased with age. This is in contrast to the microbial diversity of babies, which was found to be extremely low: only nine species were detected within each clone's library.

Hopkins et al. (23) studied fresh fecal samples obtained from seven adults, five elderly individuals, and four geriatric patients diagnosed with *C. difficile*-associated diarrhea. Selected fecal bacteria were investigated using viable counting procedures, 16S rRNA abundance measurements and community CFA profile. The principal micro-biological differences between adults and the elderly were the occurrence of higher numbers of enterobacteria and a lower number of anaerobe populations in the elderly group. Another important finding of this study was the lower number of bifidobacteria observed in the group of elderly patients.

Hopkins and Macfarlane (24) isolated bacteria from fecal samples of healthy young adults, elderly subjects, and elderly patients with *Clostridium difficile*-associated diarrhea (CDAD). The isolated bacteria were identified to species level on the basis of their CFA profiles with Microbial Identification System (MIDI, Inc., Newark, DE) (MIDI). While *Bacteroides* species diversity increased in the feces of the elderly individuals, bifidobacteria diversity dramatically decreased with age.

These observations indicate that aging may diminish bifidobacteria, and significantly increase clostridia, including *C. perfringens*, and allow a slight increase of lactobacilli, streptococci, and enterobacteriacea. The total bacterial counts and anaerobes/aerobes in the intestinal microbiota of the elderly are relatively lower than those of the healthy infants and young adults.

Historically, bifidobacteria have been considered to be the most important organisms for infants while lactobacilli, especially *L. acidophilus*, were considered the predominant beneficial bacterium for adults (5). However, bifidobacteria have recently been suggested to be more important throughout life as beneficial intestinal bacteria than lactobacilli (25,26). These ecological studies on the intestinal microbiota of the elderly indicate that bifidobacteria rather than lactobacilli are often decreased upon age. Although some changes also happen upon age in other groups of bacteria, they are non-specific, not constant, and very individual. Furthermore, a decrease in bifidobacteria has often been observed in the intestinal microbiota of various young patients (23,27,28). Therefore, a decrease in bifidobacteria in the intestinal microbiota could be considered as an important hallmark for aging and disease of the human GI tract.

BIFIDOBACTERIA IN THE ELDERLY

Taxonomic Species Placement of Bifidobacterial Microbiota with Age

Bifidobacteria have been known since Tissier (5) first described a species from the feces of breast-fed infants, which was later named as *Lactobacillus bifidus* by Orla-Jensen (29). Since that time numerous studies have been published concerning the ecology and importance of bifidobacteria in the intestine of humans, especially in infants.

A new taxonomic system was established by creating the genus *Bifidobacterium* with the description of several new species besides *B. bifidum*, which was the only existing species at that time (30,31). This was followed by an increasing number of new species isolated from humans and animals (32–35). The new concept of the genus *Bifidobacterium* taxonomy including 24 taxonomic species was summarized in special chapters of Bergey's Manual (1986). Currently, a total of 26 well-established taxonomic species have been described, among which are nine species which have been found to be exclusive residents of the human intestine. These are *B. bifidum, B. longum, B. infantis, B. breve, B. adolescentis, B. angulatum, B. catenulateum, B. pseudocatenulatum,* and *B. dentium.* Bifidobacteria appear between the 2nd and 5th day of life and continue to be one of the most numerous bacteria, amounting to about 10^{10}/g of wet feces. Many studies indicated that each healthy adult has and maintains its own specific composition of *Bifidobacterium* microbiota during his/her life (33).

Interestingly, the bifidobacterial composition of a human can progressively vary with aging, both qualitatively as well as quantitatively. The predominant species of bifidobacteria in the human GI tract can be differentiated to indicate various stages of life. However, lactobacilli, another of the important genus of endogenous bacteria considered to contribute to host health and well being, are changing only quantitatively, and do not express an apparent species diversity upon aging of the host.

Bifidobacterium was found as one of the predominate bacteria in the intestinal microbiota of both infants and adults (19,32). However, the species and biotype composition of the fecal bifidobacteria progressively varied with increasing age. Species typical for infants were *B. bifidium, B. infants, B. breve,* and *B. parvulorum.* Typical for adults were four different biovars of *B. adolescentis. B. bifidum* and *B. longum* could often be found in both age groups, but in lower numbers. *B. adolescentis* biovar b was the most common *Bifidobacterium* in the microbiota of the elderly. The frequency of the occurrence of *B. longum* was 71% for infants, 62% and 33% for adults and the elderly, respectively. *B. adolescents* occurred 100%, 91%, and 79% in the elderly, adults, and infants, respectively. These results have been supported by studies conducted by other research groups (36,37). It was found that *B. adolescentis* and *B. longum* dominated the bifidobacteria of healthy adults, which is different from the bifidobacteria composition of infants.

Mitsuoka (20) consistently observed that *B. adolescentis* biovar. b was significantly higher in the elderly, even when *Bifidobacterium* counts were similar among children, adults, and the elderly. The number of *B. adolescentis* and *B. longum* in healthy adults was significantly higher than those in aged persons. From 1829 fecal bacterial isolates from 15 healthy adults, *B. adolescentis, B. longum* and *B. bifidum* were found to be the predominant species of bifidobacteria of these healthy adults (21).

He and coworkers (38) isolated 51 *Bifidobacterium* strains from the feces of healthy adults (30–40 years old) and seniors (older than 70 years of age). The isolates were identified to species level based on the phenotypic characteristics. The isolates from the adults belonged to *B. adolescentis, B. longum, B. infantis, B. breve,* while those from the elderly were *B. adolescentis* and *B. longum.*

Studies with molecular methods indicate a similar distribution of bifidobacteria species in the various stages of life. In a study using a non–culture-based method using PCR and denaturing gradient gel electrophoresis, *B. adolescentis* was found to be the most common species in feces of adult subjects as earlier indicated in the studies with traditional culture-based methods (39).

Fecal bacteria from healthy young adults, elderly subjects, and elderly patients with CDAD were identified to species level on the basis of their CFA profiles with MIDI. Species diversity was found to decrease with age. *B. angulatum* was the most common bifidobacterial isolate in the healthy young adults. *B. bifidum, B. catenulateum, B. pseudocatenulatum* and *B. infantis* were not detected in the feces of the elderly subjects (24).

Human *Bifidobacterium* species were identified by Mullie and coworkers (40) with three multiplex PCRs. *B. bifidum, B. longum* and *B. breve* species were commonly recovered in infants, while *B. adolescentis B. catenulateum/B. pseudocatenulatum* and *B. longum* were predominant in adults.

Matsuki and coworkers (41) applied species- and group-specific PCR directly to fecal samples and found *B. catenulatum* (*B. catenulatum* and *B. pseudocatenulatum*) in 92% of adult fecal samples and *B. longum, B. adolescentis* and *B. bifidum* in 65, 60, and 38% of the samples from adults, respectively. Comparison of species-species PCR method with the classical culture method revealed that some species, most frequently *B. adolescentis,* were detected by the direct PCR method but not by culturing followed by species-specific PCR of the isolates.

The bifidobacteria in the intestinal microbiota of the healthy elderly is characterized by a reduced species diversity as well as quantitative decrease. The bifidobacteria in the elderly are characterized by *B. adolescentis* as the predominant species as well as a quantitative decrease within the whole intestinal microbiota.

Mucus Adhesion of Bifidobacteria

The reason for the age-related decrease in bifidobacteria numbers is still not well understood. Adhesion to the intestinal mucosa is regarded as a prerequisite for colonization by microbes and induction of the healthy promotion by them. It has therefore been proposed as one of the selection criteria for probiotic strains (42–45).

Ouwehand and coworkers (46) tested four *Bifidobacterium* strains for adhesion to mucus isolated from subjects of different age groups including healthy newborns, 2- and 6-month-old infants, adults (25–52 years) and elderly (74–93 years). The tested bifidobacteria adhered less to the mucus isolated from the elderly subjects compared to those from healthy infants and adults. The results suggest that the physiological condition of the mucus could be altered by aging, which can reduce the affinity spectrum of the mucus to bifidobacteria from various origins. This may be a factor involved in the decreased colonization of the elderly subjects by bifidobacteria and fewer species of *Bifidobacterium* present.

Twenty-four *Bifidobacterium* strains were examined for their ability to bind to immobilized human and bovine intestinal mucus glycoproteins (47). Each of the tested bacteria exhibited its characteristic adhesion to human and bovine fecal mucus. No significant differences were found among the taxonomic species. Among the tested bacteria, *B. adolescentis, B. angulatum, B. bifidum, B. breve, B. catenulatum, B. infantis, B. longum* and *B. pseudocatenulatum* adhered to human fecal mucus better than bovine fecal mucus, while the binding of *B. animalis and B. lactis* was not preferential. These results suggest that the mucosal adhesive properties of bifidobacteria may be a strain dependent feature, and the mucosal binding of the human bifidobacteria may be more host specific.

Fifty-one *Bifidobacterium* strains were isolated from the feces of healthy adults (30–40 years old) and seniors (older than 70 years of age) and were tested for their ability to adhere to the mucus isolated from the healthy adults (30–40 years of age) (38). The strains isolated from healthy adults, and especially *B. adolescentis,* bound better to intestinal mucus than those isolated from seniors. These results indicate that the bifidobacteria isolated from the healthy elderly may pose a reduced affinity to the intestinal mucus from healthy adults. These results suggest that the poor colonization of bifidobacteria in the intestinal microbiota of the elderly may also be related to the development of a less adherent *Bifidobacterium* population as well as the reduced ability of mucus from this age group to facilitate *Bifidobacterium* adhesion.

Laine and coworkers (48) investigated 30 *Bifidobacterium* strains isolated from the feces of the healthy elderly (>80 years of age) Japanese and Finnish subjects. These strains were tested for their ability to adhere to the mucus only isolated from their own feces. The better mucus adhesion was observed in the combination of *Bifidobacterium* from the elderly and their fecal mucus rather than that of probiotic bifidobacteria from adults and the mucus from the elderly.

The enhanced adhesion of *B. adolecentis* from the elderly to their mucus may, at the least, partly explain that *B. adolescentis* is a predominant species in the fecal *Bifidobacterium* microbiota. Therefore, there may be an advanced symbiotic relationship between *B. adolescentis* and the elderly. The replacement of the predominant species of bifidobacteria upon aging of the host may be one of the important events by which the intestinal microbes affect the homeostasis of physiological functions on the basis of the important contribution of bifidobacteria to human health and well being.

Influence of Age-Related Decline in Immune Function and Influence on Intestinal Bifidobacterial Microbiota

Immunosenescence is defined as the state of deregulated immune function that contributes to the increased susceptibility of the elderly to infection and, possibly, to autoimmune diseases and cancer (49,50). When immunosenescence appears, the functional capacity of the immune system of the host gradually declines with age. The most dramatic changes in immune function with age occur within the T cells compartment, the arm of the immune system that protects against pathogens and tumors (51–54). The fact that T lymphocytes are more severely affected than B cells or antigen-presenting cells is mainly due to the involution of the thymus, which is almost complete at the age of 60. The host is then dependent on the T cells of various specificities, which eventually leads to changes in the T cell repertoire. CD45RA + "native" cells are replaced by CD45RA − "memory" cells, and a T cell receptor oligoclonality develops. At the same time, T cells with signal transduction defects accumulate. Age-related T cell alterations lead to a decreased clonal expansion and a reduced efficiency of T cell effectors functions, such as cytotoxicity or B cell functionality. Decreased antibody production and a shortened immunological memory are the consequence. Efficient protection of elderly individuals by suitable vaccination strategies is therefore a matter of great importance (51,55). Perhaps of greater consequence to interpretation of immunosenescence in the elderly is the decline in cell-mediated immunity (CMI). This is particularly important with respect to combating infectious disease, but also to tumor surveillance, since anti-tumor effects of the immune system are almost exclusively governed by the cell-mediated component.

Interleukin (IL)-12 is a cytokine produced by mononuclear phagocytes and dendritic cells that serves as a mediator of the innate immune response to intracellular microbes and is a key inducer of cell-mediated immune responses towards microbes (56). IL-12 activates natural killer (NK) cells, promotes interferon (IFN)-γ production by NK cells and T cells, enhances cytotoxic activity of NK cells and cytolytic T lymphocytes, and promotes the development of TH1 cells. IL-10 and IL-12 are two cytokines secreted by monocytes/macrophages in response to bacterial products which have largely opposite effects on the immune system. IL-12 activates cytotoxicity and IFN-γ secretion by T cells and NK cells, whereas IL-10 inhibits these functions.

Many studies indicate that Gram-positive bacteria and their cell wall components are potent inducers of IL-12 for human monocytes, while the Gram-negative bacteria can promote more IL-10 (57–60). Karlsson and coworkers (61) also reported that Gram-positive bacteria *B. adolescentis*, *Enterococcus fecalis*, *Lactobacillus plantarum*, *Streptococcus mitis* can induce more IL-12 production by mononuclear cells from cord and adult blood compared to the gram-negative bacteria, *Bacteroides vulgatus*, *Escherichia coli*, *Pseudomonas aeruginosa*, *Veillonella parvula* and *Nerisseria sicca*. In contrast, more IL-10 was secreted by the stimulation of mononuclear from cord and adults with Gram-negative bacteria instead of gram-positive bacteria.

Furthermore, He and coworkers (62,63) characterized the ability of bifidobacteria to affect the production of macrophage-like derived cytokines with a murine macrophage-like cell line, J774.1 (Fig. 1 and Table 1). *B. adolescentis* and *B. longum*, known as adult-type bifidobacteria, induced significantly more pro-inflammatory cytokine secretion, IL-12, and TNF-α by the macrophage-like cells than did infant-type bifidobacteria,

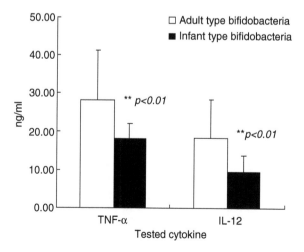

Figure 1 Cytokine production by a murine macrophage cell line J774.1 after exposure to adult-type bifidobacteria (*B. adolescentis* and *B. longum*) and infant-type bifidobacteria (*B. bifidum* and *B. breve, B. infantis*). *Abbreviations*: IL, interleukin; TNF, tumor necrosis factor.

Table 1 Cytokines Secretion by J774.1 Cells After Exposure to *Bifidobacterium* Strains[a]

No.	Species name	Strain no.	Origin	TNF-α (ng/ml)	IL-1 β (pg/ml)	IL-6 (ng/ml)	IL-10 (pg/ml)	IL-12 (ng/ml)
1	*B. adolescentis*	TMC 2704	Healthy adult	13.16±6.4	—	73.69±6.4	—	6.56±3.7
2	*B. adolescentis*	TMC 2705	Healthy adult	9.71±1.8	—	20.92±8.3	—	3.93±3.3
3	*B. adolescentis*	TMC 2718	Allergic infant	29.63±4.1	2.5±2.5	74.82±8.2	—	8.82±1.5
4	*B. adolescentis*	TMC 2720	Allergic infant	38.33±2.1	8.5±1.4	144.09±11.2	—	24.38±2.3
5	*B. adolescentis*	TMC 2721	Allergic infant	39.06±2.8	7.7±1.0	149.32±4.9	—	24.60±2.0
6	*B. adolescentis*	TMC 2723	Allergic infant	40.82±2.9	8.2±1.5	152.59±5.9	—	25.97±1.9
7	*B. adolescentis*	TMC 2736	Allergic infant	38.99±3.1	2.7±2.6	133.26±5.5	—	21.49±2.3
8	*B. adolescentis*	TMC 2737	Allergic infant	28.03±2.1	—	30.07±5.4	—	6.44±1.7
9	*B. adolescentis*	TMC 2938	Allergic infant	4.06±2.1	6.8±0.8	152.84±3.0	—	25.93±2.1
10	*B. adolescentis*	TMC 2739	Allergic infant	40.88±0.8	10.1±1.7	157.69±5.8	—	24.81±1.7
11	*B. animalis*	TMC 5101	Dairy food	14.02±2.9	—	49.93±6.3	6.15±10.1	6.19±51
12	*B. bifidum*	TMC 3101	Dairy food	11.99±5.3	—	15.32±3.1	65.8±77.3	5.71±3.1
13	*B. bifidum*	TMC 3108	Healthy infant	22.47±4.3	5.4±5.1	108.58±1.5	26.05±20.4	13.40±8.6
14	*B. bifidum*	TMC 3115	Healthy infant	21.75±4.8	3.3±2.8	120.02±12.9	19.26±21.0	12.53±4.8
15	*B. bifidum*	TMC 3116	Healthy infant	20.36±4.0	8.5±8.5	117.27±2.3	28±22.1	16.75±9.1
16	*B. bifidum*	TMC 3117	Healthy infant	23.14±4.0	11±3.9	102.23±10.3	18.7±15.1	14.81±7.6
17	*B. breve*	TMC 3207	Dairy food	18.05±3.3	19.4±19.4	94.38±7.1	18.48±19.4	8.73±3.4
18	*B. breve*	TMC 3217	Dairy food	12.91±3.0	—	35.60±7.1	—	2.24±0.3
19	*B. breve*	TMC 3218	Dairy food	17.20±3.9	3.3±5.6	86.36±15.6	16.21±14.7	7.37±3.2
20	*B. infantis*	TMC 2906	Healthy infant	18.99±4.6	14.3±13.2	86.19±2.6	15.11±14.5	8.65±2.7
21	*B. infantis*	TMC 2908	Healthy infant	19.63±4.5	4.2±3.7	87.30±7.7	23.41±21.5	7.16±2.8
22	*B. longum*	TMC 2607	Dairy food	21.83±4.4	3.8±3.8	103.34±18.9	45.75±36.7	16.41±8.3
23	*B. longum*	TMC 2608	Healthy adult	16.46±0.9	1.7±2.9	74.04±8.9	23.3±22.3	7.14±2.3
24	*B. longum*	TMC 2609	Healthy adult	15.16±2.4	—	63.47±4.3	24.56±25.3	7.49±4.6
25	*B. longum*	TMC 2614	Allergic infant	37.51±2.6	13.7±2.4	178.68±9.8	27.15±25.7	29.96±1.4
26	*B. longum*	TMC 2615	Allergic infant	39.18±1.1	16.1±3.2	185.72±12.8	36.13±35.5	33.47±1.2
27	*B. longum*	TMC 3524	Allergic infant	40.89±3.0	13.9±1.8	160.76±12.2	—	26.94±0.7
	Control[b]			—		—	—	—

[a]Results were expressed as Mean (SD); —: Detectable limit for IL-1β, 6 10, 12 and TNF-α <3 pg/ml, 7 pg/ml, 12 pg/ml, 12 pg/ml and 12 pg/ml.
[b]J774.1 cells without the tested bacteria.
Abbreviations: IL, interleukin; TNF, tumor necrosis factor.

Table 2 Cytokines Secretion by J774.1 Cells After Exposure to Various Species of *Bifidobacteria*[a]

Species	Strains	TNF-α (ng/ml)	IL-1 β (pg/ml)	IL-6 (ng/ml)	IL-10 (pg/ml)	IL-12 (ng/ml)
B. adolescentis	10	28.26±14.2	4.65±4.0	108.92±53.8	—	17.29
B. animalis	1	14.02	—	49.93	6.15	6.19±9.5
B. bifidum	5	19.94±4.6	5.64±4.3	92.71±43.8	31.562±19.6	12.64
B. breve	3	16.05±2.8	7.56±10.4	72.11±31.9	11.56±10.1	6.11±3.4
B. infantis	2	19.31±0.5	9.25±7.1	86.75±0.8	19.26±5.9	7.90±1.1
B. Iongum	6	28.50±12.0	8.2±7.1	127.66±54.1	26.15±15.4	20.23±11.5

[a]Results were expressed as Mean (SD); —: Detectable limit for IL-1β, 6 10, 12 and TNF-α <3 pg/ml, 7 pg/ml, 12 pg/ml, 12 pg/ml and 12 pg/ml.
Abbreviations: IL, interleukin; TNF, tumor necrosis factor.

Table 3 Adhesion of Lactic Acid Bacteria to Caco-2 Cells and IL-6 and IL-8 Secretion by Caco-2 Cells after Exposure to Lactic Acid Bacteria

Strain no.[a]	Species name	Origin	Adhesive ability[b]	IL-6 secretion (pg/ml)[c]		IL-8 secretion (pg/ml)[d]	
				Viable cells	Inactivated cells	Viable cells	Inactivated cells
TMC 0313	Lactobacillus acidophilus	Dairy food	±	—	—	—	—
TMC 0356	L. acidophilus	Dairy food	±	—	—	—	—
TMC 0402	L. casei	Dairy food	±	—	—	—	—
TMC 0409	L. casei	Dairy food	++	—	—	—	—
TMC 0503	L. rhamnosus	Dairy food	±	—	—	—	—
TMC 0510	L. rhamnosus	Dairy food	±	—	—	—	—
TMC 0517	L. rhamnosus	Dairy food	±	—	—	—	—
ATCC 53103	L. rhamnosus GG	Human intestine	+++	—	—	—	—
TMC 1001	L. casei	Human intestine	±	—	—	—	—
TMC 1002	L. casei	Human intestine	±	—	—	—	—
TMC 1003	L. casei	Human intestine	±	0.7±0.1	—	—	—
ATCC 15706	Bifidobacterium adolescentis	Adult intestine	±	—	—	—	—
TMC 2704	B. adolescentis	Adult intestine	±	—	—	—	—
TMC 2705	B. adolescentis	Adult intestine	±	—	—	—	—
TMC 5101	B. animalis	Dairy food	++	0.7	—	31.2±36.3	—
TMC 2906	B. infantis	Infant intestine	±	—	1.3±1.6	120±75.1	131.6±135.0

(Continued)

Table 3 Adhesion of Lactic Acid Bacteria to Caco-2 Cells and IL-6 and IL-8 Secretion by Caco-2 Cells after Exposure to Lactic Acid Bacteria (*Continued*)

Strain no.[a]	Species name	Origin	Adhesive ability[b]	IL-6 secretion (pg/ml)[c]		IL-8 secretion (pg/ml)[d]	
				Viable cells	Inactivated cells	Viable cells	Inactivated cells
TMC 2908	*B. infantis*	Infant intestine	±	−	−	94.5±71.6	32.1±33.6
TMC 2607	*B. longum*	Dairy food	±	−	−	−	−
TMC 2608	*B. longum*	Adult intestine	±	−	−	19.9±31.3	−
TMC 2609	*B. longum*	Adult intestine	++	−	−	−	−
TMC 3101	*B. bifidum*	Dairy food	++	−	−	−	−
TMC 3108	*B. bifidum*	Infant intestine	±	−	−	22.8±47.0	−
TMC 3115	*B. bifidum*	Infant intestine	++	−	−	62.2±74.8	−
TMC 3116	*B. bifidum*	Infant intestine	−	−	−	−	−
TMC 3117	*B. bifidum*	Infant intestine	±	−	−	−	−
TMC3207	*B. breve*	Infant intestine	±	−	−	−	−
TMC 3217	*B. breve*	Dairy food	±	−	−	119.5±42.0	−
TMC 3218	*B. breve*	Dairy food	±	−	−	28.7±47.0	75.8±115.1
TMC 3219	*B. breve*	Dairy food	±	−	−	−	−
JCM 1200T	*B. pseudocatenulatum*	Infant intestine	−	−	−	−	−

[a]TMC, Culture Collection of Takanashi Milk Products Co., Ltd.; ATCC, American Type Culture Collection; JCM, Japanese Collection of Microorganisms.
[b]Mean of bacteria cells bound to Caco-2 cell in 20 randomly chosen microscopic fields (−); 0; (±): 1–20; (++): 21–50; (+++): 100–200.
[c]IL-6 secretion by Caco-2 cells was expressed as Mean (SD); −: <0.7 pg/ml (detection limit).
[d]IL-8 secretion by Caco-2 cells was expressed as Mean (SD); −: <10 pg/ml (detection limit).
Abbreviation: IL, interleukin.

B. bifidum, B. breve, and *B. infantis.* In contrast, *B. adolescentis* did not stimulate the production of anti-inflammatory IL-10 as the other tested bacteria did. At the same time, neither the adult-type nor the infant-type bifidobacteria were found to be likely to trigger inflammatory responses in human enterocytes (Table 2) (64). The results suggest that the adult-type bifidobacteria, especially, *B. adolescentis,* may be more potent to amplify but less able to down-regulate the inflammatory responses (Table 3).

The intestinal microbiota of elderly people is in general different from those in infants; the former is more diverse and stable (8). One of the distinct age-related events in intestinal microbiology is the increase in numbers of facultative anaerobic Gram-negative bacteria (8); these bacteria may be opportunistic infective agents. They have been found to be the triggers of anti-inflammatory cytokine-production by macrophages and monocytes (60,61). The anti-inflammatory effects of these bacteria are believed to be one of the strategies required for their successful colonization of the host's intestine, overcoming the natural defense barrier, including inflammation. Therefore, an increase in bacteria, including bifidobacteria, which can enhance the intestinal inflammatory response in aged people, can be considered beneficial to counterbalance the age-related changes in their intestinal microecology. This may contribute to the homeostasis of the local immunity by preventing local inflammation from being oversuppressed. These results suggest that the dominance of the intestinal bifidobacteria by *B. adolescentis* may be one of the events in the intestinal environment in response to the aging of the host.

These results can lead to a hypothesis that the age-related changes of the predominant species of bifidobacteria in the human intestine is a kind of well-acquired adaptation of the host to the changes in the intestinal microbiota, localizing the beneficial microbes such as *B. adolescentis* to enhance the colonization resistance against the exogenous infectious agents. For more information on the influence of the normal intestinal microbiota on the immune system, see the chapter by Moreau.

Effects of Bifidobacterial Probiotics on Immunosenescence

Probiotics have been defined as a live microbial food ingredient that are beneficial to the health of the host (65,66). Most current probiotics are lactic acid bacteria, especially *Lactobacillus* and *Bifidobacterium* species (66). Among the proposed health-promoting effects of the probiotic strains are the enhancement of cell-mediated immune responses of the host by stimulating the pro-inflammatory cytokine, particularly IL-12 (67,68). The cell-mediated immune response, enhanced by the pro-inflammatory cytokine IL-12, has been considered as one of the most important underlying mechanisms contributing to the self-defense of the host a against tumors and allergy (56). Therefore, probiotics strains with the ability to stimulate IL-12 secretion can exhibit apparent anti-tumor and anti-allergic effects (69–71). Considering the fact that the reduced cell-mediated immune response is the main component of immunosensense of the elderly, such probiotics can be expected to benefit the elderly. After consumption of a probiotic *B. lactis,* increase in the proportion of the total CD +4 and CD25 T lymphocytes and NK cells in the blood were observed (72,73). The ex vivo phagocytic and tumoricidal activity capacity of polymorphonuclear and mononuclear cells were increased by an average of 101 and 62%, respectively. These increases were significantly correlated with age, with volunteers older than 70 years experiencing significantly greater improvement than those younger than 70. In sight of the fact that the intestinal bifidobacteria are usually dominated by *B. adolescentis* with an advanced affinity specific to the mucus from the elderly and the ability to promote IL-12 production, *B. adolescentis* from the intestine of the healthy elderly may be a more reasonable candidate for use as probiotics to help the seniors to combat immunosenescence. Compared to other predominant species of bifidobacteria in infants and young adults, *B. adolescentis* is usually less quantitatively. Therefore, a strategy to increase senior-specific bifidobacteria, including *B. adolescentis* in the elderly could be a more practical way to improve the immunomodulatory effect of the intestinal flora. For more information on probiotics, see the chapter by Ouwehand and Khedkar.

CONCLUSION

With the progress in nutrition and medicine, the life-expectancy of people has increased. In industrialized societies this has led to increasing costs and spending for health care and medical treatment of their senior citizens. Growing scientific evidence suggests that aging alters the intestinal microbiota qualitatively and quantitatively, generating a different microbial community with an aberrant structure. The intestinal microbiota in the elderly is colonized by fewer bifidobacteria, and more potentially infectious microbes compared to infants and young adults. Furthermore, there is a decrease in the species diversity in bifidobacterial population of the elderly which is dominated by *Bifidobacterium adolescentis* and *B. longum*. The advanced affinity of *B. adolescentis* to mucus both isolated from the elderly suggests a deep symbiotic relationship between this microbe and host. The elevated ability of *B. adolescentis* to enhance the production of pro-inflammatory cytokine, particularly IL-12, by macrophages and monocytes suggests that this endogenous bacterium may play an important role in the maintenance of the CMI which can be impaired by age-related immunosenescence. This evidence can be used as the basis to consider *B. adolescentis* from the healthy elderly as a reasonable probiotic candidate for targeting the elderly, a growing subpopulation more prone to infection and autoimmune disease.

REFERENCES

1. Saunier K, Dore J. Gastrointestinal tract and the elderly: functional foods, gut microflora and healthy ageing. Dig Liver Dis 2002; 34:19–24.
2. Kalache A. Active ageing makes the difference. Bull. WHO 1999; 77:299.
3. Schneeman BO. Gastrointestinal physiology and function. Br J Nutr 2002; 88:159–163.
4. Savage DC. Microbial ecology of the gastrointestinal tract. Annu Rev Microbiol 1997; 31:107–133.
5. Tissier H. Recherches sur la flora intestinal des nourrisson. Theses Paris 1900; 1–253.
6. Benno Y, Sawada K, Mitsuoka T. The intestinal microflora of infants: composition of fecal flora in breast-fed and bottle-fed infants. Microbiol Immunol 1984; 28:975–986.
7. Yuhara T, Isojima S, Tsuchiya F, Mitsuoka T. On the intestinal flora of bottle-fed infant. Bifidobact Microflora 1983; 2:33–39.
8. Mitsuoka T. A Color Atlas of Anaerobic Bacteria. Tokyo: Sobunsya, 1980.
9. Mitsuoka T. Intestinal flora and aging. Nutr Rev 1992; 50:438–446.
10. Mitsuoka T. Significance of dietary modulation of intestinal flora and intestinal environment. Biosci Microflora 2000; 19:15–25.
11. Russell RM. Changes in gastrointestinal function attributed to aging. Am J Clin Nutr 1992; 55:1203S–1207S.
12. Gill HS, Darragh AJ, Cross ML. Optimizing immunity and gut function in the healthy. J Nutr Health Aging 2001; 5:80–91.
13. Holt PR. Gastrointestinal disease in the elderly. Curr Opin Clin Nutr Metab Care 2003; 6:41–48.

14. Haenel H. Über die Mikroökologie alter Menschen. Zentralbl Bakteriol Hyg I Orgi 1963;
    188:219–230.
15. Haenel H. Gesetzmäßigkeiten in der Zusammensetzung der fakalen Mikroflora. Eubiose und
    Dysbiose der menschlichen Darmbesiedlung. Ernährungsforschung 1965; 10:289–301.
16. Jantea F, Nicolae D, Bad-Oprisescu D, Voina P. Beitrag zur Darmmikroflora bei Menschen im
    Alter von 45-100 Jahren. Ernährungsforschung 1965; 10:352–362.
17. Gorbach SL, Nahas L, Lerner P, Weinstein L. Studies of intestinal microflora 1. Effects of diet,
    age, and periodic sampling on number of fecal microrganisms in man. Gastroenterology 1967;
    53:845–855.
18. Mitsuoka T, Hayagawa K. Die Faecalflora bei Menschen I: Mitteilung: Die Zusanmmenset-
    zung der Faecalflora der verschiedenen Altersgruppen. Zbl Bakt Hyg I Abt Org 1972;
    223:333–342.
19. Mitsuoka T, Hayagawa K, Kimura N. Die Faekalflora bei Menschen II: Mitteilung: Die
    Zusammensetzung der Bifidobakterienflora der verschiedenen Altersgruppen. Zbl Bakt Hyg I
    Abt Org 1973; 226:469–478.
20. Mitsuoka T, Ohno K, Benno Y, Suzuki K, Namba K. Die Faekalflora bei Menschen IV.
    Mitteilung: Vergleich des neu entwickelten Verfahrens mit dem bisherigen üblichen Verfahren
    zur Darmfloraanalyse. Zbl Bakt Hyg I Abt Org 1976; 224:219–233.
21. Benno Y. Mitsuoka Development of intestinal microflora in human and animal. Bifdobact
    Microflora 1986; 5:13–25.
22. Blaut M, Collins MD, Welling GW, Dore J, van Loo J, de Vos W. Molecular biological
    methods for studying that gut microbiota: the EU human gut flora project. Br J Nutr 2002;
    87:S203–S211.
23. Hopkins MJ, Sharp R, Macfarlane GT. Age and disease related changes in intestinal bacterial
    population assessed by cell culture, 16S rRNA abundance, and community cellular fatty acid
    profiles. Gut 2001; 48:198–205.
24. Hopkins MJ, Macfarlane GT. Changes in predominant bacterial population in human faces
    with age and with *Clostridium difficile* infection. J Med Microbiol 2002;448–454.
25. Hopkins MJ, Sharp R, Macfarlane GT. Variation in human intestinal microbiota with age. Dig
    Liver Dis 2002; 34:512–518.
26. Mitsuoka T. Intestinal Bacteria and Health. Japan: Harcourt Brace Jovanovich, 1978.
27. Björkstén B, Naaber P, Sepp E, Mikelsaar M. The intestinal microflora in allergic Estonian and
    Swedish 2-year-old children. Clin Exp Allergy 1999; 29:342–346.
28. Kuvaeva IB, Orlova NG, Veselova OL, Kuznezova GG, Borovik TE. Microecology of the
    gastrointestinal tract and the immuneological status under food allergy. Nahrung 1984;
    28:689–693.
29. Orla-Jensen S. La classification des bactéria lactiques. Lait 1924; 4:468–474.
30. Reuter G. Vergleichende Untersuchungen über die Bifidus-Flora im Sauglings- und
    Erwachsenenstuhl. Zbl Bakt I Orig 1963; 191:486–507.
31. Reuter G. Designation of type strains for *Bifidobacterium*-species. Int J Syst Bact 1971;
    21:273–275.
32. Mitsuoka T. Vergleichende Untersuchungen über die Bifidobakterien aus dem Verdauung-
    strakt von Menschen und Tieren. Zbl Bakt I Orig 1969; 210:52–64.
33. Mitsuoka T. Taxonomy and ecology of bifidobacteria. Bifidobacteria Microflora 1984;
    3:11–28.
34. Lauer E, Kandler O. DNA–DNA homology, murein type and enzyme patterns in the type
    strains of the genus *Bifidobacterium*. Syst Appl Microbiol 1983; 4:42–62.
35. Scardovi V. Genus *Bifidobacterium*. Orla-Jensen 1924. In: Sneath et al., eds. Bergey's Manual
    of Systematic Bacteriology: Vol. 2 Section 15, Philadephia: The Williams & Wilkins
    Company, 1986; 1418–1434.
36. Moore WE, Holdeman LV. Human fecal flora: the normal flora of 20 Japanese-Hawaiians.
    Appl Microbiol 1974; 27:961–979.
37. Finegold SM. Normal human intestinal flora. Ann 1st Super Sanita 1986;731–737.

38. He F, Ouwehand AC, Isolauri E, Hosoda M, Benno Y, Salminen S. Differences in composition and mucosal adhesion of bifidobacteria isolated from healthy adults and healthy seniors. Curr Microbiol 2001; 43:351–354.

39. Satokari RM, Vaughan EE, Akkermans ADL, Saarela M, de Vos WM. Bifidobacterial diversity in human feces detected by genus-speciic PCR and denaturing gradient gel electrophoresis. Appl Environ Microbiol 2001; 67:504–513.

40. Mullie C, Odou MF, Singer E, Romond MB, Izard D. Multiplex PCR using 16 S rRNA gene-targeted primers for the identifyication of bifidobacteria from human origin. FEMS Microbiol Lett 2003; 222:129–136.

41. Matsuki T, Watanabe K, Fujimoto J, et al. Development of 16 S rRNA-gene-targeted group-species primers fro the detection and identification of predominant bacteria in human feces. Appl Environ Microbiol 2002; 68:5445–5451.

42. Isolauri E, Sütas Y, Kankaanpää P, Arvilommi H, Salminen S. Probiotics: effects on immunity. Am J Clin Nutr 2001; 73:444S–450S.

43. Ouwehand AC, Tuomola EM, Lee YK, Salminen S. Microbial interactions to intestinal mucosal models. Methods Enzymol 2001; 337:200–212.

44. Kohler H, McCormick BA, Walker WA. Bacterial-enterocyte crosstalk: cellular mechanisms in health and disease. J Pediatr Gastroenterol Nutr 2003 Feb; 36:175–185.

45. Salminen S, Laine M, von Wright A, Vuopio-Varkila J, Korhonen T, Mattila-Sandholm T. Development of selection criteria for probiotic strains to acess their potential I n functional food. A Nordic and European approach. Biosci Microflora 1996; 2:23–28.

46. Ouwehand AC, Isolauri E, Kirjavanen PV, Salminen SJ. Adhesion of four *Bifidobacterium* strains to human intestinal mucus from subjects in different age groups. FEMS Microbiol Lett 1999; 172:61–64.

47. He F, Ouwehand AC, Hashimoto H, Isolauri E, Benno Y, Salminen S. Adhesion of *Bifidobacterium* spp. to human intestinal mucus. Microbiol Immunol 2001; 45:259–262.

48. Laine R, Salminen S, Benno Y, Ouwehand AC. Performance of bifidobacteria in oat-based media. Int J Food Microbiol 2003; 83:105–109.

49. Pawelec G. Immunosenescence: impact in the young as well as the old? Mech Ageing Dev 1999; 108:1–7.

50. Mishto M, Santoro A, Bellavista E, Bonafe M, Monti D, Franceschi C. Immunoproteasomes and immunosenescence. Ageing Res Rev 2003; 2:419–423.

51. Ginaldi L, Martinis MD, D' Ostilio A, Marini L, Loreto MF, Quaglino D. Immunological changes in the elderly. Ageing Clin Exp Res 1999; 11:281–286.

52. Lesourd B, Mazari L. Nutrition and immunity in the elderly. Proc Nutr Soc 1999; 58:685–695.

53. Castle SC. Clinical relevance of age-related immune dysfunction. Clin Infect Dis 2000; 31:578–585.

54. Effros RB. Ageing and immune system. Novartis Found Symp 2001; 235:130–145.

55. Grubeck-Loebenstein B. Changes in the ageing immune system. Biologicals 1997; 25:205–208.

56. Peakman M, Vergani D. Basic and Clinical Immunology. Hong Kong: Churchill Livingstone, 1997:1–388.

57. Fujimoto T, Duda RB, Szilvasi A, Chen X, Mai M, O'Donnell MA. Streptococcal preparation OK-432 is a potent inducer of IL-12 and T helper cell 1 dominant state. J Immunol 1997; 158:5619–5629.

58. Haller D, Blum S, Bode C, Hammes WP, Schiffrin EJ. Activation of human peripheral blood mononuclear cells by nonpathogenic bacteria in vitro: evidence of NK cells as primary targets. Infect Immun 2000; 68:752–759.

59. Hessle C, Honson LA, Wold AE. Lactobacilli from human gastrointestinal mucosa are strong stimulators of IL-12 production. Clin Exp Immunol 1999; 116:276–282.

60. Hessle C, Andersson B, Wold AE. Gram-positive bacteria are potent inducers of monocytic interleukin-12 (IL-12) while gram- negative bacteria preferentially stimulate IL-10 production. Infect Immun 2000; 68:3581–3586.

61. Karlsson H, Hessle C, Rudin A. Innate immune responses of human neonatal cells to bacteria from normal gastrointestinal flora. Infect Immun 2002; 70:6688–6696.

62. He F, Morita H, Ouwehand AC, et al. Stimulation of the secretion of pro-inflammatory cytokines by *Bifidobacterium* strains. Microbiol Immunol 2002; 46:781–785.

63. He F, Isolauri E, Morita H, et al. Bifidobacteria isolated from allergic and healthy infants: differences in taxonomy, mucus adhesion and immune modulation effects. Microecol Ther 2002; 29:103–108.

64. Morita H, He F, Fuse T, et al. Adhesion of lactic acid bacteria to Caco-2 cells and their effects on cytokine secretion. Microbiol Immunnol 2002; 46:293–297.

65. Fuller R. Probiotics in man and animals. J Appl Bacteriol 1989; 66:365–378.

66. Salminen S, Ouwehand A, Benno Y, Lee YK. Probiotics: how should they be defined? Trends Food Sci Technol 1999; 10:107–110.

67. Gill HS. Stimulation of the immune system by lactic culture. Int Dairy J 1998; 8:535–544.

68. Borchers AT, Keen CL, Gershwin ME. The influence of yogurt/*Lactobacillus* on the innate and acquired immune response. Clin Rev Allergy Immunol 2002; 22:207–230.

69. Matsuzaki T, Yamazaki R, Hashimoto S, Yokokura T. The effect of oral feeding of *Lactobacillus casei* strain Shirota on immunoglobulin E production in mice. J Dairy Sci 1998; 81:48–53.

70. Aso Y, Akaza H, Kotake T, Tsukamoto T, Imai K, Naito S. BLP Study Group. Preventive effects of a *Lactobacillus casei* preparation on the recurrence of superficial cancer in a double-blind trial. Eur Urol 1995; 27:104–109.

71. Cross ML, Gill HS. Can Immunoregulatory lactic acid bacteria be used as dietary supplements to limit allergies? Int Arch Allergy Immunol 2001; 125:112–119.

72. Gill HS, Rutherfurd KJ, Cross ML. Dietary probiotics supplementation enhances natural killer cell activity in the elderly: an investigation of age-related immunological changes. J Clin Immunol 2001; 21:264–271.

73. Gill HS, Rutherfurd KJ, Cross ML, Gopal PK. Enhancement of immunity in the elderly supplementation with the probiotic *Bifidobacterium lactis* HN019. Am J Clin Nutr 2001; 74:833–839.

5

Immune Modulation by the Intestinal Microbiota

Marie-Christiane Moreau
French Institute of Agronomical Research (INRA), Nancy, France

INTRODUCTION

Today, there is a growing interest in the intestinal microbiota and its relationship with the host's immunity. This is mainly due to two causes: first, the results obtained with probiotics, which have been defined as live micro-organisms that confer a health benefit on the host when consumed in adequate amounts (1), have shown interesting immunomodulatory properties in humans (1–3). Second, the studies by Dutchmann and coworkers (4) demonstrated for the first time, some years ago, that we are tolerant to our own digestive flora. A breakdown of this state leads to inflammatory bowel diseases (IBD). Consequently, the digestive flora can be considered as an organ belonging to the host's just as the spleen, heart, or brain. It plays an important role in the host's protection, especially by its actions on the immune system.

The overall importance of the intestine, relating to health, is still not completely understood. It is an extremely complex organ, which has to assure the function of digestion of foods and absorption of nutrients. In addition to this, the intestine is the largest lymphoid organ in the body by virtue of lymphocyte number and quantity of immunoglobulin produced. It also harbors a huge reservoir of bacteria that colonize it very early after birth and which is called "the commensal or resident or autochthonous digestive microflora," and more recently the "intestinal microbiota." The relationships between the intestinal microbiota and intestinal immune system (IIS), described in some reviews (5–7) can be viewed in terms of "symbiosis" or "mutualism," which is the association of symbiosis and commensalism as explained by Hooper and Gordon (8). Indeed the IIS does not mount immune responses toward the intestinal microbiota that, in turn, exert many effects on the immune system. These effects can be characterized as activation, modulation, and regulation of immune responses and are effective at both intestinal and peripheral levels.

In this chapter, effects of the intestinal microbiota on the host's immunity will be described, and in some cases effects of probiotic bacteria will also be discussed.

BRIEF REVIEW OF THE INTESTINAL MICROBIOTA

From birth to death, the gut is colonized by a diverse, complex, and dynamic bacterial ecosystem that constitutes the intestinal microbiota. In newborns, it develops sequentially according to the maturation of intestinal mucosa and dietary diversification. In healthy conditions, the human baby's intestine is sterile at birth but, within 48 hours, 10^8 to 10^9 bacteria can be found in 1 g of feces (9–11). The bacteria colonizing the baby's intestine come from the environment, where maternal vaginal and fecal microbes represent the most important source of bacterial contamination. However, the infant conducts an initial selection, since, out of all the bacteria present, only the facultative anaerobic bacteria such as *Escherichia coli* and *Streptococcus* will be able to colonize the intestinal tract, whatever the diet. Conditions under which this initial selection is operated have yet to be fully elucidated. They are related to endogenous factors, such as maturation of intestinal mucosa, mucus, growth promoters or inhibitors present in the meconium, or exogenous factors such as delivery conditions (natural childbirth, caesarean section), mother's status (antibiotic intake), and quality of the bacterial environment. Subsequently, obligate anaerobes such as *Bacteroides, Clostridium*, and *Bifidobacterium* colonize over the first week of life, following a second selection in which the diet factor plays a fundamental role. It has long been known that *Bifidobacteria* are predominant in exclusively breast-fed babies, while in bottle-fed babies it is not always present, or present at fluctuating levels and, in contrast to breast-fed babies, often associated with other anaerobic bacteria such as *Bacteroides* and *Clostridia*. Breast milk contains oligosaccharides enabling development of *Bifidobacterium* and may also function as receptor analogues of the mucus influencing the strains able to colonize the intestinal tract (12). A bacterial balance is obtained towards the end of the second week of life in which *Bifidobacterium* and *E. coli* predominate in exclusively breast-fed infants, while a more diverse microbiota, rich in *E. coli, Bacteroides* and, possibly *Clostridium, Bifidobacterium, Staphylococcus*, and other *Enterobacteriaceae*, is found in formula-fed infants. Thus the bacterial balance of the infantile microbiota mainly depends on two important factors: bacterial environment at birth and diet. During the last decade, some modifications of the microbiota balance in babies whatever the feeding have been observed, namely, dominance of *Staphylococcus*, low levels of *E. coli*, delayed colonization with anaerobic bacteria and absence or low levels of bifidobacteria (MJ Butel, personal communication). Excessive aseptic conditions present at birth, maternal antibiotic intake immediately before parturition or during childbirth could be, among other factors, responsible for such differences (13). Because of the fragility of the baby's digestive microbiota, which is poorly diversified, with about 10 bacterial species of micro-organism versus over 400 in adults, the consequences of its modification have to be considered in terms of health. For example, recent studies suggest that some infancy pathologies, such as food allergy, could be due to the modifications of the intestinal microbiota of newborns. The latter will be discussed in chapter 10.

Thereafter, according to dietary diversification, the digestive microbiota, enriched by the development of other strictly anaerobic bacteria, becomes more and more complex. It is considered to have assumed adult characteristics toward the age of 2 years (9–11).

In adults, a complex and diverse digestive microbiota is present, mainly in the distal parts of the gut. In the duodenum, the number of bacteria is approximately 10^4 bacteria/g of intestinal content while in the ileum, the number reaches 10^7–10^8. The large intestine is the most densely colonized region (10^{10}–10^{11} bacteria/g of content), essentially because of digestive stasis. Bacterial species established at levels over 10^7–10^8 bacteria/g characterize the predominant microbiota, whereas those below such a threshold compose the subdominant microbiota. In fact, it is proposed that only predominant bacteria are able to exert a measurable function. The dominant microbiota of human feces is mainly composed of strict anaerobic and extremely oxygen-sensitive bacteria. According to several authors, 30% to 70% of the microbiota is not identified because it is uncultivable with current techniques. The predominant species commonly isolated from the human feces belong to the genera *Bacteroides, Eubacterium, Bifidobacterium, Ruminococcus*, and *Clostridium*, and subdominant species include enterobacteria, particularly *Escherichia coli* and streptococci (14). Lactobacilli are frequently subdominant in humans or cannot even be detected. Some studies suggest that they may be abundant in the ileum.

Very little data exists on the evolution of intestinal microbiota in the elderly. Nonetheless, bifidobacteria have been reported to decrease at old age, which may be related to a reduced adhesion to the intestinal mucus (15).

Currently available molecular biology techniques should bring additional and complementary approaches to those offered by the usual culture techniques. Recent molecular methods have shown that every individual has his/her own gut microbial balance, which has been described to be stable (studies over a period of 6–9 months) (16).

In conclusion, depending on the intestinal sites (duodenum, ileum, and colon) and the various periods of life, childhood, adulthood, aging, the human's intestinal microbiota also varies. This is discussed further in more detail in chapter 3 by McCartney and Gibson.

BRIEF REVIEW OF IMMUNE RESPONSES

Innate Immunity

Cells responsible for the innate immunity provide the first line of host defense: monocytes/macrophages, dendritic cells (DC), natural killer cells (NK), and neutrophils. These are the body's sentinels, able to detect danger and signal it to other cells, by the synthesis of molecules, such as NO (nitric oxide) which displays antibacterial activity, cytokines, and chemokines, which are small peptides acting by means of specific receptors expressed at the surface of targeted cells. Some of them have pro-inflammatory properties, and increase expression of surface markers on some cells allowing migration into neighboring lymphoid organs. DCs and macrophages are able to display a phagocytic activity, and by production of inflammatory chemokines and cytokines, to modulate other cells such as neutrophils, polymorphonuclear cells and eosinophils in the case of hypersensitivity, which increase the inflammatory action, and ultimately B and T cells, which will set up an "acquired" immune response (see below). NK cells contribute to antitumor activity.

Innate immunity is fast, non-specific, and not endowed with memory. It also plays an important role in acquired immunity by the process of antigen (Ag) presentation to T cells and through the synthesis of some cytokines, which play an important role in the orientation of the specific immune responses. Thus, innate immunity is the first to intervene following exposure to an Ag. It also plays a fundamental role in acquired immunity, as described below.

Macrophages and DCs are able to recognize "danger" by way of receptors called Toll-like receptors (TLRs), which respond to several bacterial components (17,18). To date, at least 10 TLRs have been found. TLR2 and TLR4 recognize cell wall structures: peptidoglycan of Gram-positive bacteria and lipopolysaccharides (LPS) of Gram-negative bacteria, respectively. TLR3 is found specifically in DCs, TLR5 is reported to recognize bacterial flagella and TLR9 recognizes pro-inflammatory CpG dinucleotide (cytosine phosphoryl guanine non-methylated) only found in the bacterial genome (19). Another surface receptor that binds LPS, CD14, is expressed on the surface of monocytes and macrophages. In addition to macrophages and DCs, mucosal epithelial cells also express TLR2 and TLR4 (20). TLRs play an important role in the initiation of innate responses and hence in acquired immunity. The binding of bacterial molecules such as LPS, peptidoglycan and CpG motifs to TLRs results in the activation of the nuclear factor κβ (NF-κβ) pathway. NF-κβ is a transcriptional factor that intervenes in the synthesis of the pro-inflammatory cytokines, TNF-α, IL-1 and IL-6, by cells of the innate immune system and intestinal enterocytes; it further stimulates phagocytosis and adhesion molecule expression, NO production and synthesis of IL-12 (21). In addition, NF-κβ activation has an important role in regulating the expression of anti-apoptotic proteins and affecting the susceptibility of cells to apoptosis (22). Because of the importance of inhibiting the NF-κβ pathway in certain circumstances, such as those found in the gut, this pathway is regulated by several processes as described elsewhere (21–25).

Acquired Immunity

Acquired responses consist of the Ag-specific humoral and cell-mediated immune responses. They express by synthesis of antibodies (Abs) and cellular responses, respectively. They involve three kinds of cells: Ag-presenting cells (APC) (mainly macrophages and DCs), T cells, and B cells. For cellular responses only APC and mainly CD8 + T cells are involved, while another population of T cells, CD4 + T cells, and B cells are needed for Ab synthesis. Antibodies can belong to several kinds of immunoglobulin isotypes: IgD, IgM, IgG, IgE, and IgA, and different subclasses including IgG1, IgG2, IgG3, and IgG4 or IgA1, and IgA2 in humans. After an initial contact with the Ag, acquired responses are slowly established (7–10 days) but are endowed with memory enabling a very rapid response after a further contact with the same Ag (within one day). The first step in the induction of the immune response is the presentation by APC, and recognition of the epitope (small part of an Ag), associated with major histocompatibility complex (MHC) molecules of class I or II, to the epitope-specific T cell receptor. In addition, the binding of co-stimulation molecules and equivalent receptors expressed by both APCs and T cells (CD40 and CD40 ligand, B7.2 and CD28, etc.), leads to the full activation and proliferation of naive T cells. Subsequently, a large proportion of those activated cells will die of apoptosis, the others surviving in the form of memory T cells.

All those "lock-and-key" mechanisms are important and greatly contribute to modulate the immune responses. It has been shown that DCs play a key role in the acquired immune responses. They exist in an immature form in tissues. The mature form is obtained following contact with an Ag and phagocytosis. Mature DCs are able to synthesize cytokines and migrate into the neighboring draining lymph nodes in order to supply Ag information to the T cells. DC populations are heterogeneous (26) and, as will be further described, subsets of intestinal DCs display specific properties in terms of Ag presentation and cytokine secretion.

Th1/Th2 Balance

Several years ago Mosmann and coworkers (27) described different subsets of CD4 + T cells that differ by the cytokine profiles produced after activation (Fig. 1). Three kinds of T cells are now described from progenitor type 0 helper T cells (Th0). The type 1 helper T cells (Th1) mainly secrete IFN-γ, a pro-inflammatory cytokine, and IL-2. Th1 induce a weak synthesis of Abs by B cells (subclass IgG2a) and are recruited more in the event of a cell-mediated response. In contrast, activation of type 2 helper T cells (Th2) induces synthesis of cytokines IL-4, IL-5, IL-10, and IL-13, which have anti-

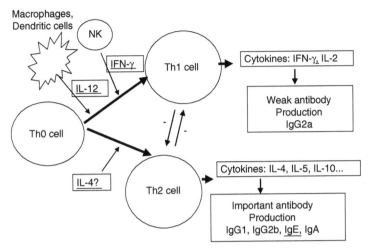

Figure 1   Schematic representation of the Th1/Th2 balance. *Abbreviations*: IFN, interferon; Ig, immunoglobulin; IL, interleukin; NK, natural killer cells; Th, T-helper cell.

inflammatory properties. They induce a large production of Abs by B cells belonging to the isotypes and subclasses of IgG1, IgG2b, IgA, and IgE, the latter being involved in allergy. Activation of one population inhibits that of another. One of the major determinants of the Th1/Th2 differentiation is the cytokine environment at initial sensitization. Indeed the transition from Th0 to Th1 or Th2 depends on environmental factors, among which the innate immune cells, macrophages, DCs, and NK cells, play a considerable role through synthesis of some cytokines, especially IL-12, and IFN-γ, acting on the orientation toward a Th1 profile (Fig. 1).

Another subset of T helper cells has been described in mice, the Th3 cells. They could play an important role in tolerance by suppressing the immune response through production of transforming growth factor-β (TGF-β) after Ag-specific triggering (28).

The Th1/Th2 balance is an example of the complexity of the host's immune system, which has to respond to various immune stimuli by an appropriate immune response. In fact, according to the situation, an inflammatory immune response involving Th1 and/or CD8 + T cells will be activated in intracellular infections needing cell-mediated responses. In contrast, a Th2 response producing a low inflammatory response with marked synthesis of IgG1 or IgA Abs, will be more activated in other situations. With regard to the IgE response (Th2 response), it must remain moderate in order to not give rise to adverse allergic reactions. A balance between Il-4 and IL-10 may intervene in that regulation, in which IL-10 is believed to play a very important anti-inflammatory role (29).

THE INTESTINAL IMMUNE SYSTEM

The IIS is a particular immune system anatomically and functionally distinct from that present at the peripheral level (30–32). It is in contact with an enormous quantity of Ags, food proteins and intestinal bacteria, and does not mount an inflammatory response against them. At the same time, it has to protect against enteric pathogens and toxins.

The IIS is mainly located in the small intestine and colon with differences in the anatomical patterning and physiological functions. It is important to be aware of the compartmentalization of the intestine even if the IIS associated with the small intestine has been subject to more studies and is the most widely described. According to the compartment, differences in immune regulations in response to local Ags can be easily understood: food Ags are more numerous in the small intestine while in the ileum and colon they have essentially been digested and absorbed. In contrast, commensal microbes are scarce in the duodenum but more numerous in the ileum and above all in the colon.

Three lines of defense are present: (i) natural defenses: stomach acidity, bile salts, mucus, motility, permeability, (ii) innate immune responses: Ag capture, cytokine secretion, TLRs, and (iii) acquired immune responses namely oral tolerance (OT) and secretory IgA (sIgA) response. All of them interact together.

Many results presented in this review are derived from studies with mice. Note for some results, it is not certain whether they reflect what is happening in humans (33,34).

Anatomy

The immune system associated with the small intestine is currently described according to two compartments: (i) the inducing sites, named the gut associated lymphoid tissue (GALT), consisting of organized aggregated lymphoid tissue, scattered small nodules, Peyer's patches (PPs), and mesenteric lymph nodes (MLN); and (ii) the effector sites, i.e., the lamina propria tissue where numerous mature B and T small lymphocytes (60% CD4+T cells), plasma cells of which about 90% synthesize IgA are present, and the epithelium richly endowed with intra-epithelial lymphocytes (IEL) (CD8+T cells) present between the tight junctions of some enterocytes (Fig. 2) (30–32,35).

PPs are the first important inductive sites. They are macroscopic lymphoid aggregates that are found in the submucosa along the length of the small intestine. They consist of large B-cell follicles and intervening T-cell areas which are separated from the single layer of intestinal epithelial cells, known as the follicle-associated epithelium (FAE), by the subepithelial dome region where APCs are numerous (31). An important feature of the FAE is the presence of microfold (M) cells, which, in contrast to enterocytes, lack the surface microvilli, the normal thick layer of mucus, and cellular lysosomes. Thus, M cells are distinctive epithelial cells that occur only in the FAE. It is believed that they play a central role in the initiation of mucosal immune responses by transporting Ags, and microorganisms, to the underlying organized lymphoid tissue within the mucosa.

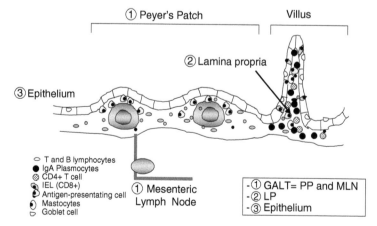

Figure 2 Schematic representation of the intestinal immune system. *Abbreviations*: GALT, gut-associated lymphoid tissue; IEL, intraepithelial lymphocytes; LP, lamina propria; MLN, mesenteric lymph nodes; PP, Peyer's patches.

Most of the mature cells found in the effector sites, T cells, plasma cells, epithelium CD8 α-β thymus-dependent IEL, derive from PPs. After oral Ag stimulation, the Ag-activated immature T and B cells present in PPs leave the PP, and migrate into the systemic compartment via the MLN, and the lymph, then enter the bloodstream via the thoracic duct. Subsequently the expression of α4β7 integrin, expressed at the surface of cells, allows them to bind the gut-specific vascular addressin, MadCAM-1, which is expressed at high levels by the vasculature of mucosal surfaces, inducing the cells to migrate across the endothelium into the lamina propria. Within the intestinal lamina propria, B cells differentiate into IgA-secreting plasma cells with a half-life of about 4½ days, and most of the T cells undergo apoptosis. This fact has been suggested to be important to maintain the gut homeostasis preventing immune responses to luminal Ags (36). This cellular traffic, between the PP and lamina propria, has been particularly described for IgA plasmocytes. After antigenic stimulation at the PP, B cells undergo Ig class switching from expression of IgM to IgA which is under the influence of several factors, including cytokines, TGF-β, IL-4, and IL-10, and cellular signals delivered by DC and T cells present in PPs. After returning to the lamina propria, IgA plasmocytes synthesize and assemble two IgA units and the J chain. Then, a polymeric-Ig receptor (pIgR) expressed by enterocytes allows selective transcytosis through the epithelial cells, and dimeric IgA are excreted in the lumen associated with the secretory component, a protein derived from the pIgR, which confers to sIgA interesting properties such as resistance to proteolytic enzymes present in the intestinal lumen (37).

The physiological significance of the entero-enteric cell circulation is important. The induction of an immune response at a PP level propagates distally relative to the induction site, not only throughout the intestine but also to other mucosa. It has been shown that T and B cells, which have been activated in the GALT, are able to reach other mucosal surfaces, which together compose the mucosa-associated lymphoid tissue (MALT; vagina, breast during pregnancy and lactation, respiratory tract) by the way of homing receptors. This is known as the "common immune system of the mucosa." The cycle also shows that there are relationships between the IIS and systemic compartment, even though they have, as yet, not been fully elucidated.

The APCs play a crucial role in the initiation and regulation of the immune responses, and are present in all the parts of IIS. In PPs we found immature DCs located in close proximity to M cells, which have the capacity to migrate into the interfollicular areas of the PP (T areas) and also via the lymphatic to T-cell areas of MLN, thereby stimulating T cells in both locations. In all these locations where T cells are stimulated by a given gut Ag, the resulting blasts have the capacity to move via the lymph to the thoracic duct and the blood to finally home in the gut wall. Ag-specific effector and memory cells thereby become disseminated along the whole gut wall, in the lamina propria, and the epithelium.

Several and unusual subsets of DCs have been described in the murine PP. They are located either in the subepithelial dome (CD11b+/CD8−), or in the intrafollicular regions (CD11b−/CD8+) or at both sites (CD11b−/CD8−) (38). It has been described in mice that CD11b+/CD8− DC, present in the subepithelial dome of PP, have unusual functional characteristics and differ from their peripheral counterparts. Upon antigenic stimulation, they secrete IL-10 and induce naive T cells to differentiate into Th2 with IL-4 and IL-10 production (38). In contrast, in the spleen, the same DC subset secretes IL-12 after antigenic stimulation under the same experimental conditions, and consequently drives the immune response to a Th1 orientation with production of IFN-γ. However the authors showed that the double negative population CD11b−/CD8− of DCs, is capable of secreting Il-12 upon recognition of microbial stimuli. These functional differences in the different PP DC populations may come from the type of Ag stimulation. Indeed, T cells in PP of mice immunized orally with live *Salmonella typhimurium* secrete large quantities of IFN-γ (39). In these studies, it also seems to be important to consider the intestinal site from which PPs originate. In fact recent studies have shown that the presence of intestinal bacteria in the ileum influence the cytokine profile secreted by DCs in PPs (40).

DCs are also found in the intestinal villi at the subepithelial level in lamina propria. When activated, they can penetrate the epithelium, and send dendrites to the epithelium surface, thus being able to directly sample luminal Ags and to present them to IEL and lamina propria lymphocytes (31,35). Unusual subsets of DCs are also found, including some that are similar to the IL-10 inducing DCs that have been described in PPs. This characteristic constitutes a particularity of DCs present both in PPs, lamina propria, and epithelium, with functional consequences as presented below for the section on oral tolerance.

Another characteristic of the IIS is the presence of large numbers of activated memory CD4+ and CD8+T cells throughout the lamina propria, expressing the chemokine receptor CCR5, probably because of the continuous exposure to environmental antigens (41). By contrast, the majority of CD4+T cells in the peripheral blood and nodes are naive T cells (lack of CCR5 expression). It has been reported that PP contains naive T cells, expressing chemokine receptor CXCR4, but also activated and memory T cells, a phenomenon which is not found in other inductive lymphoid tissue such as MLN or peripheral lymph nodes (42). The reasons for this are unknown.

Physiology

The IIS generates two important immune functions. First is a suppressive function, also termed oral tolerance (OT), characterized by regulatory mechanisms avoiding local and peripheral immune responses to harmless environmental Ags present in the intestine, such as dietary proteins and bacterial Ags of the intestinal microbiota. Second is the immune exclusion performed by sIgA Abs to protect the mucosa against pathogen microorganisms but also against bacterial translocation of commensal bacteria. Now, it is still unclear whether OT induction is accompanied by local sIgA production or not. The knowledge of regulatory mechanisms that govern the IIS functions are important to understand. When the IIS is not functioning well, diseases can develop: enteric and/or systemic infections, hypersensitivities to dietary proteins, and IBD.

Tolerance to Soluble Proteins: Oral Tolerance

In healthy conditions, the IIS does not mount immune responses to food proteins and commensal bacterial Ags. Because two kinds of studies have been reported in the literature dealing with the mechanisms involved in OT either to food proteins or to intestinal microbiota, we distinguish the mechanisms described, and postulate that they may be different according to soluble proteins, such as food proteins, or to bacterial component Ags present in the intestinal microbiota.

Oral tolerance is defined by the state of both systemic and mucosal immune unresponsiveness induced after soluble protein feeding. It is a long-lasting phenomenon, which affects suppression of both cellular and humoral Ag-specific immune responses. Despite the absence of direct evidence in infants, it is believed that OT, which has been shown to exist in adult humans (43), certainly plays an important role in the protection against hypersensitivity reactions to food proteins [hypersensitivities type I and IV, either IgE Abs (allergy) and cellular immune responses, respectively]. Studies on mice have shown that suppression of cellular responses lasts up to 17 months after one feeding of

20-mg ovalbumin (OVA) and the suppression of the IgG antibody response lasts more than 3–6 months (44).

A number of factors affect OT induction or its persistence (7,31,45). Briefly, they are linked to the Ag (nature, doses), the host (genetic, age, inflammatory diseases which affect the permeability of intestinal mucosa), intestinal microbiota (described below), and bacterial toxins (7).

The sites where OT is generated, the different mechanisms, and the conditions in which they are operating are still a matter of debate (31). Discussion persists as to where primary immune responses are initiated: the PP, lamina propria or MLN. It has been assumed for many years that PPs are the principal site in which T cells encounter Ags derived from food and presented by several distinct potential APCs: macrophages, B lymphocytes and DCs. However, some studies suggest that M cells and PPs might not be necessary for the uptake and processing of Ags in the induction of OT. For instance, study in deficient μMT mice which do not possess B cells, M cells, and PPs because of the lack of B cells, showed that these mice are nevertheless able to induce a normal suppressive T cell response to oral Ag at the systemic level (46). They concluded that systemic T cell responses to orally administered soluble Ags requires neither the specialized Ag presentation properties by B cells, nor the microenvironment provided by M cells nor PPs, but most likely, are due to characteristics of professional APCs, especially DCs. In recent years, some in vitro studies on intestinal epithelial cell-lines have shown that Ags may be incorporated into MHC class-II positive exosomes derived from enterocytes (31). These vesicles, also called "tolerosomes," can be found in the bloodstream after Ag feeding, and are able to induce systemic tolerance when transferred into naive recipients. The mechanisms by which exosomes are able to tolerize T cells are under investigation. It has been postulated that exosomes can transmit MHC class II/peptide complexes to APCs such as DCs. Indeed, incubation of free exosomes bearing MHC class II complexes with DCs resulted in a highly efficient stimulation of specific T cells (47).

Mechanisms implicated in OT are not completely elucidated (7,30,32,45). Studies in mice have supported the important roles of intestinal regulatory T cells (reg T cells) and DCs in the OT process. The key role of Ag presentation by DCs was provided by Viney and coworkers (48). In the study, they showed, in vivo, that administration of a hemopoietic growth factor, Flt3 ligand, to mice dramatically expands the number of functionally mature DCs in intestine and other lymphoid organs, and increases the susceptibility to induction of tolerance by feeding OVA. DCs recruited by Flt3L express only low levels of co-stimulatory molecules, supporting the view that intestinal DCs may normally be in a resting state without the ability to prime T cells. This mechanism has been called "anergy" and it was postulated that only high doses of Ag given to normal mice induce this mechanism.

In addition to the OT, other active suppressor mechanisms, globally named Ag-driven suppression, or bystander suppression, have been described. They involve several subsets of reg T cells. Indeed, repeated oral administration of low-dose Ag leads to the development of Th2 CD4+T cells secreting IL-4 and IL-10 and Th3 CD4+T cells secreting TGF-β cytokines, with anti-inflammatory and suppressive properties. In addition, two other reg T cell subsets have recently been described: CD4+CD25+reg T cells, which could have an important role to prevent intestinal inflammation diseases and another reg T cell subset, named Tr1, which has been demonstrated to suppress Ag-specific immune responses and actively down-regulate a pathological immune response in vivo, through production of Il-10 (49). This last finding suggests that Ag-specific Tr1 are capable of producing suppressor cytokines which exert an effect through a local bystander suppression. It has been shown that Tr1 reg T cells can be generated from repetitive stimulation of CD4+T cells in the presence of IL-10 (49). Intestinal unusual subsets of DC-secreting-IL-10 present in both PP and lamina propria could be implicated in the genesis of some of these reg T cells. Indeed, they could drive the T cells towards suppressive Tr1 and reg Tr1 cells in the intestine and may be crucial for the induction of OT.

Tolerance to the Intestinal Microbiota

Tolerance to our intestinal microbiota is important to prevent IBD. Some of the OT mechanisms may play a role in the tolerant state, but evidence is scarce. It has been described that intestinal CD4+T cells normally recognize the local commensal bacteria, but that their responses are inhibited by local reg T cells in an IL-10 and/or TGF-β-mediated manner (50). CD4+CD25+reg T cells also play an important role to suppress immune responses to bacterial Ags. However, other regulatory mechanisms, involving the regulation of immune responses specifically directed towards bacterial components, are now suggested. They mainly concern the regulation of the NF-κB pathway, as described previously, and where several inhibitory molecular mechanisms intervene (21–25).

Recently, it has been shown that the functionality of intestinal macrophages and DCs is different from that of the peripheral compartment. In humans, and under physiological conditions, neither macrophages nor enterocytes express CD14, a surface receptor involved in the response to bacterial LPS, and CD89, the receptor for IgA (51). Consequently, they do not respond to LPS by inflammatory cytokine production. The absence of CD89 on lamina propria macrophage down-regulates IgA-mediated phagocytosis, an activity that normally induces the release of pro-inflammatory mediators including reactive oxygen intermediates, leukotrienes, and prostaglandins. This fact contributes to maintain the low inflammatory level in normal human intestinal mucosa.

Modifications of the intestinal homeostasis may modify the inhibitory factors of the NF-κB pathway leading to secretion of pro-inflammatory cytokines (20), and/or up-regulated CD14 expression. During the inflammatory process in the intestinal mucosa, CD14+blood monocytes are probably recruited to the mucosal increasing inflammatory reactions. This is the situation prevailing in IBD, in which intestinal tolerance of its microbiota has been shown to be deficient (4).

Antibody sIgA Responses

Another important function elicited by the IIS is the secretion of sIgA Abs, which represent the most prominent Ab class at the mucosal surface. Secretory IgM Abs can also contribute to surface protection in the case of selective IgA deficiency. Secretory IgA perform "immune exclusion," which is a non-inflammatory immune response playing an important protective role against enteropathogenic opportunistic microorganisms (rotavirus, *Salmonella*, *Shigella*, *Toxoplasma*, etc.) for which the intestine constitutes an important portal of entry. Thus, they prevent microbial adhesion, especially in the duodenum where some pathogenic bacteria such as enterogenic *E. coli* can adhere. Furthermore, they prevent viral multiplication in enterocytes and perform neutralization of toxins. They also prevent the translocation of pathogenic and non-pathogenic bacteria towards the systemic compartment and concomitantly prevent any damage to the epithelium (52). Recently, it has been shown in mice, that dimeric IgA, when bound to the secretory component (SC), are more efficient in protection against bacterial respiratory infection (53). This effect is due to an appropriate tissue localization of sIgA to mucus, conferred by carbohydrate residues present in SC. This feature results in an optimal protective effect of sIgA at mucosal surface by immune exclusion.

In mice, a dual origin for IgA plasma cells in the small intestine has been shown. IgA plasma cells originate from two lineages of B cells designated B-1 and B-2, which differ according to their origins, anatomical distribution, cell surface markers, Ab repertoire and self-replenishing potential. B-1 cells are maintained by self-renewal of cells resident in the peritoneal cavity, and they utilize a limited repertoire that is mostly directed against ubiquitous bacterial Ags. B-2 cells, originated from bone-marrow precursors, are present in organized follicular lymphoid tissues, within PP, as precursors of plasma cells, and use a large repertoire of Abs. Thus the sIgA response to specific proteins Ags requires a classical costimulation by Ag-specific T cells, an entero-enteric cycle as described previously, and are secreted by IgA plasma cells derived from B2 lineage precursors in the PP. By contrast, sIgA Abs against Ags from commensal bacteria are T cell independent, polyspecific, and are secreted by IgA plasma cells derived from the peritoneal cavity B-1 cells (54). They protect the host from the penetration of commensal bacteria. In mice, B-1 lineage could represent 40% of total IgA plasma cells. The contribution of peritoneal B cells to the intestinal lamina propria plasma cell population in humans is still a matter of debate (33).

In conclusion, IIS have some phenotypic and functional characteristics, which profoundly differ from those found in the peripheral immune system. An important finding, which has emerged from recent studies, is the importance of the MLN in the induction of both OT and active immunity (sIgA secretion), where trafficking of DCs from PP and lamina propria, after being loaded with Ag, could prime naive T cells. Indeed, total and specific IgA-Ag responses, as well as OT induction are absent in mice that lack MLNs (31). Many studies are, however, needed to get a better understanding of the mechanisms involved in intestinal immune responses, and the conditions in which they are elicited. They are important for the maintenance of the intestinal homeostasis, and are based on a continual cross talk between all the immune cells of both IIS (including enterocytes) and peripheral immune system and external events in which the digestive microbiota plays an important role.

RELATIONSHIPS BETWEEN THE INTESTINAL IMMUNE SYSTEM AND INTESTINAL MICROBIOTA

The intestinal microbiota has marked influences on the intestinal and peripheral host's immunity. In some cases, the effects are produced by the whole intestinal microbiota, whereas in other cases only one predominant bacterium is capable of producing a certain immunostimulatory effect that is as effective as that of the whole microbiota. Moreover, the post-natal period seems to play a crucial role in the cross talk between the intestinal microbiota and the development of some important immunoregulatory processes, especially those involved in the suppressive responses.

Most of the data come from original experimental animal models of germ-free (GF) mice and gnotobiotic mice, i.e., GF mice colonized with known bacteria. The role of intestinal microbiota in humans has largely been extrapolated from studies conducted on probiotic bacteria, mainly *Bifidobacterium* and *Lactobacillus* strains, and from epidemiological studies.

The intestinal microbiota acts on the three lines of defense of IIS. Recently, very interesting papers have been published on the role of intestinal bacteria on natural defenses, which are more or less related to innate defenses, especially on epithelium, which belong to the IIS. Thus intestinal microbiota should act on: intestinal permeability (55), production of fucosylated glycoconjugates (56), glycosylation of the intestinal cell layer which is involved in resistance or susceptibility to intestinal infections by the presence or absence of appropriately glycosylated receptors (57) and, expression of angiogenins, especially angiogenin 4 which may have microbiocidal properties (58). These results and others, showing that the intestinal microbiota influence the gene expression in epithelial cells (59), give new insights in the wonderful cross talk existing between bacteria and epithelium.

The intestinal microbiota also interacts with the other lines of defense, innate and acquired immunities. These effects can be of particular importance during the early postnatal life that is a period of high risk for intestinal disorders due to enteric pathogens and/or food hypersensitivities. During the neonatal period, mammalian species exhibit some degree of reduced immunocompetence that could be attributed to a functional immaturity in cells involved in immune intestinal responses. It could be also attributed to the lack of bacterial stimulation given by the intestinal microbiota which is absent during the fetal life. After birth, a well-balanced bacterial colonization will "educate" the IIS in a good manner allowing immunoregulatory mechanisms governing IIS functions to operate rapidly.

As already mentioned in the introduction, the activation, modulation and regulation of the IIS are the main effects exerted by the intestinal microbiota. Gnotobiotic animal models are useful in analyzing such effects of intestinal microbiota on IIS activities.

Experimental Animal Models: Gnotobiotic Mice

In experimental studies, the role of the digestive microbiota is determined by comparison between GF and conventional (CV) animals, or GF mice colonized with a human fecal microbiota, the humanized-mice. Several results show that the human fecal microbiota reproduces the same immunostimulatory effects as those produced by the mouse intestinal microbiota (60,61), and consequently, this mouse model is a very interesting tool for human studies. The first step is the demonstration of the effect of the entire intestinal microbiota on a specific immune response by comparison between GF and CV or humanized-mice. The second step is to determine the bacteria that are responsible for the immunomodulatory effect observed. For this purpose, GF mice are colonized with only one or several known bacteria originated from mice or human microbiota. These "gnotobiotic mice," such as GF mice, are reared in isolators under microbial controlled conditions. After oral colonization, the bacteria expand rapidly to colonize the intestine to a very high level within one day. A period of 3 weeks is estimated to be the time required for an optimal stimulatory effect of the intestinal microbiota. Thus gnotobiotic models allow in vivo analysis of the specific role played by the various bacteria composing the intestinal microbiota with respect to immune responses. This has enabled demonstration that the bacterial immunomodulatory effect is sometimes "strain-dependent." A more detailed discussion on the use of GF in the study of the intestinal microbiota is described in chapter 15 by Norin and Midtvedt.

Activation of the Intestinal Immune System

It has been shown that the presence of intestinal microbiota plays an important role in the development and activation of IIS even if many effects are still ignored. Its role may be of particular importance in the neonatal period and could determine many of the outcomes in later life.

As newborns, GF animals exhibit an underdeveloped IIS, which can be normalized by bacterial colonization of the intestine with the fecal microbiota from a CV animal or human, within 3 weeks. In GF mice, PPs are poorly developed, and germinal centers are absent. The absence of digestive microbiota only affects some subsets of thymus-dependent IEL, the single positive thymo-dependent CD4 + or CD8 + αβ IEL, the other thymo-independent homodimeric αα CD8 + subpopulations of IEL (all the γδ-IEL and part of the αβ IEL) being always present in GF mice (35). Cellularity of the LP is greatly reduced in GF mice and it has been demonstrated that the intestinal microbiota is the major target of the IgA plasmocyte development.

IgA-Secreting Cells

As in the neonate, the intestinal IgA-secreting cell (IgA-SC) number is much reduced in adult GF mice. Three weeks after bacterial colonization of the intestine, GF mice have an IgA-SC number equivalent to that found in CV mice. In the young, the adult number of IgA-SC is reached at the age of 6 weeks in mice and between 1 and 2 years in babies (7). This important delay might be attributed to the immaturity of the IIS of the newborn and/or the suppressive effect of Abs present in the mother's milk. However, it might also be due to the stimulatory effect of the intestinal microbiota that has been established according to a sequential manner from birth to after weaning as described previously. To test the later hypothesis, several models of adult gnotobiotic mice were colonized by the entire digestive microbiota obtained from growing CV mice from one day after birth to 25 days of age (i.e., 6 days after weaning; 62). In these experimental adult models, the effect of maternal milk, and the possible immaturity of the neonate were excluded, and only the stimulatory effect of the digestive microbiota was tested. After 4 weeks, adult recipients were sacrificed, and the immunostimulatory effect of the digestive microbiota evaluated by the IgA-SC numbers present in intestinal villi by immunohistochemical observations. Digestive microbiota of mice 3 to 21 days old exerted only a partial stimulatory effect on the intestinal IgA-SC number in gnotobiotic recipients (Table 1). However, gnotobiotic recipients colonized with the digestive microbiota of 25-day-old mice had a similar IgA-SC number to that found in adult CV mice.

These results obviously show the important role played by the sequential establishment of the digestive microbiota in full development of the intestinal IgA-SC number and the pivotal role played by the bacterial diversification present after weaning in this process. Results have been confirmed by other studies (7). Moreover, taking into account the 3-week delay between the bacterial stimulus and the intestinal IgA-SC response, these results showed that the neonate is capable of developing a sIgA response at birth, the intensity of which depends on the stimulatory capacity of the intestinal bacteria present in the intestine. It is tempting to project such results onto infants where the full development of the intestinal IgA-SC number observed at 2 years of age is correlated to the stabilization of the intestinal microbiota.

Table 1 Effect of the Sequential Establishment of Intestinal Microbiota of Growing CV Mice on the Maturation of Intestinal IgA Plasma Cells Measured in Gnotobiotic Mice

Gnotobiotic mice harboring the digestive flora of:	IgA plasma cell number/villus
Adult conventional mice	41 ± 1
Adult germ-free mice	4 ± 0.5
Growing conventional mice 1–4 days old	15 ± 2
Growing conventional mice 7–23 days old	23 ± 1
Growing conventional mice 25 days old	43 ± 1

Source: From Refs. 62, 63.

Attempts have been made to elucidate the role played by individual bacterial strains present in the digestive microbiota of CV growing mice (63). Results showed that some Gram-negative bacteria such as *E. coli* or *Bacteroides* play an important adjuvant role on this immunological non-specific effect, probably due to the LPS present in the cell wall of these bacteria (7). These studies have shown the importance of the intestinal microbiota diversification on the complete development of IIS in young. They promote insight into the close correlation between dietary modification and intestinal microbiota diversification and consequently its effect on the infantile IIS. Excessively early or late dietary modification may have consequences on quality of the intestinal microbiota equilibrium and, consequently, may affect the development of the IIS.

Dendritic Cells

As described above, the intestine is populated by some characteristic subsets of DCs, which are believed to play a pivotal role in the orientation of the acquired immune responses towards tolerance. Is the intestinal microbiota the main factor that determines such characteristics? Currently, only few studies exist in this field.

From some studies, it appears that inflammatory stimuli are very important for maturation of DCs in GF mice as well as in neonates, and the intestinal microbiota could afford such stimuli. It has been demonstrated that the rapid and constitutive trafficking of DCs from the IIS to the MLNs can be increased by the presence of inflammatory stimuli, such as LPS (64). Other studies have shown that it is possible to increase the rate of postnatal development of the intestinal DC population in rats by intra-peritoneally administration of IFN-γ (65). We can conclude that these inflammatory factors are physiologically important to maintain activation of DCs and the intestinal microbiota may have an important part in this process.

Another question concerns the specific functions of intestinal DCs. Are there specific distinct lineages of DCs attracted into the intestinal mucosa under the control of specific chemokines or adhesion molecules, or are precursor DCs modified after their arrival in the tissue? In his interesting review (31), Mowat explains that, given the plasticity of DCs in other tissues, it is reasonable to believe the latter hypothesis, and mucosal DCs are the cells that integrate the genetic and environmental factors to shape T-cell responses to local Ags in ways such that homeostasis is maintained. Intestinal epithelial cells, by the ability to constitutively produce TGF-β and by the regulatory factors controlling inflammatory cytokine secretion, could be the first level of regulatory control. Moreover, recent studies have shown that lamina propria stromal cells constitutively produce cyclo-oxygenase 2 (COX2)-dependent protaglandin E2 (PGE2) under the influence of the physiological levels of LPS that are absorbed from intestinal microbiota. These metabolites act as down-regulators of the immune response to dietary Ags (66). Moreover, DCs themselves might also express COX2 and produce PGE2 in response to LPS. As PGE2 is known to polarize DC differentiation towards an IL-10-producing inhibitory phenotype, this would explain the prevalence of such DCs in the normal gut (67).

The subunit p40 is present in IL-12 and IL-23, which are both Th1-inducing cytokines. In a elegant study, Becker and coworkers (40) using transgenic mice expressing a reporter under the control of the IL-12p40 subunit promoter, showed that some subsets of lamina propria DCs, present in the small intestine but not in the colon, constitutively exhibited transgene expression. This expression was restricted to the ileum, associated with the intracellular nondegraded bacteria as revealed by fluorescent in situ hybridization (FISH), and was not found in the ileum of GF mice. In addition to supporting literature elsewhere (68), these results obviously show how the presence of the intestinal microbiota, which become more abundant in the ileum, can influence the immune responses elicited at this specific area of the intestine. They afford new data on the compartmentalization of the IIS, which have to be considered carefully to avoid erroneous conclusions.

In conclusion, GF, and gnotobiotic animal models are very useful tools to gain new insight into the fundamental role played by the intestinal microbiota on the complete activation of the IIS, with functional consequences. In certain aspects, adult GF mice, in which the IIS is poorly developed, may be considered as similar to that of the neonate and immunological immaturity of neonates can be questioned.

Modulation of Specific Immune Responses: The IgA Anti-Rotavirus Response

Little information is available regarding the role of intestinal microbiota composition on the modulation of the specific sIgA Ab response against enteropathogens. Indeed, it can be assumed that, according to the composition of the digestive microbiota and the presence, or not, of some bacteria in the dominant microbiota, the specific immune responses might be different.

This fact is of particular importance in babies where the poorly diversified intestinal microbiota is strongly influenced by the type of milk. Indeed, it is well known that breast-fed babies are more resistant to enteric infections than formula-fed babies (69,70). Human breast milk contains abundant bioactive components that may provide direct protective effects to infants against enteric pathogens (71), but breast-feeding also influences the intestinal microbiota composition enhancing *Bifidobacterium* development. To test the influence of the intestinal microbiota on the modulation of a specific intestinal sIgA-Ab response, a sIgA anti-rotavirus response was established in a mouse model. This involved an original model of adult gnotobiotic mice colonized with the fecal microbiota of a breast- or a bottle-fed infant and then orally inoculated with a heterologous simian rotavirus strain SA-11. As previously described, the adult mouse model described here excluded breast milk effects and the possible immaturity of the neonate immune system [(72) and manuscript in preparation].

Bacterial strains found in the dominant fecal microbiota of a breast- or formula-fed baby were isolated and inoculated in the digestive tract of the gnotobiotic mice. They established in a similar manner as in babies. "Breast-fed mice" were colonized with *Bifidobacterium*, *Escherichia coli* and *Streptococcus*, while only two Gram-negative bacteria, *E. coli* and *Bacteroides*, colonized the digestive tract of "formula-fed" mice. The two groups of gnotobiotic mice were similar in all respects except for the intestinal microbiota and especially by the presence or absence of *Bifidobacterium*. They were orally inoculated with rotavirus 3 weeks after bacterial colonization to allow the bacteria time to affect the immune system of the host. The kinetics of sIgA anti-rotavirus Ab responses were measured in feces by enzyme linked immunosorbent assay (ELISA) over a one month period of time and at sacrifice, numbers of sIgA-anti-rotavirus secreting cells were evaluated in the small intestine by solid phase enzyme-linked immunospot (ELISPOT) assay.

Kinetics of the sIgA anti-rotavirus response were similar in the two groups of gnotobiotic mice, but the maximal level, that was reached 20 days after viral inoculation, was approximately 4-fold higher in "breast-fed" than in "formula-fed" gnotobiotic mice (Fig. 3). The same difference was measured for the sIgA-anti-rotavirus secreting cell numbers. To assess the respective immunomodulatory role of two bacteria present in the baby's intestine, *Bifidobacterium bifidum* (Gram-positive bacteria) and *E. coli*

Gnotobiotic mice : « Breast-fed baby »
(Bifidobacterium, Escherichia coli,
Streptococcus)

Gnotobiotic mice « Formula-fed baby »
(Escherichia coli, Bacteroides)

Oral Inoculation with rotavirus

Three weeks later; level of anti-rotavirus sIgA antibodies in feces

18 AU   5 AU

Abbreviation: AU, arbitrary units

Figure 3 Adjuvant effect of the fecal microbiota of breast-fed babies on the intestinal anti-rotavirus antibody response measured in gnotobiotic mice. *Source*: From Ref. 72.

(Gram-negative bacteria), two other groups of gnotobiotic mice were created. Results presented in Table 2 obviously show the adjuvant capacity of the strain of *Bifidobacterium bifidum* on the intestinal sIgA anti-rotavirus response while, in contrast, *E. coli* exerted a suppressive effect, as compared with GF mouse response. These results show how the presence of *Bifidobacterium bifidum* in the fecal microbiota of babies modulates the suppressive effect exerted by the presence of *E. coli*. Given the importance of rotavirus infections as a cause of infantile diarrhea worldwide, the presence of *Bifidobacterium* in the intestinal microbiota of babies is of great interest to stimulate this protective Ab sIgA response. These results can be compared to those found previously, which showed that a strain of *Lactobacillus rhamnosus* GG, used as a probiotic, and given to babies suffering from rotavirus diarrhea, shortened the diarrhea duration, and stimulated the specific IgA anti-rotavirus response (73,74). Other studies have shown an enhancement of serum or intestinal Ab response to orally administered Ags by Gram-positive bacteria (75), especially lactic acid producing bacteria used as probiotics (76).

These results also showed that GF mice are able to mount a sIgA anti-rotavirus response while its IIS is poorly developed suggesting a lack of correlation between the non-specific IgA response induced after bacterial colonization and the specific anti-rotavirus Ab response. The latter findings confirm previous results from Cebra and coworkers (77). Such data have also been described in humans where one-week-old babies are capable of developing protective immunity following oral vaccination with poliovirus or hepatitis B virus while the complete development of natural sIgA is only achieved several months later (78). Consequently, the ability to give a highly specific sIgA anti-rotavirus Ab response could be correlated with the modulatory effect of intestinal bacteria rather than with the development of IIS. Mechanistic studies are required to clarify the molecular basis upon which some digestive bacteria modulate the sIgA Ab response to enteric pathogens.

The adjuvant effect of *Bifidobacterium* sp. may be strain-dependent. In a recent study we have shown that four different species of *Bifidobacterium* isolated from the fecal

Table 2 The Gut Colonization of Different Bacterial Strains Modulates the Intestinal Anti-rotavirus IgA Antibody Response Measured in Gnotobiotic Mice

Intestinal microflora of gnotobiotic mice	Anti-rotavirus sIgA antibody level (AU/g of feces)
Bifidobacterium bifidum (from baby)	$31 \pm 7^a \uparrow$
Bifidobacterium DN 173 010 (a commercial strain)	$21 \pm 3^a \uparrow$
Germ-free (control)	11 ± 2
Bifidobacterium infantis + B. pseudocatenulatum + B. angulatum + B. sp (from human adult)	$4 \pm 1^a \downarrow$
E. coli (from infants) or *Bacteroides vulgatus* (from human adult)	$4 \pm 1^a \downarrow$

[a] Significant difference with germ-free mice ($p < 0.01$).
Abbreviation: AU, arbitrary units.
Source: From Refs. 72, 79.

microbiota of an adult human lacked the adjuvant ability to stimulate the sIgA anti-rotavirus response in gnotobiotic mice but, on the contrary, exerted a suppressive effect as do *E. coli* (Table 2) (79). Thus, the modulating effect of *Bifidobacterium* is strain-dependent, as it has also been described for different *Lactobacillus* strains used as probiotics in other mice studies (80). Taken together, these data suggest that it is important to define the modulatory effect of the strains of bifidobacteria either normally colonizing the digestive tract of babies after birth or given as probiotics, to modulate in a good protective way a specific intestinal immune response.

In conclusion, and on the basis of the experimental and clinical data, we may consider that the presence of certain bacterial strains in the infantile intestinal microbiota, namely some strains of *Bifidobacterium,* or some transiting strains of probiotics, enable activation of the mechanisms that result in optimization of the anti-rotavirus protective IgA Ab response. Elucidation of the immunomodulatory mechanisms must now be pursued.

Regulation of the Immune Responses

Tolerance to Soluble Proteins: Oral Tolerance

The role of the intestinal microbiota on the OT process has been demonstrated by various experimental studies using GF mice. Results depend on the immune response considered, oral Ag, and experimental schedule used. In these experiments, immune responses to a specific Ag are compared in two groups of mice: the tolerant group where mice are fed with an Ag prior to the peripheral immunization with the same Ag, and the control group fed with only the buffer before the same peripheral immunization. Specific immune responses to the Ag used are then evaluated (Ab responses in serum or cellular response by delayed-type hypersensitivity) in both groups. The tolerant state is present when peripheral immune responses to the Ag are abolished or significantly decreased in the group Ag-fed as compared with the control group.

In an initial study, Wannemuehler and coworkers (81) showed that, in contrast to what is observed with the CV mice, gavage of GF mice with a particular antigen, sheep red blood cells (SRBC), does not enable suppression of immune responses to SRBC in serum. However, the OT process was re-established when LPS was administered orally prior to gavage. The authors concluded that Gram-negative bacteria play a fundamental role in the mechanisms responsible for OT. Subsequently, other experiments using adult GF mice fed with a soluble protein, OVA, in order to study the immune suppression of anti-OVA serum IgG response, demonstrated that it was possible to induce OT in GF mice. However, in contrast to what is observed with CV mice, the suppression was of very short duration, about 10–15 days, versus more than 5 months in CV mice (82). Similar results were obtained in human-microbiota-associated gnotobiotic mice (60). Colonization of the intestinal tract with *E. coli* alone prior to gavage was sufficient to restore lasting suppression (83), and the same results were obtained with another Gram-negative bacteria, *Bacteroides* (unpublished personal data), while in our experimental conditions, adult GF colonized with the strain of *Bifidobacterium bifidum* isolated from a baby's feces, had no effect on the serum IgG anti-OVA suppression (83).

Recently, in their experimental conditions, Sudo and coworkers (84) showed that in OVA-fed mice, the GF state does not allow suppression of the systemic anti-OVA IgE response in serum in contrast to what is observed with CV mice. Colonization of the intestinal tract by a strain of *Bifidobacterium infantis* restored the suppression but only when the strain colonized the intestinal tract of the mouse from birth. The importance of the presence of intestinal bacteria from birth in the optimization of the immune processes has also been suggested in a more recent study (60).

It is interesting to compare these experimental results to those described in human neonates by Lodinova-Zadnikova and coworkers (85). In their study, they colonized the digestive tract of babies just after birth with a given strain of *E. coli*. In these conditions *E. coli* is able to establish durably in the digestive tract of newborns as described previously (86). After 10 years (preterm infants) and 20 years (full-term infants), differences in occurrence of food allergies between colonized and control subjects were statistically significant; 21% versus 53%, and 36% versus 51% respectively. Furthermore, recent clinical trials using ingestion of a strain of probiotic, *Lactobacillus rhamnosus* GG, during the last month of pregnancy to women and after birth to babies during 6 months, reduced the incidence of atopic eczema in at-risk children during the first 4 years of life (87). However, in this case, IgE levels were not decreased in the treated group as compared with the placebo group. The protective mechanisms of these interventions are not elucidated.

All these experimental data show the importance that a single bacterial strain present in the intestinal digestive microbiota of infants may have with respect to the establishment of tolerance mechanisms. Are there *E. coli*, *Bacteroides* or some strains of *Bifidobacterium* which play this important role? First, as suggested by previous studies, it is not sure whether the mechanisms are the same for suppression of the various isotypes IgG and IgE (45,88), and consequently that the same bacteria are operating on them. Secondly, as described previously, all the strains belonging to the same bacterial genus have not the same immunoregulatory properties and it is conceivable that some *Bifidobacterium* strains may have regulatory properties on suppressive immune processes.

The cellular ways by which the bacteria are acting, and the exact bacterial components involved are not known. However, from an ecological point of view, it is important to note that some experimental data point out the importance of the neonatal period with respect to the ability to recognize bacterial messages.

Tolerance to the Intestinal Microbiota

An important question is why the intestinal microbiota does not mount an inflammatory response in the gut while this state is broken in pathologic conditions such as IBD?

The mechanisms by which commensal and non-pathogenic bacteria are tolerated by the IIS is beginning to be understood and may result from a cross-talk between bacteria, epithelium, and immune cells. In an interesting experimental study, Neish and co-workers (89) demonstrated, using an in vitro model of cultured human intestinal epithelial cells, that a non-pathogenic strain of *Salmonella* directly influenced the intestinal epithelium to limit inflammatory cytokine production. They showed that the immunosuppressive effect was due to the inhibition of the NFκ-B activation pathway by blockage of IκB-α degradation. Another interesting conclusion from this study was that non-pathogenic bacteria, which do not belong to the commensal intestinal microbiota, are unable to induce inflammatory responses. Another study converges to an opposite conclusion (90). In several intestinal epithelial cell lines, the authors demonstrated that a commensal bacterial strain, *Bacteroides vulgatus*, was able to activate the NF-κB signaling pathway through IκB-α degradation and ReIA phosphorylation. However, the presence of TGF-β1 cytokine inhibits *B. vulgatus*-mediated NF-κB transcriptional activity showing that the responsiveness of intestinal epithelial cells to luminal enteric bacteria depends on a network of communication between immune and epithelial cells and their secreted mediators.

Recently, it was shown in vivo in mice, that the intestinal microbiota itself plays a regulatory role with respect to inhibition of the NFκ-B activation pathway, by the way of another inhibitory factor, the peroxisome proliferator-activated receptor (PPARγ) (61). The latter is highly expressed in the colon and its activation has anti-inflammatory effects, with protection against colitis. PPARγ activators are able to limit inflammatory cytokine production through the inhibition of the NF-κB pathway. It has been suggested that PPARγ could play an important role in homeostasis of the gut, especially in the colon. In patients with IBD, impaired expression of PPARγ in colon epithelial cells was observed (61). In the same work, in vivo observations showed that the intestinal microbiota and TLR-4 regulates PPARγ expression by epithelial cells of the colon. Indeed, it is highly expressed in CV mice while it is barely detectable in GF mice. When TLR-4 transfected CaCo-2 cells were incubated with LPS, an increase of PPARγ expression was observed showing the involvement of TLR-4 in this process and suggesting that PPARγ may be a regulatory factor able to shut down the TLR-4 signaling given by bacterial LPS abundant in the colon (61).

Taken together, these data provide evidence that the cross-talk existing between the IIS and intestinal microbiota pass through regulatory processes preventing inflammatory responses induced by activation of some nuclear factors, such as NF-κB, which could be different, or predominant, according to the intestinal site. They are mediated through the actions of commensal bacteria, but also through exogenous non-pathogenic bacteria action and this data is of importance in terms of nutrition. Indeed, we can ingest billions of exogenous bacteria in some foods such as fermented milks and some cheeses, without detrimental consequences. In terms of pathology, a lot of other questions concerning the mechanisms and origin of IBD have yet to be answered. Why is an activation of the NF-κB pathway observed in IBD? Is it due to some subsets of the intestinal microbiota, which are suddenly dominant in an unbalanced microbiota? Is it due to enteropathogens which can interact with the NF-κB pathway during infection? Or, is it due to a decrease and modification of mucus secretion allowing excessive adhesion of commensal bacteria? All these factors, and others, may be responsible.

It is interesting to give recent clinical results concerning oral administration of probiotics on the maintenance of the remission phase in IBD, either the use of a mixture of 8 strains of lactic-acid bacteria used as probiotics (VSL#3) in chronic pouchitis (91), or a yeast strain, *Saccharomyces boulardii* (92) or the *E. coli* Nissle 1917 (93) in ulcerative colitis. The mechanisms underlying such beneficial effects are still not known and they are multifactorial. From experimental data it has been suggested that a stimulation of the non-inflammatory IL-10 cytokine production by ingestion of probiotics may be involved in such protective effect (94). Further experimental and clinical studies need to be conducted to further elucidate the mechanisms involved in the epithelium-bacterial cross talk.

RELATIONSHIPS BETWEEN THE PERIPHERAL IMMUNE SYSTEM AND INTESTINAL MICROBIOTA

Activation of the Immune System

Innate immunity plays a very important role in the activation of the immune system and the ability to develop specific acquired immune responses. Through their Ag-presenting activity and the synthesis of numerous pro-inflammatory chemokines and cytokines (IL-8, IL-1, IL-6, TNF-α, and IL-12), macrophages, and DCs play a key role in the regulation of immune responses. They are the gatekeepers of the host, generating innate resistance to pathogens, and specific immune responses by the stimulation of T-cell-acquired immunity and regulation of the TH1/Th2 balance.

It has been postulated that the immune defects in neonates may result from a developmental immaturity of APC functions (78), and bacterial components resulting from intestinal colonization could be an important factor for maturation of APCs (95). Recently, Sun and coworkers (96) investigated the ontogeny of peripheral DCs and their capacity to provide innate responses to microbial stimuli in early life. They show that neonatal murine spleen DCs have intrinsic capacity to produce bioactive IL-12. Moreover, after microbial stimulation given in vitro by LPS, they are able to up-regulate MHC and costimulatory molecule expression required for productive interaction with naive T cells. Thus, neonatal DCs could be fully competent in their innate functions but they need to be activated, through TLR recognition as described previously, by bacterial stimuli afforded by the intestinal microbiota. Another interesting study supports this hypothesis. Nicaise and coworkers (97) demonstrated that the presence of the intestinal microbiota underlies IL-12 synthesis by macrophages derived from splenic precursors.

On the basis of those experimental data, one can wonder whether the first bacteria colonizing the intestinal tract, E. coli, rich in LPS, and subsequently bifidobacteria rich in peptidoglycan and CpG dinucleotides, do not play such crucial activating roles? It is conceivable that in newborns, the abrupt colonization of the intestinal tract by the microbiota may induce a physiological inflammatory reaction with, as a consequence, an increase in intestinal permeability, bacterial translocation and systemic activation of immune cells, especially APCs. Experimental evidence supports that hypothesis. Studies in mice have shown that the presence of the intestinal microbiota induces the synthesis of pro-inflammatory cytokines IL-1, IL-6, and TNF-α by peritoneal macrophages. Such effects can be reproduced in gnotobiotic mice colonized with E. coli alone while a Bifidobacterium bifidum strain isolated from baby's feces had no effect (Table 3) (98).

Other non-specific resistance factors play an important role in host defense mechanisms to infection. GF and gnotobiotic animal models have showed that some functional parameters involved in innate immunity, phagocytosis, complement system, and opsonins, are expressed to a lesser extent than in CV animals (99).

Table 3 Influence of Intestinal Bacteria on the Inflammatory Cytokine Production by Peritoneal Macrophages

Gnotobiotic mice	Cytokines (units/ml)		
	IL-1	IL-6	TNF-α
Conventional	18200	6,33	72
Germ-free	8300[a]	2,62[a]	<50[a]
Bifidobacterium bifidum	8000[a]	2,46[a]	<50[a]
Escherichia coli	15350[b]	7,24[b]	108[b]

[a] Significant difference with conventional mice ($p < 0.01$).
[b] Not significant.
Abbreviations: IL, interleukin; TNF, tumor necrosis factor.
Source: From Ref. 98.

Modulation and Regulation of Immune Responses

Balance Th1/Th2

Experimental results, epidemiological studies and clinical trials strongly argue for the fact that bacterial environment plays a crucial role in the Th1/Th2 balance via different mechanisms of which cytokine synthesis by innate immune cells, especially IL-12, and IFN-γ, could play a decisive role.

The prenatal period and early childhood are considered to be critical for the establishment and maintenance of a normal Th1/Th2 balance. It has been described that the immune context at birth is mainly Th2, while Th1 responses are partially suppressed, enabling non-rejection of the fetus during gestation. After birth, neonates must rapidly restore the balance by developing the potential to induce Th1-type responses (100). Various studies have shown that, in atopic infants, the switch does not occur, and the infant is in a context of an imbalance toward Th2 with a predisposition to development of IgE responses (101,102). The neonatal period is thus considered to be extremely important in enabling regulation of the Th1/Th2 balance to become operative, and the switch could occur during the first 5 years of life especially during the first year of life (103).

The Th2→Th1 switch is dependent on multiple factors whose relative importance has yet to be elucidated. Bacterial stimuli are considered to play a considerable role, and some years ago it had been claimed that infections might prevent the development of atopic diseases. This is referred to as the "hygiene hypothesis" (13), but it is now a matter of debate. From a recent study (104), authors did not find any evidence that exposure to infections in infants reduces the incidence of allergic disease, but, in contrast, exposure to antibiotics may be associated with an increased risk of developing allergic disease. Today, accumulating evidence suggests that rather than infections, alteration of the composition of the intestinal microbiota early in life may be an important determinant of atopic status (13,105). Experimental studies have supported this hypothesis. Thus, in one-week-old rats, peripheral immunization leads to a Th2-biased memory response. However, when the rats are concomitantly administered a bacterial extract by the oral route with immunization, the memory response switches to both Th1 and Th2 (106). Another study showed how, in three-week-old mice, the disturbance in intestinal bacterial equilibrium following ingestion of an antibiotic, kanamycin, promoted a shift in the Th1/Th2 balance toward a Th2-dominant immunity, while it became Th1 and Th2 in non-treated growing CV mice (107). Ingestion of intestinal bacteria such as *Enterococcus faecalis* five days after antibiotic treatment again permitted the shift back towards the Th1/Th2 balance (108).

From an epidemiological point of view, very interesting studies argue in favor of the important role of the bacterial environment in the first year of life in order to ensure the good orientation of immune responses preventing the short- and long-term development of atopic diseases (13,101,103,109–111). Recent comparative studies have been conducted in children living in the same allergenic environment but under different life-style conditions, urban and farming environments. Results showed that substantial protection against development of asthma, hay fever, and allergic sensitization was seen only in children exposed to stables, farm raw milk, or both in their first year of life (103). Authors also found that prenatal exposure of women had a substantial protective effect.

Bacteria that are responsible for such effects are not known. Gram-negative bacteria rich in LPS have been suggested to be important in that phenomenon (85,109,112) but it is also possible that Gram-positive bacteria, such as bifidobacteria and *Lactobacillus*, are involved. The comparative study between Swedish and Estonian children (105) has suggested a specific role of the intestinal microbiota, regarding its nature, diversity and changes with time. Besides genetic factors, which are known to play an important role in the development of allergic diseases, all these data suggest that the infant intestinal microbiota normally rich in Gram-negative (LPS-producing) and Gram-positive bacteria may not be well-balanced in atopic children. Depending on the microbial environment associated with the life-style, especially during the first year of life, a restoration of the normal balance could be achieved.

Clinical trials using probiotics to treat or prevent atopic eczema in infants have also generated arguments suggesting that the infantile intestinal microbiota balance plays an important role in the good orientation of immune responses. In a recent double-blind trial, Kalliomäki and coworkers (87) have shown that the supplementation of pregnant women one month before delivery followed by 6 months post-parturition (mother or baby) with a probiotic strain, *Lactobacillus rhamnosus* GG, lead to a significant decrease in the incidence of atopic eczema in babies with a family history of atopic disease. At two years of age, atopic eczema was diagnosed in 23% of treated babies versus 46% in the placebo group. The preventive effect of *L. rhamnosus* GG extends to the age of 4 years follow-up treatment (87). The mechanisms involved in such a protection are unknown. Indeed, the frequencies of positive skin-prick test reactivity (measuring the specific IgE levels) were comparable between treated and placebo groups. Further studies are necessary to elucidate the mechanisms responsible for these interesting protective effects.

On the basis of all the above data, questions arise with respect to delivery conditions, infant feeding, and antibiotic treatments to be administered during infancy in order to enable and optimally establish and maintain integrity of the intestinal microbiota. Probiotics may also be considered as good palliative agents with respect to impaired equilibrium of the intestinal microbiota. Knowledge of the immunoregulatory mechanisms driven by the intestinal microbiota of infants, as well as the bacterial components which are involved, are crucial to prevent some pathologies which are dramatically increasing today.

Natural IgG

In the absence of immunization, there is a natural level of immunoglobulins (Ig) in serum named "natural Ig" or "natural Abs." The roles of those Abs in the immune responses have yet to be completely elucidated but it is known that they play important regulatory roles in humoral immune responses, especially in immune responses to self-Ag (113). It has also been demonstrated in mice that they intervene with the development of the B repertoire at peripheral level (spleen), enabling expansion of the Ab response towards thymo-dependant Ags (114,115). In man, the role of these natural Abs is under investigation in the context of research on certain autoimmune disease (116).

Intrinsic and extrinsic factors, especially the intestinal microbiota, act on the natural Ig levels, depending on isotypes and sub-classes. Thus, GF mice had normal serum IgM levels, but IgG, and IgA levels are approximately 5% of conventionally reared littermates (114). It has been established in mice that one of the roles of the natural IgG is to expand B cell repertoire. The latter can be evaluated through the expression of some genes coding for the variable part of the heavy chain of Ig (VH gene) using probes. Analysis of a VH gene expression has provided a quantitative tool for the global assessment of Ab repertoire, and a preferential use of the gene means that the repertoire is poorly diversified.

Early in ontogeny, a high frequency of B cells could bind to multiple Ags, among which auto-Ags are found, in neonatal CV mice. This fact has been correlated with preferential use of VH gene family, namely VH7183. In CV adult mice these multi-reactive B cells are much less frequent coinciding with a random usage of VH genes, as seen by the decreased utilization of VH7173 gene family, showing a diversified repertoire. Thus, there is a maturation of the immune system of adult CV mice. This fact is not present in adult GF mice where a high percentage of B cells expressing VH 7138 genes is found as in neonatal CV mice (115). The injection of purified natural IgG Ig from serum adult CV mice into GF mice reduced the use of the VH7183 gene family in the peripheral B-cells, as in CV mice (115). From these data authors concluded that if a genetic program leading to non-random position-dependent preference of rearrangement and expression initially controls the establishment of the VH repertoire, a broader utilization of the B-cell repertoire is thereafter stimulated by environmental Ags and Igs. The finding that GF mice maintain a "fetal-like" VH repertoire that can be modified by the administration of pooled Igs from normal unimmunized CV mice establishes the crucial role of the intestinal microbiota in this function.

This data may have clinical relevance. Many reports have described the beneficial results of intravenous injection of normal human IgG in treatment of autoimmune disease (116).

The mechanism by which exogenous antigenic stimulation can influence the expression of VH gene remains unclear. Exogenous Ags may play an important role in the final modulation of the expressed repertoires either by direct stimulation of Ag-specific clones or indirectly by idiotype interactions mediated by the Abs produced in those responses (113–115).

Autoimmune Diseases

One example of the regulatory effect exerted by intestinal microbiota on an autoimmune disease has been reported by Van der Broek and co-workers (117). Streptococcal cell wall (SCW)-induced arthritis is a chronic erosive polyarthritis, which can be induced in susceptible rats by a single intra-peritonal injection of a sterile aqueous suspension of SCW. The acute phase of the disease develops within a few days, the second, chronic phase, which mainly involves peripheral joint inflammation, develops from 10 days after. The second phase is dependent on functional T lymphocytes. F344 rats are genetically described as resistant to the second chronic phase, while in contrast another strain of rats, Lewis rats, are described as susceptible. These data suggest that a T-cell unresponsiveness due to immune tolerance to SCW may be the mechanism underlying resistance to SCW-induced arthritis of F344 rats, while Lewis rats are defective in their tolerance. When F344 rats are reared in GF conditions, they become susceptible to SCW-induced arthritis as are Lewis rats. There was a correlation between the susceptibility of the disease and the

T cell proliferation response to SCW measured in vitro. In CV Lewis and GF-F344 rats, a proliferation was measured while it was not present in CV F-344 rats. This concept that disease might result from a similarity between naturally occurring cell surface Ags of the host and those expressed on some commensal or pathogenic micro-organisms have been referred to as the "molecular mimicry hypothesis." Mono-association of GF F344 rats with *E. coli* resulted in resistance, which equaled that in CV F344 rats whereas mono-association with a *Lactobacillus* strain did not really affect susceptibility. Thus, in CV F-344 rats, a state of tolerance to arthritogenic epitopes is induced during the neonatal period of life and maintained through life by the bacterial microbiota, resulting in resistance to SCW-induced arthritis. In Lewis rats, this tolerant state is deficient and/or easily broken.

Bacterial effects have been suggested in other autoimmune diseases. Thus, oral antibiotic treatment after adjuvant-induced arthritis (AIA) induction in rats significantly decreased clinical symptoms of AIA while, concomitantly, *E. coli* levels increased in the distal ileum of antibiotic-treated rats (118). In addition, it has been described that *Mycobacterial* infections profoundly inhibit the development of diabetes in non-obese diabetic (NOD) mice (119).

CONCLUSION

From all the experimental epidemiological and clinical results presented here, the digestive microbiota can be considered as an organ: it is specifically tolerated by the host and in turn, it exerts many continuous regulatory effects on intestinal and peripheral host's immune responses. Consequently, it plays fundamental roles in health. It is very important to develop knowledge about its composition, the bacterial components and metabolites that participate to such immunoregulatory effects, and the exact mechanisms involved.

Studies from GF animals have demonstrated the importance of the digestive microbiota on intestinal and peripheral immune systems. In some cases, the entire digestive microbiota is needed to obtain the complete effect while other immunoregulatory effects can be reproduced with only one bacterium and sometimes with only specific strains. Because the intestinal microbiota is a dynamic community which modifies from birth to old age in predominant bacteria composition, specific targeted interests have to be defined for the study of relationships between the intestinal microbiota and the host, according to age. Indeed, bacterial species found in the predominant microbiota are not constantly the same throughout life and several studies have demonstrated the strain-dependant immunomodulatory effect of bacteria. For instance, some strains of bifidobacteria, such as *B. breve,* are more commonly found in infants but less in adults (120). Other studies from adult GF animals have demonstrated that some bacterial effects are only obtained when the bacteria colonized the intestinal tract from birth indicating that the bacterial effects need some characteristics of the neonate immune system. A number of indirect findings converge toward the idea that the neonatal period is crucial for the infant with respect to setting up the regulatory mechanisms which will play an important role in the good orientation of immune responses throughout life. Because of the long-term consequence of the establishment of appropriate immunoregulatory networks, it is very important to develop knowledge on the cross-talk between the intestinal microbiota and immune system early in life. In this context, recent studies of the innate responses to bacterial constituents should generate decisive information in support of the role of the intestinal microbiota.

In adults, regulation of immune responses seems to be constantly reshaped by persistent interactions between the host and its digestive microbiota.

Today, an increasing challenge for researchers studying immunity (IIS as well as oral or peripheral immune responses after Ag vaccination, pro-, and prebiotic effects) is that the intestinal microbiota of experimental rodents used is not defined and can differ between breeders because of the great variety in housing conditions. Since the development of knock-out mice, which are very sensitive to infections, the microbial status required by experimenters has led to the production of highly clean animals which carry a commensal microbiota with reduced diversity. This fact has probably a significant impact on the development of the immune responses. Thus, because results could not reflect the exact conditions of microbial stimulation, the interpretation of experiments may be completely different according to different laboratories. Some controversial results obtained in mice and humans might also be explained by such paucity of mouse microbiota existing in pathogen-free mouse breeding-care units. Now, it is crucial to develop animal models in which the commensal microbiota will be better defined and designed to allow the maintenance of biological features relevant in the field of immunological investigations.

A more comprehensive understanding of the relationships between the intestinal microbiota and innate and acquired immune systems should offer new approaches for the therapy of some diseases such as allergies and IBD and for the design of oral vaccinations, and the maintenance of health. Beneficial micro-organisms such as probiotics, and dietary ingredients such as prebiotics, that act on the digestive microbiota, show promise for treatment in these immune-related intestinal disorders. Researchers addressing those subjects have to consider the digestive microbiota in their investigations.

All of the studies presented here clearly indicate the close relationship between the prokaryotic and eucaryotic worlds, and the intricacy and complexity of the relationships. Much work remains to be done and much is left to discover about our intestinal microbiota and immunity. It is to be hoped that the current enthusiasm with respect to the interest in the action of intestinal microbiota on immunity will continue to increase. The practical applications that can emerge in terms of human health can be highly significant.

REFERENCES

1. Sanders ME. Probiotic bacteria: implications for human health. J Nutr 2000; 130:384S–390S.
2. Isolauri E, Sutas Y, Kankaanpaa P, Arvilommi H, Salminen S. Probiotics: effects on immunity. Am J Clin Nutr 2001; 73:444S–450S.
3. Erickson KL, Hubbard NE. Probiotic immunomodulation in health and disease. J Nutr 2000; 130:403S–409S.
4. Dutchmann R, Kaiser I, Hermann E, Mayet W, Ewe K, Meyer zum Buschenfelde KH. Tolerance exists towards commensal intestinal flora but is broken in active inflammatory bowel disease. Clin Exp Immunol 1995; 102:448–455.
5. Guaner F, Malagelada JR. Gut flora in health and disease. The Lancet 2003; 360:512–519.
6. Umesaki Y, Setoyama H. Structure of the intestinal flora responsible for development of the gut immune system in a rodent model. Microbes Infect 2000; 2:1343–1351.
7. Moreau MC, Gaboriau-Routhiau V. Influence of commensal intestinal microflora on the development and functions of the gut-associated lymphoid tissue. Microb Ecol Health Dis 2001; 13:65–86.
8. Hooper LV, Gordon JI. Commensal host-bacterial relationships in the gut. Science 2001; 292:1115–1118.
9. Rubaltelli FE, Biadaioli R, Pecile P, Nicoletti P. Intestinal flora in breast-fed and bottle-fed infants. J Perinat Med 1998; 26:186–191.

10. Raibaud P. Factors controlling the bacterial colonization of the neonatal intestine. In: Hanson LA, ed. Biology of human milk. New-York Nestec Ltd.: Vevey/Raven Press, 1988:205–219.

11. Favier CF, Vaughan EE, DeVos WM. Akkermans ADL. Molecular monitoring of succession of bacterial communities in human neonates. Appl Environ Microbiol 2002; 68:219–226.

12. Kelly D, Coutts AGP. Early nutrition and the development of immune function in the neonate. Proc Nutr Soc 2000; 59:177–185.

13. Wold AE. The hygiene hypothesis revised: is the rising frequency of allergy due to changes in the intestinal flora ? Allergy 1998; 46:20–25.

14. Salminen S, Bouley C, Boutron-Ruault MC, et al. Functional food science and gastrointestinal physiology and function. Br J Nutr 1998; 80:S147–S171.

15. Hopkins MJ, Sharp R, Macfarlane GT. Age and disease related changes in intestinal bacterial populations assessed by cell culture, 16S rRNA abundance, and community cellular fatty acid profiles. Gut 2001; 48:198–205.

16. Tannock GW. Molecular assessment of intestinal microflora. Am J Clin Nutr 2001; 73:410S–414S.

17. Vasselon T, Detmers PA. Toll receptors: a central element in innate immune responses. Inf Immun 2002; 70:1033–1041.

18. Uderhill DM, Ozinsky A. Toll-like receptors: key mediators of microbe detection. Curr Opin Immunol 2002; 14:103–110.

19. Hemmi H, Takeuchi O, Kawai T, et al. A toll-like receptor recognizes bacterial DNA. Nature 2000; 408:740–745.

20. Cario E, Rosenberg IM, Brandwein SL, Beck PL, Reinecker HC, Podolsky DK. Lipolopysaccharide activate distinct signaling pathways in intestinal epithelial cell lines expressing toll-like receptors. J Immunol 2000; 164:966–972.

21. Neish AS. The gut microflora and epithelial cells: a continuing dialogue. Microbes Infec 2002; 4:309–317.

22. Wu MX, Ao Z, Prasad KV, Wu R, Schlossman SF. IEX-1L, an apoptosis inhibitor involved in NF-κB-mediated cell survival. Science 1998; 281:998–1001.

23. Xavier RJ, Podolsky DK. How to get along-friendly microbes in a hostile world. Science 2000; 289:1483–1484.

24. Ricote M, Li AC, Willson TM, Kelly CJ, Glass CK. The peroxisome proliferator-activated receptor gamma is a negative regulator of macrophage activation. Nature 1998; 391:79–82.

25. Tato CM, Hunter CA. Host-pathogen interactions: subversion and utilization of the NF-kappa B pathway during infection. Inf Immun 2002; 70:3311–3317.

26. Rescigno M, Granucci F, Cittero S, Foti M, Ricciardi-Castagnoli P. Coordinated events during bacteria-induced DC maturation. Immunol Today 1999; 20:200–204.

27. Mosmann TR, Cherwinski H, Bond W, Giedlin A, Coffman RL. Two types of murine helper T cell clone. Definition according to profiles of lymphokine activities and secreted proteins. J Immunol 1986; 136:2348–2357.

28. Miller A, Lider O, Roberts AB, Spom MB, Weiner HL. Suppressor T cells generated by oral tolerization to myelin basic protein suppress both in vitro and in vivo immune reponses by the release of transforming growth factor β after antigen-specific triggering. PNAS USA 1992; 89:421–425.

29. Pretolani M, Goldman M. IL-10: a potential therapy for allergic inflammation? Immunol Today 1997; 6:277–280.

30. Mowat AMc, Viney JL. The anatomical basis of intestinal immunity. Immunol Rev 1997; 156:145–166.

31. Mowatt AMc. Anatomical basis of tolerance and immunity to intestinal antigens. Nat Rev Immunol 2003; 3:331–341.

32. Brandtzaeg P. Development and basic mechanisms of human gut immunity. Nutr Rev 1998; 56:S5–S18.

33. Boursier L, Farstad IN, Mellembakken JR, Brandtzaeg P, Spencer J. IgVH gene analysis suggests that peritoneal B cells do not contribute to the gut immune system in man. Eur J Immunol 2002; 32:2427–2436.

34. MacDonald TT, Monteleone G. IL-12 and Th1 immune responses in human Peyer's patches. Trends Immunol 2001; 22:244–247.

35. Guy-Grand D, Vassali P. Gut intraepithelial lymphocyte development. Curr Opin Immunol 2002; 14:255–259.

36. Bu P, Keshavarzian A, Stone DD, et al. Apotosis: one of the possible mechanisms that maintains unresponsiveness of the intestinal mucosal immune system. J Immunol 2001; 166:6399–6403.

37. Johansen FE, Braathen R, Brandtzaeg P. The J chain is essential for polymeric Ig recepetor-mediated epitheliall transport of IgA. J Immunol 2001; 167:5185–5192.

38. Iwasaki A, Kelsall BL. Unique functions of CD11b+CD8+, and double negative Peyer's patch dendritic cells. J Immunol 2001; 166:4884–4890.

39. George A. Generation of gamma interferon responses in murine Peyer's patches following oral immunisation. Infect Immun 1996; 64:4606–4611.

40. Becker C, Wirtz S, Blessing M, et al. M.F. Constitutive p40 promoter activation and IL-23 production in the terminal ileum mediated by dendritic cells. J Clin Invest 2003; 112:693–706.

41. Veazey RS, Marx PA, Lacker AA. The mucosal immune system primary target for HIV infection and AIDS. Trends Immunol 2001; 22:626–633.

42. Jump RL, Levine AD. Murine Peyer's patches favor development of an IL-10 secreting, regulatory T cell population. J Immunol 2002; 168:6113–6119.

43. Husby S, Mestecky J, Moldoveanu Z, Holland S, Elson CO. Oral tolerance in humans—T cell but not B cell tolerance after antigen feeding. J Immunol 1994; 152:4663–4670.

44. Strobel S, Ferguson A. Persistence of oral tolerance in mice fed ovalbumin is different for humoral and cell-mediated immune responses. Immunology 1987; 60:317–318.

45. Strobel S, Mowat AMc. Immune responses to dietary antigens: oral tolerance. Immunol Today 1998; 19:173–181.

46. Alpan O, Rudomen G, Matzinger P. The role of dendritic cells, B cells, and M cells in gut-oriented immune responses. J Immunol 2001; 166:4843–4852.

47. Vincent-Schneider H, Stumptner-Cuvelette P, Lankar D. Exosomes bearing HLA-DR1 molecules need dendritic cells to efficiently stimulate specific T cells. Int Immunol 2002; 14:713–722.

48. Viney JL, Mowat AM, O'Malley JM, Williamson E, Fanger NA. Expanding dendritic cells in vivo enhances the induction of oral tolerance. J Immunol 1998; 160:5815–5825.

49. Groux H, O'Garra A, Bigler M, et al. A CD4+T-cell subset inhibits antigen-specific T-cell responses and prevent colitis. Nature 1997; 389:737–742.

50. Khoo UY, Proctor IE, Macpherson AJ. CD4- Tcells down-regulation in human intestinal mucosa: evidence for intestinal tolerance to luminal bacterial antigens. J Immunol 1997; 158:3626–3634.

51. Smith PD, Smythies LE, Mosteller-Barnum M, et al. Intestinal macrophages lack CD14 and CD89 and consequently are down-regulated for LPS- and IgA-mediated activities. J Immunol 2001; 167:2651–2656.

52. Macpherson AJ, Hunziker L, McCoy K, Lamarre A. IgA responses in the intestinal mucosa against pathogenic and non-pathogenic microorganisms. Microbes Infect 2001; 3:1021–1035.

53. Phalipon A, Cardona A, Kraehenbuhl JP, Edelman L, Sansonetti PJ, Corthesy B. Secretory component: a new role in secretory IgA-mediated immune exclusion in vivo. Immunity 2002; 17:107–115.

54. Macpherson AJ, Gatto D, Sainsbury E, Harriman GR, Hengartner H, Zinkernagel RM. A primitive T cell-independant mechanism of intestinal mucosal IgA response to commensal bacteria. Science 2000; 288:2222–2226.

55. Madsen K, Cornish A, Soper P, et al. Probiotic bacteria enhance murine and human intestinal epithelial barrier function. Gastroenterology 2001; 121:580–591.

56. Bry L, Falk PG, Midtvedt T, Gordon JI. A model of host-microbial interactions in an open mammalian ecosystem. Science 1996; 273:1380–1383.

57. Freitas M, Axelsson LG, Cayuela C, Midtvedt T, Trugnan G. Microbial-host interactions specifically control the glycosylation pattern in intestinal mouse mucosa. Histochem Cell Biol 2002; 118:149–161.

58. Hooper LV, Stappenbeck TS, Hong CV, Gordon JI. Angiogenins: a new class of microbicidal proteins involved in innate immunity. Nature Immunol 2003; 4:269–273.

59. Hooper LV, Wong MH, Thelin A, Hansson L, Falk PG, Gordon JI. Molecular analysis of commensal host-microbial relationships in the intestine. Science 2001; 291:881–884.

60. Gaboriau-Routhiau V, Raibaud P, Dubuquoy C, Moreau MC. Colonization of gnotobiotic mice with human microflora at birth protects against *Escherichia coli* heat-labile enterotoxin-mediated abrogation of oral tolerance. Pediatr Res 2003; 54:739–746.

61. Dubuquoy L, Jansson EA, Deeb S, et al. Impaired expression of peroxisome proliferator-activated recepetor γ in ulcerative colitis. Gastroenterology 2003; 124:1265–1276.

62. Moreau MC, Raibaud P, Muller MC. Relation entre le développement du système immunitaire intestinal à IgA et l'établissement de la flore microbienne dans le tube digestif du souriceau holoxénique. Ann Immunol (Inst Pasteur) 1982; 133D:29–39.

63. Moreau MC, Ducluzeau R, Guy-Grand D, Muller MC. Increase in the population of duodenal IgA plasmocytes in axenic mice monoassociated with different living or dead bacterial strains of intestinal origin. Infect Immun 1978; 21:532–539.

64. Macpherson GG, Liu MM. Dendritic cells and Langherans cells in the uptake of mucosal antigens. Curr Top Microbiol Immunol 1999; 236:33–53.

65. MacWilliam AS, Holt PG. Mucosal dendritic cells in the respiratory tract. Mucosal Immunol Update 1997; 5:21–25.

66. Newberry RD, Stenson WF, Lorenz RG. Cyclooxygenase-2-dependent prostaglandin E2 production by stromal cells in the murine small intestine lamina propria: directing the tone of the intestinal immune response. J Immunol 2001; 166:4465–4472.

67. Harizi H, Jusan M, Pitard V, Moreau JF, Gualde N. Cyclooxygenase-2- issued prostaglandine E2 enhances the production of endogenous IL-10, which down-regulates dendritic cell functions. J Immunol 2002; 168:2255–2263.

68. Uhlig HH, Powrie F. Dendritic cells and the intestinal bacterial flora: a role for localized mucosal immune responses. J Clin Invest 2003; 112:648–651.

69. Wold AE, Hanson LA. Defense factors in human milk. Curr Opin Gastroenterol 1994; 10:652–658.

70. Mastretta E, Longo P, Laccisaglia A, et al. Effect of *Lactobacillus* GG and breast-feeding in the prevention of rotavirus nosocomial infection. J Pediatr Gastroenterol Nutr 2002; 35:527–531.

71. Goldman AS. Modulation of the gastrointestinal tract of infants by human milk. Interfaces and interactions. An evolutionary perspective. J Nutr 2000; 130:426S–431S.

72. Moreau MC. Effet immunomodulateur des bactéries intestinales: rôle des bifidobactéries. J Pédiatr Puériculture 2001; 14:135–139.

73. Isolauri E, Juntenen M, Rautanen T, Sillanaukee P, Koivula T. A human *Lactobacillus* strain (*Lactobacillus casei* sp strain GG) promotes recovery from acute diarrhea in children. Pediatrics 1991; 88:90–97.

74. Kaila M, Isolauri E, Soppi E, Virtanen V, Laine S, Arvilommi H. Enhancement of the circulating antibody secreting cell response in human diarrhea by a human *Lactobacillus* strain. Pediatr Res 1992; 32:141–144.

75. Herias MV, Midtvedt T, Hanson LA, Wold AE. Increased antibody production against gut-colonizing. E. coli in the presence of the anaerobic bacterium *Peptostreptococcus*. Scand J Immunol 1998; 48:277–282.

76. Flo J, Goldman H, Roux ME, Massoud E. Oral administration of a bacterial immunomodulator enhances the immune response to cholera toxin. Vaccine 1996; 14:1167–1173.

77. Cebra JJ, Bos NA, Cebra ER, Kramer DR, Kroese FGM, Schrader CE. Cellular and molecular biologic approaches for analyzing the in vivo development and maintenance of gut mucosal IgA responses. In: Mestecky et al. eds. Advances in Mucosal Immunology. New-York: Plenum press, 1995:429–434.

78. Bona C, Bot A. Neonatal immunoresponsiveness. Immunologist 1997; 5:5–9.

79. Moreau MC, Gaboriau-Routhiau V, Guiard G, Bouley. Strain-dependent immunomodulatory properties of lactic acid bacteria: experimental data from *Bifidobacterium* strains and *Lactobacillus* strains. 17th International Congress of Nutrition, Modern aspects of nutrition, Vienna, 27–31 August 2001.

80. Maassen CBM, Van Holten-Neelen C, Balk F, et al. Strain-dependent induction of cytokine profiles in the gut by orally administered *Lactobacillus* strains. Vaccine 2000; 18:2613–2623.

81. Wannemuehler MJ, Kiyono H, Babb JL, Michalek SM, McGhee JR. Lipopolysaccharide (LPS) regulation of the immune response: LPS converts germfree mice to sensitivity to oral tolerance induction. J Immunol 1982; 129:959–965.

82. Moreau MC, Gaboriau-Routhiau V. The absence of gut flora, the doses of antigen ingested, and aging affect the long-term peripheral tolerance induced by ovalbumin feeding in mice. Immunol Res 1996; 147:49–59.

83. Moreau MC, Gaboriau-Routhiau V, Dubuquoy C, Bisetti N, Bouley C, Prevoteau H. Modulating properties of intestinal bacterial strains, *Escherichia coli*, and *Bifidobacterium*, on two specific immune responses generated by the gut, i.e., oral tolerance to ovalbumin, and intestinal IgA anti-rotavirus response, in gnotobiotic mice. The 10th International Congress of Immunology, New-Dehli. In: Talwar GP, Nath I, Ganguly NK, Rao KVS, eds. Bologna: Monduzzi Editore, 1998:407–411.

84. Sudo N, Sawamura SA, Tanaka K, Aiba Y, Kubo C, Koga Y. The requirement of intestinal bacterial flora for the development of an IgE production system fully susceptible to oral tolerance induction. J Immunol 1997; 159:1739–1745.

85. Lodinova-Zadnikova R, Cukrowska B, Tlaskalova-Hogenova H. Oral administration of probiotic *Escherichia coli* after birth reduces frequency of allergies and repeated infections later in life (after 10 and 20 years). Int Arch Allergy Immunol 2003; 131:209–211.

86. Duval-Iflah Y, Ouriet MF, Moreau MC, Daniel N, Gabilan JC, Raibaud P. Implantation précoce d'une souche de *Escherichia coli* dans l'intestin de nouveau-nés humains: effet de barrière vis-à-vis de souches de *E. coli* antibiorésistantes. Ann Microbiol (Inst Pasteur), 133A 1982; 133A:393–408.

87. Kalliomaki M, Salminen S, Poussa T, Arvilommi H, Isolauri E. Probiotics and prevention of atopic disease: 4-year follow-up of a randomised placebo-controlled trial. Lancet 2003; 361:1869–1871.

88. McMenamin C, McKersey M, Kuhnlein P, Hunig T, Holt PG. Gamma-delta T cells down-regulate primary IgE responses in rats to inhaled soluble protein antigens. J Immunol 1995; 154:4390–4394.

89. Neish AS, Gewirtz A, Zeng H, et al. Procaryotic regulation of epithelial responses by inhibition of IκB-αubiquition. Science 2000; 289:1560–1563.

90. Haller D, Russo MP, Sartor RB, Jobin C. IKK beta and phosphatidylinositol 3-kinase/Akt in non-pathogenic Gram negative enteric bacteria-induced ReIA phosphorylation and NF-κB activation in both primary and intestinal epithelial cell lines. J Biol Chem 2002; 277:38168–38178.

91. Gionchetti P, Amadini C, Rizello F, Venturi A, Poggioli G, Campieri M. Probiotic for the treatment of postoperative complications following intestinal surgery. Best Pract Res Clin Gastroenterol 2003; 17:821–831.

92. Guslandi M, Giollo P, Testoni PA. A pilot trial of *Saccharomyces boulardii* in ulcerative colitis. Eur J Gastroenterol Hepatol 2003; 15:697–698.

93. Rembacken BJ, Snelling AM, Hawkey P, Chalmers DM, Axon TR. Non pathogenic *Escherichia coli* versus mesalazine for the treatment of ulcerative colitis: a randomised trial. Lancet 1999; 354:635–639.

94. Madsen K, Doyle JS, Jewell LD, Tavernini M, Fedorak RN. *Lactobacillus* species prevents colitis in interleukin-10 gene-deficient mice. Gastroenterology 1999; 116:1107–1114.

95. Ridge JP, Fuchs EJ, Matzinger P. Neonatal tolerance revisited: turning on newborn T cells with dendritic cells. Science 1996; 271:1723–1726.

96. Sun CM, Fiette L, Tanguy M, Leclerc C, Lo-Man R. Ontogeny and innate properties of neonatal dendritic cells. Blood 2003; 102:585–591.

97. Nicaise P, Gleizes A, Sandre C, et al. The intestinal microflora regulates cytokine production positively in spleen-derived macrophages but negatively in bone marrow-derived macrophages. Eur Cytokine Net 1999; 10:365–372.

98. Nicaise P, Gleizes A, Forestier F, Sandre C, Quero AM, Labarre C. The influence of *E. coli* implantation in axenic mice on cytokine production by peritoneal and bone marrow-derived macrophages. Cytokine 1995; 7:713–719.

99. Podoprigora G. The role of microbial factors in non-specific resistance of the host to infection. Microecol Ther 1996; 24:207–217.

100. Adkins B, Bu YR, Guevara P. Murine neonatal CD4+ lymph nodes are highly deficient in the development of antigen-specific Th1 function in adoptive adult hosts. J Immunol 2002; 169:4998–5004.

101. Renz H, vonMutius E, Illi S, Wolkers F, Hirsh T, Wieland SK. T(H)1/T(H)2 immune responses profiles differ between atopic children in eastern and western Germany. J Allergy Clin Immunol 2002; 109:338–342.

102. Marodi L. Down-regulation of Th1 responses in human neonates. Clin Exp Immunol 2002; 128:1–2.

103. Riedler J, Braun-Fahrländer C, Eder W, et al. The ALEX study team. Exposure to farming in early life and development of asthma and allergy: a cross-sectional survey. Lancet 2001; 358:1129–1133.

104. McKeever TM, Lewis SA, Smith C, et al. Early exposure to infections and antibiotics and the incidence of allergic disease: a birth cohort study with the West Midlands general practice research database. J Allergy Clin Immunol 2002; 109:43–50.

105. Bjorksten B, Naaber P, Sepp E, Mikelsaar M. The intestinal microflora in allergic Estonian and Swedish 2-year-old children. Clin Exp Allergy 1999; 29:342–346.

106. Bowman LM, Holt PG. Selective enhancement of systemic Th1 immunity in immunologically immature rats with an orally administered bacterial extract. Infect Immun 2001; 69:3719–3727.

107. Oyama N, Sudo N, Sogawa H, Kubo C. Antibiotic use during infancy promotes a shift in the Th1/Th2 balance towards Th2-dominant immunity in mice. J Allergy Clin Immunol 2001; 107:153–159.

108. Sudo N, Yu XN, Aiba Y, et al. An oral introduction of intestinal bacteria prevents the development of a long-term Th2-skewed immunological memory induced by neonatal antibiotic treatment in mice. Clin Exp Allergy 2002; 32:1112–1116.

109. Braun-Fahrlander C, Riedler J, Herz U, et al. Environmental exposure to endotoxin and its relation to asthma in school-age children. N Engl J Med 2002; 347:869–877.

110. Strannegard O, Strannegard IL. The causes of the increasing prevalence of allergy: is atopy a microbial privation? Allergy 2001; 56:91–102.

111. Von Mutius E. Environmental factors influencing the development and progression of pediatric asthma. J Allergy Clin Immunol 2002; 109:525S–532S.

112. Cukrowska B, Lodinova-Zadnikova R, Enders C, Sonnenborn U, Schulze J, Tlaskalova-Hogenova H. Specific proliferative and antibody responses of premature infants to intestinal colonization with non-pathogenic E. coli strain Nissle 1917. Scand J Immunol 2002; 55:204–209.

113. Avrameas S. Natural antibodies: from "horror autotoxicus" to "gnothi seauton". Immunol Today 1991; 123:154–159.

114. Bos NA, Meeuswen CG, Wostmann BS, Pleasants JR, Benner R. The influence of exogenous stimulation on the specificity repertoire of background immunoglobulin-secreting cells of different isotypes. Cell Immunol 1988; 112:371–380.

115. Freitas AA, Viale AC, Sunblad A, Heusser C, Coutinho A. Normal serum immunoglobulins participate in the selection of peripheral B-cell repertoires. PNAS 1991; 88:5640–5644.

116. Kaveri SV, Lacroix-Desmazes S, Mouthon L, Kazatchine MD. Human natural antibodies: lessons from physiology and prospects for therapy. Immunologist 1998; 6:227–233.

117. Van der Broek MF, Van Bruggen MCJ, Koopman JP, Hazenberg MP, Van den Berg WB. Gut flora induces and maintains resistance against streptococcal cell wall-induced arthritis in F344 rats. Clin Exp Immunol 1992; 88:313–317.

118. Nieuwenhuis EES, Visser MR, Kavelaars A, et al. Oral antibiotics as novel therapy for arthritis. Evidence for a beneficial effect of intestinal E. coli. Arthritis Rheum 2000; 43:2583–2589.

119. Martins TC, Aguas A. Mechanisms of *Mycobacterium avium*-induced resistance against insulin-dependant diabetes melitus (DDM) in non-obese diabetic (NOD) mice: role of Fas and Th1 cells. Clin Exp Immunol 1999; 115:248–254.

120. Gavini F, Cayuela C, Antoine JM, et al. Differences on the distribution of bifidobacterial and enterobacterial species in human faecal microflora of three different (children, adults, elderly) age groups. Microb Ecol Health Dis 2001; 13:40–45.

6

Mucosal Interactions and Gastrointestinal Microbiota

Wai Ling Chow and Yuan-Kun Lee
National University of Singapore, Department of Microbiology,
Faculty of Medicine, Singapore

INTRODUCTION

The human gut harbors a complex and diverse microbiota. The numbers of microorganisms in the upper gastrointestinal (GI) tract are kept low by the actions of gastric acid, pancreatic enzymes, bile, and a propulsive motor pattern. The colonic population of microbes is estimated to be 10^{12} organisms/gram with at least 400 possible species. The above figure was obtained by traditional culture-based methods. Modern molecular methods such as 16S ribosomal RNA clone libraries that are discussed in Chapter 1 indicate that the number of species will be even higher. The composition of the intestinal microbiota varies from human to human. These differences in the composition of the microbiota are affected by physiological, chemical, and environmental factors. The common intestinal microbiota in humans includes predominantly members of genera *Clostridium*, *Eubacterium*, *Bacteroides*, *Atopobium* and *Bifidobacterium* spp. and many others to a lesser extent. There is an approximation that almost 90% of the cells in our body are microbial, whereas only 10% are human.

The bacteria that colonize the gut must be able to proliferate at a rate that resists washout. Adherence to the intestinal mucosal surface is an important factor in intestinal bacterial colonization. In healthy individuals, a layer of mucus is found to line the gut. It is composed mostly of glycoproteins and serves as a lubricant and a protective lining over the mucosa. Microbiota degradation of the mucin polymeric glycoprotein results in the release of monosaccharides such as N-acetylglucosamine and fucose amongst others, which the microbiota use to support their growth (2). Furthermore, under the mucus the surfaces of intestinal epithelial cells are covered with an abundance of terminally fucosylated glycoproteins and glycolipids which are induced by members of the intestinal microbiota (3). In particular, it was demonstrated that *Bacteroides thetaiotaomicron* cleaves L-fucose moieties from the host's surface and internalizes them for use as an energy source. This commensal microbe modulates the production of the fucose by the host with its requirement needs, which gives it a competitive colonization advantage within the intestinal niche (68). Thus, the interaction of microorganisms with the mucosa is a complex one, which involves cross-talk between the microbes, and between the microbes and the host.

In this chapter, we provide some insights about the development and regulation of the gastrointestinal microbiota as well as the interaction of the microbes with the intestinal mucosal layer. The majority of research on the molecular interactions between microbes and the mucosa relate to pathogen-enterocyte interaction, and consequently, this field is also occasionally referred to.

FEATURES OF THE GASTROINTESTINAL TRACT

Structure and Function of the Small Intestine

The small intestine is the principal site of food digestion, nutrient absorption as well as endocrine secretion. It is the longest component of the alimentary tract, measuring over 6 meters, and is divided into three anatomic regions: duodenum, jejunum and ileum. The duodenum begins at the pylorus of the stomach and is the proximal 20–25 cm of the small intestine. The jejunum spans about 2.5 meters in length. The ileum is approximately 3.5 meters long and an extension of the jejunum.

The absorptive surface area of the small intestine is greatly increased by tissue and cell specializations such as plicae circulares, villi and microvilli (Fig. 1). Plicae circulares are permanent transverse folds of the mucosa, forming semicircular or spiral elevations. They are abundant in the distal duodenum and beginning of the jejunum. Intestinal villi are finger-like outgrowths of mucosa protruding into the lumen of the small intestine. Microvilli are protrusions of the apical plasmalemma of the epithelial cells covering the intestinal villi, increasing the surface area of the small intestine 20 times. Therefore, these modifications immensely amplify the absorptive and interactive (with intestinal content, including the microbiota) surface area of the small intestine.

The mucosa comprises the lining epithelium, a lamina propria that houses glands and muscularis mucosa. There are at least 5 types of cells found in the intestinal mucosal

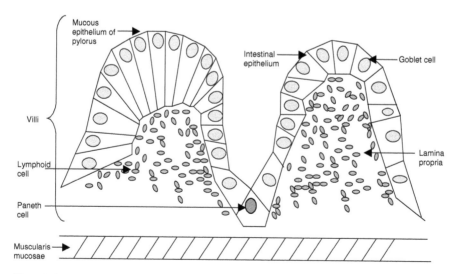

Figure 1   Schematic diagram of the mucosa, villi, and component cells of the small intestine.

epithelium. They include enterocytes, goblet cells, Paneth cells, enteroendocrine cells and M cells (microfold cells). Both the enterocytes and the goblet cells line the villus and are the major cell types in the epithelium. The enterocytes are columnar in shape and have brush borders composed of microvilli which help to enhance the water ions and nutrient absorbing surface area. Goblet cells are unicellular mucin-secreting glands which produce mucinogen and mucin, a component of mucus. The number of goblet cells increases progressively down the gastrointestinal tract from the duodenum, to jejunum, ileum and colon, where they are most abundant. The Paneth cells' role is to maintain the innate immunity by secreting antimicrobial substances such as α-defensins (4,69). Enteroendocrine cells are present only in small numbers (~ 1%) and their functions include the production of panacrine and endocrine hormones (5). M cells are modified enterocytes overlying the enlarged lymphatic nodules in the lamina propria. Their function is to phagocytose and transport antigens present in the intestinal lumen to the underlying macrophages and lymphoid cells, which then migrate to other compartments of the lymphoid nodes, where immune responses to foreign antigens are initiated (5).

The lamina propria is rich in lymphoid cells, which will protect the intestinal lining from bacterial invasion. The loose connective tissue of lamina propria forms the main part of the villi, extending down to the muscularis mucosa. The epithelium may invaginate into the lamina propria between the villi to form glands, termed the crypts of Lieberkühn. These tubular glands consist of enterocytes, goblet cells, regenerative cells, enteroendocrine cells and Paneth cells. The rate at which the regenerative cells proliferate is high and they are capable of replacing other cell types in the intestinal epithelium. As mentioned above, the pyramidal-shaped Paneth cells secrete antibacterial agents, such as lysozyme and α-defensins or cryptdins, and internalized extracellular matter such as bacteria and immunoglobulin. Therefore, it is postulated that these cells help in regulation of the bacterial microenvironment in the small intestine.

Structure and Function of the Large Intestine

The large intestine is a continuation of the ileum and is usually divided into three regions: the colon, rectum and anal canal. The colon accounts for nearly the full length of the large intestine. The colon absorbs water and electrolytes (approximately 1400 ml per day). It also compacts and eliminates feces (about 100 ml per day). Feces are composed of water (75%), dead bacteria (7%), roughage (5%), inorganic substances (5%), and undigested protein, dead cells and bile pigment (1%). Bacterial products, including the vitamins riboflavin, thiamin, vitamin B12 and vitamin K, are also excreted in the feces (5).

The colonic mucosal membrane does not have any folds due to an absence of villi (Fig. 2). The intestinal glands are long and characterized by a great abundance of goblet and absorptive cells, and a small number of enteroendocrine cells. The large intestinal epithelium is specialized for mucos secretion, salt and water absorption.

The histology of the rectum is identical to that of the colon except that the crypts of Lieberkühn are deeper and fewer in number. The rectum is about 12–18 cm in length and is continuous with the anal canal, which spans about 3 to 4 cm. The mucosa of the anal region displays a series of longitudinal folds, the rectal columns of Morgagni. These rectal columns meet one another to form pouch-like outpocketings, the anal valves with intervening anal sinuses. The anal valves assist in supporting the column of feces (5). The epithelial cells of the entire gastrointestinal tract are constantly shed. They are replaced with stem cells that have undergone mitosis. The high turnover rate of the epithelial cells may explain why the small intestine is affected rapidly by the administration of

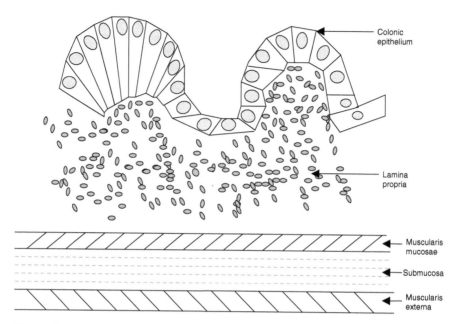

Figure 2   Schematic diagram of the colonic epithelium and associated cells.

anti-mitotic drugs, as in cancer chemotherapy. The epithelial cells continue to be lost at the tip of the villi, but drugs inhibit cell proliferation (6).

Mucus

The gastrointestinal tract contains tremendous numbers of microorganisms and some of these microorganisms are pathogenic in nature under certain conditions. Therefore a function of the mucus is to protect the underlying epithelial cells by keeping the microbes and toxins at bay, on the outer mucosal surfaces. The mucus layer is comprised of various mucosal secretions including mucins, trefoil peptides, and surfactant phospholipids.

Mucus occurs in two distinct physical forms: (1) a thin layer of stable, water insoluble mucus gel firmly adhering to the gastroduodenal mucosal surface, (2) and as soluble mucus which is quite viscous but mixes with the luminal juice (7).

The layer of mucus that is bound to the surface of the gastrointestinal tract is resistant to its removal from the mucosa. It is approximately 50–450 μm thick in humans and about twofold less in rats. This adherent mucus functions to support and define the mucosal ecosystem since it is the outermost sensory "organ" of the mucosal immune system. The mucus gel plays a role in providing surface neutralization by having the HCO_3^- barrier to the gastric acid. The surfactant lipids maintain surface hydrophobicity on the mucus. The adherent mucus also serves as a stable protective barrier that prevents the entry of luminal pepsin to the underlying epithelial cells.

The soluble mucus plays a role in maintaining the protective barrier because it is not physically attached to the mucosa and can be removed from the mucosa by gentle washing. Due to the viscous nature of the soluble mucus, the soluble mucus makes an excellent lubricant which allows easy movement of solid material in the lumen. This helps to prevent the damage to the underlying epithelial cells as well as minimize the tearing of the adherent layer of mucus from the mucosal surface (7).

The main structural component of the mucus layer are the mucins or glycoproteins of molecular weight ranging from one to several million daltons. When concentrated, these glycoprotein macromolecules ($M_r \geq 2 \times 10^6$) polymerise to form gels. Mucin molecules consist of carbohydrate side chains (70–80%) bound to a protein skeleton. The O-linked oligosaccharide chains contain a restricted number of monosaccharides, including galactose, fucose, N-acetylgalactosamine, N-acetylglucosamine and often terminated with sialic acids or sulfate groups, which account for the polyanionic nature of mucins at a neutral pH (7,8). Oligosaccharides chains are successively added on to mucins specifically by membrane bound glycosyltransferases. The biochemistry of the intestinal mucins confers their protective nature: the protein backbone has a high O-linked oligosaccharide content (>80% carbohydrate by mass) that provides lectin-binding capacity, whereas the ability of the protein core to form multimers (through disulphide bonds) causes polymerization into gels and bestows viscoelasticity and lubrication (9). The trefoil peptides also facilitate the mucins to confer visoelasticity on the mucus (10).

The composition of the mucus is constantly regulated by the varying secretion rates of the mucin types, ions, lipids, proteins and water. The variation in the composition of the mucus is also dependent on the development stage of the host as well as the host's diet and the interaction of the commensals and pathogens (10). Commensals rapidly colonize the individual soon after birth and some play a role in inhibiting the growth of pathogenic bacteria. However, many commensals are capable of becoming opportunistic pathogens by overgrowing when the stable gastrointestinal ecosystem is disturbed. Thus, the mucus has to be continuously secreted and then shed, discarded, digested or recycled. This form of protective mechanism keeps the numbers of both pathogens and commensals in check by blocking the bacterial adherence to the epithelial cells.

MICROBIOTA AND GASTROINTESTINAL SYSTEM

Distribution of Microbiota

The mucosal surface of the human body, including the gastrointestinal tract, the respiratory tract and the urogenital tract, has a total surface area of more than 400 m^2 (11). The gastrointestinal tract's surface area is about 200–300 m^2 and is colonized by 10^{13-14} bacteria with hundreds of bacterial species and subspecies.

The normal microbiota of the gastrointestinal tract has been grouped and defined into two categories, the autochthonous (indigenous) and the allochthonous (nonindigenous) species (12). The autochthonous microbes (1) are always present in the normal adult's gastrointestinal tract, (2) play a role in maintaining the stable bacterial populations in the gastrointestinal tract, (3) colonize particular parts of the tract, (4) can grow anaerobically, (5) colonize their habitats in succession in infants, and (6) often associate intimately with the gastrointestinal mucosal epithelium.

On the other hand, allochthonous species are not characteristic of the normal habitat. Allochthonous microbiota is defined as transient microbes which will not be established but would just be passing through, having arrived in the habitat in food, in water, from another habitat in the gastrointestinal tract, or from elsewhere in the body. These microbes either cannot or find it very challenging to establish themselves since they cannot compete in the various niches or may be killed by host or bacterial factors.

However, the allochthonous microorganisms might colonize the habitats vacated by the autochthonous microbes in the disturbed gastrointestinal system (13). This was evidently seen in the administration of antibiotics which caused severe disturbance in the gastrointestinal microbiota leading to undesirable effects, such as the overgrowth and superinfection with allochthonous microorganisms like yeast (14,15); see also chapter 18 by Sullivan and Nord in this book.

Thus, the main difference between autochthonous and allochthonous species is that an autochthonous microbe naturally colonizes the habitat, whereas an allochthonous one cannot colonize it except under abnormal or atypical situations (13).

In a steady gastrointestinal ecosystem, all the niches are probably occupied by indigenous microbes. The number of microorganisms in the stomach and the upper two-thirds of the small intestine is very scarce: a maximum of 10^4 per milliliter of intestinal contents. The relatively low number of microbes is due to the low pH (approximately pH 2) of the intestinal contents resulting from gastric acid production and the relatively swift flow (transit time of 4–6 hours) of digesta through the stomach and small intestine. Culturing studies indicate that lactobacilli and streptococci are commonly found microbes in the small intestine (16). Unlike the bulk of the microbes within the gastrointestinal tract, both the lactobacilli and streptococci are acid-tolerant bacteria, and are capable of surviving the passage through the stomach.

The ileum contains larger numbers of microbes (10^8–10^9 bacteria per ml of intestinal contents) in comparison to the upper regions of the gastrointestinal tract. The higher bacterial numbers in the ileum are the result of a lower peristalsis and low oxidation-reduction potential. Therefore, lactobacilli, streptococci, enterobacteriacae and anaerobic bacteria are able to establish themselves in the distal region of the small intestine. The main site of microbial colonization in the gastrointestinal tract is the colon. The slow intestinal motility in the colon with a transit time of up to 60 hours and low oxidation-reduction potential are responsible for the large numbers of bacteria present. The colon contains 10^{11}–10^{12} bacteria per gram of intestinal contents. More than 99% of the colonic microbiota are obligate anaerobes such as *Bacteroides* spp., *Eubacterium*, *Bifidobacterium and Clostridium* spp. (17).

Enteric Pathogens

Most intestinal bacterial infections are caused by enteric pathogens. The clinical symptoms usually associated with the intestinal infections include fever, abdominal pain and diarrhea. Enteric bacteria are capable of evading host defense factors such as gastric acidity, intestinal motility, the normal indigenous microbiota, mucus secretion, and specific mucosal and systemic immune mechanisms.

In order for ingested pathogenic bacteria to infect the colon, they produce virulence factors. Enteric bacteria can be divided into four main categories based on the virulence factors that enable them to overcome the host defense. The first group of bacterial pathogens consists of *Campylobacter jejuni, Yersinia enterocolitica, Shigella* and *Salmonella* species. Their mechanism of virulence involves the mucosal invasion with intraepithelial cell multiplication resulting in cell death. The second group comprises enteric pathogens that produce cytotoxins which will in turn cause cell injury and inflammation. Microorganisms that produce cytotoxins include *Clostridium difficile*, enteropathogenic *Escherichia coli* (EPEC) and enterohemorrhagic *E. coli* (EHEC). The third class of pathogens secretes enterotoxins which will alter intestinal salt and water balance without affecting mucosal morphology. *Vibrio cholerae, Shigella* and enterotoxigenic *E. coli* produce such enterotoxins. The last category of enteric pathogens can only cause disease when they tightly adhere to the intestinal surface. The classic enteropathogenic *E. coli* as well as the enteroadherent *E. coli* is typical of this group. Both the small intestine and colon are primary sites for enteroadhesion (18).

DEVELOPMENT OF GI TRACT NORMAL MICROBIOTA IN HUMANS

The fetus in utero is sterile until birth. Colonization of the human body with a heterogenous collection of microorganisms from the birth canal begins at delivery. The *Lactobacillus* species constitute the major population of the vaginal microbiota and thus provide the initial inoculum to the infant during birth. In the case of caesarean section or premature infants, most microbes that are transferred to the newborn can be traced from the environment, i.e., from other infants via the air, equipment and nursing staff (19). Therefore, the type of delivery (passage through the birth canal versus caesarean section) as well as the type of diet (breast versus formula feeding) might affect the pattern of microbial colonization.

The general pattern observed was that the facultative microorganisms appeared first and were subsequently followed by a limited number of anaerobes during the first two weeks (20). The types of bacterial strains that are capable of populating the GI tract are regulated through the limitation of the intestinal milieu, which changes with the successive establishment of the different bacteria. Hence, bacteria that are capable of oxidative metabolism, such as enterobacteria, streptococci and staphylococci, are among the first to proliferate in the gut. As the numbers of the facultative bacteria increase, they consume oxygen and lower the redox potential to negative values. These conditions are favorable for the anaerobic bacteria to multiply and reach much higher levels than that of the first week. Populations of bifidobacteria, *Bacteroides* and clostridia, the commonly found anaerobes, increase with subsequent change of conditions in the GI tract. By the fourth week, the fecal microbiota of the breast-fed infants consists mainly of bifidobacteria and other groups to a lesser extent including enterobacteria, clostridia, and Bacteroides. However, in formula-fed infants, bifidobacteria do not beome so dominant and a more complex microbiota develops. The differences between the breast-fed and formula-fed infants gradually disappear with the intake of solid food. By the twelfth month, the number of facultative anaerobes declines as the anaerobes begin to increase and form a stable population, resembling that of adults in numbers and in composition. By the age of two, the profile resembles that of an adult (19). In adults, the ratio of anaerobic to aerobic bacteria is 1000:1 (21).

Adhesion of Bacteria

The colonization of microorganisms in various niches is dependent on their ability to adhere to surfaces and substratum. Adhesion or adherence is defined as the measurable union between a bacterium and substratum. A bacterium is considered to have adhered to a substratum when energy is required to separate the bacterium from the substratum (22).

Adhesion of a bacterium to a substratum, its colonization and finally possible invasion of the tissue is a multi-step process. It usually involves two or more kinetic steps. Firstly, the bacterium approaches the substratum via long distance interactions, such as van der Waals forces and electrostatic forces and becomes loosely attached (22). Complementary adhesion-receptor interaction leads to the formation of a bacterium-cell complex:

$$\text{Bacteria + Intestinal cell} \underset{k_{-1}}{\overset{k_1}{\rightleftarrows} } \text{Bacterium} - \text{Intestinal cell complex} \qquad (1)$$

where k_1 and k_{-1} are dissociation constants for the above reaction. At equilibrium, the concentration of the adhered bacteria (e_x) can be expressed as:

$$e_x = e_m \cdot x/(k_x + x) \qquad (2)$$

where e_m is the maximum value of e_x at saturated bacterial concentration (23). The value of e_m is equivalent to the concentration of adhesion sites on the mucosal surface and x is the concentration of bacterial cells present around the adhesion site. The dissociation constant, k_x determines the affinity the bacterial cells have for the adhesion sites on the mucosal surfaces. Thus, the adhesion of a bacterium to the substratum is determined by two major properties: the concentration of the bacterium in the vicinity of the cell receptor (x in the above equation) and the affinity of the bacterium for the receptor (k_x in the equation).

Bacterial adhesion is crucial for invasive pathogenic microbes and may be important for certain commensals, prior to colonization of the intestinal mucosa. The receptors for bacterial adhesins are found in three groups of membrane consitituents: integral, peripheral and cell surface coat components. These receptors are chemically proteins, glycoproteins or glycolipids. They fulfill the criteria of a biological receptor because they exhibit specific binding followed by physiologically relevant responses. An example would be membrane-associated fibronectin acting as a receptor molecule for streptococci (22).

Bacterial adhesion to substrata receptors could involve the specific adhesin-receptor interaction and non-specific interactions. The specific adhesion is defined as the association between the bacteria and substratum that requires rigid stereochemical constraints (22). Many bacteria have the ability to produce lectins (24), carbohydrate-specific proteins, which are usually expressed on the bacterial surfaces. Lectins are a subset of adhesins that recognize and bind to a defined carbohydrate sequence present on host glycoproteins. Previous studies reported that there were three main types of adhesin-receptor interactions. The first type was based on the carbohydrate-lectin recognition, the second kind involved protein-protein interaction and the third class, which is the least characterized, involved the binding interactions between hydrophobic moieties of proteins and lipids (25). A well-established example is the type 1 fimbriae (carrying adhesins) of *E. coli* which recognize D-mannose as the receptor site on the host mucosal surface (26). Binding of some *Lactobacillus* to human colonic cells is a mannose-specific adherence mechanism (27,28). Their similarity in binding specificity may contribute to competitive exclusion of enteropathogens by some strains of probiotic lactic acid bacteria. Lactic acid bacteria have been shown to exclude enteropathogens from the mucosal surface in in vitro studies (29–32).

On the other hand, the non-specific adhesion is also an association between a bacterium and substratum that may involve the same forces involved in the specific adhesion. However, in non-specific adhesion, a precise stereochemical fit is not necessary. Non-specific interaction comprises the physiochemical forces such as van der Waals, electrostatic forces (33), hydrogen bonding (34), and hydrophobic interactions (35).

The synthesis of adhesins can be switched on and off by the bacteria, depending on the environmental conditions, a process called phase variation (36). Phase variation has been demonstrated in Gram-negative bacteria. However, the environmental regulation of adhesin expression is likely to be present in some commensal and lactic acid bacteria also, since bacteria that are unable to regulate their adhesin expression are often inefficient colonizers (37,38). It has been suggested that the mucosal adhesive properties of the lactic acid bacteria is strain and host dependent, and the mucosal binding of human lactic acid bacteria are strain- and host specific (39,40). The adhesion and colonization of bifidobacteria have been suggested to be disease (allergy, cancer) dependent (41,42). The adhesion to the intestinal mucus of the fecal bifidobacteria from healthy infants was significantly higher than for allergic infants, suggesting a correlation between allergic disease and the composition of the bifidobacteria (41). Surprisingly, bifidobacteria, amongst other bacteria, were generally positively associated with increased risk of colon cancer in a study involving native Japanese and African patients (42). The ability of intestinal bacteria to persist on the intestinal mucosal surface may ultimately be determined by their doubling time in the intestine to maintain a high local concentration. Slowly-dividing bacteria would be expected to be out-competed or washed-out with the intestinal contents (43).

CROSS-TALK BETWEEN BACTERIA AND INTESTINAL EPITHELIAL CELLS

As discussed in chapter 5, some ingested probiotic bacteria have shown immunomodulatory properties (44–46). Both commensal and pathogenic bacteria possess recognized structures named pathogen-associated molecular patterns (PAMPS). These recognized structures are essential for the microbe, mostly constitutively expressed and shared by the same group of microorganisms. PAMPS that are characterized to date include N-formylated peptide (47), lipopolysaccharides (LPS) (48), and lipopeptides (49), more recently described PAMPS are flagellin (50) and unmethylated segments of CpG DNA (51). Even though unmethylated segments of CpG DNA are not a cell surface structure, it serves to differentiate the microorganism from the host. Therefore, they epitomize the ideal targets for the innate immune system to identify the presence of infectious agents with a limited numbers of receptors.

The best studied of the PAMPS is the glycolipid LPS, an important component of the outer membrane of Gram-negative bacteria. LPS is recognized by Toll-like receptor (TLR) 4, the first described member of the family of transmembrane TLR molecules that play a central role in the transcription activation of host defense mechanisms, such as chemokine and cytokine secretion, and the expression of costimulatory molecules (52). TLRs are transmembrane receptors defined by the presence of leucine-rich repeats in the extracellular portion of the molecule and a Toll/IL-IR/resistance (TIR) cytoplasmic domain. The extracellular leucine-rich repeats are thought to function in ligand recognition, whereas the TIR domain works in signaling. Leucine-rich repeat domains are common to proteins that are involved in the recognition of foreign proteins. There are currently 10 identified members of the mammalian TLR family (52). From recent publications (53), it has been shown that some types of intestinal epithelial cells express TLR 4.

Upon activation of TLR 4 by LPS, a series of events lead to the activation of ubiquitin ligase TRAF6 by a unique self-polyubiquitination reaction. TRAF6 then activates the TAK1 complex (54). This step leads to the phosphorylation and activation of mitogen-activated protein kinase and the inhibitor κB kinase (IKK) complex (54,55). The IKK complex comprises two kinases, IKKα and IKKβ, and one protein, NEMO. When activated, IKKβ phosphorylates IκBα, triggering its polyubiquitination and degradation (56,57). In the unstimulated state, the IκBα interacts and traps NFκβ in the cytosol. Degradation of IκBα releases the NFκβ to translocate into the nucleus and to activate proinflammatory and prosurvival gene expression. Therefore, TLR 4 activates multiple signaling pathways which will eventually lead to the production of cytokines and other factors to protect the host against infection (58). The expression level of TLR 4 in the intestine of patients with inflammatory bowel disease was found to be strongly up-regulated compared to the TLR 4 expression in healthy individuals.

As for the other PAMPS such as N-formylated peptides, the cell surface receptors that recognized them are the heterotrimeric G-protein coupled receptors (59). N-formylated peptides play an important role in recruiting and activating inflammatory cells (60). They will eventually activate the NFκβ pathway the same way as the TLR.

On the other hand, enteric pathogens have also evolved mechanisms to evade the immune recognition and defense. *Helicobacter pylori*, the etiological agent of gastritis and stomach cancer, expresses hypoacylated LPS to avoid recognition by the human TLR4/MD2 module (61). Other pathogens like *Yersinia pseudotuberculosis* have developed ways to down-regulate TLR 4 signaling by injecting proteins to abolish the signaling leading to NFκβ activation (52).

At the beginning of the chapter, we mentioned that the gastrointestinal tract is colonized by huge, complex and dynamic populations of microorganisms. Hence, the molecular pattern recognition of the epithelial cells of the gastrointestinal mucosa needs to be tightly regulated so as to avoid an extreme immune response and uncontrolled inflammatory reaction. The exact mechanism by which they do this still remains to be elucidated. However, recent studies have shed light into this area of interest. The mechanism by which one TLR, TLR 5, achieved this feat is due to the fact that gut epithelial cells express TLR 5 only on their basolateral surfaces. Therefore only those bacteria that breached the epithelial cells or have translocated flagellin across the epithelia will activate the receptor (62).

Using a gnotobiotic mouse model it was shown that *Bacteroides thetaiotaomicron* is able to induce the production of α-L fucose on intestinal epithelial cells via a regulator, FucR, as a molecular sensor of L-fucose availability (3,68). FucR coordinates expression of an operon encoding enzymes in the L-fucose metabolic pathway in the bacteria with expression of another locus that regulates production of fucosylated glycans in the intestinal enterocytes. By tightly coordinating presentation of host-derived fucose with the rate of fucose utilization, an excess of epithelial fucose is avoided. This may minimize the risk of encroachment by pathogens that use fucosylated glycans as receptors for their adhesins (69).

Certain pathogenic bacteria require intimate contact with the host to cause disease. *E. coli* (EPEC) is one such pathogen which requires intimate attachment to the host cells for maximum virulence to occur. There are a few factors which facilitate the cross-talk between the microorganism and the host epithelial cells and this involves the EPEC-secreted proteins, the type-three secretion system and the expression of outer membrane protein, intimin (64,65). The release of extracellular protein via the type-three secretion system is necessary for the formation of attaching lesions by EPEC. The attachment of bacteria is by means of intimin binding to a 90 kDa tyrosine phoshorylated protein in the host membrane. This receptor is known as translocated intimin receptor (Tir) and is of bacterial origin; it is translocated on to the host membrane where its tyrosine residues become phosphorylated and binds to intimin. Subsequent signal transduction events that occur within the host cells are the activation of protein kinase C, inositol triphosphate and calcium release. This leads to the formation of an actin-rich pedestal that forms a dome-like anchoring site for the bacteria which is an essential feature of EPEC pathogenesis (63).

There is evidence to suggest that in some strains of *Lactobacillus reuteri*, mucus-binding adhesion could be induced by the presence of mucin glycoproteins and solid substratum (66).

CONCLUSION

The gastrointestinal tract is a highly dynamic ecosystem where interaction of the microbiota with the host mucosa plays an important role. Thus, it not only functions to digest food and absorb nutrients; it is also the major site where communication between microbes, and also between microbiota and their host takes place.

Probiotics and prebiotics offer dietary means to support the balance of intestinal microbiota. They may be used to counteract local immunological dysfunctions, to stabilize the gut mucosal barrier function, to prevent infectious succession of pathogenic microorganisms or to influence intestinal metabolism. However, many of the proposed mechanisms still need to be validated in human clinical trials (67). Future research on commensal microbiota interactions with mucosal surfaces of the host should focus on the cross-talk and determining the signaling mechanisms involved.

REFERENCES

1. Mitsuoka T. The human gastroinstestinal tract. In: Brian JB Wood, ed. In: Lactic Acid Bacteria, Vol. 1. London and New York: Wood Publisher Elseivier Applied Science, 1992:69–114.
2. Hoskins LC. Mucin degradation in the human gastrointestinal tract and its significance to enteric microbial ecology. Eur J Gastroenterol Hepatol 1993; 5:205.
3. Bry L, Falk PG, Midtvedt T, Gordon JI. A model of host-microbial interactions in an open mammalian ecosystem. Science 1996; 273:1380–1383.
4. Ross MH, Kaye GI, Pawlina W. Digestive system III: esophagus and gastrointestinal tract. Histology: A Text and Atlas. Philadelphia: Lippincott Williams & Wilkins, 2003 pp. 474–531.
5. Gartner LP, Hiatt JL. Digestive system III— alimentary canal. Color Textbook of Histology. Philadelphia: W.B. Saunders Company, 1997 pp. 325–335.
6. Junqueria LC, Carneiro J, Contopoulos AN. Digestive tract. In: Junqueria LC, Carneiro J, Contopoulos AN, eds. Basic Histology: Text and Atlas. Los Altos. CA: Lange Medical Publications, 1998 pp. 272–303.
7. Allen A, Carroll NJ. Adherent and soluble mucus in the stomach and duodenum. Dig Dis Sci 1985; 30:55S–62S.
8. Forstner JF, Oliver MG, Sylvester FA. Production, structure, and biologic relevance of gastrointestinal mucins. In: Blaser MJ, Smith PD, Ravdin JI, Greenberg HB, Guerrant RL, eds. Infections of the Gastrointestinal Tract. New York: Raven Press, 1995:71–88.
9. Belley A, Keller K, Gottke M, Chadee K, Gottke M. Intestinal mucins in colonization and host defense against pathogens. Am J Trop Med Hyg 1999; 60:10–15.
10. Cone RA. Mucus. In: Orga PL, Mestecky J, Lamm ME, Strober W, Bienenstock J, McGhee JR, eds. Mucosal Immunology. New York: Academic Press, 1999:43–60.
11. Kraehenbuhl JP, Neutra MR. Molecular and cellular basis of immune protection of mucosal surfaces. Physiol Rev 1992; 72:853–879.
12. Savage DC. Microbial ecology of the gastrointestinal tract. Ann Rev Microbiol 1977; 6:155–178.
13. Alexander M. Microbial Ecology. New York: Wiley, 1971 pp. 3–21.
14. Nord CE, Hermdah IA, Kager L. Antimicrobial induced alteration of the human oropharyngeal and intestinal microflora. Scand J Infect Dis Suppl 1986; 49:64–72.
15. Trenschel R, Peceny R, Runde V, et al. Fungal colonization and invasive fungal infections following allogeneic BMT using metronidazole, ciprofloxacin and fluconazole or ciprofloxacin and fluconazole as intestinal decontamination. Bone Marrow Transplant 2000; 26:993–997.
16. Tannock GW. Normal Microflora: An Introduction to Microbes Inhabiting the Human Body. U.K.: Chapman and Hall, 1995.
17. Herbert MK, Holzer P, Roewer N. Problems of the Gastrointestinal Tract in Anesthesia, the Perioperative Period, and Intensive Care. New York; Berlin: Springer, 1999.
18. Cohen MB, Giannella RA. Bacterial infections: pathophysiology, clinical features and treatment. In: Philips SF, Permberton JH, Shorter RG, eds. The Large Intestine: Physiology, Pathophysiology and Disease. New York: Raven Press, 1991:395–428.
19. Conway PL. Microbial ecology of the human large intestine. In: Gibson GR, Macfarlane GT, eds. Human Colonic Bacteria: Nutrition, Physiology, and Pathology. Boca Raton: CRC, 1995:1–24.
20. Stark PL, Lee A. The microbial ecology of the large bowel of the breast-fed and formula-fed infants during the first year of life. J Med Microbiol 1982; 15:189–203.
21. Berg RD. The indigenous gastrointestinal microflora. Trends Microbiol 1996; 4:430–435.

22. Ofek I, Doyle R. Recent developments in bacterial adhesion to animal cells. In: Ofek I, Doyle R, eds. Bacterial Adhesion to Cells and Tissues. London: Chapman and Hall, 1994:321–512.

23. Lee YK, Lim CY, Teng WL, Ouwehand AC, Tuomola EM, Salminen S. Quantitative approach in the study of adhesion of lactic acid bacteria to intestinal cells and their competition with Enterobacteria. Appl Environ Microbiol 2000; 66:3692–3697.

24. Mirelman D, Ofek I. Microbial lectins and agglutinins. In: Mirelman D, ed. Microbial Lectins and Agglutinins. New York: J. Wiley and Sons, 1986:1–19.

25. Abraham SM, Sharon N, Ofek I. Adhesion of bacteria to mucosal surfaces. In: Orga PL, Mestecky J, Lamm ME, Strober W, Bienenstock J, McGhee JR, eds. Mucosal Immunology. New York: Academic Press, 1999:31–42.

26. Schembri MA, Kjaergaard K, Sokurenko EV, Klemm P. Molecular charcterisation of the *Escherichia coli* FimH adhesion. J Infect Dis 2001; 183:S28–S31.

27. Adlerberth I, Ahrné S, Johansson ML, Molin G, Hanson LÅ, Wold AE. A mannose-specific adherence mechanism in *Lactobacillus plantarum* conferring binding to the human colonic cell line HT-29. Appl Environ Microbiol 1996; 62:2244–2251.

28. Neesen J-R, Granato D, Rouvet M, Servin A, Teneberg S, Karlsson K-A. *Lactobacillus johnsonii* La1 shares carbohydrates-binding specificities with several enteropathogenic bacteria. Glycobiology 2000; 10:1193–1199.

29. Bernet MF, Brassart D, Neeser JR, Servin AL. *Lactobacillus acidophilus* LA1 binds to cultured human intestinal cell lines and inhibits cell attachment and cell invasion by enterovirulent bacteria. Gut 1994; 35:483–489.

30. Chauviere G, Coconnier MH, Kerneis S, Darfeuille-Michaud A, Joly B, Servin AL. Competitive exclusion of diarrheagenic Escherichia coli (ETEC) from human enterocyte-like Caco-2 cells by heat-killed *Lactobacillus*. FEMS Microbiol Lett 1992; 70:213–217.

31. Gopal PK, Prasad J, Smart J, Gill HS. In vitro adherence properties of *Lactobacillus rhamnosus* DR20 and *Bifidobacterium lactis* DR10 strains and their antagonistic activity against an enterotoxigenic *Escherichia coli*. Int J Food Microbiol 2001; 67:207–216.

32. Tuomola EM, Ouwehand AC, Salminen SJ. The effect of probiotic bacteria on the adhesion of pathogens to human intestinal mucus. FEMS Immunol Med Microbiol 1999; 26:137–142.

33. Busscher HJ, Weerkamp AH. Specific and non-specific interactions in bacterial adhesion to solid substrata. FEMS Microbiol Rev 1987; 46:165–173.

34. Pimentel GC, McClellan AL. In: Freeman WH, ed. The Hydrogen Bond. San Francisco; New York: Reinhold Pub. Corp, 1960.

35. Reid G, Servin AL, Bruce AW, Busscher HJ. Adhesion of three lactobacillus strains to human urinary and intestinal epithelial cells. Microbios 1993; 75:57–65.

36. Klemm P. Fimbrial adhesions of *Escherichia coli*. Rev Infect Dis 1985; 7:321–340.

37. Adlerberth I. Establishment of a normal intestinal microflora in the newborn infant. In: Hanson L, Yolken RH, eds. Probiotics, Other Nutritional Factors, and Intestinal Microflora. Philadelphia: Vevey/Lippincott-Raven Publishers, 1999:63–78.

38. Duguid JP, Old DC. Adhesive properties of enterobacteriaceace. In: Beachey EC, ed. In: Bacterial Adherence, Receptors and Recognition, Vol. 6. London: Chapman and Hall, 1980:185–216.

39. He F, Ouwehan AC, Hashimoto H, Isolauri E, Benno Y, Salminen S. Adhesion of *Bifidobacterium* spp. to human intestinal mucus. Microbiol Immunol 2001; 45:259–262.

40. Kirjavainen PV, Ouwehand AC, Isolauri E, Salminen SJ. The ability of probiotic bacteria to bind to human intestinal mucus. FEMS Microbiol Lett 1998; 167:185–189.

41. He F, Ouwehand AC, Isolauri E, Hashimoto H, Benno Y, Salminen S. Comparison of mucosal adhesion and species identification of bifidobacteria isolated from healthy and allergic infants. FEMS Immunol Med Microbiol 2001; 30:43–47.

42. Moore WE, Moore LH. Intestinal floras of populations that have a high risk of colon cancer. Appl Environ Microbiol 1995; 61:3202–3207.

43. Lee YK, Ho PS, Low CS, Arvilommi H, Salminen S. Permanent colonization by *Lactobacillus casei* is hindered by low rate of cell division in mouse gut. Appl Environ Microbial 2004; 70:670–674.

44. Link-Amster H, Rochat F, Saudan KY, Mignot O, Aeschlimann JM. Modulation of a specific humoral immune response and changes in intestinal flora mediated through fermented milk intake. FEMS Immunol Med Microbiol 1994; 10:55–63.

45. Schiffrin EJ, Rochat F, Link-Amster H, Aeschlimann JM, Donnet-Hughes A. Immunomodulation of human blood cells following the ingestion of lactic acid bacteria. J Dairy Sci 1995; 78:491–497.

46. Haller D, Blum S, Bode Ch, Hammes WP, Schiffrin EJ. Activation of human PBMC by non-pathogenic bacteria in vitro: evidence of NK cells as primary targets. Infect Immun 2000; 68:752–759.

47. Anton P, O' Conell J, O'Conell D, et al. Mucosal binding sites for the bacterial chemotactic peptide, formyl-methionyl-leucyl-phenylalanine (FMLP). Gut 1998; 42:374–379.

48. Poltorak A, He X, Smirnova I, et al. Defective LPS signaling in C3H/HeJ and C57BL/10ScCr mice: mutations in Tlr4 gene. Science 1998; 282:2085–2088.

49. Takeuchi O, Sato S, Horiuchi T, et al. Cutting edge: role of Toll-like receptor 1 in mediating immune response to microbial lipoproteins. J Immunol 2002; 169:10–14.

50. Gerwitz AT, Simon PO, Jr., Schmitt CK, et al. *Salmonella typhimurium* translocates flagellin across intestinal epithelia inducing a proinflammatory response. J Clin Invest 2001; 107:99–109.

51. Di Martino P, Girardeau JP, Der Vartanian M, Joly B, Darfeuille-Michaud A. The central variable V2 region of the CS31A major subunit is involved in the receptor-binding domain. Infect Immun 1997; 65:609–616.

52. Bäckhed F, Hornef M. Toll-like receptor 4-mediated signaling by epithelial cell surfaces: necessity or threat? Microbes Infect 2003; 5:951–959.

53. Abreu MT, Arnold ET, Thomas LS, et al. TLR4 and MD-2 expression is regulated by immune-mediated signals in human intestinal epithelial cells. J Biol Chem 2002; 277:20431–20437.

54. Wang C, Deng L, Hong M, Akkaraju GR, Inoue J, Chen ZJ. TAK1 is a ubiquitin-dependent kinase of MKK and IKK. Nature 2001; 412:346–351.

55. Ninomiya-Tsuji J, Kishimoto K, Hiyama A, Inoue J, Cao Z, Matsumoto K. The kinase TAK1 can activate the NIK-I kappa B as well as MAP kinase cascade in the IL-1 signaling pathway. Nature 1999; 398:252–256.

56. Delhase M, Hayawaka M, Chen Y, Karin M. Positive and negative regulation of IkappaB kinase activity through IKKbeta subunit phosphoryation. Science 1999; 284:309–313.

57. Hu Y, Baud V, Delhase M, et al. Abnormal morphogenesis but intact IKK activation in mice lacking the IKKalpha subunit of IkappaB kinase. Science 1999; 284:316–320.

58. Zhang Y, Bliska JB. Role of toll-like receptor signaling in the apoptotic response of macrophages to *Yersinia* infection. Infect Immun 2003; 71:1513–1519.

59. Radel SJ, Genco RJ, De Nardin E. Structural and functional characterization of the human formyl peptide receptor ligand-binding region. Infect Immun 1994; 62:1726–1732.

60. Snyderman R, Fudman EJ. Demonstration of a chemotactic factor receptor on macrophages. J Immunol 1980; 124:2754–2757.

61. Chaudhary PM, Ferguson C, Nguey V, et al. Cloning and characterization of two Toll/Interleukin-1 receptor-like genes TIL3 and TIL4: evidence for multi-gene receptor family in human. Blood 1998; 91:4020–4027.

62. Gewirtz AT, Liu Y, Sitareman SV. Madara. Intestinal epithelial pathobiology: past, present, future. Best Pract Res Clin Gastroenterol 2002; 16:851–867.

63. Lu L, Walker WA. Pathologic and physiologic interactions of bacteria with the gastrointestinal epithelium. Am J Clin Nutr 2001; 73:1124S–1130S.

64. Luo Y, Frey EA, Pfuetzner RA, et al. Crystal structure of enteropathogenic *Escherichia coli* intimin-receptor complex. Nature 2000; 405:1073–1077.

65. Sinclair J, O' Brien AA. Cell surface-localized nucleolin is a eukaryotic receptor for the adhesion intimin-γ of enterohemorrhagic *Escherichia coli* O157: H7. J Biol Chem 2002; 277:2876–2885.

66. Jonsson H, Ström E, Ross S. Addition of mucin to the growth medium triggers mucus-binding activity in different strains of *Lactobacillus reuteri* in vitro. FEMS Microbiol Lett 2001; 204:19–22.

67. Holzapfel WH, Haberer P, Snel J, Schillinger U, Huis in't Veld JHJ. Overview of gut flora and probiotics. Int J Food Microbiol 1998; 41:85–101.

68. Hooper LV, Xu J, Falk PG, Midtvedt T, Gordon JI. A molecular sensor that allows a gut commensal to control its nutrient foundation in a competitive ecosystem. Proc Natl Acad Sci U.S.A. 1999; 96:9833–9838.

69. Ayabe T, Satchell DP, Wilson CL, Parks WC, Selsted ME, Ouellette AJ. Secretion of microbicidal alpha-defensins by intestinal Paneth cells in response to bacteria. Nat Immunol 2000; 1:113–118.

7

The Metabolism of Nutrients and Drugs by the Intestinal Microbiota

Barry R. Goldin
Department of Public Health and Family Medicine, Tufts University School of Medicine, Boston, Massachusetts, U.S.A.

INTRODUCTION

The intestinal microbiota of humans is comprised of a complex ecosystem of metabolically active microorganisms that reside close to the mucosal surface of the intestine. The bacteria of the intestine can interact with substrates introduced orally or compounds entering the intestinal lumen via the bile, mucosal secretions, or systemically from the circulatory system. This chapter will review the bacterial reactions performed on nutrients and drugs entering the intestine. The composition and distribution of the intestinal microbiota will not be discussed, and the readers are referred to other chapters in this book and review articles that address this topic (1–4). It is, however, important to note that the intestinal microbiota at any given time weighs approximately 110 to 200 grams and consists of at least 400 different species. The number of bacterial cells is approximately ten times greater than the total number of cells comprising the human body. Although the mass of the intestinal microbiome is equivalent to that of a single kidney, the number and diversity of species affords the microbiota a diverse metabolic role in the human body. This chapter will review some of these reactions and implications of these transformations to the host; however, no attempt will be made to exhaustively review all known reactions carried out by the microorganisms that inhabit the gastrointestinal tract of humans and animals.

GENERAL METABOLISM AND FUNCTION OF THE MICROBIOTA

The bacteria of the intestinal microbiota are predominantly anaerobic with a small percentage of facultative anaerobes. Therefore, intestinal bacteria do not use oxygen as a terminal election acceptor, and derive their energy from anaerobic respiration or substrate level phosphorylation. The magnitude of energy derived is the difference in redox potential between the substrate, and the products formed (5,6). The major overall balance of the intestinal microbiota derives from the ability to convert available substrates, principally originating from oral ingestion by the host of nutrients, fiber, and intestinal secretions or endogenously host synthesized compounds entering the intestine via the bile into the biomass that makes up the microorganisms in the intestine. The total biomass is principally controlled by space constraints, transit time of the digesta, and substrate availability. In general, approximately 50% of the fecal mass is composed of intestinal microorganisms. In addition to the utilization of substrates derived from the host, the intestinal microbiota can provide the host with energy mainly in the form of short chain fatty acids, and nutritive benefit by producing certain vitamins.

Metabolic Reactions of the Intestinal Microbiota

In Table 1 the major chemical reactions performed by the microbiota are listed. Most of the bacterial reactions can be classified as reductive, hydrolytic, or removal of functional groups such as dehydroxylation and decarboxylation. These reactions are often catalyzed by specific bacterial enzymes.

NUTRIENTS AND DIETARY PLANT COMPOUNDS

Fermentation of Carbohydrates

Carbohydrate fermentation is a major source of energy for the intestinal microbiota. It has been estimated based on the biomass of the microbiota in the intestine that 20–70 grams of carbohydrate or equivalent substrates based on similar energy density would be required to be fermented to provide a biomass steady state (5–8). This calculation takes into account

Table 1 Reactions Performed by the Intestinal Microflora

Types of reaction	Example of substrate
Hydrolytic reactions	
Glucuronides	Phenolphthalein-glucuronide
Glycosides	Cellobiose
Amides	Methotrexate
Esters	Acetyldigoxin
Sulfamates	Amygdalin
Nitrates	Pentaerythritol trinitrate
Reductive reactions	
Nitrocompounds	1-nitropyrene
Azocompounds	Direct red 2
Double bonds	Polyunsaturated fatty acids
Aldehydes	Benzaldhydes
N-oxides	4-Nitroquinoline-1-oxide
Nitrosation	
Amines	Dimethylamine
Removal of functional group	
C-hydroxy	Bile acids
N-hydroxy	N-hydroxyfluorenyl-acetamide
Carboxyl	Amino acids
Methyl	Biochanin A
Amine	Amino acids
Chlorine	DDT

Abbreviation: DDT, dichloro-diphenyl-trichloroethane.

that bacteria are excreted daily, and that the normal intestinal transit time varies between 48 and 72 hours. In "Western societies," such as the United Kingdom, the major intestinal bacterial carbohydrate substrates available are non-starch polysaccharide 12 grams, oligosaccharides 5 grams, simple sugars less than 5 grams, resistant starch 4 grams, and fermentable polysaccharides from intestinal mucus unknown (5,9). The amount of carbohydrate derived from colonic mucus available for bacterial fermentation is limited based on the fact that elemental diets support a very low bacterial biomass (10,11). All of these sources of carbohydrates are not readily digested and absorbed by humans, and thus arrive intact in the colon.

Bacteroides, the most abundant bacterial genus in the ileum and colon can degrade, and ferment a number of different polysaccharides, including xylan, psyllium hydrocolloid, and numerous other plant polysaccharides (12,13). *Bacteroides* can also degrade host derived glycans such as chondroitin sulfate, mucin, heparin, hyaluronate, and glycosphingolipids.

The fact that non-absorbable polysaccharides would not provide energy for the host adds a function to intestinal bacterial carbohydrate fermentation, namely salvaging energy. The major end products of bacterial fermentation in the intestine are the short chain fatty acids, acetate, propionate, and butyrate (14). For humans approximately 20–70 grams of carbohydrate would normally be fermented by intestinal flora per day. This would translate into 30–105 kcal per day or between 1.5 and 5% of typical human caloric intake. This percentage of caloric requirements varies greatly with the amount of fiber and other non-absorbable polysaccharides consumed per day. In developing countries, populations may derive larger benefits from bacterial metabolism in the intestine, as a result of greater consumption of plant fiber.

Intestinal Bacterial Protein, Amino Acid, and Nitrogen Metabolism

In monogastric animals there are several sources of nitrogen containing compounds that enter the large intestine, and thus are substrates for metabolic action by the microflora. The sources include incompletely digested dietary protein, protein from intestinal epithelial cells, and digestive secretions including digestive enzymes, glycoprotein mucins, free amino acids, and peptides including those derived from a bacterial origin. In addition, ammonia, urea, and nitrate are found in the ileal effluent. In terms of amounts and composition of nitrogen containing compounds entering the large intestine, it has been estimated in humans that 12–18 grams of protein enter the cecum from the ileum per day, and 2–3 grams per day of nitrogen (15). The approximate relative amount of nitrogen containing compounds in the large intestine is 48–51%, 34–42% peptides, and 10–15% urea/ammonia/nitrate, and free amino acids (15). The nitrogen sources in ileal effluent are primarily pancreatic enzyme protein and dietary protein residue. In contrast, in the feces the nitrogen compounds are more than 50% of bacterial origin (16). Therefore, although the balance of nitrogen is relatively well maintained between the amount entering and leaving the colon, the bacteria change the nature of the nitrogen containing compounds by utilizing these nitrogen compounds, and to large extent converting them into bacterial protein, which is found in the feces as intact bacteria, and as products of the lysed microorganisms.

There are five major bacterial pathways for deaminating amino acids; four are designated as direct pathways, and one is considered an indirect pathway. The direct pathways are: reduction resulting in saturated fatty acid production; oxidation resulting in the formation of keto acids; hydrolysis causing the formation of an alpha-hydroxy fatty acid; and removal of the elements of ammonia, producing an unsaturated fatty acid (17). A fifth deamination pathway is known as the Strickland reaction, and is carried out by clostridia that have little or no capacity to degrade single amino acids. As consequence the clostridia degrade amino acids in pairs by a coupled oxidation-reduction reaction forming a keto acid and a saturated fatty acid. Reduction reactions are the major pathway for the degradation of amino acids in the intestine. The reduction products of the action of intestinal anaerobic organisms include: acetic, propionic, butyric, and isovaleric, isobutyric, and 2-methylbutyric acids (18). Other reductive products are ammonia, amines, carbon dioxide, and hydrogen (19). Some of the products that result from reductive degradation of aromatic amino acids include phenol, p-cresol, phenylactic acid, phenylpropionic acid, indole, indoleacetic acid, and indolepropionine acid.

Decarboxylation is a second class of reactions that the intestinal microbiota perform in the course of the intestinal amino acid degradation (20). Bacterial decarboxylases act on amino acids to form amines and carbon dioxide. Many of these decarboxylases are specific, acting only on a single amino acid. There are a number of different genera of intestinal bacteria that have decarboxylase activity including: enterobacteria, enterococci, lactobacilli, clostridia, *Bacteroides*, and bifidobacteria (19). Some of the specific products formed from bacterial decarboxylation are the formation of cadavarine from lysine, putrescine from ornithine, histamine from histidine, and tyramine from tyrosine.

Intestinal bacteria can assimilate ammonia from the surrounding environment (20), and incorporate it into cell structures. Bacteria are also capable of ammonia production from peptides and amino acids (21).

Bacterial Intestinal Lipid Metabolism

In healthy humans, the vast majority of free fatty acid formed from dietary lipids is absorbed in the small intestine. The anaerobic bacterial microflora have the capability to hydrate, and hydrogenate double bonds found in unsaturated fatty acids (22,23). This is evidenced by the presence of 10-hydroxystearic acid in human feces. The limited amounts of fatty acids that are transported to the lower intestinal tract relegate intestinal bacterial metabolism to minor significance in humans.

Short Chain Fatty Acids

Short chain fatty acids are not an important dietary nutrient, however, they are being discussed at this point because they are a significant end product of carbohydrate and amino acid bacterial metabolism. Short chain fatty acids are readily absorbed from the human colon, and facilitate the absorption of salt and water by the colon. Colonic epithelium derives 60–70% of its energy from short chain fatty acids with butyrate being the most important in this regard (24). Short chain fatty acids also stimulate mucosal growth in the colon. As stated previously, the major short chain fatty acids produced by intestinal bacterial fermentation are acetate, butyrate, and propionate. Additional end acid products include: lactate, succinate, and formate (25). The fate of these bacterially produced acid end products has been studied to varying extents. In humans, acetate is always found at a concentration of 50 micromolar in fasting venous blood. After a carbohydrate rich meal, these blood levels rise to 100 to 300 micromolar (5). The half-life of acetate in the blood is only a few minutes, and is taken up and metabolized in skeletal and cardiac muscle, brain, and adipocytes for lipogenesis (5). Acetate spares fatty acid oxidation but has only a small influence on glucose metabolism, and has no effect on insulin release in humans.

Intestinal Bacterial Synthesis and Metabolism of Vitamins

The human intestinal bacteria can synthesize vitamin K, a member of the naphtoquinone family. The liver cannot synthesize the prothrombin complex, a blood-clotting factor, unless menaquinone, a substituted naphthoquinone, is present. The peptides that become the glycopeptides of the prothrombin complex require menaquinone for synthesis from the appropriate RNA codon.

Bacteria found in the intestine can also synthesize homologues of menaquinone-7 (vitamin K_2). The synthesized homologues range from the 6-isoprene unit side chain containing menaquinone-6 to menaquinone-13 (26,27). The vitamin K bacterial reactions occur, in part, in the ileum, where the menaquinone is absorbed. The importance of bacterial synthesis of vitamin K has been demonstrated in human studies (28). Adult subjects maintained on a low vitamin K diet for several weeks did not develop a deficiency. When these subjects were treated with antibiotics such as neomycin that reduce the bacterial population of the intestine, a significant decrease in plasma prothrombin levels was noted (28,29).

Most of the vitamin B_{12} (cyanocobalamin) required by humans comes indirectly from the meat and milk of ruminants. The synthesis of B_{12} in ruminants is exclusively of bacterial origin. The human intestinal microflora also synthesize vitamin B_{12} as evidenced by the fecal secretion of approximately 5 micrograms per day. However, it appears most of the bacterially formed B_{12} in humans occurs in the large bowel where absorption most likely does not occur due to lack of B_{12} mucosal receptors. However, there is a study of healthy subjects from Southern India that reported the synthesis of vitamin B_{12} in the jejunum and ileum, an area where absorption of the vitamin can occur (30). It was demonstrated that *pseudomonas* and *klebsiella* were two of the bacteria that synthesized B_{12} in the small intestine.

Biotin is synthesized by the human intestinal microflora. The administration of antibiotics can lower human urinary biotin levels. The importance of bacterial involvement in biotin synthesis has been demonstrated in germfree rats. The germfree animals require biotin in their diet; in contrast conventional rats can thrive without dietary biotin (22).

Folic acid and thiamine B complex vitamins are also synthesized by bacteria in the intestinal tract. This synthesis does not solely provide for human requirements, and dietary sources of these vitamins are required to prevent deficiencies (31).

Intestinal Bacterial Metabolism of Isoflavones and Lignans

Dietary plant sources, such as vegetables, fruit, and cereals contain in addition to nutrients a large number of physiological active compounds. Many of these orally consumed compounds are transformed by the intestinal bacteria, which can result in either biological activation or deactivation of these substances. There are many plant-derived substances. In this section, two of these compounds of recent interest, the bacterial metabolism of the phytoestrogen compounds isoflavones and lignans are discussed.

Isoflavones have weak estrogenic and antiestrogenic activities. Soybeans contain the highest levels of isoflavones in the human food chain. Other plant foods that contain isoflavones are pinto beans, navy beans, and chick peas, which have approximately two orders of magnitude lower levels. For populations consuming soy-based foods, the amount of isoflavones eaten daily is between 30 and 150 mgs. For daidzein, one of three major soy isoflavones, the intestinal bacteria can convert the parent compound into several end products. Among the end products are o-desmethylangolensin, equol, cis-4-equol, and dihydrodaidzein (32). There is a large individual variation in the ability of intestinal bacteria to metabolize daidzein. Studies have shown that production of equol does not occur in 30–40% of people fed soy isoflavones, and the remainder are active equol producers (33). The conversion of daidzein to equol can be of physiological importance since equol is a more potent estrogenic substance. The factors that control the extent of bacterial conversion of isoflavones in the intestine are unknown.

Genistein, the isoflavone with the highest concentration in soy, is converted by intestinal bacteria to dihydrogenistein, and p-ethyl phenol. These reactions most likely lower or destroy the estrogenic activity of genistein. Glycetein, the third most prevalent isoflavone contained in the soybean, is bacterially converted to 5-hydroxy-, and 5-methoxy-o-desmethylangolensin. There are other bacterial metabolites of isoflavones, and new end products are still being isolated.

Lignans are found in relatively high concentrations in flaxseed, whole-grain products, vegetables, and sesame seeds (32). Lignans also exhibit weak estrogen and anti-estrogen activity, although these activities are lower than those found in isoflavones (32).

The plant lignan precursors secoisolariciresinol and matainresinol are converted by the intestinal microflora to enterodiol and enterolactone, respectively (32). The physiological importance of these bacterial conversions are not clear.

INTESTINAL BACTERIAL METABOLISM OF HOST ENDOGENOUSLY SYNTHESIZED COMPOUNDS

Bacterial Cholesterol Metabolism

The intestinal tract has a major impact on cholesterol metabolism (34–36). A major source of intestinal cholesterol comes from the de novo synthesis of the sterol compound. Cholesterol also can enter the intestine from dietary sources. It has been estimated that 34–57% of dietary cholesterol is absorbed from the intestine (37). In humans cholesterol synthesized by the intestinal cells is introduced into the lumen by exfoliation of these cells. An additional source of intestinal cholesterol is via biliary excretion.

The fecal excretion of total neutral sterols in humans ranges from 350–900 mg/day, with a mean of 700 mg/day (38). Cholesterol accounts for about 20% of the total neutral sterols excreted in the feces, or about 150 mg/day. The normal range of cholesterol excreted by humans in the feces is between 75–200 mg/day. As discussed above, there are three sources of intestinal and fecal cholesterol: unabsorbed cholesterol from the diet which contributes 20%, bile which contributes 67%, and sloughed intestinal epithelial cells which contribute 13% of the total fecal cholesterol (39).

The cholesterol that enters the intestine can be metabolized by bacterial microflora. Cholesterol is converted to 4-cholesten-3 one which is an intermediate formed by the oxidation of the 3 beta-hydroxyl group to a ketone, and isomerization of the 5–6 double bond to the 4–5 position. Coprostonone is formed by the reduction of the 4–5 double bond. The final reaction is the formation of coprostonal by reduction of the 3-beta to a hydroxyl group (34).

The amounts of cholesterol and its metabolites found in feces are approximately 20% cholesterol, 65% coprostonal, and 10% coprostanone (40). An additional 5% of fecal neutral sterols are made up of cholesterol, cholestanone, and epicoprostanol (40).

Studies in Americans have shown that the majority of this population metabolizes cholesterol in the intestine (41). The distribution of intestinal bacterial conversion was bimodal. The majority of subjects converted 70–99% of cholesterol in their feces to metabolites, and a smaller group of individuals converted 0–19% of cholesterol (42).

Intestinal Bacterial Metabolism of Bile Acids and Bile Pigments

Cholesterol is a precursor of bile acids, and both are synthesized in the liver from two carbon units. Bile acids synthesized in the liver are conjugated through an amide bond to either glycine or taurine. The conjugated bile acids are deposited in the bile, and excreted into the upper small intestine. The bacterial conversion of bile acids primarily occurs in the distal ileum and colon. The bacterial reactions on bile acids include: the hydrolysis of the amide bond to release free bile acids from their corresponding glycine and taruine conjugates; an oxidoreduction of the hydroxyl groups at C3, C7, and C12 to form either oxo bile acids or alpha hydroxyl groups after the reduction of the beta groups (inversion products); and dehydroxylation at C7, and to a smaller extent at the C3 and C12 positions (43). The consequence of these reactions is the conversion of primary to secondary bile acids, and the re-absorption of free bile acids from the ileum, and to a lesser extent, from the colon. Only approximately 5% of bile acids are lost in the feces in each cycle as a result of bacterial deconjugation of bile acids (44).

Bacterial Metabolism of Androgens and Estrogens

Estrone, estradiol, and estriol are the three major estrogens that are excreted into the bile. These estrogens are conjugated to glucuronic acid and/or sulfate. Upon excretion of these conjugated estrogens from the bile into the small intestine the conjugates are available substrates for bacterial metabolism. The bacteria of the lower small intestine and colon can hydrolyze the estrogen conjugate releasing free estrogens (45). The nonconjugated estrogens are then subject to additional bacterial action. A major reaction involves oxidoreduction of the C17 position. Bacteria can convert estrone to estradiol, and the fecal flora can also convert 16 alpha hydroxyestrone to estriol (46).

The intestinal bacteria can also modify androgens. The intestinal bacteria can reversibly oxidize and reduce the 3-hydroxy group, and reduce steroid nuclear double bonds at the one and four positions. The latter reactions can result in several interconversions of androgens (47).

Other Steroid Hormone Bacterial Conversions

Studies have shown that fecal organisms can modify corticosteroids. The corticosteroids undergo reduction in ring A, and undergo side-chain dehydroxylation separately or sequentially with the reduction (48). Cortisol is converted to 21-deoxycortisol, tetrahydrocortisol, and tetrahyro-21-deoxycortisol (48). Corticosterone is metabolized to tetrahydrocorticosterone, 21-deoxycorticosterone, and 3-alpha-hydroxy or 3-beta-hydroxy epimers of tetrahydro-21-deoxycorticosterone (48).

The intestinal bacteria can also transform progesterone similar to the reactions described above. Bacterial reduction of ring A can occur, as well as 16-alpha dehydroxylation, which can cause epimersation of the side chain (48).

OTHER BACTERIAL REACTIONS

Sulphate Metabolism

The human colon contains Gram-negative anaerobes capable of reducing sulphates. The process is referred to as dissimilatory sulphate reduction, and results in the conversion of sulphates and sulphites to sulphides (49,50). The major bacterial genus that performs this reaction in the human colon is *Desulfovibrio*. Hydrogen gas in the colon is used as an electron donor in the formation of sulphides (50). The source of sulphates for bacterial reduction can come from food preservatives and drugs, and the levels of sulphides are highest in the sigmoid colon and rectum (51). Less than half of the human population appears to actively reduce sulphate in the large bowel (52).

Aromatization

Quinic acid is found in food products such as coffee, tea, fruits, and vegetables. Quinic acid has an aliphatic cyclic structure. Quinic acid is excreted in the urine as hippuric acid, an aromatic ring containing compound (53). Evidence that the intestinal bacteria are involved in the aromatization comes from the observation that hippuric acid is not formed when Quinic acid is given parenterally, and the formation of hippuric acid is inhibited when the antibiotic neomycin is given to humans (53). These findings strongly support the hypotheses that aromatization occurs as a result of intestinal bacterial action.

Bacterial Carbon–Carbon Bond Cleavage

The human intestinal flora has been shown capable of breaking the carbon bond between two of the rings of the product sennidin (54), which is found in senna and rhubarb. The product formed from this cleavage is rhein anthrone. The carbon-carbon cleavage is of physiological importance since this reaction is required for the observed laxative action of plant sennosides.

BACTERIAL INTESTINAL FORMATION OF MUTAGENS

In Table 2 are shown some of the mutagens formed as a result of intestinal bacterial reactions. The bacterial enzymes that catalyze the reactions that potentially can produce mutagens, carcinogens, and tumor promoters are also presented in Table 2. Some of the reactions discussed in this section also act on various drugs, and will again be discussed in the section on drug metabolism.

Intestinal bacterial enzymes that have been implicated in the formation of carcinogens, mutagens, and tumor promoters include: beta-glucuronidase, beta-glucosidase, beta-galactosidase, nitroreductase, azoreductase, sulfatases, nitrosation, tryptophanase, 1-alpha-steroid dehydrogenase, and 7-alpha-hydroxysteroid dehydroxylase (55).

Table 2 Substrates Converted into Mutagens as a Result of Intestinal Bacterial Reactions

Substrate	Bacterial enzyme
2-Nitrofluorene	Nitroreductase
Metronidazole	Nitroreductase
Trypan blue	Azoreductase
Ponceau 3R	Azoreductase
Cycasin	Beta-glucosidase
1-Nitropyrene	Beta-glucuronidase
Cyclamate	Sulfatase
Dimethylamine	Nitrosation
Tryptophan	Tryptophanase

Glycosidase

A classic example of the role of the intestinal flora in generating carcinogens is illustrated by the action of this bacterial enzyme on the plant derived compound cycasin (56). Cycasin is a naturally occurring beta-glucoside of methylazoxymethanol, extractable from the seeds and roots of cycad plants. It was observed that when Cycasin was fed to infant rats a number of different tumors developed. The Cycasin-induced tumors included hepatomas, renal sarcomas, squamous-cell carcinomas of the ear duct, and most frequently large bowel and duodenal adenocaricomas (56). The genetic strain of the rat did not appear to have a major influence on tumor development. It was, however, noted that the intestinal flora was required for tumorgenesis, since when Cycasin was given orally to germfree rats no tumors were observed (57). The discovery of the carcinogenicity of Cycasin led to experiments to test the precursor aglycones of Cycasin azoxymethane, azomethane, and dimethylhydrazine. These compounds were carcinogenic in conventional and germfree rats (58). The route of administration was not critical, and tumors developed after oral or subcutaneous administration (56). These results confirmed that the hydrolysis by the intestinal flora of the glycosidic bond was required for the activation of Cycasin. It was also observed that infant but not adult rats developed tumors when given Cycasin by intraperitoneal injection confirming the observation that tissue b-glucosidase disappeared in rats after 3 weeks of life (56).

Many other plant natural products occur as glycosides. These glycosides do not demonstrate mutagenicity when tested in the Salmonella test; however, upon hydrolysis of the glucosidic linkages they become mutagenic. There have been several studies showing mixed fecal cultures or fecal isolates of *Streptococcus faecium* can convert non-mutagenic rutin (quercetin-3-D-beta-D-glucose-alpha-L-rhamnose) to quercetin (59). Quercetin has been shown to be mutagenic in the Ames salmonella assay. Red wine and tea contain glycosides of quercetin.

Beta-Glucuronidase

The formation of glucuronides in the liver is an important mechanism for detoxifying and enhancing excretion of a large number of orally ingested nutrients and their end products, other dietary compounds, and drugs, as well as endogenously synthesized compounds, such as estrogens. In humans many of these glucuronides depending on the structure of the aglycone, are excreted in the bile, and subsequently enter the duodenum. The glucuronides are then subject to bacterial deconjugation primarily in the ileum and colon. As a consequence of this bacterial deconjugation physiologically active, toxic, and carcinogenic compounds are regenerated. In addition to their formation in the intestine these compounds can be reabsorbed into the portal blood system. This results in recycling of these hydrolyzed glucuronides, and this process is referred to as the enterohepatic circulation.

Several studies have shown that intestinal beta-glucuronidase can alter or amplify the biological activity of exogenous and endogenous compounds.

The metabolism of the carcinogen N-hydroxyflourenylacetamide administered parenterally to conventional and germfree rats was studied by Weisburger et al. (60). Germfree rats excreted larger amounts of the glucuronides of N-hydroxyflourenylace-tamide in their feces compared to conventional animals. The cecal and fecal contents of conventional rats contained mostly unconjugated N- hydroxyflourenylacetamide, and its metabolites; in contrast most of these metabolites were glucuronide or sulfate conjugates in germfree animals.

It has been shown that cell-free extracts derived from a number of different bacteria residing in the intestinal tract, including *Bacteroides fragilis*, *Bacteroides vulgatus*, *Bacteroids thetaiotamicron, Eubacterium eligens, Peptostreptoccus, and Escherichia coli*, were capable of increasing the mutagenic activity of bile from rats fed 1-nitropyrene via stomach tube. These extracts had beta-glucuronidose activity. Cell-free extracts of bacteria that were not able to enhance the mutangenicity of the bile did not possess beta-glucuronidose activity (61). These data support the hypothesis that glucuronides of 1-nitropyrene metabolites entering the bile can be hydrolyzed by intestinal bacterial beta-glucuronidase to produce active deconjugated mutagenic products.

Bacterial Azoreductase

Azoreductase activity, which is of exclusively bacterial origin in the lumen of the intestine, catalyzes the reduction of the azo bond to cause the formation of aromatic amines. The highly reactive intermediates and end products have been shown to be mutagenic and carcinogenic. Azo dyes are used for coloring in the food industry, and as dyes and stains in textiles and other products. Water-soluble azo dyes are degraded by the intestinal microflora in the gastrointestinal tract (62). There is a 90% correlation between carcinogenicity and mutagenicity for aromatic amines and azo dyes tested by the Ames Salmonella test (63). The need for bacterial azoreductase and nitroreductase to activate mutagens, such as azocompounds in combination with intestinal mucosal microsomal enzymes has been demonstrated (64,65).

The reduction of azocompounds by azoreductase is mediated through a free radical mechanism that produces intermediates that react with nucleic acids and proteins. The action of azoreductase on food dyes results in the release of phenyl-, and naphthyl-substituted amines. The amines generated in the lower intestine by bacterial action are probably oxidized by microsomal enzymes in the intestinal mucosa to carcinogens.

Bacterial generation of mutagens from a number of azodyes has been demonstrated. Trypan blue, a widely used biologic stain, is converted to the mutagen O-toluidine by cell free extracts of *Fusobacterium*, an anaerobic organism found in the large intestine (66). Ponceau 3R, another biologic stain, is reduced *Fusobacterium* to 2, 4, 5-trimethylaniline which is mutagenic (67). Other azo dyes that have been shown to be transformed by bacterial reduction to mutagenic or carcinogenic products are direct black 38, direct red 2, and direct blue 15. Congo red lacks mutagenic activity, however, preincubation of dye with cecal bacteria generates mutagen-positive products (68).

Bacterial Nitroreductase

Nitroreductase similar to azoreductase is exclusively of bacterial origin in the lumen of the intestine. The enzyme is required for the mutagenic activity of nitrocompounds (64). Nitroreductase generates reactive nitroso and N-hydroxyintermediates in the course of converting aromatic amines. 1-nitropyrene is formed by the reaction of nitrogen oxides with the combustion product pyrene. The presence of 1-nitropyrene in diesel exhaust makes exposure to this compound a real risk. 1-nitropyrene is mutagenic in bacterial test systems, and carcinogenic when administered to the rat. When 1-nitropyrene was fed to conventional rats 5% to 6% of the dose was detected in the feces as 1-aminopyrene (69). When the same feeding experiment was performed with germfree rats no 1-aminopyrene was detected in the feces. The reduction of 1-nitropyrene to 1-aminopyrene is a carcinogen activation process, and the results cited above indicate that the intestinal microflora are important in the activation of 1-nitropyrene.

Mixed bacterial fecal specimens obtained from humans have been shown to reduce 6-nitrochrysene to 6-aminochrysene, a compound that causes cancer in mice (70). The intermediate nitrosopolychic aromatic hydrocarbons generated in the conversion of the nitro groups to an amine are highly reactive compounds that can alter DNA.

Bacterial Nitrosation

Since the first report of the induction of liver cancer in rats fed dimethylnitrosamine (71), more than 80 different nitroso compounds have been identified as cancer-causing agents. The formation of nitrosamines results from the reaction of secondary amines with nitrite at acid pH. Nitrite is commonly added to cured meat and fish, and nitroso compounds have been measured in these foods (72).

Bacteria have been implicated in the formation of N-nitroso compounds. Nitrite can be produced by the bacterial reduction of nitrate. High levels of nitrate are often present in leafy vegetables. The oral microbial flora of humans can reduce nitrate with the formation of nitrite. This reaction can raise nitrite levels in saliva to 6–10 ppm (73).

It has been shown that when dimethylamine and sodium nitrite are incubated at pH 7.0 under anaerobic conditions with rat intestinal microflora, the formation of dimethylnitrosamine was detected (74). These findings indicate that nitrosamines could be generated in the intestine, where the pH is nearly neutral, and the reaction would occur extremely slowly without bacterial enzyme catalysis.

Bacterial Metabolism of Tyrosine and Tryptophan

Tyrosine and tryptophan are amino acids that can be converted by bacterial reactions in toxins and carcinogens. The tryptophanase containing *Bacteroides thetaiotamicron,* an organism found in the intestine, can convert tryptophan to indole a compound with carcinogenic activity (75).

Tyrosine is converted to phenol by aerobic intestinal bacteria, and to p-cresol by intestinal anaerobic bacteria. These metabolites of tyrosine are not found in the urine of germfree mice. Phenol and cresol have been shown to be tumor promoters in mice.

BACTERIAL INTESTINAL DRUG METABOLISM

In Table 3 are shown some representative natural and synthetic compounds that are or have been used as drugs that have been shown to be metabolized by the intestinal microflora. A description of the bacterial reactions involved for some of these drugs is given below.

DOPA

DOPA (3, 4-Dihydroxyphenylalanine) is used for the treatment of Parkinson's disease. DOPA replaces dopamine lost to Parkinson's disease because dopamine itself cannot cross the blood-brain barrier. Intestinal microbial metabolism of DOPA influences the dose required for the pharmacological action of this drug. The bacterial modification involves a dehydroxylation resulting in the removal of the hydroxyl group at the para position of the aromatic ring of phenylalanine (76). The product of this reaction, meta-hydroxylphenyl-acetic acid, is not active in the treatment of Parkinson's disease. In addition DOPA can be decarboxylated by intestinal bacteria forming inactive amines which can be detected in

Table 3   Drugs, Supplements, and Additives Metabolized by the Intestinal Microflora

Digoxin
Diethylstilbesterol
Estrogens
Cyclamate
Azulfidine
3, 4-Dihydroxyphelalanine
Amygdalin
Metronidazole
Caffeine
Propachlor
Morphine
Buprenophine
Oxazepum
Phenolphthalein
Warfarin

urine. As a consequence of these bacterial reactions the dose of DOPA required to influence the symptoms associated with Parkinson's disease is greatly elevated.

Salicylazosulfapyridine (Azulfidine)

Azulfidine has been shown to be beneficial for the treatment and prevention of recurrence of ulcerative colitis. The drug structurally has sulfapyridine and aminosalicylate moieties attatched via an azo bond. The drug was originally designed to deliver the anti-inflammatory action of aminosalicylate, and the antimicrobial activity of sulfapyridine. The introduction of the azo bond linkage produced an unsymmetrical molecule that was non-absorbable in the upper intestine.

It has been demonstrated that the azo bond of azulfidine is reductively cleaved by fecal bacterial cultures and that conventional but not germfree animals can also perform this cleavage reaction (77). The resultant products of the bacterial cleavage have been shown to have a different distribution (78). 5-aminosalicylate, because of its dual positive and negative charge is not absorbed from the colon, and is found almost exclusively in the feces. Sulfapyridine is readily absorbed from the intestine, and is excreted in the urine. This observation has been noted in humans and in rats (78). The evidence suggests that aminosalicylate is the active component for treating ulcerative colitis, and that the azo bond linkage affords an effective delivery system to the large intestine by being non-absorbable in the upper gastrointestinal tract, and then being slowly released by the action of bacteria in the lower ileum and large intestine.

Metronidazole

Metronidazole is an antibiotic which has a specificity against pathogenic anaerobes (79). Metronidazole structurally is a 5-nitroimidazole. The compound has been shown to be mutagenic in the Ames assay. This activity is lost when tester strains deficient in nitroreductase are used in the assay. The nitro group is reduced to amine group prior to ring cleavage which yields acetamide and N-(2-hydroxyethyl) oxamic acid. Therefore the amine intermediate generated by bacterial action is not stable, and breaks down to simpler metabolites.

Cyclamate

Cyclamate (cyclohexylamine- N- sulfonate) was used as an artificial sweetening agent until it was banned. It had been reported that the intestinal flora can hydrolyze c-sulfonates, o-sulfonates, and N-sulfonates (54). Initially it was reported that Cyclamate could not be metabolized in the body. It was however, shown that Cyclamate could be converted to the bladder carcinogen cyclohexylamine as a result of the action of intestinal bacterial catalyzed N-sulfate ester hydrolysis (80). Cyclohexylamine was absorbed from the intestine, and excreted in the urine. Prolonged feeding of Cyclamate to rats increased the hydrolysis to the amine, and withholding cyclamate from the diet caused a decline in hydrolytic activity within 5 days (81).

Digoxin

The role of intestinal bacterial metabolism is important in the action of the cardiac glycoside drug digoxin (82). In order to form a pharmacologically active drug, the bacterial flora has to remove a trisacchride from the parent compound, releasing digoxigenin. The bacterial intestinal flora can further reduce the double bond in the lactone ring to form dihydrodigoxigenin (82). This compound is pharmacologically inactive. It was found that 36% of Americans in New York city given digoxin had the capability to reduce the double bond forming the inactive metabolite of digoxin (83). A total of 14% of New Yorkers excreted large amounts of metabolites of digoxin. These findings indicate at least 14%, and possibly a greater percentage of the population receiving digoxin will not achieve predicted serum levels resulting from the action of the intestinal microflora. Studies on a population residing in southern India indicated only 13.7% of those tested could reduce digoxin, and only 1% excreted large amounts of metabolites (83). These studies indicate that there are interethnic variations in the metabolic capacity of the intestinal microflora to reduce the double bond in the lactone ring of digoxin. This finding is not surprising based on the observation that *Eubacterium lentum* is exclusively responsible for the reductive reaction (82).

Diethylstilbesterol

Diethylstilbesterol is a highly active synthetic estrogen. This compound had been used prior to its being banned as a drug to prevent spontaneous abortions during pregnancy. It was subsequently discovered that this compound had serious side-effects, including reproductive problems, and vaginal cancer in the daughters of mothers given diethylstilbesterol during pregnancy. The metabolic fate of this compound has been studied (84). When diethylstilbesterol glucuronide was given orally to germfree rats, the compound was rapidly recovered in the feces. This results from poor absorption of the glucuronide from the intestine. In conventional rats the fecal recovery of diethylstilbes-terol is significantly reduced. The explanation for this finding is based on the ability of the beta-glucuronidase produced by intestinal microflora to generate the free compound from its glucuronide. Free diethylstilbesterol is more readily absorbed from the intestine. In conventional animals diethylstilbesterol makes approximately 1.5 passes through the enterohepatic circulation. The increased exposure resulting form the enterohepatic circulation can enhance the pharmacologic action, as well as the side-effects of diethylstilbesterol.

Estrogens: Hormone Replacement Therapy and Birth Control

Estrogens are used as a drug in a number of different human conditions. The most common are in hormone replacement therapy for treating menopausal symptoms and other consequences of aging in postmenopausal women, and for preventing conception in premenopausal women.

The metabolism of estrogens involves an enterohepatic circulation that is dependent on intestinal bacterial deconjugation, and intestinal re-absorption similar to those of bile acids. Approximately 60% of circulating estrogens are conjugated in the form of glucuronides or sulfates, and are excreted in the bile (85–87). Deconjugation, a required step to cause intestinal mucosal cell re-absorption, is catalyzed by bacterial beta-glucuronidase and sulfatase. Approximately 97% of the estrogens excreted in the feces are in the deconjugated form, although virtually all of the estrogens in bile are conjugated.

Another indicator of the involvement of the intestinal microflora in estrogen metabolism and pharmacokinetics is the observation that oral antibiotics exert an effect on the enterohepatic circulation of estrogens. It has been observed that urinary estriol concentration is decreased following oral administration of penicillin, ampicillin or neomycin (88). When antibiotics were given, fecal excretion of estrogens increased 60-fold, and unconjugated estrogens increased 3-fold.

These findings have a clinical significance. Failures of oral contraception pills have been associated with the use of oral antibiotics. Five pregnancies were reported among 88 women receiving rifampin at the same time they were on oral contraceptive pills (89). Other antibiotics associated with birth-control failures are ampicillin, chloramphenicol, and sulfamethoxy-pyridazine (90).

CONCLUSION

This chapter has reviewed some of the important intestinal bacterial interactions with nutrients, endogenously synthesized hormones and other compounds, and orally ingested drugs. Since there is no available human germfree model to compare the magnitude of the importance of the intestinal flora in the various reactions cited in this chapter it is difficult to quantitatively evaluate. Based on animal models, the intestinal microflora are not an absolute requirement for survival, however, they do influence nutrient requirements, drug responses, and the effectiveness of various endogenously produced substances. Therefore, the metabolic potential of the intestinal microflora has to be considered in human biochemical and physiological activities and responses.

REFERENCES

1. Finegold SM, Atteberg HR, Sutter VL. Effect of diet on human fecal flora: comparison of Japanese and American diets. Am J Clin Nutr 1974; 27:1456–1469.
2. Moore WEC, Holderman LV. Human fecal flora: the normal flora of 20 Japanese-Hawaiians. Appl Microbiol 1974; 27:961–979.
3. Finegold SM, Sutter VL, Mathisen GE. In: Hentges DJ, ed. Human Intestinal Microflora in Health and Disease. New York: Academic Press, 1983:3–31.
4. Goldin BR. Intestinal microflora: metabolism of drugs and carcinogens. Ann Med 1990; 22:43–48.
5. Cummings JT, Macfarlane GT. Role of intestinal bacteria in nutrient metabolism. J Parenter Enteral Nutr 1997; 21:357–365.

6.   Hooper LV, Midtvedt T, Gordon JI. How host-microbial interactions shape the nutrient environment of the mammalian intestine. Ann Rev Nutr 2002; 22:283–307.

7.   McNeil NI. The contribution of the large intestine to energy supplies in man. Am J Clin Nutr 1984; 39:338–342.

8.   Bergman EN. Energy contributions of volatile fatty acids from the gastrointestinal tract in various species. Physiol Rev 1990; 70:567–590.

9.   Cummings JH, Englyst HN. Fermentation in the large intestine and the available substrates. Am J Clin Nutr 1987; 45:1243–1255.

10.   McCamman S, Beyer PL, Rhodes JB. A comparison of three defined formula diets in normal volunteers. Am J Clin Nutr 1977; 30:1655–1660.

11.   Winitz M, Seedman DA, Groff J. Studies in metabolic nutrition employing chemically defined diets. Am J Clin Nutr 1970; 23:525–545.

12.   Salyers AA, Gheruidini F, O'Brien M. Utilization of xylan by two species of human colonic bacteroides. Appl Environ Microbiol 1981; 41:1065–1068.

13.   Salyers AA, Harris CJ, Wilkins TD. Breakdown of psyllium hydrocolloid by strains of bacteroides ovatus from human intestinal tract. Can J Microbiol 1978; 24:336–338.

14.   Hoverstad T, Fausa O, Bjorneklatt A, Bohmer T. Short-chain fatty acids in the normal human feces. Scand J Gastroenterol 1984; 19:375–381.

15.   Chacko A, Cummings JH. Nitrogen losses from the human small bowel: obligatory losses and the effect of physical form of food. Gut 1988; 29:809–815.

16.   Stephen AM, Cumings JH. The microbial contribution to fecal mass. J Med Microbiol 1980; 13:45–56.

17.   Barker HA. Amino acid degradation by anaerobic bacteria. Ann Rev Biochem 1981; 50:23–40.

18.   Macfarlane GT, Allison C. Utilization of protein by human gut bacteria. FEMS Microbiol Ecol 1986; 38:19–24.

19.   Macfarlane S, Macfarlane GT. In: Gibson GR, Macfarlane GT, eds. Human Colonic Bacteria. Boca Raton: CRC Press, 1995:75–100.

20.   Metges CC, Petzke KJ, El-Khoury AE, Henneman L, Grant I. Incorporation of urea and ammonia nitrogen into ileal and fecal microbial proteins and plasma five amino acid in normal men and ileostomates. Am J Clin Nutr 1999; 70:1046–1058.

21.   Goldin BR, Lichtenstein AH, Gorbach SL. In: Shils ME, Olson JA, Moshe S, eds. Modern Nutrition in Health and Disease. 85th ed. Philadelphia: Lea and Febiger, 1994:569–582.

22.   Coates ME, Fuller R. In: Clarke RTJ, Bauchop T, eds. Microbial Ecology of the Gut Flora. New York: Academic Press, 1977:311–342.

23.   Cummings JH. Short chain fatty acids in the human colon. Gut 1981; 22:763–779.

24.   Cummings JH. In: Gibson GR, Macfarlane GT, eds. Human Colonic Bacteria. Boca Raton, FL: CRC Press, 1995:101–130.

25.   Hill MJ. In: Hill MR, ed. Microbial Metabolism in the Digestive Tract. Boca Raton, FL: CRC Press, 1986:31–41.

26.   Fernandez F, Collins MD. Vitamin K composition of anaerobic bacterial. FEMS Microbiol Lett 1987; 41:175–180.

27.   Ramotor K, Conly JM, Chubb H, Louie TJ. Production of menoquinones by intestinal anaerobes. J Infect Dis 1984; 150:213–218.

28.   Frick PG, Riedler G, Brogli H. Dose response and minimal daily requirement for vitamin K in man. J Appl Physiol 1967; 23:387–389.

29.   Udall JA. Human sources and absorption of Vitamin K in relation to anticoagulation stability. J Am Med Assoc 1965; 194:127–129.

30.   Albert MJ, Mathan VJ, Baber SJ. Vitamin B12 synthesis by human small intestinal bacteria. Nature 1980; 283:781–782.

31.   Wostmann BS. The germfree animal in nutritional studies. Ann Rev Nutr 1981; 1:257–279.

32.   Adlercreutz H, Mousovi Y, Loukovaara M, Hanaloimen E. In: Hochberg RB, Naftolin F, eds. The New Biology of Steroid Hormones, Vol. 72. New York: Raven Press, 1991:145–154.

33. Rowland IR, Wiseman H, Saners TAB, Adlercreutz H, Bowery EA. Interindividual variation in metabolism of soy isoflavones and lignans: influence of habitual diet on equol production by the gut microflora. Nutr Cancer 2000; 36:27–32.
34. Lichtenstein AH. Intestinal cholesterol metabolism. Ann Med 1990; 22:49–52.
35. Eyssen HJ, Parmantier GG. Biotransformation of sterols and fatty acids by the intestinal microflora. Am J Clin Nutr 1974; 27:1329–1340.
36. McNamara DJ, Prosa A, Miettinen TA. Thin layer and gas-liquid chromatographic identification of neutral steroids in human and rat feces. J Lipid Res 1981; 22:474–484.
37. Grundy SM, Mok HYL. Determination of cholesterol absorption in man by intestinal perfusion. J Lipid Res 1977; 18:263–271.
38. Grundy SM, Ahrens EH, Davignon J. The interaction of cholesterol absorption and synthesis in man. J Lipid Res 1969; 10:103–115.
39. Ferezou J, Coste T, Chevaller F. Origins of neutral sterols in human feces studied by stable isotope labeling. Digestion 1981; 21:232–243.
40. Reddy BS, Martin CW, Wynder EL. Fecal bile acids and cholesterol metabolites of patients with ulcerative colitis, a high risk group for development of colon cancer. Cancer Res 1977; 37:1697–1701.
41. Reddy BS, Wynder EL. Large-bowel carcinogenesis: fecal constituents of populations with diverse incidence rates of colon cancer. J Natl Cancer Inst 1973; 50:1437–1442.
42. Wilkins TD, Hackman AS. Two patterns of neutral sterol conversion in the feces of normal North Americans. Cancer Res 1974; 34:2250–2254.
43. Owen RW. In: Hill MR, ed. Microbial Metabolism in the Digestive Tract. CRC Press: Boca Raton, 1986:51–65.
44. Plaa GL. In: Gillette JR, ed. Handbook of Experimental Pharmacology, Vol. 28. New York: Springer, 1975:139–140.
45. Adlercreutz H, Martin F, Lehtinen T, Tilskanen MJ, Pulkkinen MO. Effect of ampicillin administration on plasma. Conjugated and unconjugated estrogen and progesterone levels in pregnancy. Am J Obstet Gynecol 1977; 128:266–271.
46. Adlercreutz H, Martin F, Pulkknen M, et al. Intestinal metabolism of estrogens. J Clin Endocrinol Metab 1976; 43:497–505.
47. Lombardi P, Goldin B, Boutin E, Gorbach SL. Metabolism of androgens and estrogens by human fecal microorganisms. J Steroid Biochem 1978; 9:795–801.
48. Bolskenheuser VD, Winter J. In: Hentges DJ, ed. Human Intestinal Microflora in Health and Disease. New York: Academic Press, 1983:215–239.
49. Gibson GR, Macfarlane S, Macfarlane GT. Metabolic interactions involving sulphate-reducing methanogenic bacteria in the human large intestine. FEMS Microb Ecol 1993; 12:117–125.
50. Gibson GR, Cummings JH, Macfarlane GT. Use of a three-stage continuous culture system to study the effect of mucin on dissimilatory sulfate reduction and methanogenesis by mixed populations of human gut bacteria. Appl Environ Microbiol 1988; 54:2750–2755.
51. McFarlane GT, Gibson GR, Cummings JH. Comparison of fermentation reductions in different regions of the human colon. J Appl Bacteriol 1992; 72:57–64.
52. Pitcher MCL, Beatty ER, Gibson GR, Cummings JH. Incidence and activities of sulphate-inducing bacteria in patients with ulcerative colitis. Gut 1995; 36:A63.
53. Asatoor MA. Aromatization of quinic acid and shikimic acid by bacteria and the production of urinary hippurate. Biochem Biophy Acta 1965; 100:200–202.
54. Scheline RR. Metabolism of foreign compounds by gastrointestinal microorganisms. Pharmacol Rev 1973; 25:451–523.
55. Goldin BR. In situ bacterial metabolism and colon mutagens. Ann Rev Microbiol 1986; 40:367–393.
56. Laqueur GL, Spatz M. Oncogenicity of cycasin and methylazoxymethanol. Gann Monograph Cancer Res 1975; 17:189–204.
57. Laqueur GL, McDaniel EG, Matsumoto H. Tumor induction in germ free rats with methylazoxymethanol (MAM) and synthetic MAM acetate. J Natl Cancer Inst 1967; 39:355–371.

58. Druckery H, Pressmann R, Matzkies F, Ivanovic S. Selective erzengung von darmkrebs bei ratten durch 1,2-dimethylhydrozin. Naturwissenschoften 1967; 54:285–286.

59. Macdonald IA, Brussard RG, Hutchinson DM, Holdeman LV. Rutin induced beta-glucosidase activity in *Streptococcus faecuim* VGH-1 and streptococcus sp strain FRP-17 isolated from human feces: formation of the mutagen quercetin from rutin. Appl Environ Microbiol 1984; 47:350–355.

60. Weisburger JH, Grantham PM, Horton RE, Weisburger EK. Metabolism of the carcinogen N-hydroxy-N-2-fluorenylacetamide in germfree rats. Biochem Pharmacol 1970; 19:151–162.

61. Morotomi M, Nanno M, Watanbe T, Sakurai T, Mutai M. Mutagenic activitaton of biliary metabolites of 1-nitropyrene by intestinal microflora. Mutat Res 1985; 149:171–178.

62. Chung KT. The significance of azo-reduction in the mutagenesis and carcinogenesis of azo dyes. Mutat Res 1983; 114:269–281.

63. Ames BN, McCann J, Yamoski E. Methods for detecting carcinogens and mutagens with salmonella/mammalian-microsome mutagenicity test. Mutat Res 1975; 31:347–364.

64. McCoy EC, Rosenkranz HS, Mermelstein R. Evidence for the existence of a family of nitroreductase capable of activating nitrated polycyclics to mutagens. Environ Mutogen 1981; 3:421–427.

65. Karpinsky GE, Rosenkranz HS. The anaerobic-mediated mutagenicity of 2-nitrofluorene and 2-aminofluorene for *Salmonella typhimurium*. Environ Mutogen 1980; 2:353–358.

66. Hartman CP, Falk CE, Andrews AW. Azo reduction of trypan blue to a known carinogen by a cell-free extract of a human intestinal anoerobe. Mutat Res 1978; 58:125–132.

67. Hartman CP, Andrews AW, Chung KT. Production of mutagen from ponceau 3R by a human intestinal anaerobe. Infect & Immunity 1979; 23:686–689.

68. Reid TM, Morton KC, Wang CY, King CM. Conversion of congo red and 2-azoxyfluorence to mutagens following in vitro reduction by whole-all rat cecal bacteria. Mutat Res 1983; 117:105–112.

69. El-Bayoumy K, Fharma C, Louis Y, Reddy B, Hecht SS. The role of the intestinal microflora in the metabolic reduction of 1-nitropyrene to 1-aminopyrene in conventional and germfree rats and in humans. Cancer Lett 1983; 19:311–316.

70. Manning BW, Campbell WL, Franklin W, Delclos KB, Cerniglia CE. Metabolism of 6-nitrochyfesene by intestinal microflora. Appl Environ Microbiol 1988; 54:197–203.

71. Magee PH, Barnes JM. The production of malignant hepatic tumors in the rat by feeding dimethylnitrosamine. Br J Cancer 1956; 10:114–122.

72. Ender F, Ceh L. Occurrence of nitrosamines in foodstuffs for human and animal consumption. Food Cosmet Toxicol 1968; 6:569–571.

73. Tannenbaum SR, Sinskey AJ, Weisman M, Bishop W. Nitrate in human saliva, its possible relationship to nitrosamine formation. J Natl Cancer Inst 1974; 53:79–84.

74. Klubes P, Cerna I, Rabinowitz AD, Jondorf WR. Factors effecting dimethrylnitrosamine formation from simple precursors by rat intestinal bacteria. Food Cosmet Toxicol 1972; 16:307–330.

75. Chung KT, Fulk GE, Slein MW. Tryptophanase of fecal flora as a possible factor in the etiology of colon cancer. J Natl Cancer Inst 1975; 54:1073–1078.

76. Goldin BR, Peppercorn MA, Goldman P. Contribution of host and intestinal microflora in the metabolism of L-DOPA by the rat. J Pharmacol Exp Ther 1973; 186:160–166.

77. Peppercorn MA, Goldman P. The role of intestinal bacteria in the metabolism of salicylazosulfapyridine. J Pharmacol Exp Ther 1972; 181:151–162.

78. Peppercorn MA, Goldman P. Distribution studies of salicylazosulfapyridine and its metabolities. Gastroenterology 1973; 64:240–245.

79. Goldman P. Drug therapy: Metronidazole. N Engl J Med 1980; 303:1212–1218.

80. Renwick AG, Williams RT. The fate of cyclamate in man and other species. Biochem J 1972; 129:869–879.

81. Wallace WC, Lethco ET, Brower EA. The metabolism of cyclamate in rats. J Pharmacol Exp Ther 1970; 175:325–330.

82. Lindenbaum J, Rund DG, Butler VP, Tse-Eng D, Sala JR. Inactivation of digoxin by the gut flora: revised by antibiotic therapy. N Engl J Med 1981; 305:789–794.

83. Muthan VI, Wiederman J, Dobkin JF, Lindenbaum J. Geographic differences in digoxin inactivation, a metabolic activity of the human anaerobic gut flora. Gut 1989; 30:971–977.

84. Fisher LJ, Millburn P, Smith RL. The fate of ^{14}C stilleslerol in the rat. Biochem J 1966; 100:69.

85. Eriksson H, Gustafsson JA. Excretion of steroid hormones in adults: steroids in feces from adults. Eur J Biochem 1971; 18:146–150.

86. Sandberg AA, Slaunwhite WR, Jr. Studies on phenolic steroids in human subjects. II. The metabolic fate and hepato-biliary-enteric circulation of ^{14}C-estrone and ^{14}C-estradiol in women. J Clin Invest 1957; 36:1266–1278.

87. Sandberg AA, Slaunwhite WR, Jr. Studies on phenobic steroids in human subjects. VII. Metabolic fate of estriol and its glucuronide. J Clin Invest 1965; 44:694–702.

88. Adlercreutz H, Martin F, Tikkanen MJ, Pulkkinen MO. Effect of ampicillin administration on the excretion of twelve oestrogens in pregnancy urine. Acta Endocrinol 1975; 80:551–557.

89. Reimer D. Rifampicin "pill" do not go well together. J Am Med Assoc 1974; 227:608.

90. Dossetar J. Drug interaction with oral contraceptives. BMJ 1975; 4:467–468.

8

The Metabolism of Polyphenols by the Human Gut Microbiota

Max Bingham
Unilever Research and Development, Vlaardingen, The Netherlands

SUMMARY

Polyphenols are considered to be key active constituents of fruits and vegetables and responsible for many of the health protective effects of diets rich in these foods. While their structure varies considerably, following ingestion, most ($\sim 95\%$) persist to the colon where they encounter the human gut microbiota. Here they may undergo considerable structural alteration to compounds that may have enhanced biological properties or possibly degraded into inert metabolites and excreted. As such, the human gut microbiota may have a significant influence on the final outcomes of polyphenol ingestion. Moreover, inter-individual variation in the composition of the microbiota means that certain compounds are metabolized in different ways, and this is reflected in the considerable variability seen in excreted polyphenol metabolites. Consequently, polyphenols as active ingredients in functional foods may turn out to be beneficial for only a certain proportion of the population. Clearly, this may further have an impact on disease risk and health protection. This chapter considers the potential role of the human gut microbiota in polyphenol metabolism and highlights the level of current understanding of this process.

INTRODUCTION

Interest in the role of polyphenols in health has never been greater. Responsible for much of the flavor, texture, and appearance of fruits, vegetables, pulses, and grains, polyphenols are also considered to be largely responsible for many of the positive health effects of a diet rich in these particular food groups. In particular, epidemiologic studies suggest a protective effect of fruits and vegetables against cancer and coronary heart disease (1–3). In addition to antioxidant properties, polyphenols show a number of interesting activities in animal models and within in vitro systems. These effects include scavenging free radicals, nitric oxide regulation, apoptosis induction, inhibition of cell proliferation and angiogenesis and phytoestrogenic activity (4–7). As such these effects may contribute to their potentially protective role in cancer and coronary heart disease. And yet, the question remains of whether these types of studies are relevant in humans since substantial proportions of ingested polyphenols persist to the colon and may undergo extensive metabolism in the gut prior to absorption (8,9). This may further explain the failure of

155

many studies that sought to detect increases in antioxidant capacity in plasma following diets rich in polyphenols.

Significantly, our understanding of these processes is limited but a considerable focus of attention on this issue has recently been established. This is following the realization that achievable concentrations of polyphenols in circulation may be significantly affected by the metabolic activities of the human gut microbiota (9,10). Overall, such processes may represent a significant factor in determining the final health outcomes of a diet rich in fruits, vegetables, pulses, and grains. We consider how the human gut microbiota can influence the bioavailability of polyphenols and establish the extent to which variations in microbiota composition between individuals may affect such processes.

TYPES OF POLYPHENOLS AND MICROBIAL METABOLISM

Polyphenol compounds are ubiquitous in the plant kingdom. They are secondary plant metabolites since they are not required in their primary metabolism. Rather, these compounds are essential for appearance, taste, stability, and often the protection of plant tissue. They have a wide variety of structures, chemical characteristics and to date several thousand compounds of this nature have been identified in higher plants (3). A more limited proportion of these compounds are present in edible food crops. The wide variety of compound characteristics means they are often separated into different classes according to their structural properties (Fig. 1).

An important issue in the study of polyphenols in the diet is that the most commonly consumed ones are not necessarily the most active within the human body. There are several possible reasons for this: they may have low intrinsic activity, they may be poorly absorbed from the intestine, extensively metabolized in the intestine or rapidly eliminated. In addition, the metabolites circulating in blood or reaching target organs, and those that result from hepatic or digestive processes may differ from their original substances in biological activity. It is therefore crucial to have an extensive knowledge of the bioavailability of polyphenols if the true health effects of these compounds are to be understood.

Even though polyphenols exhibit a large structural diversity, the metabolism of these compounds occurs via a common pathway (11). A limited proportion of polyphenols (mostly as aglycones) is absorbed intact in the small intestine. The balance, mostly present as glycosides, persists to the colon where they may undergo extensive metabolism and structural alteration by the colonic microbiota. A diverse range of smaller molecular weight compounds result and these can be detected in plasma, urine, and feces in various forms (10,12,13). A consistent observation in studies of polyphenol metabolism is that considerable inter-individual variation is seen in both the types and amounts of polyphenol metabolites that result from polyphenol ingestion (14–16). In many cases it is thought that compositional variations in the colonic microbiota are responsible for this. Many factors can influence the development and composition of the microbiota, including the overall diet, drugs, age, xenobiotics, and host factors such as gastric secretions and luminal pH (17–19). By proxy, the composition of the microbiota may have significant influence on the metabolism of polyphenols and thus the final health outcomes of diets rich in fruits, vegetables, cereals, and grains. In the light of this we consider the fate of a number of representative polyphenols classes (phenolic acids, flavonoids, anthocyanins, and proanthocyanidins) to illustrate an important aspect of the extent to which the colonic microbiota activity can impact on health.

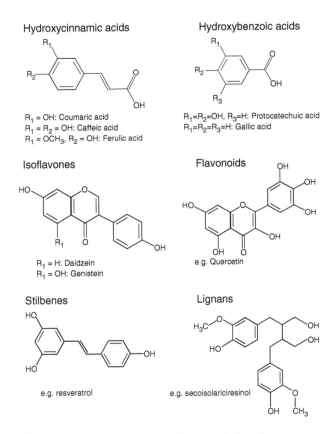

Hydroxycinnamic acids

$R_1 = OH$: Coumaric acid
$R_1 = R_2 = OH$: Caffeic acid
$R_1 = OCH_3$, $R_2 = OH$: Ferulic acid

Hydroxybenzoic acids

$R_1=R_2=OH$, $R_3=H$: Protocatechuic acid
$R_1=R_2=R_3=H$: Gallic acid

Isoflavones

$R_1 = H$: Daidzein
$R_1 = OH$: Genistein

Flavonoids

e.g. Quercetin

Stilbenes

e.g. resveratrol

Lignans

e.g. secoisolariciresinol

Figure 1 Example structures of main polyphenol groups.

PHENOLIC ACIDS—HYDROXYCINNAMATES AND HYDROXYBENZOATES

The phenolic acids are generally the lowest molecular weight polyphenol compounds. Hydroxycinnamates (Fig. 2) are a core class of polyphenols and are central to the biosynthetic pathways of polyphenols. Caffeic and quinic acids combine to form chlorogenic acid, which is found in many types of fruit and in high concentrations in coffee (20). Ferulic acid is the most abundant phenolic acid found in grains and may constitute the main dietary source of this compound. Hydroxybenzoic acids (Fig. 2) are less abundant in plants but are often found in red fruits, black radishes and onions. Tea is an important source of gallic acid containing 4.5 g/kg fresh weight tea leaves (21), whilst ellagic acid is a major polyphenol in some berry fruits. Hydroxybenzoic acids are also important components of complex hydrolysable tannins such as gallotannins in mangoes and ellagitannins in red fruits, hazelnuts, walnuts, pomegranates, and oak aged wines (from the barrels) (22–25).

Human bioavailability studies for hydroxycinnamates reveal that between 0.3 and 25% of ingested dose is excreted in urine (12). Chlorogenic acid (ingested as coffee) has been detected at low concentrations in urine samples (26,27) along with a range of smaller molecular weight secondary metabolites including ferulic acid, isoferulic acid, dihydroferulic acid, vannilic acid, 3,4-dihydroxyphenylpropionic acid, 3-hydroxyhippuric acid, and hippuric acid (27–29). One third of ingested chlorogenic acid is absorbed in the

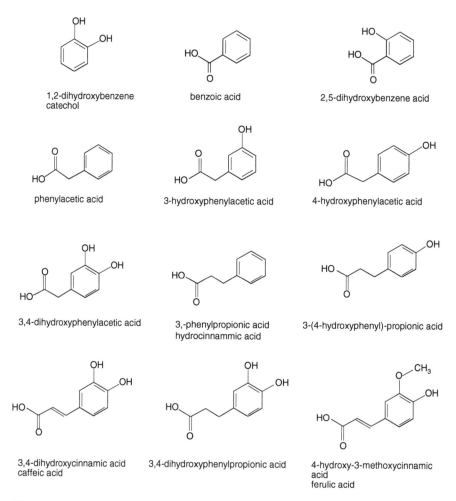

Figure 2   Examples of phenolic acids and flavonoid metabolites (e.g., hydroxycinnamic acids and hydroxybenzoic acids).

small intestine, leaving the balance to persist to the colon where it is exposed to the gut microbiota (26,30). Inter-individual variations in excretion profiles in these studies suggest that hydroxycinnamates are substantially metabolized by the colonic microbiota. In vitro studies have also revealed that chlorogenic acid is extensively metabolized by the colonic microbiota (31). Using inocula from different volunteers, it was clearly demonstrated that degradation rates of chlorogenic acid and the production of the metabolites 3,4-dihydroxyphenylpropionic acid and 3-hydroxyphenylpropionic acid varied considerably between volunteers. Meanwhile in studies with healthy human volunteers, over 50% of the ingested dose was excreted as hippuric acid (a potential microbial metabolite of chlorogenic acid), whilst in the same study individuals without a colon excreted a much smaller amount of aromatic acids. The balance was not metabolized and excreted as chlorogenic acid in the latter volunteers. The absence of a colon and therefore a substantial colonic microbiota in the volunteers and the apparent excretion of intact chlorogenic acid effectively demonstrate the necessity for and the metabolic capacity of the colonic microbiota (27).

Successful attempts have been made to identify colonic microbiota species capable of metabolizing hydroxcinnamates. Inter-individual differences in excretion profiles of volunteers imply that the composition of the resident microbiota may be important in determining this final profile. Given that some of these metabolites are considered to be potentially protective of health, knowledge of the identity of species responsible for such metabolic activity is valuable. It has been demonstrated that at least three colonic microbiota species (*Bifidobacterium lactis*, *Lactobacillus gasseri*, and *Escherichia coli*) can release hydroxycinnamates from chlorogenic acid in the gut (32) as well as diferulic acid being released in the colon as a result of metabolism by esterase activity of the colonic microbiota (33,34). Given that free hydroxycinnmates (including ferulic, caffeic, and p-coumaric acids) exhibit antioxidant and anticarcinogenic properties in vitro and in animal models, and that various microbial metabolites can be absorbed readily (35), this supports the notion that some beneficial effects of hydroxycinnamtes can be ascribed to the metabolic activities and products of the colonic microbiota.

A more limited set of studies has been carried out for hydroxybenzoates. Ellagitannins are polyphenols made up of subunits of ellagic acid (a hydroxybenzoate) and are thought to possess chemopreventative properties that might contribute to health benefits in humans (36–38). Their fate has been studied in 40 volunteers consuming a variety of foodstuffs known to contain high levels of ellagitannins (39). In all cases the ellagitannin microbial metabolite urolithin B (which may be antiangiogenic) could be identified although inter-individual variation in excretion rates was large. Furthermore they were able to identify high and low excretors of this compound in much the same way that consumers of soya can be differentiated by their ability to excrete equol (40,41). Again this observation indicates that the gut microbiota is likely to be important in the bioavailability of these potentially health-promoting compounds and that variations in the composition of the microbiota may dictate the production of a potentially health-promoting metabolite. At present, we are unaware of any studies designed to identify components of the colonic microbiota that are potentially responsible for the metabolism of ellagitannins.

FLAVONOIDS

Flavonoids are the most important class of polyphenols in plants. Over 6000 flavonoids have been identified so far (3) and their structural variety is based on the flavan or 2-phenyl-benzo-dihydropyrane skeleton. Flavonoids are further differentiated into subclasses (Fig. 3). The metabolism of two of these classes are discussed here—flavonols and flavan-3-ols.

Flavonols

Flavonols are the most ubiquitous flavonoids in plants, with the main representatives being quercetin, kaempferol, myricetin, and isohamnetin, which are predominantly present as glycosides bound to a variety of sugar moieties. The richest sources are onions, curly kale, leeks, broccoli, and blueberries and are present at levels of approximately 30 mg/kg fresh weight although in certain circumstances can reach in excess of 1.2 g/kg fresh weight. Red wine and tea are also rich sources.

In contrast to other classes of polyphenols, flavonols such as quercetin and kaempferol have received a larger amount of attention in terms of bioavailability over the past few years. This is largely because of their ubiquitous nature in food crops but also because a great deal of their apparent in vitro effects on health parameters have failed to be

Anthocyanidins

Flavanones

R$_1$=R$_2$=H: Perlargonidin
R$_1$=OH, R$_2$=H: Cyanidin
R$_1$=R$_2$=OH: Delphinidin
R$_1$=OCH$_3$, R$_2$=OH: Petunidin
R$_1$=R$_2$=OCH$_3$: Malvidin

R$_1$=H: R$_2$=OH: Narigenin
R$_1$=R$_2$=OH: Eridocytol
R$_1$=OH: R$_2$=OCH$_3$: Hesperetin

Flavanols

Flavonols

R$_1$=R$_2$=OH: R$_3$=H: Catechins
R$_1$=R$_2$=R$_3$=OH: Gallocatechins

R$_2$=OH: R$_1$=R$_3$=OH: Kaempferol
R$_1$=R$_2$=OH: R$_3$=H: Quercetin
R$_1$=R$_2$=R$_3$=OH: Myricetin

Trimeric procyanidin

Flavones

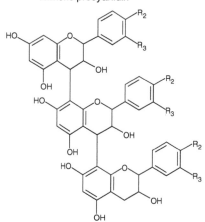

R$_1$=H: R$_2$=OH: Apigenin
R$_1$=R$_2$=OH: Luteolin

Figure 3 Example structures of various flavonoid groups.

repeated in vivo (13). The majority of these compounds exist as glycosides in their original food matrices and thus reach the colon intact following ingestion. Here they can serve as substrate for the microbiota. Associations between the urinary excretion of simple phenolics such as hydroxyhippuric acid, hydroxyphenylacetic acid and 3-(hydroxyphenyl)-propionic acid, and a high flavonoid intake have been observed in a number of human studies (28,29,40–48) indicating that a substantial proportion of polyphenols undergo metabolism in the gut. In addition, the microbiota has also been confirmed as the major site for the release of free flavonol aglycones from their conjugated forms following cleavage of ester or glycosidic forms (49). Quercetin derivatives are deconjugated and converted to hydroxyphenylacetic acids by the colonic microbiota in vitro (50). Confirming these observations, recent in vitro studies revealed the production of 3-hydroxyphenylacetic acid and 3-(hydroxyphenyl)-propionic acid from rutin (a representative glycoside of quercetin) in human gut microbiota fermentation studies (31). An important observation in these studies was that the pattern of degradation varied considerably between donor fecal microbiota samples and with concentration of the initial substrate. This is significant since many of the compounds produced in this degradative process may have enhanced biological properties. 3,4-dihydroxyphenylacetic acid and 4-hydroxyphenylacetic acid have more effective antiplatelet aggregation activity than their precursors rutin and quercetin (51).

Compositional variations in the microbiota may have a significant impact on the final metabolic products of flavonol metabolism. Indeed reports of studies designed to confirm these observations are now appearing. *Eubacterium ramulus* is capable of metabolizing quercetin both in vitro and in rats associated with the organism (8). In both cases, the isolate was capable of releasing quercetin from its glycosidic form and was then able to cleave the ring system of quercetin and produce mainly 3,4-dihydroxyphenylacetic acid. Further studies in humans revealed that *E. ramulus* is a common member of the human gut microbiota (52); its resident population level is dependant on flavonoid intake and the production of secondary metabolites of flavonoids (such as 3,4-dihydroxyphenylacetic acid) was greatest when *E. ramulus* populations where increased (15). Meanwhile *E. ramulus* has also been tested for its abilities to degrade other structurally related flavonoids including other flavonols, flavones, flavan-3-ols, and flavonones, and in certain cases, significant metabolism can occur (53). *Clostridium orbiscindens*, which is an obligate anaerobe commonly found in the intestinal tract, is also capable of cleaving the C3-C4 bond of quercetin to give 3,4-dihydroxyphenylacetic acid (54). In recent studies, it was also shown to degrade a range of other flavonols and flavanones in vitro and that it was present in 8 of 10 volunteers at levels of 1.87×10^8 to 2.5×10^9 cells/ g (55). At present, these are the most extensively published reports on the influence of microbiota composition in polyphenol metabolism and set the benchmark for future studies in other polyphenol classes.

Flavan-3-ols

Flavan-3-ols are found in most plants and the stereo isomers (+)-catechin and (−)-epicatechin are the most common monomeric flavan-3-ols in fruits. (+)-gallocatechin and (−)-epigallocatechin are their corresponding O-3 gallates and are rarer but found in certain seeds of leguminous plants, in grapes and in tea. Catechins are found in many types of fruit and red wine but by far the most abundant sources are green tea and chocolate (56).

The bioavailability of flavan-3-ols differs markedly among the different catechins and appears to be related substantially to structure and degree of galloylation (12). Again due to their structure, a substantial proportion of ingested flavan-3-ols may persist to the colon where they encounter the colonic microbiota. The colonic degradation of flavan-3-ols such as catechin, epicatechin, and epicatechin gallate have been investigated previously (14,44,57–59) revealing that in contrast to other similar structures the heterocyclic C-ring is not cleaved per se. The hydroxylation pattern of flavan-3-ols (5,7,3′,4′-) has instead been suggested to enhance the opening of the heterocyclic ring after hydrolysis (60,61) and this results in the production of a large number of metabolites from the colonic microbiota: 3,4-dihydroxyphenylacetic acid, 3-hydroxyphenylacetic acid, homovanillic acid and their conjugates are derived from the B-ring and phenolic acids from the C-ring (61). In animal studies, phenylvalerolactones, and phenylpropionic acids have also been identified as degradation products (61).

Antibiotic treatment in rats significantly alters the metabolism of catechin and decreases the urinary elimination of many of the compounds of flavan-3-ol metabolism indicating that an intact microbiota is necessary for the production of many of these compounds (57). At present there are limited studies (described by 8; see Flavonols) that have investigated the specific species that may be responsible for the conversion of flavan-3-ols, and data is limited on any possible inter-individual variation in final metabolic profiles. However, green tea catechins have been shown to cause a shift in bacterial populations in humans (62), pigs (63) and chickens (64), and this may have relevance to the overall polyphenol metabolic capabilities of the resident microbiota. Structurally, these compounds may exhibit substantial anti-oxidant activities and thus the influence of the composition of the resident microbiota and associated metabolic variations could impact on the overall health impact of flavan-3-ol ingestion.

ANTHOCYANIDINS

The red to purple colored anthocyanidins are responsible for a good portion of color in fruits and flowers. They are only present as glycosides or anthocyanins and their color is pH dependent. In the human diet, anthocyanidins are present in red wine, certain varieties of cereals, certain leafy and root vegetables (e.g., aubergines, cabbage, beans, onions, and radishes) and most abundantly in fruit. The content is generally proportional to the color intensity and may reach values of 2–4 g/kg fresh weight in blackberries and black currants. They are found mainly in the skin, except where the flesh is also colored.

Anthocyanidins and anthocyanins have been reported previously as having several positive effects on health (35,65–72). Much of this evidence has been derived in vitro and very little is known about their bioavailability in vivo. Previous human and rat studies have reported very low recoveries of intact anthocyanins in urine (73). Very little is known of the specific fate of the balance of these compounds. Given their structure, it is likely that they will undergo substantial metabolism by the human gut microbiota in much the same way as any other flavonoid structure. And yet, studies performed in the 1970s indicated that degradation of anthocyanins by the microbiota occurs to a much more limited extent than with other flavonoid structures (61). However recent studies investigated in vitro whether the anthocyanin glycosides, cyanidin-3-glucoside, and cyanidin-3-rutinoside were deglycosylated and whether the resulting aglycones were degraded further to smaller phenolic compounds by colonic bacteria (74). Cyanidin-3-glucoside and cyanidin aglycone were identified as intermediary metabolites of cyanidin-3-rutinoside. Protocatechuic acid was identified as a major metabolite at early stages of the fermentations along with a variety of other low molecular weight metabolites suggesting that the anthocyanins were converted by the gut microbiota. However, protocatechuic acid was also formed in vitro with the simple incubation of cyanidin with rat plasma in the absence of colonic microbiota (75). These experiments, although far from conclusive indicate that bacterial metabolism of anthocyanins can occur and is likely to involve the cleavage of glycosidic links and the breakdown of the anthocyanidin heterocycle—thus having a potential impact on the bioavailability of these compounds in vivo. However, significantly more investigation is needed before the real extent of the involvement of the microbiota is uncovered in terms of metabolism and the bioavailability of anthocyanins.

PROANTHOCYANIDINS

Proanthocyanidins are dimers, oligomers, and polymers of flavan-3-ols and are formed by enzymatic or chemical condensation. These so-called "condensed tannins" contribute to astringent tastes in fruits (e.g., grapes, peaches, apples, pears, berries etc.), beverages (e.g., wine, cider, tea, beer etc.) and chocolate. At a lower degree of polymerization they are colorless and bitter to taste, but with greater polymerization the taste becomes astringent and the color yellow to brown. Proanthocyanidins purely consisting of catechin and epicatechin monomers are called procyanidins, which are the most common type of proanthocyanidins. Less abundant are the prodelphinidins, which include both epicatechin and gallocatechin monomers.

Previous studies in rats have indicated that the bioavailability of procyanidins is low and characterized by a very low urinary recovery (0.5% ingested dose) (76). Procyanidin consumption in rats and in humans is associated with the production of several aromatic compounds including derivatives of phenylpropionic, phenylacetic, and benzoic acids (77,78). More recent studies have also established that consumption of proanthocyanidins from grape seed extract can result in a consistent increase in urinary excretion of 3-hydroxyphenylpropionic acid and 4-O-methylgallic acid. Inter-individual variation in excretion of 3-hydroxyphenylproionic acid was significant (79). The microbial metabolism of proanthocyanidins has never been studied in humans but the microbial origin of these compounds was established in vitro following incubation of proanthocyanidins with rat cecal contents (80) and human fecal microbiota (81). These studies utilized ^{14}C labeled proanthocyanidin oligomers and led to the formation of m-hydroxyphenylpropionic acid, m-hydroxyphenylacetic acid and their p-hydroxy isomers, m-hydroxyphenylvaleric acid, phenylpropionic acid, phenyl acetic acid and benzoic acid. Attempts have been made in the past to identify intestinal bacteria that can degrade proanthocyanidins (82,83) although these studies actually failed. The impact of proanthocyanidins on colonic microbiota populations has been investigated in rat studies and revealed that there was a shift in the predominant bacteria present towards Gram-negative Enterobacteriaceae and Bacteroides species (84). Furthermore, proanthocyanidin intestinal absorption and microbial metabolism of some of the above metabolites fell as the degree of polymerisation increased (77,81). Thus studies on antioxidant and biological effects of proanthocyanidins are only useful when targeted at compounds with a low degree of polymerization. Larger compounds do not appear to be able to reach systemic circulation or be available for microbial metabolism that would result in significant production of readily absorbable phenolic acid metabolites. However, this does highlight that at least some of the purported health effects of proanthocyanidin-rich diets may be due to secondary metabolites rather than the original ingested compounds.

PERSPECTIVES

The role of dietary polyphenols in health and disease continues to be the focus of much academic and commercial research. Consumption of diets rich in polyphenols is generally thought to be beneficial to health and this has led to great excitement over the potential of diets, supplements, and pro-drugs based on polyphenol compositions. As we have discussed, much of the latest research surrounds the question of bioavailability since they must reach target tissues in a form that is viable and can have the desired effects. A major obstacle for these compounds, though, is the microbial mass in the colon since in many cases they persist intact to the colon, and are structurally ideal for metabolism by the human gut microbiota. This review highlights the extent to which certain polyphenol classes undergo metabolism and structural alteration in the colon, and suggests that much of the prescribed in vivo health benefits of polyphenols may be due to secondary metabolites of polyphenols rather than the original compounds.

Well-designed studies have evaluated the need for an intact microbiota in polyphenol metabolism, although this is not so for all classes. Perhaps one of the most consistent observations in human bioavailability studies of dietary polyphenol compounds is how striking the inter-individual variations are in the types and amounts of metabolic breakdown products seen following polyphenol ingestion. The reasons for this have at present not been rigorously investigated, but it seems very likely that variations in composition of the resident microbiota between individuals is key. Numerous factors can influence the composition of the microbiota (19) and this in turn may affect the overall metabolic capabilities of this system.

Only a limited amount of research has been targeted at specifically trying to identify actual species of the human gut microbiota that are responsible or capable of metabolizing polyphenols. This is perhaps a reflection of the difficulties encountered in undertaking such an effort. Simply isolating single strains of bacteria anaerobically and carrying out suitable fermentation assays when presented with such a complex mixed culture of bacteria is an extreme challenge in terms of the laboratory time required. Furthermore much of the microbial mass in the colon remains to be described or cultured (85,86). In terms of the metabolic pathways that polyphenols follow during their degradation in the gut, they are fairly complex and multistaged, suggesting (although unconfirmed at this stage) that more than one species/strain may be required for the complete degradation of the original substrate. The application of culture independent molecular microbiology techniques (such as fluorescent in situ hybridization) (87) and modern analytical chemistry techniques (such as metabonomics in combination with pattern recognition techniques) means that an understanding can be gained of whether variable levels of target populations present in the gut are related to the production of specific metabolites. This may in turn have an impact on specific health outcomes (such as cardiovascular markers of health or the development of cancers). This is particularly so given that a number of the microbial metabolites are now thought to have specific activities related to health.

REFERENCES

1. World Cancer Research Fund. Vegetables and fruit. In: Food, Nutrition and the Prevention of Cancer: A Global Perspective. Washington, D.C.; World Cancer Research Fund/ American Institute for Cancer Research 1997; 436–446.
2. Law MR, Morris JK. By how much does fruit and vegetable consumption reduce the risk of ischemic heart disease? Eur J Clin Nutr 1998; 52:549–556.
3. Arts IC, Hollman PC. Polyphenols and disease risk in epidemiologic studies. Am J Clin Nutr 2005; 81:317–325.
4. Higdon JV, Frei B. Tea catechins and polyphenols: health effects, metabolism and antioxidant functions. Crit Rev Food Sci Nutr 2003; 43:89–143.
5. Yang CS, Landau JM, Huang MT, Newmark HL. Inhibition of carcinogenesis by dietary polyphenolic compounds. Annu Rev Nutr 2001; 21:381–406.
6. Nijveldt RJ, van Nood E, van Hoorn DEC, Boelens PG, van Norren K, van Leeuwen PAM. Flavonoids: a review of probable mechanisms of action and potential applications. Am J Clin Nutr 2001; 74:418–425.
7. Adlercreutz H, Mazur W. Phytoestrogens and western diseases. Ann Med 1997; 29:95–120.

8. Blaut M, Schoefer L, Braune A. Transformation of flavonoids by intestinal microorganisms. Int J Vitam Nutr Res 2003; 73:79–87.

9. Clifford MN. Diet-derived phenols in plasma and tissues and their implications for health. Plant Med 2004; 70:1103–1114.

10. Halliwell B, Rafter J, Jenner A. Health promotion by flavonoids, tocopherols, tocotrienols and other phenols: direct or indirect effects? antioxidant or not? Am J Clin Nutr 2005; 1:268–276.

11. Spencer JP. Metabolism of tea flavonoids in the gastrointestinal tract. J Nutr 2003; 1:3255–3261.

12. Manach C, Williamson G, Morand C, Scalbert A, Remesy C. Bioavailability and bioefficacy of polyphenols in humans I. Review of 97 bioavailability studies. Am J Clin Nutr 2005; 1:230–242.

13. Williamson G, Manach C. Bioavailability and bioefficacy of polyphenols in humans II. Review of 93 intervention studies. Am J Clin Nutr 2005; 1:243–255.

14. Mulder TP, Rietveld AG, van Amelsvoort JM. Consumption of both black tea and green tea results in an increase in the excretion of hippuric acid into urine. Am J Clin Nutr 2005; 1:256–260.

15. Simmering R, Pforte H, Jacobasch G, Blaut M. The growth of the flavonoids-degrading intestinal bacterium, Eubacterium ramulus, is stimulated by dietary flavonoids in vivo. FEMS Microb Ecol 2002; 40:243–248.

16. Jenner AM, Rafter J, Halliwell B. Human fecal water content of phenolics: the extent of colonic exposure to aromatic compounds. Free Rad Biol Med 2005; 38:763–772.

17. Rowland IR. Factors affecting metabolic activity of the intestinal microbiota. Drug Metab Rev 1988; 19:243–261.

18. Kirjavainen PV, Gibson GR. Healthy gut microbiota and allergy: factors influencing development of the microbiota. Ann Med 1999; 31:288–292.

19. Mallett AK, Rowland IR. Factors affecting the gut microbiota. In: Rowland I, ed. Role of the Gut Flora in Toxicity and Cancer. London: Academic Press, 1998:347–382.

20. Clifford MN. Chlorogenic acids and other cinnamates—nature, occurance and dietary burden. J Sci Food Agric 1999; 79:362–372.

21. Tomas-Barberan FA, Clifford MN. Dietary hydroxybenzoic acid derivatives and their possible role in health protection. J Sci Food Agric 2000; 80:1024–1032.

22. Clifford MN, Scalbert A. Elagitannins—nature, occurence and dietary burden. J Sci Food Agric 2000; 80:118–125.

23. Laszlavik M, Gla L, Misik S, Erdei L. Phenolic compounds in two hungarian red wines matured in quercus robur and quercus petrea barrels: HPLC analysis and diode array detection. Am J Enol Vitic 1995; 46:67–74.

24. Haslam E. Plant Polyphenols, Vegetable Tannins Revisitied. Cambridge, U.K.: Cambridge University Press, 1989.

25. Gil MI, Tomas-Barberan FA, Hess-Pierce B, Holcroft DM, Kader AA. Antioxidant activity of pomegranate juice and its relationship with phenolic composition and processing. J Agric Food Chem 2000; 48:4581–4589.

26. Olthof MR, Hollman PC, Katan MB. Chlorogenic acid and caffeic acid are absorbed in humans. J Nutr 2001; 131:66–71.

27. Olthof MR, Hollman PC, Buijsman MN, van Amelsvoort JM, Katan MB. Chlorogenic acid, quercetin-3-rutinoside and black tea phenols are extensively metabolized in humans. J Nutr 2003; 133:1806–1814.

28. Rechner AR, Spencer JP, Kuhnle G, Hahu U, Rice-Evans CA. Novel biomarkers of the metabolism of caffeic acid derivatives in vivo. Free Rad Biol Med 2001; 30:1213–1222.

29. Rechner AR, Pannala AS, Rice-Evans CA. Caffeic acid derivatives in artichoke extract are metabolized to phenolic acids in vivo. Free Rad Res 2001; 35:192–202.

30. Gonthier MP, Verney MA, Besson C, Remesy C, Scalbert A. Chlorogenic acid bioavailability largely depends on the its metabolism by the gut microbiota in rats. J Nutr 2003; 133:1853–1859.

31. Rechner AR, Smith MA, Kuhnle G, et al. Colonic metabolism of dietary polyphenols: influence of structure on microbial fermentation products. Free Rad Biol Med 2004; 36:212–225.

32. Couteau D, McCartney AL, Gibson GR, Williamson G, Faulds CB. Isolation and characterisation of human colonic bacteria able to hydrolyse chlorogenic acid. J Appl Microbiol 2001; 90:873–881.

33. Andreasen MF, Kroon PA, Williamson G, Garcia-Conesa MT. Intestinal release and uptake of phenolic antioxidant diferulic acids. Free Rad Biol Med 2001; 31:304–314.

34. Konishi Y, Kobayashi S. Microbial metabolites of ingested caffeic acid are absorbed by monocarboxylic acid transporter (MCT) in intestinal Caco-2 cell monolayers. J Agric Food Chem 2004; 52:6418–6424.

35. Tsuda T, Watanabe M, Ohshima K, et al. Antioxidative activity of the anthocyanin pigments cyanidin 3-O-β-D-glucoside and cyanidin. J Agric Food Chem 1994; 42:2407–2410.

36. Gali HU, Perchellet EM, Klish DS, Johnson JM, Perchellet JP. Antitumour-promoting activities of hydrolyzable tannins in mouse skin. Carcinogenesis 1992; 13:715–718.

37. Castonguay A, Gali HU, Perchellet EM, et al. Antitumourigenic and antipromoting activities of ellagic acid, ellagitannins and oligomeric anthocyanin and procyanidin. Int J Oncol 1997; 10:367–373.

38. Kresty LA, Morse MA, Morgan C, et al. Chemoprevention of esophageal tumourogenesis by dietary administration of lyophilised black rasperberries. Cancer Res 2001; 61:6112–6119.

39. Cerda B, Tomas-Barberan FA, Espin JC. Metabolism of antioxidant and chemopreventative ellagitannins from strawberries, raspberries, walnuts and oak aged wine in humans: identification of biomarkers and individual variability. J Agric Food Chem 2005; 53:227–235.

40. Rechner AR, Kuhnle G, Bremner P, Hubbard GP, Moore KP, Rice-Evans CA. The metabolic fate of dietary polyphenols in humans. Free Rad Biol Med 2002; 33:220–235.

41. Setchell KDR, Brown NM, Lydeking-Olsen E. The clinical importance of the metabolite equol. A clue to the effectiveness of soy and its isoflavones. J Nutr 2002; 132:3577–3584.

42. Booth AN, Emerson OH, Jones FT, DeEds F. Urinary metabolites of caffeic and chlorogenic acids. J Biol Chem 1957; 229:51–59.

43. Booth AN, Murray CW, Jones FT, DeEds F. The metabolic fate of rutin and quercetin in the animal body. J Biol Chem 1956; 223:251–257.

44. Bravo L, Abia R, Eastwood MA, Saura-Calixto F. Degradation of polyphenols (catechin and tannic acid) in the rat intestinal tract: effect on colonic fermentation and fecal output. Br J Nutr 1994; 71:933–946.

45. DuPont MS, Bennett RN, Mellon FA, Williamson G. Polyphenols from alcoholic apple cider are absorbed, metabolized and excreted by humans. J Nutr 2002; 132:172–175.

46. Graefe EU, Veit M. Urinary metabolites of flavonoids and hydroxycinnamic acids in humans after application of a crude extract from *Equisetum arvense*. Phytomedicine 1999; 6:239–246.

47. Gross M, Pfeiffer M, Martini M, Cambell D, Slavin J, Potter J. The quantitation of metabolites of quercetin flavonols in human urine. Cancer Epidemiol Biomarkers Prev 1996; 5:711–720.

48. Pietta PG, Gardana C, Mauri PL. Identification of ginko biloba flavonol metabolites after oral administration to humans. J Chromatogr B 1997; 693:249–255.

49. Plumb GW, Garcia-Conesa MT, Kroon PA, Rhodes M, Ridley S, Williamson G. Metabolism of chlorogenic acid by human plasma, liver, intestine and gut microbiota. J Sci Food Agric 1999; 79:390–392.

50. Aura AM, O'Leary KA, Williamson G, et al. Quercetin derivatives are deconjugated and converted to hydroxyphenylacetic acids but not methylated by human fecal flora in vitro. J Agric Food Chem 2002; 50:1725–1730.

51. Kim DH, Jung EA, Sohng IS, Han JA, Kim TH, Han MJ. Intestinal bacterial metabolism of flavonoids and its relation to some biological activities. Arch Pharm Res 1998; 21:17–23.

52. Simmering R, Kleessen B, Blaut M. Quantification of the flavonoid-degrading bacterium *Eubacterium ramulus* in human fecal samples with a species specific oligonucleotide hybridisation probe. Appl Environ Microbiol 1999; 65:3705–3709.

53. Schneider H, Blaut M. Anaerobic degradation of flavonoids by *Eubacterium ramulus*. Arch Microbiol 2000; 173:71–75.

54. Winter J, Popoff MR, Grimont P, Bokkenheuser VD. *Clostridium orbiscindens* sp. Nov., a human intestinal bacterium capable of cleaving the flavonoid C-ring. Int J Syst Bacteriol 1991; 41:355–357.

55. Schoefer L, Mohan R, Scwiertz A, Braune A, Blaut M. Anaerobic degradation of flavonoids by *Clostridium orbiscindens*. Appl Environ Microbiol 2003; 69:5849–5854.

56. Labenbrink C, Lapczynski S, Maiwald B, Engelhardt UH. Flavonoids and other polyphenols in consumer brews of tea and other caffeinated beverages. J Agric Food Chem 2000; 48:1752–1757.

57. Gott DM, Griffiths LA. Effects of antibiotic pretreatments on the metabolism and excretion of [U14C](+)-catechin [U14C](+)-cyanidanol-3 and its metabolite, 3′-0-methyl-(+)-catechin. Xenobiotica 1987; 17:423–434.

58. Meselhy MR, Nakamura N, Hattori M. Biotransformation of (−)-epicatechin-3-O-gallate by human intestinal bacteria. Chem Pharm Bull 1997; 45:888–893.

59. Wang LQ, Meselhy MR, Li Y, et al. The heterocyclic ring fission and dehydroxylation of catechin and related compounds by *Eubacterium* sp. Strain SDG-2, a human intestinal bacterium. Chem Pharm Bull 2001; 49:1640–1643.

60. Spencer JP, Scroeter H, Rechner AR, Rice-Evans CA. Bioavailability of flavan-3-ols and procyanidins: gastrointestinal tract influences and their relevance to bioactive forms in vivo. Antioxid Redox Signal 2001; 3:1023–1039.

61. Scheline RR. CRC Handbook of Mammalian Metabolism of Plant Compounds. Boca Raton, FL: CRC Press, 1991.

62. Okubo T, Ishihara N, Oura A, et al. In vivo effects of tea polyphenol intake on human intestinal microbiota and metabolism. Biosci Biotechnol Biochem 1992; 56:588–591.

63. Hara H, Orita N, Hatano S, et al. Effect of tea polyphenols on fecal flora and fecal metabolic products of pigs. J Vet Med Sci 1995; 57:45–49.

64. Hara Y. Influence of tea catechins on the digestive tract. J Cell Biochem 1997; 27:52–58.

65. Morazzoni P, Bombardelli E, *Vaccinium myrtillus* L. Fitoterapia 1996; 67:3–29.

66. Renaud S, de Longeril M. Wine, alcohol, platelets, and the french paradox for coronary heart disease. Lancet 1992; 339:1523–1526.

67. Havsteen B. Flavonoids, a class of natural products of high pharmacological potency. Biochem Pharmacol 1983; 32:1141–1148.

68. Mian E, Curri SB, Lietti A, Bombardelli E. Anthocyanosides and the walls of microvessels: further aspects of the mechanism of action of their protective effect in syndromes due to abnormal capillary fragility. Minerva Med 1977; 68:3565–3581.

69. Kadar A, Robert L, Miskulin M, Tixier JM, Brechemier D, Robert AM. Influence of anthocyanoside treatment on the cholesterol-induced atherosclerosis in the rabbit. Paroi Artérielle 1979; 5:187–205.

70. Wang H, Nair MG, Strasburg GM, et al. Antioxidant and anti-inflammatory activities of anthocyanins and their aglycon, cyanidin, from tart cherries. J Nat Prod 1999; 62:294–296.

71. Kamei H, Kojima T, Hasegawa M, et al. Suppression of tumor cell growth by anthocyanins in vitro. Cancer Invest 1995; 13:590–594.

72. Wang H, Cao G, Prior RL. Oxygen radical absorbing capacity of anthocyanins. J Agric Food Chem 1997; 45:304–309.

73. Felgines C, Texier O, Besson C, Fraisse D, Lamaison JL, Remesy C. Blackberry anthocyanins are slightly bioavailable in rats. J Nutr 2002; 132:1249–1253.

74. Aura AM, Martin-Lopez P, O'Leary KA, et al. In vitro metabolism of anthocyanins by human gut microbiota. Eur J Nutr 2004; 44:133–142.

75. Tsuda T, Horio F, Osawa T. Absorption and metabolism of cyanidin 3-O-β-D-glucoside in rats. FEBS Lett 1999; 449:179–182.

76. Baba S, Osakabe N, Natsume M, Terao J. Absorption and urinary excretion of procyanidin B_2 [epicatechin-(4β-8)-epicatechin] in rats. Free Rad Biol Med 2002; 33:142–148.

77. Gonthier MO, Donovan JL, Texier O, Felgines C, Remesy C, Scalbert A. Metabolism of dietary procyanidins in rats. Free Rad Biol Med 2003; 35:837–844.

78. Rios LY, Gonthier MP, Remesy C, et al. Chocolate intake increases urinary excretion of polyphenol-derived phenolic acids in healthy human subjects. Am J Clin Nutr 2003; 77:912–918.

79. Ward NC, Croft KD, Puddey IB, Hodgson JM. Supplementation with grape seed polyphenols results in increased urinary excretion of 3-hydroxyphenylpropionic acid, an important metabolite of proanthocyanidins in humans. J Agric Food Chem 2004; 52:5545–5549.

80. Groenewoud G, Hundt HK. The microbial metabolism of condensed (+)-catechins by rat - cecal microbiota. Xenobiotica 1986; 16:99–107.

81. Deprez S, Brezillon C, Rabot S, et al. Polymeric proanthocyanidins are catabolised by human colonic microbiota into low molecular weight phenolic acids. J Nutr 2000; 130:2733–2738.

82. Makkar HP, Becker K. Degradation of condensed tannins by rumen microbes exposed to quebrancho tannins (QT) in rumen simulation technique (RUSITEC) and effects of (QT) on fermentative processes in the RUSITEC. J Sci Food Agric 1995; 69:495–500.

83. Makkar HP, Blummel M, Becker K. In vitro effects of and interactions between tannins and saponins and fate of tannins in the rumen. J Sci Food Agric 1995; 69:481–493.

84. Smith AH, Mackie RI. Effect of condensed tannins on bacterial diversity and metabolic activity in the rat gastrointestinal tract. Appl Environ Microbiol 2004; 70:1104–1115.

85. McCartney AL. Application of molecular biological methods for studying probiotics and the gut flora. Br J Nutr 2002; 88:29–37.

86. Manguin I, Bonnet R, Seksik P, et al. Molecular inventory of fecal microbiota in patients with Chrohn's disease. FEMS Microb Ecol 2004; 50:25–36.

87. Namsolleck P, Thiel R, Lawson P, et al. Molecular methods for the analysis of gut microbiota. Microb Ecol Health Dis 2004; 16:71–85.

9

Molecular Analysis of Host-Microbe Interactions in the Gastrointestinal Tract

Peter A. Bron
Wageningen Centre for Food Sciences and NIZO Food Research, BA Ede, Wageningen, The Netherlands

Willem M. de Vos
Wageningen Centre for Food Sciences and Laboratory of Microbiology, Wageningen University, Wageningen, The Netherlands

Michiel Kleerebezem
Wageningen Centre for Food Sciences and NIZO Food Research, BA Ede, Wageningen, The Netherlands

INTRODUCTION

From birth to death, the human gastrointestinal tract (GI tract) is colonized by a vast and complex consortium of mainly bacterial cells that outnumbers our somatic and germ cells (1). The microflora in this niche is estimated to be composed of at least 500 different species. However, this number is likely to represent a large underestimate, since it has been based on culturing studies that are known to be selective and notably underestimate the large number of Gram-positive intestinal bacteria. Molecular approaches, such as broad-range sequencing of 16S ribosomal RNA genes, have been used to monitor the composition of the dominant GI-tract microbiota in different individuals at different points in their lives (see chapter 1). These approaches revealed a relatively stable composition in individual adults, but they appeared to be considerably variable when different individuals were compared (2,3). Moreover, host development (4,5), host genotype (6), and environmental factors (7) influence the composition of the microbiota, emphasizing how challenging it is to define and compare bacterial communities within and between specified intestinal niches of a given individual at a particular time point in his or her life. The fact that we have not yet been able to culture the majority of the members of this bacterial community further complicates studies on the activity of individual members of the GI-tract consortium. An important development in this respect are the sophisticated enrichment strategies that have led to the isolation of new bacterial species from fecal samples [(8) and see chapter 1].

Several biological barriers are met by bacteria during residence in and travel through the different parts of the host's GI tract, such as the gastric acidity encountered in the stomach, the presence of bile salts in the duodenum and stress conditions associated with oxygen gradients that are steep at the mucosal surface, while the colon lumen is virtually anoxic. Moreover, considerable bacterial competition is encountered throughout the intestinal tract and is most prominent in the colon where bacterial density is highest. There are many functions that can be ascribed to the bacterial GI-tract communities, including the processing of undigested food, the stimulation of the host's immune system, and providing colonization resistance to pathogens (9). However, it seems that we are only beginning to understand the dimensions of these interactions. This is evident from the major impact that bacterial colonization seems to have on the host and the presently known response of intestinal bacteria that are reviewed below.

BACTERIAL RESPONSES TO THE HOST

In Vitro Approaches

Due to the complex nature of host-specific and chemical stress conditions that are met by bacteria in the GI tract many studies describe the in vitro response of intestinal bacteria to a simplified model that mimics (a component of) the stress encountered in the host's GI tract.

Historically, these studies have been performed in pathogens, including studies describing the response towards acid stress in enteropathogenic bacteria such as Salmonella and *Escherichia coli*, which revealed that RpoS, Fur, PhoP, and OmpR are important pH-response regulators (10). More recent studies describe food-grade bacteria and their tolerance to acid stress. These studies have focused mainly on physiological aspects such as determination of levels of acid-tolerance (11,12). Changes in protein synthesis during acid adaptation have been studied in *Propionibacterium freudenreichii* using 2D-gel electrophoresis, indicating an important role in the early acid tolerance response for a biotin carboxyl carrier protein and enzymes involved in DNA synthesis and repair, as well as a role in the late response for the universal chaperones GroEL and GroES (13).

Several studies describe the defense mechanisms of Gram-negative enteric bacteria towards bile acids, which include the synthesis of porins, transport proteins, efflux pumps and lipopolysaccharides (14). In addition, a few genome-wide approaches aiming at the identification of proteins important for bile salt resistance in Gram-positive bacteria have been described. In *Propionibacterium freudenreichii, Listeria monocytogenes* and *Enterococcus faecalis* differential proteome analysis using 2D-gel electrophoresis led to the identification of several proteins that were expressed at a higher level in the presence of bile salts relative to control conditions lacking bile salts (15–17). In *Propionibacterium freudenreichii* these bile-induced proteins were further analyzed by N-terminal sequencing and peptide mass fingerprinting, leading to the identification of 11 proteins important in bile stress response. The induced proteins include general stress proteins such as ClpB and the chaperons DnaK and Hsp20 (16). Analogously, a subset of the proteins identified in *E. faecalis* appeared to be inducible by multiple sublethal stresses, including heat, ethanol, and alkaline pH (18). The fact that these general stress proteins are induced by bile is in agreement with the cross protection against bile after thermal or detergent pre-treatment that has been observed in several bacteria, including *Enterococcus faecalis*, *Listeria monocytogenes* and *Bifidobacterium adolescentis* (15,19,20). Moreover, in *Escherichia coli* an *rpoS* mutant failed to develop starvation-mediated cross protection after in vitro mimicking of osmotic, oxidative, and heat stresses (21). Two other bile-induced proteins in *Propionibacterium freudenreichii* are the superoxide dismutase and cysteine synthase, which could be involved in the protection against the oxidative stress imposed on *Propionibacterium freudenreichii* by bile. In addition, other studies describe the oxidative stress response of GI-tract organisms, including *Campylobacter coli*,

Escherichia coli and several *Shigella* species (21–23). A deletion mutant in the gene encoding superoxide dismutase in *Campylobacter coli* displayed poor survival and colonization during infection of an animal model (23). Moreover, proteins involved in signal sensing and transduction, and an alternative sigma factor appeared to be bile-inducible (16). Next to these proteomic approaches, random gene disruption strategies have been applied to *Listeria monocytogenes* and *Enterococcus faecalis*, resulting in strains that are more susceptible to bile salts than the wild-type strains. Subsequent genetic analysis of the mutants revealed that the disrupted genes encode diverse functions, including an efflux pump homologue (19) and genes involved in oxidative stress response, and cell wall and fatty acid biosynthesis (24). In *Lactobacillus plantarum* a genetic screen resulted in the identification of 31 genes of which the expression appeared to be induced by bile. In analogy with the random gene disruption strategies applied in other species, this genetic screen in *L. plantarum* led to the observation that efflux pumps and changes in the architecture in the cell envelope are important for bile resistance of these bacteria (25). Moreover, these findings are in agreement with several physiological studies in GI-tract bacteria such as *L. plantarum*, *Propionibacterium freudenreichii* and *L. reuteri* that demonstrated that bile salts induce severe changes in the morphology of the cell membrane and/or cell wall of these organisms (Fig. 1) (16,25,26).

Overall, the aforementioned in vitro experiments have provided insight in the response of specific bacteria towards components of the complex mixture of stress conditions that is met by these bacteria during residence in or transit through the GI tract of their hosts. Although these approaches have helped to unravel the response of specific micro-organisms towards certain GI-tract conditions, they will not suffice to describe their behavior in the GI tract. The full response repertoire will only be triggered in vivo, where all physicochemical conditions are combined with specific host-microbe and microbe-microbe interactions. Therefore, more sophisticated approaches have aimed at the development of tools that allow the in vivo identification of genes that are important in the GI tract.

Overview of In Vivo Strategies

Three main strategies have been developed for the identification of genes that are either highly expressed, differentially expressed or specifically required in vivo (Fig. 2). These

(A)   **(B)**

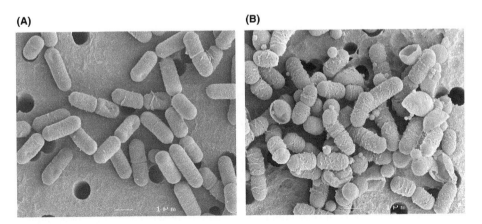

Figure 1 Exemplary representation of the morphological changes induced by bile. *L. plantarum* cells were grown on laboratory media without (**A**) or with (**B**) 0.1% of porcine bile and the bacterial cells were investigated by scanning electron microscopy. *Source*: From Ref. 25.

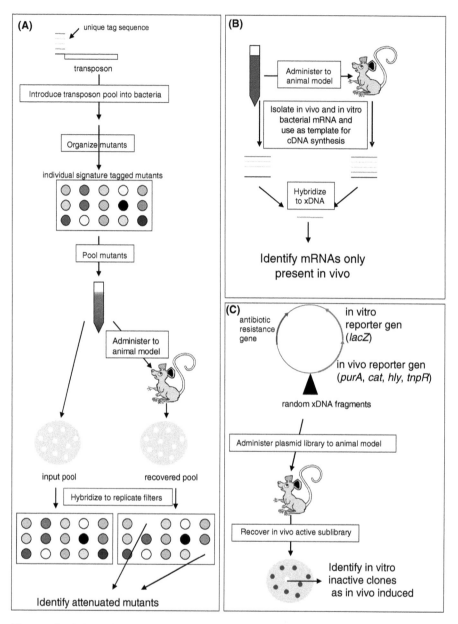

Figure 2 Schematic representation of the basic principles of STM (**A**), SCOTS (**B**) and (R-)IVET (**C**). *Abbreviations*: IVET, in vivo expression technology; SCOTS, selective capture of transcribed sequences; STM, signature tagged mutagenesis.

strategies have mainly been applied for the identification of genes from pathogens which are important during infection of their animal host. Signature tagged mutagenesis (STM) utilizes a negative selection strategy in which an animal host is infected with a pool of sequence-tagged insertion mutants. Mutated genes represented in the initial inoculum but not recovered from the host are essential for growth in the host (27,28). A major advantage of STM is that this type of screen provides direct proof for the importance of the mutated genes in the relevant niche. Unfortunately, only limited numbers of mutants can be screened per animal model. Therefore, large scale animal experiments are required for genome-wide mutant screens and for this reason STM screens are labor-intensive. In addition, mutants that are slow-growing, contain mutations in genes encoding redundant functions, or that can be complemented in a mixed population remain undetected or are at least underrepresented (29). Moreover, mutants for genes that are essential in the laboratory can never be obtained and, therefore, their importance for persistence in vivo cannot be investigated using this technique. Nevertheless, the STM strategy has been applied successfully to identify genes important in GI-tract colonization by at least six enteric pathogens, including *Klebsiella pneumoniae, Vibrio cholerae*, and *Escherichia coli* (27). Lipopolysaccharides have been recognized as an important factor in GI-tract persistence and colonization of several Gram-negative bacteria, as they have emerged as a common theme in the STM-based studies. In addition, the importance of the global regulator of anaerobic metabolism Fnr was highlighted by several STM screens, which is not surprising considering the low oxygen tension in the colon. Moreover, the alternate sigma factor RpoN was found in several of the STM screens and is likely to associate with RNA polymerase to promote the transcription of genes that are specifically required in the GI-tract niche. Finally, STM studies revealed the importance of specific adhesins, including the type IV pili of *Vibrio cholerae* and *Citrobacter rodentium* (27).

A second strategy that has been applied for the identification of in vivo transcribed genes is selective capture of transcribed sequences (SCOTS). cDNA is prepared from total RNA isolated from infected cells, or tissue samples. cDNA mixtures obtained are then enriched for sequences that are transcribed preferentially during growth in the host, using hybridizations to biotinylated bacterial genomic DNA in the presence of cDNA similarly prepared from bacteria grown in vitro. This strategy is very effective for the identification of highly abundant genes in situ which are also expressed to a lower level in the laboratory. In contrast to the STM strategy, genes that are essential in the laboratory can be investigated for their importance in GI-tract colonization. Nevertheless, major disadvantages of SCOTS are the instability of bacterial mRNA for the construction of cDNA libraries, the low abundance of mRNA from transiently or lowly expressed genes, and the technical difficulty in isolation of sufficient high-quality mRNA from small populations of bacteria in vivo (29). SCOTS has only been applied in a limited number of studies and the majority of these screens was performed to identify bacterial genes expressed within macrophages (30–33). More recently, the first SCOTS strategy utilizing an animal model to identify genes important during infection was performed (34). This approach resulted in the identification of *Escherichia coli* genes of which the expression is either relatively abundant or induced in vivo. Similar to the STM approaches described above, this SCOTS approach revealed the induction of expression of genes involved in pilus formation and lipopolysaccharide (LPS) biosynthesis. Other genes identified included iron-responsive and plasmid- and phage-encoded genes (34).

The third strategy that has been used to identify genes that are specifically induced or required during infection is in vivo expression technology (IVET). Similar to SCOTS, the IVET strategy is capable of identifying genes that are non-essential or redundant, while in an STM approach genes are only identified that are essential in vivo. An important difference between IVET and SCOTS lies in the fact that SCOTS is capable of identifying genes that are active in the laboratory, but, nevertheless, are induced in the host, while IVET only identifies in vivo induced genes that are very lowly or not expressed in the laboratory. The IVET approach relies on the generation of transcriptional fusions of genomic sequences to a reporter gene encoding an enzymatic activity. Nowadays, four variations of IVET utilizing different reporter genes have evolved as discussed in the section below.

In Vivo Expression Technology Approaches

The original IVET approach involves a tandem set of two promoterless reporter genes, namely *purA* and *lacZ*, which were used to identify promoters that are specifically switched on in *Salmonella typhimurium* during infection (35). Purine auxotroph mutants (Δ*purA*) of *Salmonella typhimurium* were only able to survive in a mouse model system when complemented *in trans* with a plasmid encoded *purA* copy. The promoterless *purA* gene was thereby utilized as a reporter for the identification of chromosomal fragments that are capable to complement the mutants, thereby strongly selecting for chromosomal fragments which harbor promoter elements that are active in the mouse model system. Subsequently, the in vivo active promoters are tested for the absence of promoter activity in vitro utilizing the second reporter gene (*lacZ*). The second variation of IVET is based on selection of an antibiotic resistance gene as selectable marker. One obvious disadvantage of this second variation of IVET is that the antibiotic must be administered to the host animal, which will certainly disturb the naturally occurring microflora in the GI tract. Therefore, the screening conditions assessed with this variant of IVET significantly differ from the native, in vivo situation. On the other hand, the addition of different levels of the selective antibiotic allows for selection of in vivo induced genes in a wider range of promoter activities. The third type of IVET selection uses a single gene as a dual reporter. The first example of such a dual reporter was *hly*, encoding the pore-forming haemolysin listeriolysin O (LLO) of *Listeria monocytogenes* (36). LLO mediates lysis of the phagosomal membrane in macrophages following infection. This reporter provides an in vivo selection for active fusions that allow for escape from the phagosomal compartment and subsequent multiplication. Moreover, a convenient screen on blood agar plates can be performed to identify inactive fusions in vitro, since clones harboring such fusions do not display haemolysis on these plates. The major drawback of the three aforementioned IVET variations is that the experimental set-up is designed in such a way that gene activity is required throughout the residence of the bacteria in the host. Hence, genes that are weakly expressed in the laboratory or transiently expressed only in a specific compartment of the host's GI tract slip through the selection procedure without being noticed. The fourth IVET variation circumvents this disadvantage by using the irreversible enzymatic activity of resolvases as reporter gene. Recombination-based IVET (R-IVET) is the only IVET approach that functions as a genetic screen. An antibiotic resistance marker flanked by two resolvase-recognition sites is integrated into the chromosome of the bacterium of interest. Subsequently, a promoterless copy of a resolvase-encoding gene, typically the *tnpR* gene from Tnγδ, is introduced on a plasmid and used to trap transcriptional activation by monitoring changes in the antibiotic resistance phenotype. Importantly, this approach does not rely on selective pressure during the animal experiments, as promoter activations are irreversibly trapped by the excision of the antibiotic resistance marker and can be identified after recovery of the bacterium under investigation from the host.

In the first decade, (R-)IVET was extensively utilized for the identification of genes important during infection of at least 15 different pathogens, including *Klebsiella pneumoniae*, *Salmonella enterica*, and *Listeria monocytogenes* (29,37). Thereby, (R)-IVET is the most extensively applied screen for the identification of in vivo-induced genes during infection in animal models. The number of genes that are identified with an individual (R)-IVET screen varies strongly and ranges from 1 to approximately 100 genes (37).

Several of these screens identified genes that were already known to be involved in virulence and this observation was considered an intrinsic validation of these (R-)IVET screens (29). An exemplary finding along these lines is the identification of *agrA* using R-IVET in *Staphylococcus aureus* (38). This gene encodes a quorum-sensing transcriptional activator and *agrA* mutants constructed in this organism prior to the R-IVET screen had already been shown to display a virulence defective phenotype (39). In general, regulators are one of the predominant classes of genes identified with (R-)IVET (29). Another frequently encountered class of in vivo induced genes in pathogenic bacteria are involved in the uptake of divalent cations, including many examples of Fe^{2+} transporters (29). The harsh conditions these pathogens encounter when they transit from rich laboratory media to the host's GI tract apparently results in the induction of this group of genes. This suggestion is further supported by the observation that several in vivo induced genes were demonstrated to be similarly regulated under low Fe^{2+} concentrations in vitro (40–42). Other genes that frequently arise from (R-)IVET screens have functions in a variety of generally recognized functional categories, including cell metabolism, DNA repair and general stress response.

Recently, the first two reports appeared that describe the utilization of (R-)IVET strategies in food-grade or commensal micro-organisms in order to determine the specific induction of gene expression in these bacteria after introduction in the GI tract of animal models. In *L. reuteri* an IVET strategy based on in vivo selection of an antibiotic resistant phenotype (the aforementioned second variation of IVET) led to the identification of three genes important for this organism during colonization of the GI tract of *Lactobacillus*-free mice (43). One of these genes encodes a peptide methionine sulfoxide reductase (*msrB*) which has previously been identified using IVET in the non-food-associated *Streptococcus gordonii* during endocarditis (44). Although not noticed by the authors at that time, this was an important clue suggesting an overlap in the genetic response triggered in the pathogenic and non-pathogenic world following contact with the host. The second report dealing with in vivo induction of genes in food-associated microbes describes a R-IVET approach in *L. plantarum* (45). Previously, the resolvase-encoding *tnpR-res* system (46) has been applied to trap promoter activities in R-IVET experiments in several pathogenic bacteria. Therefore, initial attempts aimed at implementation of this system in *L. plantarum*. A *res-ery-res* cassette was successfully integrated into the chromosome of this bacterium and a promoterless copy of the *tnpR* gene was cloned on a low-copy plasmid. Despite the successful cloning of the endogenous, highly active *ldhL1* promoter upstream of *tnpR*, excision of the *ery* gene from the chromosome of *L. plantarum* was never observed (Bron et al. unpublished data). These experiments strongly suggest that the *tnpR* resolvase is not functional in *L. plantarum* under the conditions applied during the experiments. Therefore, an alternative strategy was chosen to implement R-IVET in *L. plantarum*, which involved the *cre-loxP* system (47). This system was previously demonstrated to be functional in another lactic acid bacterium (LAB), *Lactococcus lactis* (48). Hence, a *loxP-ery-loxP* cassette was integrated into the chromosome of *L. plantarum* and a promoterless copy of *cre* was cloned on a low-copy vector. This system appeared to be functional in *L. plantarum*, as *ldhL1*-promoter driven expression of the *cre* gene led to the irreversible excision of the *loxP-ery-loxP* cassette from the chromosome. Subsequently, a library containing *L. plantarum* chromosomal fragments upstream of *cre* was constructed and administered to mice. The library was recovered from fecal samples and analyzed for *L. plantarum* colonies that had lost their erythromycin resistant phenotype during passage through the animal model. These erythromycin sensitive colonies potentially harbor chromosomal fragments of which the expression was in vivo induced. Using this strategy, 72 *L. plantarum* genes were identified as being in vivo induced (*ivi* genes) during host GI-tract transit (45). The distribution over the generally recognized classes of main biological functions appeared to be random. A slight overrepresentation of R-IVET genes is observed around the origin of replication as compared to the rest of the genome (Fig. 3). However, the significance of the latter

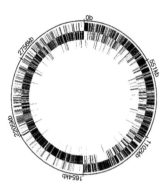

Figure 3 Using R-IVET 72 *L. plantarum* genes could be identified as in vivo induced (*ivi*) during passage of the mouse GI tract. The chromosomal localization of these *ivi* genes is represented in the inner circle, while the outer two circles represent the ORFs on the positive (*outer circle*) and negative (*middle circle*) chromosomal DNA strand.

observation is unclear. Nine of the 72 *ivi* genes appeared to encode sugar-related functions, including genes involved in ribose, cellobiose, sucrose, and sorbitol transport. Another nine genes encode functions involved in acquisition and synthesis of amino acids, nucleotides, cofactors, and vitamins, indicating their limited availability in the GI tract. Four genes involved in stress-related functions were identified, reflecting the harsh conditions that *L. plantarum* encounters in the GI tract. Another four genes encoding extracellular proteins were identified that could mediate interactions with host GI-tract epithelial cells. Remarkably, the protein encoded by one of the hypothetical proteins identified in this study in *L. plantarum* is a homologue (32% identity) of the only conserved hypothetical protein that was identified with IVET in *L. reuteri* (43). Moreover, a large number of the functions and pathways identified in *L. plantarum* have previously been identified in pathogens as being important in vivo during infection (45). This striking amount of parallels between the pathogenic and non-pathogenic in vivo response suggests that survival rather than virulence is the explanation for the importance of these genes during host residence. Recently, nine of the *L. plantarum ivi* genes were selected, mainly focusing on genes that encode proteins with a predicted role in cell envelope functionality, stress response and regulation, for the construction of isogenic gene replacement mutants. Quantitative polymerase chain reaction (PCR) experiments were performed to monitor the relative population abundance of the group of *L. plantarum* replacement mutants in fecal samples after competitive passage through the GI tract of mice. These experiments revealed that after GI-tract passage the relative abundance of three of the *ivi* gene mutants was 100- to 1000-fold reduced as compared to other mutant strains, suggesting an important role for these three *ivi* genes, encoding the IIC transport component of a cellobiose phosphotransferase system (PTS), an extracellular protein that contains an LPQTNE motif, and a copper transporting ATPase, in the functionality of *L. plantarum* during passage of the GI tract (49).

INSIGHTS FROM GENOMICS

Nowadays more and more bacteria are undergoing genome sequencing and as a result over 130 completed bacterial genomes have become available in the public domain. Following the first example of *Haemophilus influenzae* in 1995 (50) the major focus of these efforts has initially been on pathogenic bacteria and includes the completion of several genome sequences of food-borne pathogens, including *Bacillus cereus* (51), *Salmonella typhimurium* (52), and *Listeria monocytogenes* (53). Over the last years sequencing of the genomes of food-associated, non-pathogenic bacteria has received considerable attention, including the elucidation of the complete genome sequence of *Bacillus subtilis* in 1997 (54). Moreover, the first complete LAB genome sequence published was that of *Lactococcus lactis* subspecies *lactis* strain IL1403 (55). To date, only two other high-fidelity genome sequences of LAB, *L. plantarum* strain WCFS1 (56) and *L. johnsonii* strain NCC533 (57), have been published. An additional number of LAB genomes is nearing completion and draft genome information has become available in the public domain in 2002 with the publication and appearance of genome sequences for LAB provided by the Joint Genome Institute (http://genome.jgi-psf.org/microbial/) in collaboration with the lactic acid bacteria genomics consortium (58,59). Next to this large amount of sequence data from food-associated LAB, successful efforts have been put in determination of the (complete) genome sequences of members of our normal colonic microbiota, in particular *Bacteroides thetaiotaomicron* (60) and *Bifidobacterium longum* (Fig. 4) (61).

L. plantarum is a versatile and flexible organism that is able to grow on a wide variety of sugar sources. This phenotypic trait is reflected in the genome sequence of *L. plantarum*, which harbours a remarkably high number of 25 complete PTS enzyme II complexes as well as several incomplete complexes. This high number of PTS systems is far more than that found in other complete bacterial genomes, and similar only to *Listeria monocytogenes* (53) and *Enterococcus faecalis* (62). In addition to these PTS systems, the *L. plantarum* genome encodes 30 transporters that are predicted to be involved in the transport of carbon sources. This high sugar uptake flexibility has also been observed in the genomes of other LAB, such as *L. johnsonii* (57) and *L. acidophilus* (http://www.calpoly.edu/~rcano/Lacto_genome.html). Moreover, a remarkably high percentage of regulatory genes (8.5%) appeared to be

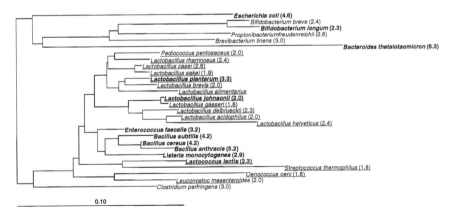

Figure 4 Phylogenetic relationship based upon the neighbor-joining method of partial 16S rDNA sequences (*Escherichia coli* positions 107 to 1434). It should be noted that for some species the genome sequence has (partially) been determined for multiple strains. LAB genomes are underscored, and published, complete genomes are shown in bold. The estimated genome sizes are indicated between brackets.

encoded in the *L. plantarum* genome. Similar percentages were found in *Listeria monocytogenes*, in which 7.3% of all the encoded genes were predicted to be involved in regulatory functions. This could be a reflection of the many different environmental conditions that these bacteria face. Moreover, these sophisticated regulatory systems enable these organisms to adapt quickly to changes in the sugar composition of the host's diet during residence in the proximal parts of the GI tract (Fig. 5).

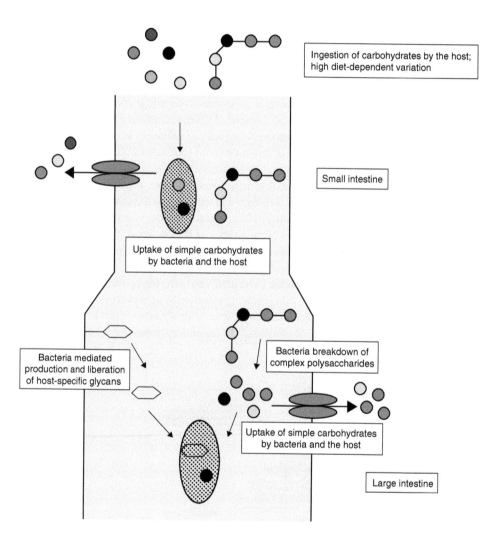

Figure 5 Molecular model of bacterial sugar utilization in the GI tract. In the small intestine mono- and disaccharides are rapidly consumed by the host. Typically, bacteria that live in this niche display highly flexible sugar utilization capacities, allowing them to quickly adapt to changes in the carbon source availability that is determined by the host's diet. This high sugar flexibility is required to compete with the host for carbon acquisition. In the large intestine more complex oligo- and polysaccharides are the only available C-source. Therefore, bacteria in this niche are usually able to hydrolyse complex dietary polysaccharides and host-derived glycoproteins and glycoconjugates. Subsequently, the released, simpler sugars are utilized as C-source by the host and the bacteria residing in the colon. *Source*: From Ref. 63.

The genomes of *B. thetaiotaomicron* and *Bifidobacterium longum* encode an elaborate apparatus for acquiring and hydrolysing otherwise indigestible dietary polysaccharides (60,61). In *B. thetaiotaomicron* this "colonic substrate dependence" is associated with an environment-sensing system consisting of a large repertoire of extracytoplasmic function sigma factors and one- and two-component signal transduction systems (60). In contrast, genes involved in sugar transport and hydrolysis in *Bifidobacterium longum* are organized in operons which are predominantly regulated by LacI-type, sugar responsive repressors (61). The tight regulation of sugar utilization observed in these bacteria allows a stringent response to environmental changes and is in accordance with the fact that *Bifidobacterium longum* and *B. thetaiotaomicron* need to adapt to wide fluctuations in substrate availability in the colon (60,61). It is speculated that the mode of regulation via repression of genes could allow a quicker response in *Bifidobacterium longum* (61). Similarly, an operon in *L. acidophilus* involved in utilization of the prebiotic compound fructooligosaccharide contains a LacI type repressor. Moreover, the expression of this operon is subject to global catabolite repression in the presence of readily fermentable sugars (64). Another interesting finding in the genome of *B. thetaiotaomicron* is that it appears to encode the capacity to use a variety of host-derived glycoproteins and glycoconjugates. Sixty-one percent of its glycosylhydrolases are predicted to be located in the periplasm, outer membrane, or extracellularly. This suggests that these enzymes are not only important for fulfilling the needs of *B. thetaiotaomicron* but may also help shape the metabolic milieu of the intestinal ecosystem in ways conducive to maintaining a microbiota that supplies the host with 10 to 15% of our daily calories as fermentation products of dietary polysaccharides (Fig. 5) (60). Similarly, the genome sequence of *Bifidobacterium longum* revealed insights into the interaction of bifidobacteria with their host, as genes encoding polypeptides with homology to glycoprotein-binding fimbriae are present in the genome. Moreover, a eukaryotic-type serine protease inhibitor is encoded in the genome and could be involved in the reported immunomodulatory activity of bifidobacteria (61).

Recently, the complete genomes of *L. plantarum* (3.3 Mbp) and *L. johnsonii* (2.0 Mbp) were compared, revealing that these genomes have only 28 regions with conservation of gene order, encompassing approximately 0.75 Mbp (65). Notably, these regions are not co-linear, indicating major chromosomal rearrangements. Moreover, metabolic reconstruction indicated many differences between these two lactobacilli, as numerous enzymes involved in sugar metabolism and the biosynthesis of amino acids, nucleotides, fatty acids and cofactors are lacking in *L. johnsonii*. Interestingly, major differences were also seen in the number and types of putative extracellular proteins, which could play a role in host-microbe interactions in the GI tract. The differences between *L. plantarum* and *L. johnsonii*, both in genome organization and gene content, are exceptionally large for two bacteria of the same genus, emphasizing the complexity and diversity of the *Lactobacillus* genus (65).

Overall, the availability and comparison of bacterial genome sequences and their annotated functions provides valuable clues towards the survival strategy of these bacteria during their residence in the human GI tract. Additionally, these complete genome sequences are powerful tools for the convenient and effective interpretation of the data generated by the in vitro and in vivo screening procedures described above. Moreover, comparative genomics can provide important insight in diversity, evolutionary relationship and functional variation between bacteria, which might eventually generate a comprehensive view of the behavior of microbes during residence in the human GI-tract.

IN SITU PROFILING OF TRANSCRIPTION IN THE GI TRACT

As soon as sequence data is available for a few genes in a bacterium of interest, one could think of several sophisticated tools that allow investigation of the in situ expression levels of specific genes. One example of such an approach is the implementation of quantitative reverse transcriptase polymerase chain reaction (qRT-PCR) in the gram-negative bacterium *Helicobacter pylori* (66). This study describes the assessment of gene expression in this pathogen within the mouse and human gastric mucosae. Three genes, encoding urease, catalase, and a putative adhesin specific for adherence to human gastric mucosa, were selected for analysis, as their role during host residence was already established. Using minute quantities of mRNA isolated from human and mouse infected mucosae, the in situ expression of these three genes could be established. Moreover, the results of this study indicate that the relative abundance of transcripts was the same in the human and mouse model system. Hence, this study demonstrates that qRT-PCR is a powerful tool for the detection and quantification of bacterial gene expression in the GI tract (66). Similar experiments were performed in *L. plantarum*. An in vitro screen and a R-IVET screen were already performed in this LAB to identify genes of which the expression is induced in vitro by bile or in situ in the GI tract of a mouse model system, respectively (25,45). Matching of the results obtained in these two screening procedures revealed two genes, encoding an integral membrane protein and an argininosuccinate synthase that appeared to be induced by bile in vitro as well as in vivo in the GI tract of a mouse model system. Therefore, the expression of these two genes was assessed using qRT-PCR followed by SYBR green fluorescence detection. As the duodenum is the site of bile release, expression in this specific region of the host's GI tract was investigated. The results confirmed that the expression levels of these two genes were significantly higher in *L. plantarum* cells isolated from the mouse duodenum relative to cells grown in standard laboratory media (25). Current studies aim at the confirmation of gene-induction of several other *L. plantarum* genes initially identified with the R-IVET screen (Marco et al. unpublished data). Moreover, transcription profiling under a variety of in vitro conditions could identify more matches with the R-IVET screen and these genes could subsequently be analyzed with qRT-PCR. These experiments might reveal the specific environmental cue involved in in situ induction in the GI tract and could eventually elucidate the regulatory mechanism(s) involved. These approaches could help to unravel the geographical differentiation of *L. plantarum* gene expression along the GI tract, i.e., specific induction in the stomach, small intestine or colon.

Another promising development is the optimization of bacterial RNA isolation protocols from fecal samples (67), and GI-tract samples from conventional mice fed with *L. plantarum* (Marco et al. unpublished data) or human cancer patients who volunteered to consume an oatmeal-based drink containing high numbers of *L. plantarum* prior to surgery (de Vries et al. unpublished data). Although it is technically difficult to isolate high quality bacterial mRNAs from these samples, such RNA samples originating from an in vivo animal tissue, could prove extremely valuable, as they should allow analysis using DNA micro-array technology, providing direct information on the in situ expression levels of thousands of bacterial genes. Moreover, comparison of bacterial responses in samples from the GI tract from animal models and of human origin could provide an indication of the overlap in the response of *L. plantarum* during residence in the GI tract of different hosts.

Studies in gnotobiotic mice have indicated that there is specific signaling between the commensal bacterium *B. thetaiotaomicron* and its host. Synthesis of host epithelial glycans is elicited by a *B. thetaiotaomicron* signal of which the expression is regulated by a fucose-binding bacterial transcription factor. This factor senses environmental levels of fucose and coordinates the decision to generate a signal for production of host fucosylated glycans when environmental fucose is limited or to induce expression of the bacteria's fucose utilization operon when fucose is abundant (68). Additional studies have evaluated the global intestinal response to colonization of gnotobiotic mice with *B. thetaiotaomicron*. This colonization dramatically affected the host's gene expression, including several important intestinal functions such as nutrient absorption, mucosal barrier fortification, and postnatal intestinal maturation (9). From the in situ global transcription profiles mentioned above and follow-up experiments it could be established that the production of a previously uncharacterized angiogenin is induced when gnotobiotic mice are colonized with *B. thetaiotaomicron*, revealing a mechanism whereby intestinal commensal bacteria influence GI-tract bacterial ecology and shape innate immunity (69). In addition, the cellular origin of the angiogenin response was investigated when different intestinal cell types were separated by laser-capture microdissection and analyzed by qRT-PCR, revealing that angiogenin-3 mRNA is specifically induced only in crypt epithelial cells. Hence, these experiments strongly suggest an intestinal tissue specific response of the host during colonization (9). Interestingly, comparison of the changes in global host gene expression in mice after colonization with *B. thetaiotaomicron*, *Bifidobacterium infantis* or *E. coli* led to the observation that part of this host response was only induced in mice by colonization with *B. thetaiotaomicron* (9). However, analysis of a broader range of members of the intestinal microbiota will reveal what the level of bacterial response specificity within the host's tissues actually is. One such study is currently performed for *L. plantarum* (Peters et al. unpublished data). Overall, the aforementioned studies on *B. thetaiotaomicron* colonization of gnotobiotic mice provided valuable information on the influence of one particular member of the microbiota on the host. However, the host response during colonization by more complex mixtures of microbes and/or the host response in other animal systems remained to be investigated at that time. Recently, it was found that conventionalization of adult gnotobiotic mice with normal microbiota harvested from the distal intestine of conventionally raised mice produced a 60% increase in body fat content and insulin resistance despite reduced food intake. Studies of gnotobiotic and conventionalized mice revealed that the microbiota promotes absorption of monosac-charides from the gut lumen, which results in induction of de novo hepatic lipogenesis. Fastin-induced adipocyte factor (Fiaf), a member of the angiopoietin-like family of proteins, is selectively suppressed in the intestinal epithelium of normal mice by conventionalization. Analysis of gnotobiotic and conventionalized, normal and *Fiaf* knockout mice established that Fiaf is a circulating lipoprotein lipase inhibitor and that its suppression is essential for the microbiota-induced deposition of triglycerides in adipocytes. These results suggest that the gut microbiota have a major impact on food-derived energy harvest and storage in the host (70). Another recent study investigated the host response during colonization of a different animal model. DNA micro-array comparison of gene expression in the digestive tracts of six days post-fertilization gnotobiotic, conventionalized, and conventionally raised zebrafish (*Danio Rerio*) revealed 212 genes regulated by the microbiota. Notably, 59 of these genes were also found to be regulated in the mouse intestine during colonization, including genes that encode functions involved in stimulation of epithelial proliferation, promotion of nutrient metabolism, and innate immune response, indicating a substantial overlap in the genetic response of mice and zebrafish towards intestinal colonization (71). Despite these recent developments, an important future challenge lies within the translation of these animal host response analyses to the human system.

CONCLUDING REMARKS AND FUTURE PERSPECTIVES

Historically, research on the bacterial flora of the GI tract has concentrated on the inhabitants that have negative effects on their hosts. More recently, research has expanded from these pathogenic to non-pathogenic bacteria, including symbionts and commensals. One obvious reason for this is the accumulating evidence that certain bacteria, especially strains from the genus *Lactobacillus* and *Bifidobacterium*, may have probiotic effects in man and animals (72). At present there is a detailed understanding on the distribution of specific microbes along the human colon and the variations that can occur between different individuals (73–75). Moreover, knowledge on the activity and response of specific species to the conditions encountered when they transit through this complex niche is starting to accumulate. Several in vitro studies mimicking specific conditions in the GI tract have been performed, which allowed the identification of the repertoire of genes and their corresponding proteins that respond to the condition applied. More recently, in vivo approaches aiming at the identification of bacterial genes that are induced during passage of the GI tract have been performed in various microbes, including food-grade species. The current knowledge on promoter elements regulating gene expression of food-grade bacteria in the GI tract could have application possibilities, as these bacteria have been shown to possess great potential to serve as delivery vehicles of health-promoting or therapeutic compounds to the human GI tract (76–84). (R-)IVET approaches have provided the required promoters that will allow the construction of LAB-based dedicated GI-tract delivery vehicles that only express certain desired functions in situ. Moreover, geographically more detailed insight in the exact site of in situ gene activation in the GI tract derived from qRT-PCR using specific tissue samples might allow the construction of highly site-specific delivery vehicles. Combination of these promoters with certain genes, e.g., bacteriophage-derived or other lytic cassettes, might generate LAB strains that release their cellular content at a specific location in the GI tract.

At present, a large part of the consortium of bacteria residing in the GI tract has not been cultured in vitro. Since most genetic approaches require the culturability of the microbe under investigation, the expansion of our knowledge of this group of bacteria is highly challenging and very limiting at this stage. Metagenomic approaches might shed light on the genetic complexity of the collective genomic material of the intestinal microbiota (85). Moreover, such studies could reveal previously unknown, critical genes for intestinal microbiota functioning. However, effective exploration of metagenomic functionality will depend on high throughput screening systems that allow function identification. Moreover, the development of effective and robust methods to assess microbiota activity in situ in a culture independent manner will be critical for our functional understanding of the large number of unculturable bacteria in the GI tract.

A promising prospect from the increasing availability of complete genome sequences is the construction of DNA micro-arrays in several laboratories working on food-associated microbes. As a consequence, the first publications presenting data from these DNA micro-arrays appeared recently (86–88). These genomics-based, global investigations of gene expression in food-grade microbes under various conditions will further detail our understanding of their behavior. However, the application of these transcriptome profiling techniques on microbe-containing GI tract samples will still have to overcome some technical hurdles (RNA extraction procedures, response validation, etc.), but will eventually lead to a more complete view of the activity of these bacteria in this complex niche. Besides the application of DNA micro-array technology to reveal the bacterial side of host-microbe interactions in the GI tract, this technology has already been used by Hooper and co-workers in several elegant studies aiming at identification of the response of gnotobiotic mice to colonization with the commensal *B. thetaiotaomicron* (9,68,69). In addition, more recent gnotobiote studies have shed light on the differences in mouse gene expression upon colonization by a more complex mixture of bacteria (70), and have provided the first steps towards the comparative analysis of host responses in different animal models (71). Nevertheless, an important question that still remains to be answered is to what extent the data obtained on host and bacterial gene expression using animal models can be extrapolated to the situation in humans.

In conclusion, the genome-wide transcript profiling approaches that have been performed to date have provided us with clues of the possible role of individual host and bacterial genes during host-microbe interactions. Combination of bacterium and host transcriptomes should allow the construction of molecular models that describe host-microbe interactions, allowing more pinpointed experiments in the future, designed on the basis of a molecular interaction hypothesis. As GI-tract bacteria like *L. plantarum* and *B. thetaiotaomicron* are genetically accessible, gene deletion and overexpression mutants can be constructed and employed to study the effect of a single bacterial gene and its corresponding function on host gene expression. After profiling of these host genes, knock-out mice and/or antisense RNA approaches might allow gene silencing on the host side of the spectrum, thereby enabling us to study the effect of single host gene mutations on the colonization of microbes. Ultimately, such studies may provide a molecular knowledge base to understand GI-tract colonization of commensals or symbionts, and could lead to the molecular explanation of probiotic effects associated with LAB and related species.

ACKNOWLEDGMENTS

The authors gratefully acknowledge Jos Boekhorst and Sergey Konstantinov for the construction of Figures 3 and 4, respectively. We cordially thank our WCFS colleagues for sharing unpublished data. Part of this work was supported by the EU project LABDEL (EU-QLRT-2000-00340).

REFERENCES

1. Savage DC. Microbial ecology of the gastro intestinal tract. Annu Rev Microbiol 1977; 31:133.
2. Zoetendal EG, Akkermans AD, De Vos WM. Temperature gradient gel electrophoresis analysis of 16S rRNA from human fecal samples reveals stable and host-specific communities of active bacteria. Appl Environ Microbiol 1998; 64:3854–3859.
3. Zoetendal EG, von Wright A, Vilpponen-Salmela T, Ben-Amor K, Akkermans AD, de Vos WM. Mucosa-associated bacteria in the human GI-tract are uniformly distributed along the colon and differ from the community recovered from feces. Appl Environ Microbiol 2002; 68:3401–3407.
4. Hopkins MJ, Sharp R, Macfarlane GT. Age and disease related changes in intestinal bacterial populations assessed by cell culture, 16S rRNA abundance, and community cellular fatty acid profiles. Gut 2001; 48:198–205.
5. Favier CF, Vaughan EE, De Vos WM, Akkermans AD. Molecular monitoring of succession of bacterial communities in human neonates. Appl Environ Microbiol 2002; 68:219–226.
6. Zoetendal EG, Akkermans AD, Akkermans-van Vliet WM, de Visser JAGM, de Vos WM. The host genotype affects the bacterial community in the human gastrointestinal tract. Microb Ecol Health Dis 2001; 13:129–134.

7. Sullivan A, Edlund C, Nord CE. Effect of antimicrobial agents on the ecological balance of human microflora. Lancet Infect Dis 2001; 1:101–114.

8. Zoetendal EG, Plugge CM, Akkermans AD, de Vos WM. *Victivallis vadensis* gen. nov., sp. nov., a sugar-fermenting anaerobe from human faeces. Int J Syst Evol Microbiol 2003; 53:211–215.

9. Hooper LV, Wong MH, Thelin A, Hansson L, Falk PG, Gordon JI. Molecular analysis of commensal host-microbial relationships in the intestine. Science 2001; 291:881–884.

10. Audia JP, Webb CC, Foster JW. Breaking through the acid barrier: an orchestrated response to proton stress by enteric bacteria. Int J Med Microbiol 2001; 291:97–106.

11. Chou LS, Weimer B. Isolation and characterization of acid- and bile-tolerant isolates from strains of *Lactobacillus acidophilus*. J Dairy Sci 1999; 82:23–31.

12. Hyronimus B, Le Marrec C, Sassi AH, Deschamps A. Acid and bile tolerance of spore-forming lactic acid bacteria. Int J Food Microbiol 2000; 61:193–197.

13. Jan G, Leverrier P, Pichereau V, Boyaval P. Changes in protein synthesis and morphology during acid adaptation of *Propionibacterium freudenreichii*. Appl Environ Microbiol 2001; 67:2029–2036.

14. Gunn JS. Mechanisms of bacterial resistance and response to bile. Microbes Infect 2000; 2:907–913.

15. Flahaut S, Frere J, Boutibonnes P, Auffray Y. Comparison of the bile salts and sodium dodecyl sulfate stress responses in *Enterococcus faecalis*. Appl Environ Microbiol 1996; 62:2416–2420.

16. Leverrier P, Dimova D, Pichereau V, Auffray Y, Boyaval P, Jan G. Susceptibility and adaptive response to bile salts in *Propionibacterium freudenreichii*: physiological and proteomic analysis. Appl Environ Microbiol 2003; 69:3809–3818.

17. Phan-Thanh L, Gormon T. Stress proteins in *Listeria monocytogenes*. Electrophoresis 1997; 18:1464–1471.

18. Rince A, Le Breton Y, Verneuil N, Giard JC, Hartke A, Auffray Y. Physiological and molecular aspects of bile salt response in *Enterococcus faecalis*. Int J Food Microbiol 2003; 88:207–213.

19. Begley M, Gahan CG, Hill C. Bile stress response in *Listeria monocytogenes* LO28: adaptation, cross-protection, and identification of genetic loci involved in bile resistance. Appl Environ Microbiol 2002; 68:6005–6012.

20. Schmidt G, Zink R. Basic features of the stress response in three species of bifidobacteria: *B. longum, B. adolescentis, and B. breve*. Int J Food Microbiol 2000; 55:41–45.

21. Krogfelt KA, Hjulgaard M, Sorensen K, Cohen PS, Givskov M. rpoS gene function is a disadvantage for *Escherichia coli* BJ4 during competitive colonization of the mouse large intestine. Infect Immun 2000; 68:2518–2524.

22. Khanduja V, Kang G, Rajan DP, Balasubramanian KA. Oxidative stress response in *Shigella* & nonpathogenic gut bacteria. Indian J Med Res 1998; 108:3–7.

23. Purdy D, Cawthraw S, Dickinson JH, Newell DG, Park SF. Generation of a superoxide dismutase (SOD)-deficient mutant of *Campylobacter coli*: evidence for the significance of SOD in *Campylobacter* survival and colonization. Appl Environ Microbiol 1999; 65:2540–2546.

24. Breton YL, Maze A, Hartke A, Lemarinier S, Auffray Y, Rince A. Isolation and characterization of bile salts-sensitive mutants of *Enterococcus faecalis*. Curr Microbiol 2002; 45:434–439.

25. Bron PA, Marco M, Hoffer SM, Van Mullekom E, de Vos WM, Kleerebezem M. Genetic characterization of the bile salt response in *Lactobacillus plantarum* and analysis of responsive promoters in vitro and in situ in the gastrointestinal tract. J Bacteriol 2004; 186:7829–7835.

26. Taranto MP, Fernandez Murga ML, Lorca G, de Valdez GF. Bile salts and cholesterol induce changes in the lipid cell membrane of *Lactobacillus reuteri*. J Appl Microbiol 2003; 95:86–91.

27. West NP, Sansonetti PJ, Frankel G, Tang CM. Finding your niche: what has been learnt from STM studies on GI colonization. Trends Microbiol 2003; 11:338–344.

28. Mecsas J. Use of signature-tagged mutagenesis in pathogenesis studies. Curr Opin Microbiol 2002; 5:33–37.

29. Mahan MJ, Heithoff DM, Sinsheimer RL, Low DA. Assessment of bacterial pathogenesis by analysis of gene expression in the host. Annu Rev Genet 2000; 34:139–164.

30. Morrow BJ, Graham JE, Curtiss R, III. Genomic subtractive hybridization and selective capture of transcribed sequences identify a novel *Salmonella typhimurium* fimbrial operon and putative transcriptional regulator that are absent from the *Salmonella typhi* genome. Infect Immun 1999; 67:5106–5116.

31. Graham JE, Clark-Curtiss JE. Identification of *Mycobacterium tuberculosis* RNAs synthesized in response to phagocytosis by human macrophages by selective capture of transcribed sequences (SCOTS). Proc Natl Acad Sci USA 1999; 96:11554–11559.

32. Daigle F, Graham JE, Curtiss R, III. Identification of *Salmonella typhi* genes expressed within macrophages by selective capture of transcribed sequences (SCOTS). Mol Microbiol 2001; 41:1211–1222.

33. Hou JY, Graham JE, Clark-Curtiss JE. *Mycobacterium avium* genes expressed during growth in human macrophages detected by selective capture of transcribed sequences (SCOTS). Infect Immun 2002; 70:3714–3726.

34. Dozois CM, Daigle F, Curtiss R, III. Identification of pathogen-specific and conserved genes expressed in vivo by an avian pathogenic *Escherichia coli* strain. Proc Natl Acad Sci USA 2003; 100:247–252.

35. Mahan MJ, Slauch JM, Mekalanos JJ. Selection of bacterial virulence genes that are specifically induced in host tissues. Science 1993; 259:686–688.

36. Gahan CG, Hill C. The use of listeriolysin to identify in vivo induced genes in the gram- positive intracellular pathogen *Listeria monocytogenes*. Mol Microbiol 2000; 36:498–507.

37. Angelichio MJ, Camilli A. In vivo expression technology. Infect Immun 2002; 70:6518–6523.

38. Lowe AM, Beattie DT, Deresiewicz RL. Identification of novel staphylococcal virulence genes by in vivo expression technology. Mol Microbiol 1998; 27:967–976.

39. Gillaspy AF, Hickmon SG, Skinner RA, Thomas JR, Nelson CL, Smeltzer MS. Role of the accessory gene regulator *(agr)* in pathogenesis of staphylococcal osteomyelitis. Infect Immun 1995; 63:3373–3380.

40. Smith HE, Buijs H, de Vries RR, Wisselink HJ, Stockhofe-Zurwieden N, Smits MA. Environmentally regulated genes of *Streptococcus suis*: identification by the use of iron-restricted conditions in vitro and by experimental infection of piglets. Microbiology 2001; 147:271–280.

41. Lai YC, Peng HL, Chang HY. Identification of genes induced in vivo during *Klebsiella pneumoniae* CG43 infection. Infect Immun 2001; 69:7140–7145.

42. Janakiraman A, Slauch JM. The putative iron transport system SitABCD encoded on SPI1 is required for full virulence of *Salmonella typhimurium*. Mol Microbiol 2000; 35:1146–1155.

43. Walter J, Heng NC, Hammes WP, Loach DM, Tannock GW, Hertel C. Identification of *Lactobacillus reuteri* genes specifically induced in the mouse gastrointestinal tract. Appl Environ Microbiol 2003; 69:2044–2051.

44. Kiliç AO, Herzberg MC, Meyer MW, Zhao X, Tao L. Streptococcal reporter gene-fusion vector for identification of in vivo expressed genes. Plasmid 1999; 42:67–72.

45. Bron PA, Grangette C, Mercenier A, de Vos WM, Kleerebezem M. Identification of *Lactobacillus plantarum* genes that are induced in the gastrointestinal tract of mice. J Bacteriol 2004; 186:5721–5729.

46. Reed RR, Grindley ND. Transposon-mediated site-specific recombination in vitro: DNA cleavage and protein-DNA linkage at the recombination site. Cell 1981; 25:721–728.

47. Austin S, Ziese M, Sternberg N. A novel role for site-specific recombination in maintenance of bacterial replicons. Cell 1981; 25:729–736.

48. Campo N, Daveran-Mingot ML, Leenhouts K, Ritzenthaler P, Le Bourgeois P. Cre-loxP recombination system for large genome rearrangements in *Lactococcus lactis*. Appl Environ Microbiol 2002; 68:2359–2367.

49. Bron PA, Meijer M, Bongers R, de Vos WM, Kleerebezem M. Competitive population dynamics of *ivi*-gene mutants of *Lactobacillus plantarum* in the gastrointestinal tract of mice. *In* Bron PA, the molecular response of *Lactobacillus plantarum* to intestinal passage and conditions. PhD thesis 2004, Wageningen Centre for Food Sciences, Wageningen.

50. Fleischmann RD, Adams MD, White O, et al. Whole-genome random sequencing and assembly of *Haemophilus influenzae* Rd. Science 1995; 269:496–512.

51. Ivanova N, Sorokin A, Anderson I, et al. Genome sequence of *Bacillus cereus* and comparative analysis with *Bacillus anthracis*. Nature 2003; 423:87–91.

52. Parkhill J, Dougan G, James KD, et al. Complete genome sequence of a multiple drug resistant *Salmonella enterica* serovar Typhi CT18. Nature 2001;848–852.

53. Glaser P, Frangeul L, Buchrieser C, et al. Comparative genomics of *Listeria* species. Science 2001; 294:849–852.

54. Kunst F, Ogasawara N, Moszer I, et al. The complete genome sequence of the gram-positive bacterium *Bacillus subtilis*. Nature 1997;249–256.

55. Bolotin A, Wincker P, Mauger S, et al. The complete genome sequence of the lactic acid bacterium *Lactococcus lactis ssp. lactis* IL1403. Genome Res 2001; 11:731–753.

56. Kleerebezem M, Boekhorst J, Van Kranenburg R, et al. Complete genome sequence of *Lactobacillus plantarum WCFS1*. Proc Natl Acad Sci USA 2003; 100:1990–1995.

57. Pridmore RD, Berger B, Desiere F, et al. The genome sequence of the probiotic intestinal bacterium *Lactobacillus johnsonii* NCC 533. Proc Natl Acad Sci USA 2004; 101:2512–2517.

58. Klaenhammer T, Altermann E, Arigoni F, et al. Discovering lactic acid bacteria by genomics. Antonie Van Leeuwenhoek 2002; 82:29–58.

59. de Vos WM. Advances in genomics for microbial food fermentations and safety. Curr Opin Biotechnol 2001; 12:493–498.

60. Xu J, Bjursell MK, Himrod J, et al. A genomic view of the human-*Bacteroides thetaiotaomicron* symbiosis. Science 2003;2074–2076.

61. Schell MA, Karmirantzou M, Snel B, et al. The genome sequence of *Bifidobacterium longum* reflects its adaptation to the human gastrointestinal tract. Proc Natl Acad Sci USA 2002; 99:14422–14427.

62. Paulsen IT, Banerjei L, Myers GS, et al. Role of mobile DNA in the evolution of vancomycin-resistant *Enterococcus faecalis*. Science 2003; 299:2071–2074.

63. Hooper LV, Midtvedt T, Gordon JI. How host-microbial interactions shape the nutrient environment of the mammalian intestine. Annu Rev Nutr 2002; 22:283–307.

64. Barrangou R, Altermann E, Hutkins R, Cano R, Klaenhammer TR. Functional and comparative genomic analyses of an operon involved in fructooligosaccharide utilization by *Lactobacillus acidophilus*. Proc Natl Acad Sci USA 2003; 100:8957–8962.

65. Boekhorst J, Siezen RJ, Zwahlen MC, et al. The complete genomes of *Lactobacillus plantarum* and *Lactobacillus johnsonii* reveal extensive differences in chromosome organization and gene content. Microbiology 2004; 150:3601–3611.

66. Rokbi B, Seguin D, Guy B, et al. Assessment of *Helicobacter pylori* gene expression within mouse and human gastric mucosae by real-time reverse transcriptase PCR. Infect Immun 2001; 69:4759–4766.

67. Fitzsimons NA, Akkermans AD, de Vos WM, Vaughan EE. Bacterial gene expression detected in human faeces by reverse transcription-PCR. J Microbiol Methods 2003; 55:133–140.

68. Hooper LV, Xu J, Falk PG, Midtvedt T, Gordon JI. A molecular sensor that allows a gut commensal to control its nutrient foundation in a competitive ecosystem. Proc Natl Acad Sci USA 1999; 96:9833–9838.

69. Hooper LV, Stappenbeck TS, Hong CV, Gordon JI. Angiogenins: a new class of microbicidal proteins involved in innate immunity. Nat Immunol 2003; 4:269–273.

70. Backhed F, Ding H, Wang T, et al. The gut microbiota as an environmental factor that regulates fat storage. Proc Natl Acad Sci USA 2004; 101:15718–15723.

71. Rawls JF, Samuel BS, Gordon JI. Gnotobiotic zebrafish reveal evolutionarily conserved responses to the gut microbiota. Proc Natl Acad Sci USA 2004; 101:4596–4601.

72. Ahrne S, Nobaek S, Jeppsson B, Adlerberth I, Wold AE, Molin G. The normal *Lactobacillus* flora of healthy human rectal and oral mucosa. J Appl Microbiol 1998; 85:88–94.

73. Zoetendal EG, Collier CT, Koike S, Mackie RI, Gaskins HR. Molecular ecological analysis of the gastrointestinal microbiota: a review. J Nutr 2004; 134:465–472.

74. Zoetendal EG, Cheng B, Koike S, Mackie RI. Molecular microbial ecology of the gastrointestinal tract: from phylogeny to function. Curr Issues Intest Microbiol 2004; 5:31–47.

75. Vaughan EE, de Vries MC, Zoetendal EG, Ben-Amor K, Akkermans AD, de Vos WM. The intestinal LABs. Antonie Van Leeuwenhoek 2002; 82:341–352.

76. Grangette C, Muller-Alouf H, Hols P, et al. Enhanced mucosal delivery of antigen with cell wall mutants of lactic acid bacteria. Infect Immun 2004; 72:2731–2737.

77. Hanniffy S, Wiedermann U, Repa A, et al. Potential and opportunities for use of recombinant lactic acid bacteria in human health. Adv Appl Microbiol 2004; 56:1–64.

78. Reuter MA, Hanniffy S, Wells JM. Expression and delivery of heterologous antigens using lactic acid bacteria. Methods Mol Med 2003; 87:101–114.

79. Vandenbroucke K, Hans W, Van Huysse J, et al. Active delivery of trefoil factors by genetically modified *Lactococcus lactis* prevents and heals acute colitis in mice. Gastroenterology 2004; 127:502–513.

80. Steidler L, Neirynck S, Huyghebaert N, et al. Biological containment of genetically modified *Lactococcus lactis* for intestinal delivery of human interleukin 10. Nat Biotechnol 2003; 21:785–789.

81. Steidler L, Hans W, Schotte L, et al. Treatment of murine colitis by *Lactococcus lactis* secreting interleukin- 10. Science 2000; 289:1352–1355.

82. Steidler L. Live genetically modified bacteria as drug delivery tools: at the doorstep of a new pharmacology? Expert Opin Biol Ther 2004; 4:439–441.

83. Steidler L. Genetically engineered probiotics. Best Pract Res Clin Gastroenterol 2003; 17:861–876.

84. Grangette C, Muller-Alouf H, Geoffroy M, Goudercourt D, Turneer M, Mercenier A. Protection against tetanus toxin after intragastric administration of two recombinant lactic acid bacteria: impact of strain viability and in vivo persistence. Vaccine 2002; 20:3304–3309.

85. Schloss PD, Handelsman J. Biotechnological prospects from metagenomics. Curr Opin Biotechnol 2003; 14:303–310.

86. Bron PA, Molenaar D, de Vos WM, Kleerebezem M. DNA microarray based identification of bile-responsive genes in *Lactobacillus plantarum*. J Appl Microbiol 2006; in press.

87. Azcarate-Peril MA, McAuliffe O, Altermann E, Lick S, Russell WM, Klaenhammer TR. Microarray analysis of a two-component regulatory system involved in acid resistance and proteolytic activity *Lactobacillus acidophilus*. Appl Environ Microbial 2005; 71:5794–5804.

88. Pieterse B, Leer RJ, Schuren FH, van der Werf MJ. Unravelling the multiple effects of lactic acid stress on *Lactobacillus plantarum* by transcription profiling. Microbiology 2005; 151:3881–3894.

10

The Infant Intestinal Microbiota in Allergy

Pirkka V. Kirjavainen and Gregor Reid
Canadian Research and Development Center for Probiotics, The Lawson Health Research Institute, London, Ontario, Canada

INTRODUCTION

Allergies represent a condition where impaired immunological tolerance to common environmental allergens is the fundamental determinant of the disease. The immunopathological mechanism of the disease development is poorly understood. It is thought to involve complex genetic predisposition, which depending on environmental triggers and/or protective factors, may lead to allergic sensitization and development of allergic disease and the consequent symptoms (1–5). One environmental factor that has received particular interest in recent years is the variation in early microbial exposure, which has indisputable, although incompletely understood, effects on immunological maturation. Wider acknowledgment of the possible association between microbes and allergic diseases followed the introduction of what became known as "hygiene hypothesis" by Strachan, 1989 (6). Based upon epidemiological findings, he suggested that the rise in prevalence of allergic diseases in past decades was due to factors associated with changes in life style such as reduced family size and improved hygiene measures. He assumed that these epidemiological correlations reflected reduced opportunities for cross-infections in families with young children.

The basic idea linking microbes and allergies is that adequate microbial exposure may be able to direct the early immunological development away from allergic type responsiveness. In contrast, inadequate exposure does not provide this necessary stimulus and may even promote the development of allergic disease. The original hygiene hypothesis was based on infections, but what truly constitutes the characteristics and source of "adequate" microbial stimulus remains unknown. Intestinal microbiota are at least quantitatively the primary source of host-microbe interactions soon after birth. Moreover, the early establishment of the microbiota has been shown to be prerequisite for the formation of tolerance to mucosally encountered antigens (7–12). Arguably the best clinical evidence linking intestinal microbiota and allergies is provided by preliminary trials that have had success in preventing or treating allergic conditions by oral administration of intestinal bacterial isolates (13–18). Also, early use of antibiotics has been implicated to predispose the infant to allergic sensitization and development of allergic disease, although this view is controversial (19,20). The aim of this chapter is to summarize the current knowledge of the characteristics of gut microbiota in allergic infants and discuss their implication in allergic disease development.

ALLERGIES—AN OVERVIEW

Allergies are by definition immunological hypersensitivity reactions to substances (allergens), usually proteins, tolerated in defined dose by normal individuals (21). Allergic reactions are manifested in allergic diseases such as asthma, eczema, and rhinoconjunctivitis, each defined by a group of symptoms and signs. The life-impairing effect of these diseases varies from subtle to dominant. In addition to impairing physical health there may be an impact on social and emotional health, especially in childhood (22). Allergic symptoms can significantly disturb productivity in school and work where they are among the major causes of absenteeism. The personal and social economic burden is considerable (22–24). During the second half of the twentieth century the prevalence of allergic diseases has increased in epidemic proportions. The highest prevalence is in children and teenagers. With, on average, every fourth child affected, allergic diseases represent the most common chronic childhood illnesses in many countries (25,26). The reasons for this increase are not known (25,27).

There are many exceptions, but in most cases in established allergic disease the inflammatory cascade leading to the symptoms follows allergen contact at mucosal membranes in airways or gastrointestinal tract and is initiated through specific recognition by Immunoglobulin E (IgE) antibodies (27). Overactive T helper 2 (Th2) cells may be considered as the immunopathological cornerstone of these reactions (28). When, for example, pollen-derived aeroallergen is inhaled by a non-allergic subject the immune system reacts mildly by producing allergen-specific IgG_2 and IgG_4 antibodies. This is probably due to specific recognition and action, e.g., production of interferon (IFN)-γ by T helper 1 cells (Th1) cells (28,29). In contrast, in allergic individuals Th2 cells typically infiltrate to the affected tissue and produce cytokines such as interleukins (IL)-4, -5, -9, and, -13. These cytokines promote the production of IgE antibodies, development and accumulation of mast cells, eosinophils, and basophils (the primary effector cells in allergic inflammation) as well as overproduction of mucus and airway hyper-responsiveness in asthma. Recognition of allergens by specific IgE antibodies on the surface of mast cells and basophils triggers these cells to release pre- and newly formed proinflammatory and vasoactive molecules (e.g., histamine) that may cause tissue damage and other detrimental effects. Eosinophilic inflammation contributes to the airway hyper-responsiveness (28).

It is clear that there is a hereditary trait that predisposes to the formation of allergen-specific IgE antibodies and development of allergic disease (27). This genetic predisposition, known as atopy, affects arguably as many as 30–50% of the world population (2,25,27). Although the immunopathological mechanisms in established allergic diseases are well characterized, it is poorly understood how and why atopy leads or does not lead to allergic sensitization and why only some sensitized individuals develop symptomatic allergic disease (30). Intriguingly, the immune responses to common environmental allergens are initially dominated by Th2 cells in all newborn infants but these responses are not suppressed in atopic infants during the first year of life (31,32). This is thought to be due to defects associated with atopy, for example, impaired production of IFN-γ, which compromise the normal maturation of Th2 antagonistic Th1 responses. The major driving force for the Th1 maturation is considered to be the nature of the microbial exposure encountered after birth. Recent studies indicate that another type of T helper cells, collectively referred to as regulatory T cells (Tregs), may also be involved or even be the chief executers in natural suppression of Th2 mediated responses to environmental allergens. At least two types of Tregs have been shown to have this ability in humans: (1) $CD4^+CD25^+$ Tregs, which probably mediate their action primarily via production of immunosuppressive cytokines transforming growth factor (TGF)-β (also in membrane-bound form) and IL-10 and (2) IL-10 producing Tregs (33–35). Notably, there is indication that the numbers of allergen-specific Tregs may be lower and their suppressive ability defective in those subjects who become sensitized (36,37). Also, the mechanism of successful allergen-injection immunotherapy has been linked with induction of IL-10 Tregs that suppress Th2 responses and induce switching from IgE to IgG_4 antibody (33).

ALLERGY-ASSOCIATED COMPOSITIONAL CHARACTERISTICS OF INFANT GUT MICROBIOTA

The predominant site for host-microbe interaction is in the gut. Thus, its compositional development has been suggested to be the key determinant in whether or not the atopic genotype will be fully expressed and thereby affect the development of allergic diseases. The determination of characteristics in compositional development of intestinal microbiota in association with the expression of allergies may provide a starting point for elucidating which microbial components, if any, may have particular relevance in immunopathology of allergic diseases.

Studies by Traditional Plate Culture Methods

The first reports associating allergy with characteristic microbial composition in the gut appear to be from studies in the former Soviet Union in the early1980s (38–40). One of these studies, reported also in English, involved an assessment of 60 under one-year-old infants with food allergy and atopic eczema. It was claimed that the severity of the disease was in direct correlation with the stage of aberrancy in the fecal microbiota. This aberrancy was characterized as low prevalence of bifidobacteria and lactobacilli and high prevalence of *Enterobactericeae*, pathogenic species of staplylococci and streptococci as well as *Candida* species (39). Indication that such differences may persist beyond infancy was provided a few years later by Ionescu and co-workers (1986) who studied 10- to 45-year-old subjects. Subjects with atopic eczema (n=58) were shown to have lower prevalence of lactobacilli, bifidobacteria, and enterococci species than the healthy subjects (n=21) but higher prevalence of *Klebsiellae*, *Proteus*, *Staphylococcus aureus*, *Clostridium innocuum* and *Candida* species (41,42). Supporting findings were later published by this group from a comparison of the fecal microbiota of 30 healthy subjects and 110 subjects with atopic eczema (43).

Although these early studies have not received wider acknowledgment in the scientific community, they are well in agreement with later studies that began to accumulate a decade later. In one study *Klebsiellae* species were again found more frequently in the feces of 6-month-old infants with atopic eczema (n=27) and the presence of *Streptococcus* species was less frequent than in the healthy controls (n=10) (44). Collectively, the predominant anaerobic and facultatively anaerobic microbiota of allergic infants has been characterized by significantly lower prevalence of gram-positive species. In a study by Björkstén and co-workers (1999), colonization by lactobacilli was shown to be less common in both Estonian and Swedish two-year-old children with food allergies (n=27) than in the age compatible healthy children (n=36), whilst the opposite was true for coliforms and

S. aureus (45). In addition, their results indicated that *Bacteroides* comprised a larger proportion of the whole microbiota in healthy compared to allergic infants. They later studied the development of microbiota in a prospective follow-up. Surprisingly, lactobacilli were significantly more frequently present during the neonatal period in the feces of infants who at 2 years had atopic eczema and/or positive skin prick test (n = 18) than in the feces of infants who remained symptom free and had negative skin prick test (n = 26) (46). The rest of the characteristics that were associated with allergy were in concordance with the previous studies with less frequent presence of bifidobacteria and enterococci during the neonatal period. Later in the first year of life, a relatively high prevalence of *S. aureus* and numbers of clostridia and relatively low numbers of *Bacteroides* were associated with allergic eczema (46). The putative differences in the bifidobacterial microbiota were studied at species level by Ouwehand and co-workers (2001) and they found that the feces of 2 to 7-month-old infants with atopic eczema (n = 7) contained more frequently *B. adolescentis* and less frequently *B. bifidum* than the feces of healthy infants (n = 6) (47).

Studies by Molecular Methods

Results obtained by molecular-based culture-independent techniques are largely supportive of the findings presented above. In another prospective follow-up, the fecal microbiota in Finnish neonates was studied prior to the expression of atopy as detected by a positive skin prick test at year one (n = 12). The microbiota of these sensitized children tended to contain lower numbers of bifidobacteria and significantly higher numbers of *Clostridium histolyticum* than those in samples from infants with a negative prick test (n = 17) (48). The *Clostridium* species detectable with the oligonucleotide-probe used in that study include common infant gut colonizers such as *C. paraputrificum*, *C. butyricum* and *C. perfringens* but not *C. difficile*. However, another study indicated that relatively high fecal levels of rarely detected i-caproic acid indicative of *C. difficile* activity was associated with presence of IgE mediated allergic condition in Swedish infants at around one year of age (49). The association between low numbers of fecal bifidobacteria and subsequent allergic sensitization was confirmed in a study showing that neonatal bifidobacteria numbers were significantly lower in children who had food allergen-specific IgE antibodies in their serum at 2 years (n = 10) than in those who did not have the antibodies (n = 16) (50). In addition, the numbers of bifidobacteria present during the neonatal period correlated inversely with total IgE concentration at 2 years (n = 25). In accordance with the association suggested by the earlier studies between the high prevalence of coliforms and allergy, another study showed a direct correlation between the fecal numbers of *Escherichia coli* and total IgE concentration in infants with early onset atopic eczema at mean age of 5 months (n = 19) (18). Furthermore, at weaning around 1 year of age total bacterial cell counts correlated inversely with the severity of eczema as indicated by severity Scoring Atopic Dermatitis (SCORAD) scores (44).

Somewhat contrasting results to those presented by plate culture methods have also been reported. In a study of 6-month-old exclusively breast-fed infants the mean bifidobacterial numbers were not found to be lower in the feces of infants with early onset atopic eczema (n = 15) compared to controls (n = 10), with the exception of a small subgroup of allergic infants (n = 5) that additionally had gastrointestinal symptoms. Moreover, as opposed to studies by Björkstén and co-workers, *Bacteroides* numbers were higher in a subgroup of allergic infants (n = 6) who were later confirmed to have cow milk allergy by challenge (44). *Bacteroides* numbers were also associated with cow milk allergy in a later study where the high counts correlated directly with serum total IgE concentration in a subgroup of infants intolerant to extensively hydrolyzed whey formula (n = 7) (18).

During weaning, the numbers of *Clostridium histolyticum* correlated inversely with the severity of atopic eczema as indicated by SCORAD scores whereas lactobacilli/enterococci numbers correlated directly with the serum total IgE levels (44). It is worthy of note that although high total IgE concentration represents phenotypic characteristics associated with an atopic background, unlike allergen-specific IgE antibodies, the immunopathophysiological significance of total IgE is questionable (51).

Common Trends and Contradictions

The microbial characteristics of infants presented above are summarized in Figure 1. Although some variability exists depending upon the study, there are relatively clear trends evident. The most consistent trends associated with allergy are low numbers of bifidobacteria and high numbers of *S. aureus* and certain species of coliforms and clostridia. It should be pointed out that there are several aspects of these studies that complicate their interpretation. A fundamental downfall is the evaluation of intestinal microbiota by use of the feces, which may only be indicative of the composition of the microbial community in the lower bowel (52,53). Notably, it has been shown that the proportional quantities of specific strains in the colonic mucosa may differ from those in the feces (54). Moreover, the studies on fecal microbiota reveal little with respect to the composition of the small intestine, which immunologically may be more relevant than the large intestine. Another significant deficiency in these studies is the lack of more detailed characterization, especially at the species and strain level. It is well known that bacterial properties, including their immunological effects, vary between bacterial species

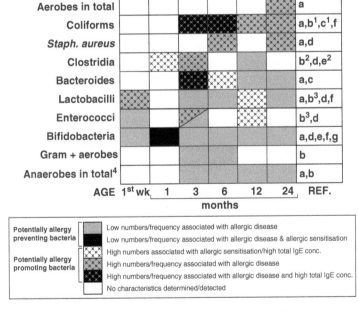

Figure 1 Map of bacterial characteristics in infant microbiota during the first 2 years of life relative to development and presence of allergic disease and IgE antibodies and total IgE concentrations. [1]*Klebsiellae* in ref. "a", *E. coli* in ref. "c", [2]*Clostridium histolyticum,* [3]lactobacilli and enterococci enumerated together, [4]Includes results from total microbial cell counts. *Abbreviation*: IgE, Immunoglobulin E. *Source*: a, From Ref. 45; b, From Ref. 44; c, From Ref. 18; d, From Ref. 46; e, From Ref. 48; f, From Ref. 39; g, From Ref. 50.

and strains of the same species (55,56). Many of the apparent contradictions in the results may therefore reflect the fact that different species or strains within the same genera may have dramatically different effects on allergies. Having said that, they could be the result of different study protocols, methodologies, and particularly differences in the study populations and their nutritional and therapeutic management. Clearly defined study populations are particularly important in studies of allergies. This applies even within allergic diseases such as atopic eczema, which rather than a single disease is an aggregation of several conditions which have certain clinical characteristics in common (57). It is difficult to state that specific microbial patterns can be generalized to be common in all allergic conditions, in part because the microbiota composition remains to be fully elucidated in all the mucosal compartments, and as human genomic and environmental exposures differ between individuals. In all the studies reported to date, the composition of infant intestinal microbiota has been assessed in relation to development of atopic eczema, food allergy or signs of allergic sensitization. This is an obvious shortcoming, albeit understandable, as these are nearly exclusive manifestations of allergy in childhood.

INTERPRETING THE GUT MICROBIOTA CHARACTERISTICS

The reason for the compositional differences in the average microbiota of allergic and healthy infants is not yet known. Undisputable conclusions regarding causal relationship cannot be drawn based on mere characterization of microbial composition relative to clinical sings and symptoms. In a few studies, characteristics of the fecal microbiota have been shown to precede the beginning of the expression of atopy, implying that these differences are not necessarily secondary to the disease. However, these, and other studies have not taken into account changes that occur in the development of the gut mucosa as these likely influence which microbes colonize and how these influence clinical signs of allergy.

Theoretically, there are a number of plausible causes for microbial compositional differences seen to date; these are listed in Table 1. Many of these factors are intertwined. Some plausible ways by which desirable microbes may protect the host from allergic sensitization and alleviate symptoms are presented in Figure 2.

Reflection of Atopic Genotype

Incomplete knowledge of the genetic characteristics of allergic diseases restricts the full understanding of their possible influence on the development of gut microbiota (58). Theoretically, microbial colonization could be directly affected for example if the atopic genotype was associated with receptor expression on epithelial cells or production of intestinal mucus. There is some indication that the atopic genotype is associated with

Table 1 Possible Causes for Microbial Compositional Differences in Atopic versus Healthy Children

---

Atopic genotype related defects in the host's ability to interact with bacteria

The role of microbial stimulus in the normal maturation of the immune system away from allergic type responsiveness

The influence of allergic symptoms and consequent inflammation on microbial colonization

The effects microbes have on allergen processing and uptake, for example, by inducing gut inflammation

Environmental factors that affect the expression of atopy in parallel with the microbiota or via the microbiota

---

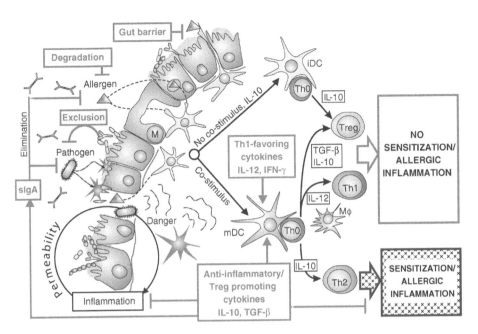

Figure 2 Mechanisms by which specific components of intestinal microbiota may protect from allergic sensitization and/or alleviate symptoms. "Adequate" microbial composition may reduce allergen uptake by providing maturational stimulus for *gut barrier* function, enhancing allergen *degradation* by production of digestive enzymes (this may also reduce allergen allergenicity), improving mucosal integrity by direct *exclusion* of pathogens that may cause epithelial damage or by enhancing *secretory IgA (sIgA)* production (possibly via inducing *TGF-β* secretion) and by inducing secretion of *anti-inflammatory cytokines*, which may break a vicious circle where inflammation increases gut permeability allowing invasion of pathogens and allergens, which then results in further inflammation. Danger signals caused by epithelial damage and inflammation promote the maturation of dendritic cells, which influence the differentiation of naïve Th cells. Presentation of allergen in absence of danger signals may promote formation of regulatory T cells (Treg) and thus formation of tolerance to the allergen. The fate of Th cells in the presence of danger signals depends on additional stimulus: presence of *TGF-β* (produced, e.g., by epithelial cells) may promote development of Treg population and again tolerance to the allergen, presence of *IL-12 and IFN-γ* (produced, e.g., by macrophages or dendritic cells) promotes development of Th1 population and non-allergic type immune responses, whereas presence of IL-10 may promote formation of allergen specific Th2 cells. In the symptomatic phase induction of *anti-inflammatory cytokines* may also directly alleviate the allergic inflammation by active suppression. *Abbreviations*: sIgA, secretory IgA; M, M-cell; iDC, immature dendritic cell; mDC, mature dendritic cell; IL, interleukin; TGF, transforming growth factor; Th, T-helper; Treg, regulatory T-cell; MΦ, macrophage.

immunological deviancies that could result in impaired recognition of specific bacterial groups and thus allow them to flourish. These defects include compromised expression of Toll-like receptor (TLR) 4 and its soluble co-receptor CD14 (sCD14), albeit the results regarding sCD14 are conflicting (59–64). However, also low breast-milk levels of sCD14 have been associated with subsequent development of eczema in children irrespective of atopy (65). TLR4 and sCD14 are pattern recognition receptors of innate immune systems that are important in detection of components in both Gram-positive and Gram-negative bacteria but especially the cell-wall lipopolysaccharides (LPS) in the latter (66,67). Notably, CD14-independent recognition of LPS would seem to be defective during the neonatal period (68). Compromised recognition may facilitate colonization by bacteria which would otherwise be cleared or reduced in numbers due to immune responses mounted against them. This could partly explain why relatively a high prevalence and numbers of potentially pathogenic Gram-negative bacteria but low numbers of Gram-positive bacteria appear to accompany atopic eczema and high levels of IgE (18,39,42–45,50).

From another perspective, microbial compositional differences may reflect their influence on allergic sensitization and disease development. If the recognition of gut colonizers is compromised, then so may be the interactions that drive the normal immunological maturation (10,32,60,69,70). Recognition of peptidoglycan, a major component of Gram-positive cell-wall, is less dependent on CD14 and TLR4 but rather on co-operation between TLRs 2 and 6 (71–73). Thereby, an atopic host, with deficient TLR4 and CD14 recognition, may have better chances to interact with Gram-positive than Gram-negative bacteria. This interaction may, on one hand, limit the ability of Gram-positive bacteria to colonize the gut, but on the other, provide maturational stimulus for the developing immune system (44,69).

Whereas the recognition of one specific bacterial component occurs primarily via one or two different pattern recognition receptors, the recognition of whole bacterium is likely to involve a set of different receptors such as TLR9 recognizing unmethylated bacterial CpG DNA and TLR5 recognizing flagella (74). Accordingly, a quantitatively strong enough exposure may compensate the poor recognition of Gram-negative bacteria, especially due to ligation of TLR9. This would be in agreement with the observation that postnatal administration of exogenous Gram-negative bacteria, namely non-enteropathogenic *E. coli* strain, was associated with reduced risk of developing allergic diseases later in life (14,15).

Reflection of Effects on Th1, Th2, and Treg Differentiation

The effects of intestinal bacteria on cytokine production, epithelia-damaging action or proinflammatory action may have a major influence on naive T-cell differentiation to Th1, Th2 or Treg cells (Fig. 2). A study in mice with compromised Toll-mediated signaling capacity indicated that antigen specific Th1 responses to food allergens are dependent on simultaneously induced Toll-mediated activities, whilst similar dependency was not observed in Th2 responses. Re-exposing the mice to the allergen enhanced the production of IL-13 by T-cells, a cytokine capable of inducing isotype class-switching of B-cells to produce IgE (75).

Th differentiation is directed by dendritic cells, which monitor the antigenic environment and presence of danger signals in the gut. Danger signals may include epithelial damage and inflammation. In the absence of maturational/inflammatory stimuli, dendritic cells aim to tolerize the immune system to what they assume to be harmless antigens. It is noteworthy that the immunological stimulus initiated may vary depending on which TLR or combination of TLRs are ligated (76). This may provide a mechanistic basis for consistent data from in vitro studies, which indicate that cytokine responses mounted by mononuclear cells in response to whole Gram-negative and whole Gram-positive bacteria are different. The induction of IL-12 is greater for Gram-positive bacteria and IL-10 for Gram-negative bacteria (77–79). IL-12 is produced by dendritic cells and macrophages and is a key cytokine promoting the Th cell differentiation into Th1 cells. IL-10 may contribute in maintaining a Th2 bias, but it may also induce tolerance by promoting the formation of Tregs and anergic T-cells (80–82).

In a study by He and co-workers (2002) bifidobacteria isolated from the feces of allergic infants tended to induce murine macrophage-like cells to produce more of IL-12, but less IL-10 than bifidobacteria from the feces of healthy infants (83). In their earlier, aforementioned, study *B. adolescentis* was associated with allergic and *B. bifidum* with healthy infants (47). Accordingly, in a recent study, Young and co-workers showed that *B. bifidum* enhanced IL-10 production by dendritic cells isolated from cord blood (84). However, *B. adolescentis*, or any other bifidobacterial strain, did not induce IL-12 production. Moderate differences were observed in the effects of bifidobacterial strains on the expression of dendritic cell activation markers. The basis for speculation on the possible significance of these findings is weak until more detailed characterization is performed. Arguably, the findings could collectively indicate that bifidobacteria in allergic infants may promote formation of tolerogenic responses but this remains to be confirmed (Fig. 2).

Also *Lactobacillus* strains have been shown to confer differential effects on cytokine production and expression of surface markers on murine dendritic cells (85). Furthermore, lactobacilli induced in vitro, in a strain dependent manner, Treg-like low proliferating Th population producing TGF-β and IL-10 (86). TGF-β is the key cytokine in induction of T-cell differentiation towards Tregs (Fig. 2) (87). In a clinical study, improvement in atopic eczema symptoms following oral administration of lactobacilli was accompanied by increased serum concentrations of TGF-β (17). Interestingly, oral supplementation of lactobacilli in breast-feeding mothers was followed by increased TGF-β concentrations in breast-milk (88). This increase may have contributed to subsequently lower prevalence of atopic eczema in children. It should be noted, however, that allergic sensitization was not affected and allergic rhinitis and asthma may have increased in frequency (89). Nevertheless, these studies are not only indicative of the influence of infant microbiota on allergy development but also of the possible influence of maternal microbiota during pregnancy and via breast-milk.

Reflection of Effects on Allergen Uptake, Processing, and Presentation

The original hygiene hypothesis implicated pathogens in an allergy-preventing role. However, their role may be two-sided (90). Whereas the host immune system may become tolerant towards commensal microbes, this should and will not happen with pathogens (91,92). Therefore, pathogens may have a greater potential to stimulate the neonatal immunity away from the allergic type responsiveness than the commensal microbes towards which tolerance has been formed (90). Conversely, potential pathogens may induce and sustain inflammation and compromise the gut barrier (18,93). This may allow greater numbers of allergens to pass the barrier and alter their presentation to lymphocytes due to the presence of danger signals. Consequently, allergic sensitization may be more likely to occur, and may be aggravated in already sensitized subjects with allergic disease (94–96). *E. coli* and *Bacteroides* bacterial groups colonizing these subjects may include strains with such detrimental properties (97–100). Such bacteria were implicated with higher serum total IgE concentrations and sensitivity to cow's milk proteins in studies referred to above (18,44). Some non-pathogenic bacteria, such as lactobacilli and bifidobacteria, may have the opposite effects by reducing gut inflammation either via excluding colonization by pathogens or inducing secretion of anti-inflammatory cytokines, reducing gut permeability, allergen antigenicity, and fortifying gut defense barrier e.g., by stimulating IgA production (101–110). Intestinal microbes are likely to affect the allergen uptake also by promoting the maturation and integrity of gut barrier but there is little information on how this ability may vary between different bacteria (111).

Reflection of Allergic Symptoms

The possibility that allergic symptoms either affect, or are affected by, the microbiota is supported by an observation that alleviation in atopic eczema and allergic inflammation following oral administration of bifidobacteria was accompanied by modified dynamics in the microbiota (i.e., restriction in the growth of *E. coli* and *Bacteroides*) (18). Also, earlier findings attest to this possibility implicating direct correlation between numbers of *Enterobacteriaceae* family bacteria and severity of atopic eczema symptoms (39). The compositional characteristics associated with the severity of symptoms may be caused by intestinal inflammation exacerbated in some allergic conditions (95,112–115).

Reflection of Environmental Factors

Amongst the best examples of factors which have been clearly shown to influence the development of the gut microbiota and have also been implicated in allergic diseases include the mode of delivery and breast-feeding (116–123). Indeed, it is plausible that the characteristics of fecal microbiota associated with atopic eczema and allergic sensitization may partly reflect dietary factors. It is well known that changes in diet may dramatically affect the microbial composition of the gut. Then again, in allergic infants the diet can reflect the child's health status due to food restrictions. In 39–63% of all infants and young children, atopic eczema is triggered by one or more challenge-confirmed food allergies (124–126). Moreover, the development of manifestations of allergic diseases in children correlates with differences in the composition and immunological characteristics of breast-milk, which on the other hand are affected by maternal gut microbiota and atopy (127–133). For example, the polyunsaturated fatty acid composition in breast-milk has been shown to correlate with the development of allergic disease in children (131,132). In vitro these compounds have been shown to selectively affect microbial growth and adhesion to intestinal cells (134). Recently, lactobacilli in breast-milk were shown to have properties in vitro that could promote the development and maintenance of gut barrier in neonates, thus warranting further studies on this area (135). Albeit the effect of caesarean delivery in promoting allergy is disputable, it is notable that colonization by *Lactobacillus*- and *Bifidobacterium*-like bacteria, the high numbers of which have mainly been associated with non-allergic phenotype, may be delayed for up to 10 days and 1 month, respectively, as compared to vaginally delivered infants (136).

Regarding our earlier discussion on pathogens and *E. coli*, it is noteworthy that in developing countries with low prevalence of allergies, the establishment of intestinal microbiota is characterized by rapid initial colonization, formation of enterobacterial microbiota predominated by *E. coli*, and frequent colonization by pathogens such as salmonellae. The *E. coli* population is characterized by a wide spectrum of strains and instability (137,138). Whether such rapid colonization and strongly variable exposure has special influence on immunological maturation and gut barrier formation and maintenance remain to be established.

CONCLUSION

It has been well established that allergic sensitization and the development of allergic disease are associated, at least in some infants, with characteristic developmental patterns in fecal microbiota composition that are atypical to healthy infants. With relative consistency these characteristics include low numbers of bifidobacteria and anaerobes in total and high numbers of clostridia, *S. aureus* and certain coliforms such as *Klebsiellae*. Data on lactobacilli, *Bacteroides* and *E. coli* are somewhat variable. How this aberrancy in fecal microbiota depicts the situation in the intestine and how it is clinically significant, remains to be known. The possibility that the characteristics are secondary to the disease cannot be excluded, but it is also feasible that they reflect their significance in the aetiology of allergy. Extensive experimental data implies that the development of atopic type immunoreactivity could be promoted by the establishment of an early gut microbiota that (1) is incapable of directing the immune system towards tolerogenic responses to, what should be, harmless environmental antigens and/or (2) induces inflammatory responses against itself, thereby increasing mucosal permeability to potential allergens.

It has been convincingly demonstrated that microbial exposure is likely to be the primary exogenous stimulus directing the immunological maturation away from allergic type immunoresponsiveness early in life. However, it is still not clear what are the qualitative or quantitative characteristics of the indigenous microbiota or other sources of microbial exposure that could protect from, or conversely promote ("allow"), the expression of allergies. Future studies should assess whether specific microbial species have particular importance in this respect or whether the "adequate" stimulus is only a matter of quantitatively high enough exposure or strongly variable exposure. More efforts should be directed to characterizing microbial composition of nasal and oral cavities and different compartments in the intestinal tract of children as well as the gut of pregnant women and the gut and breast-milk of breast-feeding mothers.

ACKNOWLEDGMENTS

Pirkka Kirjavainen gratefully acknowledges financial support from the Academy of Finland.

REFERENCES

1. Arruda LK, Sole D, Baena-Cagnani CE, Naspitz CK. Risk factors for asthma and atopy. Curr Opin Allergy Clin Immunol 2005; 5:153–159.
2. Steinke JW, Borish L, Rosenwasser LJ 5. Genetics of hypersensitivity. J Allergy Clin Immunol 2003; 111:S495–S501.
3. von Mutius E. Influences in allergy: epidemiology and the environment. J Allergy Clin Immunol 2004; 113:373–379; quiz 380.
4. von Mutius E. The environmental predictors of allergic disease. J Allergy Clin Immunol 2000; 105:9–19.
5. O'Connell EJ. Pediatric allergy: a brief review of risk factors associated with developing allergic disease in childhood. Ann Allergy Asthma Immunol 2003; 90:53–58.
6. Strachan DP. Hay fever, hygiene, and household size. BMJ 1989; 299:1259–1260.
7. Noverr MC, Huffnagle GB. Does the microbiota regulate immune responses outside the gut? Trends Microbiol 2004; 12:562–568.
8. Moreau MC, Corthier G. Effect of the gastrointestinal microflora on induction and maintenance of oral tolerance to ovalbumin in C3H/HeJ mice. Infect Immun 1988; 56:2766–2768.
9. Moreau MC, Gaboriau-Routhiau V. The absence of gut flora, the doses of antigen ingested and aging affect the long-term peripheral tolerance induced by ovalbumin feeding in mice. Res Immunol 1996; 147:49–59.
10. Sudo N, Sawamura S, Tanaka K, Aiba Y, Kubo C, Koga Y. The requirement of intestinal bacterial flora for the development of an IgE production system fully susceptible to oral tolerance induction. J Immunol 1997; 159:1739–1745.
11. Gaboriau-Routhiau V, Moreau MC. Gut flora allows recovery of oral tolerance to ovalbumin in mice after transient breakdown mediated by cholera toxin or Escherichia coli heat-labile enterotoxin. Pediatr Res 1996; 39:625–629.

12. Hooper LV, Gordon JI. Commensal host-bacterial relationships in the gut. Science 2001; 292:1115–1118.
13. Kalliomäki M, Salminen S, Arvilommi H, Kero P, Koskinen P, Isolauri E. Probiotics in primary prevention of atopic disease: a randomised placebo-controlled trial. Lancet 2001; 357:1076–1079.
14. Lodinová- Žádníková R, Cukrowská B. Influence of oral colonization of the intestine with a non-enteropathogenic *E. coli* strain after birth on the frequency of infectious and allergic diseases after 10 and 20 years. Immunol Lett 1999; 69:64.
15. Lodinová-Žádníková R, Cukrowska B, Tlaskalova-Hogenova H. Oral administration of probiotic *Escherichia coli* after birth reduces frequency of allergies and repeated infections later in life (after 10 and 20 years). Int Arch Allergy Immunol 2003; 131:209–211.
16. Majamaa H, Isolauri E. Probiotics: a novel approach in the management of food allergy. J Allergy Clin Immunol 1997; 99:179–185.
17. Isolauri E, Arvola T, Sutas Y, Moilanen E, Salminen S. Probiotics in the management of atopic eczema. Clin Exp Allergy 2000; 30:1604–1610.
18. Kirjavainen PV, Arvola T, Salminen SJ, Isolauri E. Aberrant composition of gut microbiota of allergic infants: a target of bifidobacterial therapy at weaning? Gut 2002; 51:51–55.
19. Celedon JC, Weiss ST. Use of antibacterials in infancy: clinical implications for childhood asthma and allergies. Treat Respir Med 2004; 3:291–294.
20. Voor T, Julge K, Bottcher MF, Jenmalm MC, Duchen K, Bjorksten B. Atopic sensitization and atopic dermatitis in estonian and Swedish infants. Clin Exp Allergy 2005; 35:153–159.
21. Johansson SG, Bieber T, Dahl R, et al. Revised nomenclature for allergy for global use: report of the nomenclature review committee of the world allergy organization, October 2003. J Allergy Clin Immunol 2004; 113:832–836.
22. O'Connell EJ. The burden of atopy and asthma in children. Allergy 2004; 78:7–11.
23. Blaiss MS. Important aspects in management of allergic rhinitis: compliance, cost, and quality of life. Allergy Asthma Proc 2003; 24:231–238.
24. Cisternas M, Blanc P, Yen I, et al. A comprehensive study of the direct and indirect costs of adult asthma. J Allergy Clin Immunol 2003; 111:1212–1218.
25. Jarvis D, Burney P. ABC of allergies. The epidemiology of allergic disease. BMJ 1998; 316:607–610.
26. Wickman M, Lilja G. Today, one child in four has an ongoing allergic disease in Europe. What will the situation be tomorrow? Allergy 2003; 58:570–571.
27. Johansson SG, Hourihane JO, Bousquet J, et al. A revised nomenclature for allergy. An EAACI position statement from the EAACI nomenclature task force. Allergy 2001; 56:813–824.
28. Kay AB. Allergy and allergic diseases. First of two parts. N Engl J Med 2001; 344:30–37.
29. Ebner C, Schenk S, Najafian N, et al. Nonallergic individuals recognize the same T cell epitopes of Bet v 1, the major birch pollen allergen, as atopic patients. J Immunol 1995; 154:1932–1940.
30. Prescott SL. Allergy: when does it begin and where will it end?. Allergy 2003; 58:864–867.
31. Prescott SL, Macaubas C, Holt BJ, et al. Transplacental priming of the human immune system to environmental allergens: universal skewing of initial T cell responses toward the Th2 cytokine profile. J Immunol 1998; 160:4730–4737.
32. Prescott SL, Macaubas C, Smallacombe T, Holt BJ, Sly PD, Holt PG. Development of allergen-specific T-cell memory in atopic and normal children. Lancet 1999; 353:196–200.
33. Robinson DS, Larche M, Durham SR. Tregs and allergic disease. J Clin Invest 2004; 114:1389–1397.
34. Nakamura K, Kitani A, Fuss I, et al. TGF-beta 1 plays an important role in the mechanism of CD4+CD25+ regulatory T cell activity in both humans and mice. J Immunol 2004; 172:834–842.
35. Nakamura K, Kitani A, Strober W. Cell contact-dependent immunosuppression by CD4(+)CD25(+) regulatory T cells is mediated by cell surface-bound transforming growth factor beta. J Exp Med 2001; 194:629–644.

36. Ling EM, Smith T, Nguyen XD, et al. Relation of CD4+CD25+regulatory T-cell suppression of allergen-driven T-cell activation to atopic status and expression of allergic disease. Lancet 2004; 363:608–615.

37. Akdis M, Verhagen J, Taylor A, et al. Immune responses in healthy and allergic individuals are characterized by a fine balance between allergen-specific T regulatory 1 and T helper 2 cells. J Exp Med 2004; 199:1567–1575.

38. Kuvaeva IB, Zakharova NV, Orlova NG, Veselova OL. Functional state of the immunological system and of the gastrointestinal tract in children with a food allergy. Vopr Pitan 1980;33–40.

39. Kuvaeva IB, Orlova NG, Veselova OL, Kuznezova GG, Borovik TE. Microecology of the gastrointestinal tract and the immunological status under food allergy. Nahrung 1984; 28:689–693.

40. Shaternikov VA, Kuvaeva ID, Ladodo KS, Orlova NG, Veselova OL. General and local humoral immunity and intestinal microflora in children with skin manifestations of food allergy. Vopr Pitan 1982;51–56.

41. Ionescu G, Radovicic D, Schuler R, et al. Changes in fecal microflora and malabsorption phenomena suggesting a contaminated small bowel syndrome in atopic eczema patients. Microecol Ther 1986; 16:273.

42. Ionescu G, Kiehl R, Ona L, Schuler R. Abnormal fecal microflora and malabsorption phenomena in atopic eczema paitents. J Adv Med 1990; 3:71–91.

43. Ionescu G, Kiehl R, Wichmann-Kunz F, Leimbeck R. Immunobiological significance of fungal and bacterial infections in atopic eczema. J Adv Med 1990; 3:47–58.

44. Kirjavainen PV, Apostolou E, Arvola T, Salminen SJ, Gibson GR, Isolauri E. Characterizing the composition of intestinal microflora as a prospective treatment target in infant allergic disease. FEMS Immunol Med Microbiol 2001; 32:1–7.

45. Björkstén B, Naaber P, Sepp E, Mikelsaar M. The intestinal microflora in allergic estonian and Swedish 2-year-old children. Clin Exp Allergy 1999; 29:342–346.

46. Björkstén B, Sepp E, Julge K, Voor T, Mikelsaar M. Allergy development and the intestinal microflora during the first year of life. J Allergy Clin Immunol 2001; 108:516–520.

47. Ouwehand AC, Isolauri E, He F, Hashimoto H, Benno Y, Salminen S. Differences in Bifidobacterium flora composition in allergic and healthy infants. J Allergy Clin Immunol 2001; 108:144–145.

48. Kalliomäki M, Kirjavainen P, Eerola E, Kero P, Salminen S, Isolauri E. Distinct patterns of neonatal gut microflora in infants in whom atopy was and was not developing. J Allergy Clin Immunol 2001; 107:129–134.

49. Böttcher MF, Nordin EK, Sandin A, Midtvedt T, Björkstén B. Microflora-associated characteristics in faeces from allergic and nonallergic infants. Clin Exp Allergy 2000; 30:1590–1596.

50. Kirjavainen P. The Intestinal Microbiota—A Target for Treatment in Infant Atopic Eczema, in Department of Biochemistry and Food Chemistry. Turku: University of Turku, 2003: 79.

51. Howard TD, Meyers DA, Bleecker ER. Mapping susceptibility genes for allergic diseases. Chest 2003; 123:363S–368S.

52. Mikelsaar M, Mändar R, Sepp E. Lactic acid microflora in the human microbial ecosystem and its development. In: Salminen S, Von Wright A, eds. Lactic Acid Bacteria: Microecology and Functional Aspects. New York: Marcel Dekker Inc., 1998:278–342.

53. Marteau P, Pochart P, Dore J, Bera-Maillet C, Bernalier A, Corthier G. Comparative study of bacterial groups within the human cecal and fecal microbiota. Appl Environ Microbiol 2001; 67:4939–4942.

54. Zoetendal EG, von Wright A, Vilpponen-Salmela T, Ben-Amor K, Akkermans AD, de Vos WM. Mucosa-associated bacteria in the human gastrointestinal tract are uniformly distributed along the colon and differ from the community recovered from feces. Appl Environ Microbiol 2002; 68:3401–3407.

55. Kirjavainen PV, ElNezami HS, Salminen SJ, Ahokas JT, Wright PF. Effects of orally administered viable *Lactobacillus rhamnosus* GG and *Propionibacterium freudenreichii* subsp. shermanii JS on mouse lymphocyte proliferation. Clin Diagn Lab Immunol 1999; 6:799–802.

56. Kirjavainen PV, El-Nezami HS, Salminen SJ, Ahokas JT, Wright PF. The effect of orally administered viable probiotic and dairy lactobacilli on mouse lymphocyte proliferation. FEMS Immunol Med Microbiol 1999; 26:131–135.

57. Wollenberg A, Bieber T. Atopic dermatitis: from the genes to skin lesions. Allergy 2000; 55:205–213.

58. Feijen M, Gerritsen J, Postma DS. Genetics of allergic disease. Br Med Bull 2000; 56:894–907.

59. Koppelman GH, Reijmerink NE, Colin Stine O, et al. Association of a promoter polymorphism of the CD14 gene and atopy. Am J Respir Crit Care Med 2001; 163:965–969.

60. Baldini M, Lohman IC, Halonen M, Erickson RP, Holt PG, Martinez FD. A Polymorphism* in the $5'$ flanking region of the CD14 gene is associated with circulating soluble CD14 levels and with total serum immunoglobulin E. Am J Respir Cell Mol Biol 1999; 20:976–983.

61. Zdolsek HA, Jenmalm MC. Reduced levels of soluble CD14 in atopic children. Clin Exp Allergy 2004; 34:532–539.

62. Kabesch M, Hasemann K, Schickinger V, et al. A promoter polymorphism in the CD14 gene is associated with elevated levels of soluble CD14 but not with IgE or atopic diseases. Allergy 2004; 59:520–525.

63. Kedda MA, Lose F, Duffy D, Bell E, Thompson PJ, Upham J. The CD14 C-159T polymorphism is not associated with asthma or asthma severity in an Australian adult population. Thorax 2005; 60:211–214.

64. Fageras Bottcher M, Hmani-Aifa M, Lindstrom A, et al. A TLR4 polymorphism is associated with asthma and reduced lipopolysaccharide-induced interleukin-12(p70) responses in Swedish children. J Allergy Clin Immunol 2004; 114:561–567.

65. Jones CA, Holloway JA, Popplewell EJ, et al. Reduced soluble CD14 levels in amniotic fluid and breast milk are associated with the subsequent development of atopy, eczema, or both. J Allergy Clin Immunol 2002; 109:858–866.

66. Haziot A, Ferrero E, Kontgen F, et al. Resistance to endotoxin shock and reduced dissemination of gram-negative bacteria in CD14-deficient mice. Immunity 1996; 4:407–414.

67. Miller SI, Ernst RK, Bader MW. LPS, TLR4 and infectious disease diversity. Nat Rev Microbiol 2005; 3:36–46.

68. Cohen L, Haziot A, Shen DR, et al. CD14-independent responses to LPS require a serum factor that is absent from neonates. J Immunol 1995; 155:5337–5342.

69. Kirjavainen PV. Exposure to gram-positive bacteria: the key in natural defence against atopic sensitisation?. Microecol Ther 2002;109–114.

70. Sudo N, Yu XN, Aiba Y, et al. An oral introduction of intestinal bacteria prevents the development of a long-term Th2-skewed immunological memory induced by neonatal antibiotic treatment in mice. Clin Exp Allergy 2002; 32:1112–1116.

71. Ozinsky A, Underhill DM, Fontenot JD, et al. The repertoire for pattern recognition of pathogens by the innate immune system is defined by cooperation between toll-like receptors. Proc Natl Acad Sci USA 2000; 97:13766–13771.

72. Ozinsky A, Smith KD, Hume D, Underhill DM. Co-operative induction of pro-inflammatory signaling by Toll-like receptors. J Endotoxin Res 2000; 6:393–396.

73. Dziarski R, Ulmer AJ, Gupta D. Interactions of CD14 with components of gram-positive bacteria. Chem Immunol 2000; 74:83–107.

74. Takeda K, Akira S. Toll-like receptors in innate immunity. Int Immunol 2005; 17:1–14.

75. Schnare M, Barton GM, Holt AC, Takeda K, Akira S, Medzhitov R. Toll-like receptors control activation of adaptive immune responses. Nat Immunol 2001; 2:947–950.

76. Aderem A, Ulevitch RJ. Toll-like receptors in the induction of the innate immune response. Nature 2000; 406:782–787.

77. Cross ML, Ganner A, Teilab D, Fray LM. Patterns of cytokine induction by gram-positive and gram-negative probiotic bacteria. FEMS Immunol Med Microbiol 2004; 42:173–180.

78. Karlsson H, Hessle C, Rudin A. Innate immune responses of human neonatal cells to bacteria from the normal gastrointestinal flora. Infect Immun 2002; 70:6688–6696.

79. Hessle C, Andersson B, Wold AE. Gram-positive bacteria are potent inducers of monocytic interleukin-12 (IL-12) while gram-negative bacteria preferentially stimulate IL-10 production. Infect Immun 2000; 68:3581–3586.

80. Raghupathy R. Pregnancy: success and failure within the Th1/Th2/Th3 paradigm. Semin Immunol 2001; 13:219–227.

81. Akdis CA, Blaser K. Mechanisms of interleukin-10-mediated immune suppression. Immunology 2001; 103:131–136.

82. Levings MK, Gregori S, Tresoldi E, Cazzaniga S, Bonini C, Roncarolo MG. Differentiation of Tr1 cells by immature dendritic cells requires IL-10 but not CD25 + CD4 + Tr cells. Blood 2005; 105:1162–1169.

83. He F, Morita H, Hashimoto H, et al. Intestinal *Bifidobacterium* species induce varying cytokine production. J Allergy Clin Immunol 2002; 109:1035–1036.

84. Young SL, Simon MA, Baird MA, et al. Bifidobacterial species differentially affect expression of cell surface markers and cytokines of dendritic cells harvested from cord blood. Clin Diagn Lab Immunol 2004; 11:686–690.

85. Christensen HR, Frokiaer H, Pestka JJ. Lactobacilli differentially modulate expression of cytokines and maturation surface markers in murine dendritic cells. J Immunol 2002; 168:171–178.

86. von der Weid T, Bulliard C, Schiffrin EJ. Induction by a lactic acid bacterium of a population of CD4(+) T cells with low proliferative capacity that produce transforming growth factor beta and interleukin-10. Clin Diagn Lab Immunol 2001; 8:695–701.

87. Huber S, Schramm C, Lehr HA, et al. Cutting edge: TGF-beta signaling is required for the in vivo expansion and immunosuppressive capacity of regulatory CD4 + CD25 + T cells. J Immunol 2004; 173:6526–6531.

88. Rautava S, Kalliomäki M, Isolauri E. Probiotics during pregnancy and breast-feeding might confer immunomodulatory protection against atopic disease in the infant. J Allergy Clin Immunol 2002; 109:119–121.

89. Kalliomäki M, Salminen S, Poussa T, Arvilommi H, Isolauri E. Probiotics and prevention of atopic disease: 4-year follow-up of a randomised placebo-controlled trial. Lancet 2003; 361:1869–1871.

90. Kirjavainen PV. In: Mattila-Sandholm T, Saarela M, eds. Probiotics and the management of food allergy, in functional dairy products. Cambridge, U.K.: Woodhead Publishing, 2003:108–131.

91. Neish AS, Gewirtz AT, Zeng H, et al. Prokaryotic regulation of epithelial responses by inhibition of IkappaB-alpha ubiquitination. Science 2000; 289:1560–1563.

92. Nagler-Anderson C. Man the barrier! Strategic defences in the intestinal mucosa. Nat Rev Immunol 2001; 1:59–67.

93. Kirjavainen PV, Apostolou E, Salminen SJ, Isolauri E. New aspects of probiotics–a novel approach in the management of food allergy. Allergy 1999; 54:909–915.

94. Batt RM, Rutgers HC, Sancak AA. Enteric bacteria: friend or foe?. J Small Anim Pract 1996; 37:261–267.

95. Berin MC, Yang PC, Ciok L, Waserman S, Perdue MH. Role for IL-4 in macromolecular transport across human intestinal epithelium. Am J Physiol 1999; 276:C1046–C1052.

96. Gee JM, Wal JM, Miller K, et al. Effect of saponin on the transmucosal passage of beta-lactoglobulin across the proximal small intestine of normal and beta-lactoglobulin-sensitised rats. Toxicology 1997; 117:219–228.

97. Dahlgren UI, Wold AE, Hanson LA, Midtvedt T. Expression of a dietary protein in *E. coli* renders it strongly antigenic to gut lymphoid tissue. Immunology 1991; 73:394–397.

98. Deitch EA, Specian RD, Berg RD. Endotoxin-induced bacterial translocation and mucosal permeability: role of xanthine oxidase, complement activation, and macrophage products. Crit Care Med 1991; 19:785–791.

99. Obiso RJ, Jr., Lyerly DM, Van Tassell RL, Wilkins TD. Proteolytic activity of the *Bacteroides fragilis* enterotoxin causes fluid secretion and intestinal damage in vivo. Infect Immun 1995; 63:3820–3826.

100. Duchmann R, Kaiser I, Hermann E, Mayet W, Ewe K, Meyer zum KH. Buschenfelde, Tolerance exists towards resident intestinal flora but is broken in active inflammatory bowel disease (IBD). Clin Exp Immunol 1995; 102:448–455.

101. Moreau MC, Ducluzeau R, Guy-Grand D, Muller MC. Increase in the population of duodenal immunoglobulin A plasmocytes in axenic mice associated with different living or dead bacterial strains of intestinal origin. Infect Immun 1978; 21:532–539.

102. De Simone C, Ciardi A, Grassi A, et al. Effect of *Bifidobacterium bifidum* and *Lactobacillus acidophilus* on gut mucosa and peripheral blood B lymphocytes. Immunopharmacol Immunotoxicol 1992; 14:331–340.

103. Kaila M, Isolauri E, Soppi E, Virtanen E, Laine S, Arvilommi H. Enhancement of the circulating antibody secreting cell response in human diarrhea by a human *Lactobacillus* strain. Pediatr Res 1992; 32:141–144.

104. Yasui H, Nagaoka N, Mike K, Hayakawa K, Ohwaki M. Detection of *Bifidobacterium* strains that induce large quantities of IgA. Microb Ecol Health Dis 1992; 5:155–162.

105. Majamaa H, Isolauri E, Saxelin M, Vesikari T. Lactic acid bacteria in the treatment of acute rotavirus gastroenteritis. J Pediatr Gastroenterol Nutr 1995; 20:333–338.

106. Isolauri E, Majamaa H, Arvola T, Rantala I, Virtanen E, Arvilommi H. *Lactobacillus casei* strain GG reverses increased intestinal permeability induced by cow milk in suckling rats. Gastroenterology 1993; 105:1643–1650.

107. Matsuzaki T, Yamazaki R, Hashimoto S, Yokokura T. The effect of oral feeding of *Lactobacillus casei* strain Shirota on immunoglobulin E production in mice. J Dairy Sci 1998; 81:48–53.

108. Sütas Y, Soppi E, Korhonen H, et al. Suppression of lymphocyte proliferation in vitro by bovine caseins hydrolyzed with *Lactobacillus casei* GG-derived enzymes. J Allergy Clin Immunol 1996; 98:216–224.

109. Sütas Y, Hurme M, Isolauri E. Down-regulation of anti-CD3 antibody-induced IL-4 production by bovine caseins hydrolysed with *Lactobacillus* GG-derived enzymes. Scand J Immunol 1996; 43:687–689.

110. Pessi T, Isolauri E, Sütas Y, Kankaanranta H, Moilanen E, Hurme M. Suppression of T-cell activation by *Lactobacillus rhamnosus* GG-degraded bovine casein. Int Immunopharmacol 2001; 1:211–218.

111. Hooper LV, Falk PG, Gordon JI. Analyzing the molecular foundations of commensalism in the mouse intestine. Curr Opin Microbiol 2000; 3:79–85.

112. Ogawa H, Yoshiike T. A speculative view of atopic dermatitis: barrier dysfunction in pathogenesis. J Dermatol Sci 1993; 5:197–204.

113. Majamaa H, Laine S, Miettinen A. Eosinophil protein X and eosinophil cationic protein as indicators of intestinal inflammation in infants with atopic eczema and food allergy. Clin Exp Allergy 1999; 29:1502–1506.

114. Majamaa H, Miettinen A, Laine S, Isolauri E. Intestinal inflammation in children with atopic eczema: faecal eosinophil cationic protein and tumour necrosis factor-alpha as non-invasive indicators of food allergy. Clin Exp Allergy 1996; 26:181–187.

115. Majamaa H, Aittoniemi J, Miettinen A. Increased concentration of fecal alpha1-antitrypsin is associated with cow's milk allergy in infants with atopic eczema. Clin Exp Allergy 2001; 31:590–592.

116. Xu B, Pekkanen J, Hartikainen AL, Jarvelin MR. Caesarean section and risk of asthma and allergy in adulthood. J Allergy Clin Immunol 2001; 107:732–733.

117. Kero J, Gissler M, Gronlund MM, et al. Mode of delivery and asthma—is there a connection?. Pediatr Res 2002; 52:6–11.

118. Saarinen UM, Kajosaari M, Backman A, Siimes MA. Prolonged breast-feeding as prophylaxis for atopic disease. Lancet 1979; 2:163–166.
119. Gdalevich M, Mimouni D, Mimouni M. Breast-feeding and the risk of bronchial asthma in childhood: a systematic review with meta-analysis of prospective studies. J Pediatr 2001; 139:261–266.
120. Gdalevich M, Mimouni D, David M, Mimouni M. Breast-feeding and the onset of atopic dermatitis in childhood: a systematic review and meta-analysis of prospective studies. J Am Acad Dermatol 2001; 45:520–527.
121. Schoetzau A, Filipiak-Pittroff B, Franke K, et al. Effect of exclusive breast-feeding and early solid food avoidance on the incidence of atopic dermatitis in high-risk infants at 1 year of age. Pediatr Allergy Immunol 2002; 13:234–242.
122. Kull I, Wickman M, Lilja G, Nordvall SL, Pershagen G. Breast feeding and allergic diseases in infants-a prospective birth cohort study. Arch Dis Child 2002; 87:478–481.
123. Kirjavainen PV, Gibson GR. Healthy gut microflora and allergy: factors influencing development of the microbiota. Ann Med 1999; 31:288–292.
124. Sampson HA. The immunopathogenic role of food hypersensitivity in atopic dermatitis. Acta Derm Venereol Suppl (Stockh) 1992; 176:34–37.
125. Isolauri E, Turjanmaa K. Combined skin prick and patch testing enhances identification of food allergy in infants with atopic dermatitis. J Allergy Clin Immunol 1996; 97:9–15.
126. Burks AW, James JM, Hiegel A, et al. Atopic dermatitis and food hypersensitivity reactions. J Pediatr 1998; 132:132–136.
127. Kalliomäki M, Ouwehand A, Arvilommi H, Kero P, Isolauri E. Transforming growth factor-beta in breast milk: a potential regulator of atopic disease at an early age. J Allergy Clin Immunol 1999; 104:1251–1257.
128. Järvinen KM, Laine S, Suomalainen H. Defective tumour necrosis factor-alpha production in mother's milk is related to cow's milk allergy in suckling infants. Clin Exp Allergy 2000; 30:637–643.
129. Järvinen KM, Laine ST, Jarvenpaa AL, Suomalainen HK. Does low IgA in human milk predispose the infant to development of cow's milk allergy? Pediatr Res 2000; 48:457–462.
130. Järvinen KM, Suomalainen H. Leucocytes in human milk and lymphocyte subsets in cow's milk-allergic infants. Pediatr Allergy Immunol 2002; 13:243–254.
131. Duchen K, Casas R, Fageras-Bottcher M, Yu G, Bjorksten B. Human milk polyunsaturated long-chain fatty acids and secretory immunoglobulin A antibodies and early childhood allergy. Pediatr Allergy Immunol 2000; 11:29–39.
132. Duchen K, Bjorksten B. Polyunsaturated n-3 fatty acids and the development of atopic disease. Lipids 2001; 36:1033–1042.
133. Thijs C, Houwelingen A, Poorterman I, Mordant A, van den Brandt P. Essential fatty acids in breast milk of atopic mothers: comparison with non-atopic mothers, and effect of borage oil supplementation. Eur J Clin Nutr 2000; 54:234–238.
134. Kankaanpää P, Nurmela K, Erkkila A, et al. Polyunsaturated fatty acids in maternal diet, breast milk, and serum lipid fatty acids of infants in relation to atopy. Allergy 2001; 56:633–638.
135. Martin R, Olivares M, Marin ML, Fernandez L, Xaus J, Rodriguez JM. Probiotic potential of 3 Lactobacilli strains isolated from breast milk. J Hum Lact 2005; 21:8–17 quiz 18–21, 41.
136. Grönlund MM, Lehtonen OP, Eerola E, Kero P. Fecal microflora in healthy infants born by different methods of delivery: permanent changes in intestinal flora after cesarean delivery. J Pediatr Gastroenterol Nutr 1999; 28:19–25.
137. Adlerberth I. In: Hanson LÅ, Yolken RH, eds. Establishment of Normal Intestinal Microflora in the Newborn Infant., in Probiotics, Other Nutritional Factors, and the Intestinal Flora. Philadelphia: Vevey/Lippicott-Raven Publishers, 1999.
138. Mata LJ, Urrutia JJ. Intestinal colonization of breastfed children in a rural area of low socio-economic level. Ann NY Acad Sci 1971; 176:93–109.

11

Probiotics: A Role in Therapy for Inflammatory Bowel Disease

Barbara Sheil, Jane McCarthy, Liam O'Mahony, and Malik M. Anwar
Alimentary Pharmabiotic Centre, Departments of Medicine and Surgery, Microbiology, National Food Biotechnology Centre, National University of Ireland, Cork, Ireland

Fergus Shanahan
Alimentary Pharmabiotic Centre, Departments of Medicine and Surgery, National University of Ireland, Cork, Ireland

INTRODUCTION

Hippocrates is credited with saying: "Let food be thy medicine and medicine be thy food" (1). The term "functional food" includes "any food or food ingredient that may provide a health benefit beyond the traditional nutrients it contains" (2). Probiotic bacteria are forms of functional food that are of particular relevance to gastroenterologists, with evidence for their role in the treatment of infectious and antibiotic-associated diarrhea. Their putative therapeutic role in inflammatory bowel disease (IBD) is receiving growing interest; however, it remains unproven. The Noble laureate, Elie Metchnikoff, suggested that bacteria could be of some benefit to the health of man (3). He suggested that the consumption of copious amounts of fermented dairy products, which served to introduce "beneficial" bacteria to the gastrointestinal tract, was responsible for the longevity of Bulgarian peasants. This marked the birth of probiotics, which are live microorganisms that, when consumed in an adequate amount, confer a health effect on the host (4).

The last decade has seen a resurgence of interest in probiotic research. This renewal of interest in enteric (intestinal) microbiota and gut host-microbe interactions has been generated for a number of reasons. Firstly, the gut contains a complex microbial community, the composition of which has remained elusive due to limited bacteriological culturing techniques. Molecular techniques have now been applied to accurately profile intestinal bacterial groups. Secondly, cross-talk between the gut epithelium and bacteria has been demonstrated. The mechanisms underlying this interaction, and the role of the microbiota in the development and function of the gastrointestinal tract needs further investigation. A breakdown in immune tolerance to enteric microbiota has also been implicated in the pathogenesis of inflammatory disorders, such as inflammatory bowel disease. While evidence suggests that inflammatory bowel disease is characterized by an aggressive immune response to luminal antigens, including members of the commensal microbiota, the precise role of the luminal microbiota in the pathogenesis of disease has yet to be elucidated. Finally, there is evidence suggesting a role for probiotic bacteria in ameliorating inflammatory disease. This has led to the suggestion that probiotics may be an option in the therapy of inflammatory bowel disease, the rationale being that these bacteria without proinflammatory potential might alter the intestinal microbiota balance and modulate the immune response (5–8).

Inflammatory bowel disease encompasses two major diseases, ulcerative colitis (UC) and Crohn's disease (CD). These two syndromes, while sharing similar features of gut mucosal inflammation, are distinct entities. Their pathogenesis remains incompletely understood. Both diseases are commonest in the Western, developed world, with highest incidence in northern climates (9,10).

Genetic factors are known to play a role in the pathogenesis of inflammatory bowel disease. This is demonstrated by concordance in monozygous twin studies. Also, 10–25% of affected patients have a first-degree relative with the disease. However, the incomplete concordance seen in twin studies (concordance rates are 40–50% for CD and <10% for ulcerative colitis) suggests that environmental factors also contribute to the pathogenesis of the disease. In addition, there has been a marked rise in the frequency of CD in the developed world in the past fifty years, with a prevalence of approximately 100 per 100,000 population in North America and northern Europe. This rise in incidence in CD underscores the importance of environmental factors in its etiology. The increase in the incidence of CD has occurred as countries become more developed and industrialized. With changes in lifestyle and environment, improving levels of sanitation have altered the microbial environment. This means altered patterns of exposure to microbes and infections during childhood (11). Inflammatory bowel disease may be a disorder of mucosal immune responsiveness due to lack of stimulation and education of the immune responses (12). It is interesting that parallel to an increase in CD, other chronic inflammatory disorders, including allergies, asthma, multiple sclerosis and insulin-dependent diabetes mellitus have also increased in incidence. Environmental changes associated with industrialization may alter immune system development and pose a risk factor for inflammatory bowel disease in the genetically susceptible individual (12).

THE ROLE OF THE ENTERIC MICROBIOTA IN THE NORMAL GUT

Underpinning the probiotic concept is the importance of the normal intestinal microbiota in health and disease (12). Establishment of gut microbiota begins within minutes of delivery of the newborn (13,14). During delivery the infant is exposed to bacteria in the birth canal, the environment, maternal fecal microbiota, and other sources (15). The gut is initially colonized by facultative anaerobes such as *Escherichia coli* and *Enterococcus* species, possibly due to the absence of anaerobic conditions in the intestine (16). Colonization with bifidobacteria follows, particularly in breast-fed infants, and as the environment becomes more anaerobic, Bacteroides and Clostridia.

The importance of the intestinal microbiota is suggested by the fact that the healthy adult gastrointestinal tract is home to a gut microbiota comprising over 400 different species with more bacterial cells in the gut than eucaryotic cells in the human body and with the average mass of bacteria being 1–2 kg. Commensal bacteria are present at a number of 10^{4-6} per gram of intestinal content in the small bowel, up to 10^8 per gram of ileal content in the distal ileum and up to 10^{13} cells per gram of colonic content (17).

The collective metabolic activity of the normal microbiota, of which little is known, is estimated to rival that of the liver (18–21). Up to 99% of the microbiota is comprised of 30 to 40 strains, with the most abundant populations being strict anaerobes (22,23).

Bacterial members of the genus Bacteroides are amongst the most prominent species found in human feces. Other species include bifidobacteria, clostridia, streptococci, enterococci, lactobacilli, ruminococci, and eubacteria (4,22). Information regarding the microbiota has been restricted by the limitations of bacteriological culture methodology with only 40% of bacterial communities being cultivated on non-selective media in the laboratory (24).

Effects of Enteric Microbiota in the Healthy Intestine

Experiments with germ-free and re-colonized animals demonstrate beneficial effects of the resident microbiota (20). The commensal bacteria act as a defense against infection using several mechanisms, including competition for nutrients, the production of antimicrobial factors against pathogens, such as lactic acid and bacteriocins, and blockage or antagonism of adhesion sites.

In addition, the integrity of the mucosa requires cell signaling between the microbiota, epithelium, and mucosal immune system (7). Without the microbiota, mucosal associated lymphoid tissue is underdeveloped and cell mediated immunity is defective. The enteric microbiota plays an important role in immune system education by fine-tuning T-cell repertoires and Th1/Th2 cytokine profiles (11). Compared with conventional animals, germ-free animals have reduced mucosal cell turnover, cytokine production, mucosal associated lymphoid tissue and lamina propria cellularity leading to an ineffective cell mediated immunity, decreased vascularity and less muscle wall thickness (25–27). There are also differences in intraepithelial lymphocytes (28,29). The intestinal microbiota primes the mucosal immune response and keeps it in a state of "controlled physiological inflammation" (26). Induction and/or maintenance of oral tolerance to ingested antigens also require microbial colonization of the gastrointestinal tract in early life.

Understanding the influence of the gastrointestinal microbiota has prompted interest in the therapeutic modification of the enteric microbiota with probiotics or prebiotics.

THE IMPORTANCE OF THE ENTERIC MICROBIOTA IN INFLAMMATORY BOWEL DISEASE

Considerable evidence implicates the enteric microbiota in the pathogenesis of inflammatory bowel disease (Table 1) (7,8,30,31). Firstly, mucosal inflammation occurs in areas of the gut with highest bacterial numbers. Secondly, surgical diversion of the fecal stream has been associated with clinical improvement in the distal bowel, but relapse is predictable following surgical restoration. Thirdly, putative therapeutic efficacy is seen with the use of antibiotics in colonic disease. Fourthly, immune reactivity to intestinal bacteria is detectable in patients with inflammatory bowel disease suggesting a loss of immune tolerance to components of the microbiota (32,33). Fifthly, there are reports of increased numbers of bacteria within the mucosa of patients with inflammatory bowel disease compared with controls (34,35). The highest bacterial numbers have been seen in CD patients and numbers increase with severity of disease. Finally, the description of the first susceptibility gene for CD, CARD15/NOD2, has provided a basis for explaining the interaction between bacteria and the immune response. CARD15/NOD2 encodes a protein

Table 1 Evidence Implicating the Enteric Microbiota in the Pathogenesis of IBD

The distribution of the lesions is greatest in areas of highest numbers of luminal bacteria
Interruption of the fecal stream has been associated with clinical improvement but relapse is predictable following surgical restoration
Evidence for loss of immunological tolerance to components of the commensal microbiota
Serology and cellular immune reactivity to enteric microbiota that has formed the basis of putative diagnostic tests
Efficacy of antibiotics in patients
Description of first susceptibility gene for Crohn's disease (CARD15/NOD2)
Colonization with normal enteric microbiota is required for expression of disease in animal models of colitis irrespective of the underlying defect
Attenuation of inflammation in animal models of enterocolitis
Efficacy of probiotics in animal models of colitis
Effect of probiotics in human studies of IBD

Abbreviation: IBD, inflammatory bowel disease.

that is involved in the recognition of bacterial products and initiates the inflammatory cascade via activation of the transcription factor Nuclear Factor kappaB (NFκB) (36,37).

Compelling evidence for the interactive role of genes, bacteria, and immunity has been derived from experimental animal models of both Crohn's-like and colitis-like disease (38,39). There are now about 30 different spontaneously occurring or genetically engineered (knockout or transgenic) animal models for inflammatory bowel disease (40–42). Colonization with normal enteric microbiota is required for full expression of disease. Thus, the normal microbiota is a common factor driving the inflammatory process irrespective of the genetic underlying predisposition and immunological effector mechanism (43,44). Several different microorganisms have been demonstrated to induce colitis in animal models. These include *Enterococcus faecalis*, causing colitis in the anti-inflammatory interleukin-10 (IL-10) knockout mice, and *Bacteroides vulgatus*, which induced inflammation in the HLA-B27 rat model (45,46). This evidence has prompted the therapeutic modification of the enteric microbiota in inflammatory bowel disease.

In patients with ulcerative colitis, the construction of an ileal pouch following a colectomy represents a human "model" showing the contribution of genes, bacteria, and immune mechanisms to its pathogenesis. A genetic contribution is consistent with the relative frequency of pouchitis in patients undergoing surgery for colitis compared with those having a pouch created surgically for familial polyposis coli. The contribution of bacteria to the pathogenesis of pouchitis is shown by the efficacy of both antibiotic and probiotic therapy in treating the disease (47). The immune system mediates the tissue damage and pouchitis appears to be a colitis-like process occurring in the colonized ileum.

Specific Microorganisms in Inflammatory Bowel Disease

Despite the importance of bacteria in the pathogenesis of colitis and CD, no specific microorganism has been implicated in causing the intestinal inflammation. The roles of *Mycobacterium paratuberculosis*, measles virus, *Listeria monocytogenes* and adherent *E. coli* in the pathogenesis have been examined. Strains of adherent-invasive *E. coli* have been isolated in the mucosa of patients with CD (48). *M. paratuberculosis* has been cultured from the intestine of patients with CD and detected by molecular methods in the granulomas of resected tissue from patients (49). Possible disease modifying mechanisms of transient pathogens include the disruption of the mucosal barrier (allowing increased uptake of luminal antigens), mimicry of self-antigens and activation of the mucosal immune system via modulation of transcription factors such as NFκB. However, a direct cause and effect relationship has not been established for any of these organisms. Indeed, conditions favoring transmission of infection (low socio-economic status, overcrowding, poor sanitation) appear to protect against inflammatory bowel disease, arguing against an infectious aetiology (50).

Since there is evidence for the role of luminal microbiota in the pathogenesis of inflammatory bowel disease, the alteration of the microbiota by the introduction of probiotic bacteria may result in clinical improvement of the condition. Conventional drug therapy for inflammatory bowel disease involves suppression of the immune system or modulation of the inflammatory response. Probiotics offer an alternative without the risk of side effects associated with conventional therapy.

PROBIOTICS

Probiotic Definition

Probiotics may be defined as "Live microorganisms which when administered in adequate amounts confer a health benefit on the host" (4,51). Probiotics are non-pathogenic microbial organisms which survive passage through the gastrointestinal tract and are believed to have potential beneficial health effects. The desirable properties of probiotic bacteria include having generally regarded as safe status, acid, and bile stability, adherence to intestinal cells, persistence for some time in the gut, antagonism against pathogenic bacteria and modulation of the immune response (52). Bacteria of human origin were originally required for safety reasons and because probiotic efficacy appeared to be host-specific. This stipulation may now be unnecessary as potential probiotics are fully identified and characterized by phenotypic and genotypic methods and tested for safety before use. Probiotic activity has been associated most commonly with lactobacilli and bifidobacteria, but other non-pathogenic bacteria including species of streptococci and enterococci, non-pathogenic *E. coli* Nissle 1917, and the yeast *Saccharomyces boulardii* have been used (53).

However, the current definition of a probiotic may now be too limited. Whilst the definition is one of live microorganisms, studies have demonstrated that bacterial DNA or bacterial components could themselves be responsible for any observed probiotic effects (54). Genetically modified bacteria have also been tested and a genetically engineered lactobacillus secreting the anti-inflammatory cytokine IL-10 has attenuated colitis in animals (55). Therefore, future use of the functional microbes may be outside the definition of probiotics. The definition of probiotics is likely to undergo continuing modification, and the term "pharmabiotics" may be more appropriate [(56), www.apc.ucc.ie]. This umbrella term includes live and dead organisms and constituents thereof, and encompasses genetically engineered microbes.

How Probiotics May Exert an Effect in Inflammatory Bowel Disease

The mechanisms of action of probiotic bacteria in the setting of inflammation are not completely elucidated and are likely to involve a number of factors and be strain specific. Proposed mechanisms focus on how probiotics influence the immune response. Commensal microbiota are known to contribute to immune homeostasis (7,26). There are several molecular pathways which are suggested as candidates for the site of probiotic immune effects. In the context of IBD, anti-inflammatory activity may involve signaling with the gastrointestinal epithelium and perhaps mucosal regulatory T-cells (7).

Gut Epithelium and Dendritic Cells

Within the gut, intestinal epithelial cells are the first point of contact for bacteria and play an important role in bacteria-host communication (57). The epithelial cells act as sensors of commensal and pathogenic bacteria, with discriminatory capacity to activate signaling pathways (8,58,59). Interactions with Toll-like receptors and dendritic cells in the gut are believed to be involved in this communication between host and bacteria (8,60). Dendritic cells in the gut mucosa are responsible for the stimulation of T cells and seem to have an important role in the balance between inducing TH1, TH2, and TH3 cytokine profiles (61). Gut dendritic cells are mostly immature and potentially prone to modulation by the environment, containing microorganisms. TH1/TH2/TH3 cytokine profiles induced by gut dendritic cells have been modulated by the administration of lactobacilli (62). In a further study, the probiotic bacteria *Bifidobacterium infantis* and *Lactobacillus salivarius* have induced dendritic cells to produce the anti-inflammatory cytokine IL-10 rather than pro-inflammatory IL-12 (63). In addition, intestinal dendritic cells have been shown to retain small numbers of commensal bacteria. This allows induction of protective IgA by the dendritic cells, preventing mucosal penetration by bacteria (64).

Modulation of the Cytokine Response

The ability of probiotic bacteria to induce an anti-inflammatory or regulatory cytokine profile by in vitro immunocompetent cells has been confirmed (65). In vitro studies examined the effect of probiotics on cytokine production by human intestinal mucosa. Both *Lactobacillus casei* and *Lactobacillus bulgaricus* down-regulated the production of TNF-α from normal and inflamed mucosa (66,67). The effects of various lactic acid bacteria on the cytokine profile produced by peripheral blood mononuclear cells in vitro have been studied (57,68–71). Alterations in cytokine production have been observed in the IL-10 knockout mouse model which develops colitis similar to human inflammatory bowel disease. The anti-inflammatory effects of *Lactobacillus salivarius* UCC118, and *Bifidobacterium infantis* 35624, when administered both orally and subcutaneously to IL-10 knockout mice, were accompanied by a reduction in pro-inflammatory cytokines IFN-γ, TNF-α and IL-12 from splenocytes, while levels of the regulatory cytokine TGF-β were maintained (72,73).

It is suggested that live bacteria may not be necessary for the immune responses seen with probiotics. Indeed bacterial DNA has been shown to have potent immunostimulatory effects and has reduced colitis in a number of murine models (54). The DNA sequences used are termed immunostimulatory sequences or CpG motifs. CpG DNA can activate dendritic cells and its effects are mediated via Toll-like receptors (74,75).

Nuclear Factor kappaB Pathway

The NFκB pathway, a nuclear factor involved in the transcriptional regulation of inflammatory genes, mediates responses to invasive pathogenic bacteria. Certain non-pathogenic organisms have been shown to counterbalance epithelial responses to invasive bacteria via an effect on the inhibitor kappaB / NFκB pathway (76). A recent study has demonstrated that a commensal bacterium, *Bacteroides thetaiotaomicron*, also acted on NFκB to attenuate pro-inflammatory cytokine expression, but via a unique mechanism. The mechanism involved limiting the duration of action of NFκB by promoting its nuclear export through a peroxisome proliferator activated receptor-γ-dependent (PPAR-γ) pathway (77).

Intestinal Permeability

Apart from immune mechanisms, it is also suggested that probiotic bacteria may have a beneficial effect on permeability of the gut barrier. There is evidence to suggest that the epithelial barrier function is reduced in inflammatory bowel disease (78).

Probiotic strains have demonstrated an ability to enhance the epithelial barrier function, based on measurements of intestinal permeability in excised mucosal tissue from animal models and humans (79,80). Probiotics given to IL-10 knockout mice normalized colonic physiological function and barrier integrity, along with a reduction in severity of colitis.

EFFICACY OF PROBIOTICS IN INFLAMMATORY BOWEL DISEASE

Probiotics in Animal Models of IBD

The efficacy of probiotics in attenuating colitis has been demonstrated in experimental animal models (Table 2). These models include the interleukin-10 knockout murine model (81–84), methotrexate induced colitis (85), HLA-B27 transgenic rats (86), and the CD45Rbhi transfer model (87).

The model of IL-10 knockout mice develop colitis when colonized with normal enteric microbiota but remain disease-free if kept in germ-free conditions. In a study of IL-10$^{-/-}$ mice colonization with *Lactobacillus plantarum* 299v was performed 2 weeks before transferring from a germ-free environment to a specific pathogen-free environment (84). This treatment led to a reduction in disease activity and a significant decrease in mesenteric lymph node IL-12 and IFN-γ production. A role for *Lactobacillus reuteri* in prevention of colitis in IL-10$^{-/-}$ mice was also demonstrated (81). In this study, the oral administration of the prebiotic lactulose (shown to increase the levels of *Lactobacillus* species) and rectal swabbing with *L. reuteri* restored *Lactobacillus* levels to normal in neonatal mice, originally found to have low levels of lactobacilli species. This effect was associated with the attenuation of colitis. In a placebo controlled trial, orally administered *Lactobacillus salivarius* UCC118 reduced the incidence of colon cancer and the severity of mucosal inflammation in IL-10$^{-/-}$ mice (82). *L. salivarius* was also shown to modify the gut microbiota in these animals as *Clostridium perfringens*, enterococci and coliform levels were significantly reduced in the probiotic group. A further trial confirmed the efficacy of *L. salivarius* UCC118 and demonstrated efficacy for *Bifidobacterium infantis* 35624 in attenuation of colitis in the IL-10$^{-/-}$ mouse model (83). The amelioration of disease activity in this study was associated with modulation of the gut microbiota as investigated by culture-independent 16S ribosomal RNA targeted PCR-direct gradient gel electrophoresis. In addition, mucosal pro-inflammatory cytokine production was significantly reduced. Indeed, the oral route of administration may not be essential for certain probiotic effects. Reduced inflammatory scores and reduced production of pro-inflammatory cytokines have been observed in IL-10$^{-/-}$ mice which had been injected subcutaneously with *L. salivarius* UCC118 (73).

Modified Probiotics in Animal Models

Combinations of probiotic treatment with prebiotics or antibiotics have been used to increase the beneficial effect. The combination of the prebiotic inulin, and the probiotic

Table 2 Summary of Probiotic Efficacy in Animal Models of Enterocolitis

Probiotic microorganism	Type of study	Trial outcome	Reference
Lactobacillus reuteri	IL-$10^{-/-}$ mice. N=4–8 per group. Placebo controlled trial	Prebiotic lactulose and probiotic *L. reuteri* attenuated colitis and improved mucosal barrier function.	Madsen et al. 1999 (81)
Lactobacillus salivarius UCC118	IL-$10^{-/-}$ mice. N=10 per group. Placebo controlled	Reduced incidence of colon cancer and mucosal inflammation. Modulation of fecal microbiota.	O'Mahony et al. 2001(82)
Lactobacillus salivarius UCC118 and *Bifidobacterium infantis* 35624	IL-$10^{-/-}$ mice. N=10 per group. Placebo controlled	Attenuation of disease. Modulation of gut microbiota. Reduction in in vitro production of IFN-γ, TNF-α and IL-12. TGF-β levels maintained.	McCarthy et al. 2003 (83)
Lactobacillus salivarius UCC118	L-$10^{-/-}$ mice. CIA model N=10 per group. Placebo controlled	Attenuation of colitis and arthritis following subcutaneous administration of probiotic. Reduction in proinflammatory cytokine production.	Sheil et al. (73)
Lactobacillus plantarum 299v	IL-$10^{-/-}$ mice. Placebo controlled	Attenuation of colitis. Reduction in IL-12 and IFN-γ produced by stimulated mesenteric lymph node cells.	Schultz et al. 2002 (84)
Lactobacillus rhamnosus GG	HLA-B27 transgenic rats	Prevented recurrence of colitis.	Dieleman et al. 2001 (86)
Combination of *Lactobacillus acidophilus* La-5, *L. delbrückii* subsp. *bulgaricus*, *Bifidobacterium* Bb-12, and *Streptococcus thermophilus*	HLA-B27 transgenic rats	Attenuated colitis following treatment with the prebiotic inulin and a combination of probiotic organisms.	Schultz et al. unpublished data

Abbreviations: HLA, human leukocyte antigen; IFN, interferon; IL, interleukin; N, number of animals; TGF, transforming growth factor; TNF, tumor necrosis factor.

organisms *Lactobacillus acidophilus* La-5, *Lactobacillus delbrueckii* subsp. *bulgaricus, Bifidobacterium lactis* Bb-12, and *Streptococcus thermophilus* significantly decreased inflammation in HLA-B27 rats (Schultz, unpublished data). Furthermore, genetically modified probiotics have been developed. *Lactococcus lactis* was engineered to secrete biologically active IL-10 and a significant reduction in inflammation was observed in both IL-10$^{-/-}$ and dextran sodium sulfate-induced murine colitis models (55). The investigators concluded that genetically engineered bacteria for local administration of a therapeutic agent, such as IL-10, may be a useful strategy in the treatment and prevention of IBD.

Live versus Dead Bacteria

It may not be necessary to administer live bacteria to achieve benefit. Bacterial DNA has been shown to have potent immuno-stimulatory effects. In a trial by Rachmilewitz et al. (54) bacterial DNA was used to attenuate colitis in a number of murine models suggesting an anti-inflammatory effect for bacterial DNA that warrants further study. A more recent study investigated the role of Toll-like receptors in mediating these effects of bacterial DNA (88).

Human Trials of Probiotics in Patients with Inflammatory Bowel Disease

Evidence that the enteric microbiota play a role in the pathogenesis of IBD and results from models of IBD which have demonstrated beneficial effects for probiotics has prompted clinical studies examining the effect of these organisms in patients with inflammatory bowel disease.

Trials in Ulcerative Colitis

A number of studies have examined the use of a non-pathogenic *E. coli* strain Nissle 1917, in the setting of ulcerative colitis. Kruis et al. (89) first performed in 1997 a randomized, double-blind clinical trial where 120 patients with inactive ulcerative colitis were randomized to receive oral *E. coli* strain Nissle 1917 or mesalazine. They reported that there was no difference in relapse rates in the probiotic treated group compared to patients on mesalazine. Relapse rates were 11.3% for the mesalazine treated group and 16.0% for the *E. coli* group. Life table analysis showed a relapse free time of $103 \pm$ 4 days for mesalazine and 106 ± 5 days for *E. coli*. From the results of this preliminary study, probiotic treatment appeared to offer another option for maintenance therapy of ulcerative colitis (89). Further beneficial results were described by Rembacken et al. (90) in a study where a total of 116 patients with active ulcerative colitis were recruited. Seventy-five percent and 68% of the mesalamine and *E. coli* groups achieved remission, respectively. In the second maintenance part of this study, the relapse rate in both groups was markedly higher than the investigators anticipated, 73% for the mesalamine group and 67% for the *E. coli* group. The time to relapse was not significantly different between the groups (90). These results suggested that the non-pathogenic *E. coli* was equivalent to mesalazine in maintaining remission, however these relapse rates are similar to those of placebo-treated patients. In a larger, 1-year multi-center, randomized, double-blind, remission maintenance study of 327 patients, *E. coli* was shown to be as effective as mesalazine in maintaining remission with relapse rates of 45% for the *E. coli* group and 36% in the mesalazine group, therefore offering an alternative to mesalazine in maintenance of remission in ulcerative colitis patients (Table 3) (92).

The probiotic cocktail VSL#3, a mixture of four lactobacilli (*Lactobacillus plantarum, Lactobacillus casei, Lactobacillus acidophilus, Lactobacillus delbrueckii* ssp. *Bulgaricus*), three bifidobacteria strains (*Bifidobacterium infantis, Bifidobacterium breve, Bifidobacterium longum*), and one strain of Streptococcus salivarius ssp. *thermophilus*, has been studied in ulcerative colitis. There is a high concentration of bacteria in this mixture with potential synergistic relationships to enhance suppression of potential pathogens. The effect of VSL#3 on maintenance of remission in UC patients was

Table 3 Summary of Human Trials of Probiotic Therapy in Ulcerative Colitis

Study type	Organism used	Trial outcome	Reference
Randomized controlled trial	*E. coli* strain Nissle 1917. N = 120	Patients with active colitis demonstrated similar relapse rates compared to patients on mesalazine	Kruis et al. 1997 (89)
Randomized, controlled trial	*E. coli* strain Nissle 1917. N = 116	Confirmed result from Kruis et al. 1997	Rembacken et al. 1999 (90)
Open labeled trial	VSL#3. N = 20	Maintenance of remission in patients	Venturi et al. 1999 (91)
Randomized controlled trial	*E. coli* strain Nissle 1917. N = 327	Remission maintained in patients receiving probiotic	Kruis et al. 2001 (92)
Open labeled trial	*Saccharomyces boulardii.* N = 25	Treatment given in combination with mesalamine for relapse of ulcerative colitis. Remission achieved in 17 patients	Guslandi et al. 2003 (93)

Abbreviation: N, number of subjects in trial.

evaluated using an open label design (91). In this pilot study, 20 patients in remission were treated for 12 months. At the end of the trial 15 out of 20 patients (75%) remained in remission.

A recent study has investigated the use of *Saccharomyces boulardii* in the setting of ulcerative colitis. In an open, non-placebo controlled study, 25 patients with a relapse of ulcerative colitis were treated with mesalazine in combination with *S. boulardii*. Seventeen patients achieved remission (93).

Trials in Pouchitis

Convincing evidence for beneficial probiotic effects in inflammatory bowel disease is seen in the treatment of pouchitis. In an open labeled study, patients with pouchitis were treated with *Lactobacillus* GG and fructooligosaccharide (94). The patients reported a beneficial effect when the probiotic-prebiotic mix was administered as an adjuvant to antibiotic therapy. Remission was documented by suppression of symptom scores and reversal of endoscopic findings (94). Gionchetti et al. (95) have studied VSL#3 in the setting of pouchitis and have demonstrated the efficacy of this probiotic mix in maintenance of remission in patients with chronic pouchitis. In a randomized, double-blind, placebo-controlled trial, 40 patients with pouchitis received one month of antibiotic treatment and were in clinical and endoscopic remission. Patients were then randomized to receive VSL#3 or placebo for 9 months. At the end of the study three patients (15%) had relapsed in the VSL#3 group compared to 20 (100%) in the placebo group. In a follow-up study, this group has also used VSL#3 as prophylaxis in patients after ileo-anal pouch formation surgery to prevent pouchitis. Forty patients were randomized to receive VSL#3 or placebo. At 1-year follow-up, 10% of probiotic treated patients had developed pouchitis, compared with 40% of the placebo treated group (96). A recent study has again examined the role of VSL#3 in maintaining remission following treatment of refractory or recurrent pouchitis. Thirty-six patients with recurrent pouchitis (at least twice in the past year) or requiring continuous antibiotics, in whom remission was induced by 4 weeks of antibiotics, were randomized to receive

Table 4 Summary of Human Trials of Probiotic Therapy in Pouchitis

Study type	Organism used	Trial outcome	Reference
Open labeled trial	Prebiotic fructooli-gosaccharide and probiotic. N=10	Effective in inducing remission in combination with antibiotic	Friedman et al. 2000 (94)
Randomized controlled trial	VSL#3. N=40	Maintenance of remission in chronic pouchitis after antibiotic induced remission. 15% relapse rate compared with 100% in control group	Gionchetti et al. 2000 (95)
Randomized controlled trial	VSL#3. N=40	Prevention of acute pouchitis in patients after ileo-anal pouch surgery. 10% pouchitis rate in probiotic group compared with 40% in control group	Gionchetti et al. 2003 (96)
Randomized controlled trial	VSL#3 (6 g). N=36	Maintenance of remission in recurrent or refractory pouchitis after antibiotic induced remission. 85% remained in remission at one year, compared with 6% in placebo group	Mimura et al. 2004 (97)

Abbreviation: N, number of subjects in trial.

6 gram of VSL#3 or placebo daily for one year or until relapse. Eighty-five percent of the VSL#3 treated group remained in remission at one year compared with 6% (one patient) in the placebo group (Table 4) (97).

Trials in Crohn's Disease

In CD, an early study involved the use of *Sacccharomyces boulardii* (98). In a double-blind study, 20 patients with moderately active CD were randomized to treatment with this organism or placebo for 7 weeks. The probiotic treated patients had a significant decrease in CD activity index (CDAI) compared with the control group. More recently, a double-blind trial randomized 32 CD patients in clinical remission to receive either mesalamine alone or mesalamine plus *S. boulardii*. Clinical relapse was observed in only 6.25% of patients receiving mesalamine plus *S. boulardii*, while 37.5% relapse rate was observed in the group receiving mesalamine alone (Table 5) (103).

The efficacy of *Lactobacillus rhamnosus* GG in the treatment of CD has been studied (99). Malin et al. (99) reported that in pediatric CD, consumption of *Lactobacillus* GG was associated with increased gut IgA levels which could promote the gut immunological barrier. Gupta et al. (101) also reported improved clinical scores and improved intestinal permeability in an open labeled pilot study in a small study involving four pediatric CD patients.

A double-blind study investigated the use of the *E. coli* Nissle 1917 strain in CD (100). Malchow et al. randomized 28 patients in remission to receive either *E. coli* or placebo. At 1-year follow-up, the relapse rates were significantly reduced in the group that received *E. coli* (30%) compared with 70% in the placebo group. In a large double-blind, randomized study the efficacy of VSL#3 combined with antibiotic treatment on the post-operative recurrence of CD was compared to treatment with mesalamine alone (102). Forty patients

Table 5 Summary of Human Trials of Probiotic Therapy in Crohn's Disease

Study type	Organism used	Trial outcome	Reference
Randomized controlled trial	*Saccharomyces boulardii.* N=20	Decrease in CDAI in probiotic group	Plein et al. 1993 (98)
Open labeled trial	*Lactobacillus rhamnosus* GG. N=14	Increase in gut IgA response	Malin et al. 1996 (99)
Randomized controlled trial	*E. coli* strain Nissle 1917. N=28	Remission achieved in patients on probiotics and steroids greater than with steroids alone	Malchow et al. 1997 (100)
Open labeled trial	*Lactobacillus rhamnosus* GG in children. N=4	Improved intestinal permeability and CDAI	Gupta et al. 2000 (101)
Randomized controlled trial	VSL#3 with antibiotic. N=40	Patients with CD had 20% remission when given antibiotic and VSL#3 compared to 40% in mesalamine treated group	Campieri et al. 2000 (102)
Randomized controlled trial	*Saccharomyces boulardii.* N=32	Maintenance of remission in treatment group superior as relapse observed in 6.25% of patients receiving probiotic plus mesalasine compared to 37.5% on mesalamine alone	Guslandi et al. 2000 (103)
Open labeled trial	*Lactobacillus salivarius* 118. N=25	Reduction of mean CDAI and induction of IgA in patients with relapse	McCarthy et al. 2001 (104)
Randomized controlled trial	*Lactobacillus rhamnosus GG*	No difference seen in rate of recurrence 1 year after surgery between group given probiotic or control	Prantrera et al. 2002 (105)

Abbreviations: N, number of subjects in trial; CD, Crohn's disease; CDAI, Crohn's disease activity index.

were randomized to receive rifaximin for 3 months followed by VSL#3 for 9 months or mesalamine for 12 months. At the end of the trial 20% of the patients had recurrent CD in the probiotic/antibiotic group while 40% of patients in the mesalamine group relapsed (102). In an open study of patients with mildly active CD despite 5-ASA therapy, patients were offered either steroids or a trial of *Lactobacillus salivarius* subsp. *salivarius* UCC118 for 6 weeks (104). Of the 25 patients enrolled, 19 successfully completed the study and avoided steroids for a 3-month follow-up period. The mean CDAI at enrolment was 217, falling to 150 at the end of the study period (104). Finally, in a recent study of 45 CD patients who underwent curative surgery, the recurrence rate 1 year after surgery in patients treated with *Lactobacillus rhamnosus* GG or placebo was compared. No difference was seen between the patients receiving probiotic (16% recurrence rate) and the placebo group (10%) (105).

In conclusion, while the trials for probiotics in treatment of IBD to date are promising, results have been mixed; consequently, better-designed trials are needed.

DISCUSSION

Although preliminary studies are promising, large placebo-controlled, randomized, double-blinded clinical trials are needed to clarify the role of probiotic bacteria in the treatment of inflammatory bowel disease. Studies of probiotics in inflammatory bowel disease in the future will also need to increase our knowledge of how probiotics exert their effect. Optimal dosing schedules need to be determined. Detailed comparisons of probiotic performance amongst different bacterial strains have not yet been performed, in vivo or under clinical trial conditions, and the level of scientific characterization of individual organisms has been variable. The route of administration also requires more study, in particular to determine whether the oral route is always essential. The issue of live versus dead bacteria remains unclear. The beneficial effect of bacterial DNA and other metabolites or constituents versus whole organisms needs comprehensive study.

Irrespective of the mechanism of action, however, there are reasons which might favor therapeutic usage of live over dead bacteria. Live bacteria may be more reliable for enteric transit and occupation of microbial niche. Secondly, live bacteria offer the advantage of elaborating biological molecules other than immunomodulatory DNA.

Detailed strain characterization is also required for all potential probiotic strains before the use of combinations can be recommended. The potential exists for synergistic or antagonistic effects amongst bacterial strains and this requires further study. Finally, disease-specific probiotic organisms designed to target particular patients, (the "designer probiotic"), may become a possibility as we increase our understanding of molecular mechanisms behind the anti-inflammatory effects of individual probiotics. What is already clear, is that there will be an increasing role for bacteria or bacterial products in a therapeutic setting along with conventional treatments for inflammatory bowel disease. The concept of a food influencing the health of the gastrointestinal tract is appealing to many people. Therapeutic modification of the microbiota with functional foods such as probiotics empowers patients with an enhanced sense of control in the management of their illness. Microbial therapeutics is an expanding field inviting further investigation, and we should not allow ourselves to become captive of the definition of probiotics.

REFERENCES

1. Milner JA. Functional foods and health promotion. J Nutr 1999; 129:1395S–1397S.
2. McCarthy J, Shanahan F. Functional food and probiotics: time for gastroenterologists to embrace the concept. Curr Gastroenterol Rep 2000; 2:345–346.
3. Metchnikoff E. Prolongation of life. London: William Heinmann, 1907.
4. Fooks LJ, Fuller R, Gibson GR. Prebiotics, probiotics and human gut microbiology. Int Dairy J 1999; 9:53–61.
5. Dunne C. Adaption of bacteria to the intestinal niche: probiotics and gut disorders. Inflamm Bowel Dis 2001; 7:136–145.
6. Schultz M, Sartor RB. Probiotics and inflammatory bowel disease. Am J Gastroenterol 2000; 95:519–521.
7. Shanahan F. Probiotics and inflammatory bowel disease: is there a scientific rationale. Inflamm Bowel Dis 2000; 6:107–115.
8. Shanahan F. Therapeutic manipulation of the gut flora. Science 2000; 289:1311–1312.
9. Shanahan F. Crohn's disease. Lancet 2002; 359:62–69.
10. Ekbom A. The changing faces of Crohn's disease and ulcerative colitis. In: Targan S, Shanahan F, Karp L, eds. Inflammatory Bowel Disease. From Bench to Bedside. 2nd ed. Dordrecht: Kluwer Academic Publishers, 2003:5–20.
11. Rook GAW, Stanford JL. Give us this day our daily germs. Immunol Today 1998; 19:113–116.
12. Shanahan F. The host-microbe interface within the gut. Best Pract Res Clin Gastroenterol 2002; 19:915–931.
13. Fuller R. Probiotics in man and animals. J Appl Bacteriol 1989; 66:365–378.

14. Collins JK, Thornton G, Sullivan G. Selection of probiotic strains for human applications. Int Dairy J 1998; 8:487–490.

15. Mackie R, Gaskins HR. Gastrointestinal microbial ecology. Sci. Med. 1999; 6:18–27.

16. Vanderhoof JA, Young RJ. Use of probiotics in childhood gastrointestinal disorders. J Paediatr Gastroenterol Nutr 1998; 27:323–332.

17. Berg RD. The indigenous gastrointestinal microbiota. Trends Microbiol 1996; 434:430–435.

18. Bengmark S. Ecological control of the gastrointestinal tract. The role of the probiotic flora. Gut 1998; 42:2–7.

19. Mackowiak PA. The normal microbial flora. N Engl J Med 1982; 307:83–93.

20. Midtvedt T. Microbial functional activities. In: Hanson LA, Yolken RH, eds. Intestinal Microflora Nestle Nutrition Workshop Series, Vol. 42. Philadelphia: Lippincott-Raven, 1999:79–96.

21. Bocci V. The neglected organ: bacterial flora has a crucial immunostimulatory role. Perspect Biol Med 1992; 35:251–260.

22. Alderberth I, Breslin L, Bjoerkstein B, et al. Digestive flora. In: Gorthier G, Arnaud C, eds. Mechanisms of Protection of the Digestive Tract. Paris: John Libby Eurotext, 1998:17–20.

23. Tannock GW. Probiotic properties of lactic acid bacteria: plenty of scope for fundamental R&D. Trends Biotechnol 1997; 15:270–274.

24. Tannock GW, Munro K, Harmsen HJM, et al. Analysis of the fecal flora of human subjects consuming a probiotic containing *Lactobacillus rhamnosus* DR20. Appl Environ Microbiol 2000; 66:2578–2588.

25. Hooper LV, Gordon JI. Commensal host-bacterial relationships in the gut. Science 2001; 292:1115–1118.

26. MacDonald TT, Carter PB. Requirement for a bacterial flora before mice generate cells capable of mediating the delayed hypersensitivity reaction in sheep red blood cells. J Immunol 1979; 122:2426–2429.

27. Gordon HA, Pesti L. The gnotobiotic animal as a tool in the study of host-microbial relationships. Bacterial Rev 1971; 35:390–429.

28. Kawaguchi M, Nanno M, Umesaki Y, et al. Cytolytic activity of intestinal intraepithelial lymphocytes in germ-free mice is strain dependent and determined by T-cells expressing T-cell receptors. Proc Natl Acad Sci 1993; 90:8591–8594.

29. Umesaki Y, Setomaya S, Matsumoto S, et al. Expansion of the alpha beta T-cell receptor-baring intestinal intraepithelial lymphocytes after microbial colonisation in germ-free mice, and its independence from the thymus. Immunology 1993; 79:32–37.

30. Guarner F, Casellas F, Borruel N, et al. Role of microecology in chronic inflammatory bowel diseases. Eur J Clin Nutr 2002; 56:S34–S38.

31. Campierei M, Gionchetti P. Probiotics in inflammatory bowel disease: new insight to pathogenesis or possible therapeutic alternative. Gastroenterology 1998; 116:1246–1260.

32. Macpherson A, Khoo UY, Forgacs I, et al. Mucosal antibodies in inflammatory bowel disease are directed against intestinal bacteria. Gut 1996; 38:365–375.

33. Duchmann R, Kaiser I, Hermann E, et al. Tolerance exists towards resident intestinal flora but is broken in active inflammatory bowel disease (IBD). Clin Exp Immunol 1995; 102:448–455.

34. Schultsz C, Van Den Berg F, Ten Kate FW, et al. The intestinal mucus layer from patients with inflammatory bowel disease harbors high numbers of bacteria compared with controls. Gastroenterology 1999; 117:1089–1097.

35. Swindinski A, Ladhoff A, Pernthaler A, et al. Mucosal flora in inflammatory bowel disease. Gastroenterology 2002; 122:44–54.

36. Hugot JP, Chamaillard M, Zouali H, et al. Association of NOD2 leucine rich repeat variants with susceptibility to Crohn's disease. Nature 2001; 411:599–603.

37. Hugot JP, Zouali H, Leasage S. Lessons to be learned from the NOD2 gene in Crohn's disease. Eur J Gastroenterol Hepatol 2003; 15:593–597.

38. Elson CO, Sartor RB, Tennyson GS, et al. Experimental models of inflammatory bowel disease. Gastroenterology 1995; 109:1344–1367.

39. Strober W, Fuss IJ, Ehrhardt RO, et al. Mucosal immunoregulation and inflammatory bowel disease: new insights from murine models of inflammation. Scand J Immunol 1998; 48:453–458.

40. Blumberg RS, Saubermann LJ, Strober W, et al. Animal models of mucosal inflammation and their relation to human inflammatory bowel disease. Curr Opin Immunol 1999; 11:648–656.

41. Wirtz S, Neyrath MF. Animal models of intestinal inflammation: new insights into the molecular pathogenesis and immunotherapy of inflammatory bowel disease. Int J Colorectal Dis 2000; 15:144–160.

42. Kosiewicz MM, Nast CC, Krishnan A, et al. Th1-type responses mediate spontaneous ileitis in a novel murine model of Crohn's disease. J Clin Invest 2001; 107:695–702.

43. Kuhn R, Lohler J, Rennick D, et al. Interleukin-10-deficient mice develop chronic enterocolitis. Cell 1993; 75:263–274.

44. Shanahan F. Gene-targeted immunologic knockouts: new models of inflammatory bowel disease. Gastroenterology 1994; 17:312–314.

45. Balish E, Warner T. *Enterococcus faecalis* induces inflammatory bowel disease in interleukin-10 knockout mice. Am J Pathol 2002; 160:2253–2257.

46. Rath HC, Wilson KH, Sartor RB. Differential induction of colitis and gastritis in HLA-B27 transgenic rats selectively colonized with *Bacteroides vulgatus* or *Escherichia coli*. Infect Immunol 1999; 67:2969–2974.

47. Sartor RB. Probiotics in chronic pouchitis: restoring the luminal balance. Gastroenterology 2000; 119:305–309.

48. Darfeuille-Michaud A, Neut C, Barnich N, et al. Presence of adherent *Escherichia coli* strains in ileal mucosa of patients with Crohn's disease. Gastroenterology 1998; 115:1405–1413.

49. Ryan P, Bennett MW, Aarons S, et al. PCR detection of Mycobacterium paratuberculosis in Crohn's disease granulomas isolated by laser capture microdissection. Gut 2002; 51:665–670.

50. Gent AE, Hellier MD, Grace RH, et al. Inflammatory bowel disease and domestic hygiene in infancy. Lancet 1994; 343:766–767.

51. Guarner F, Schaafsma GJ. Probiotics. Int J Food Microbiol 1998; 39:237–238.

52. Dunne C, O'Mahony L, Murphy L. In vitro selection criteria for probiotic bacteria of human origin: correlation with in vivo findings. Am J Clin Nutr 2001; 73:386S–392S.

53. Shanahan F. Inflammatory bowel disease: immunodiagnostics, immunotherapeutics, and ecotherapies. Gastroenterology 2001; 120:622–635.

54. Rachmilewitz D, Karmeli F, Takabayashi K, et al. Immunostimulatory DNA ameliorates experimental and spontaneous murine colitis. Gastroenterology 2002; 122:1428–1441.

55. Steidler L, Hans W, Schotte L, et al. Treatment of murine colitis by *Lactococcus lactis* secreting interleukin-10. Science 2000; 289:1352–1355.

56. Shanahan F. Host-flora interactions in inflammatory bowel disease. Inflamm Bowel Dis 2004; 10:S16–S24.

57. Haller D, Bode C, Hammes WP, et al. Non-pathogenic bacteria elicit a differential cytokine response by intestinal epithelial cell / leukocyte co-cultures. Gut 2000; 47:79–87.

58. Gordon JI, Hooper LV, McNevin MS, Wong M, Bry L. Epithelial cell growth and differentiation, III. Promoting diversity in the intestine: conversations between the microbiota, epithelium and diffuse GALT. Am J Physiol 1997; 273:G565–G570.

59. Hooper LV, Wong MH, Thelin A, et al. Molecular analysis of commensal host/microbial relationships in the intestine. Science 2001; 291:881–884.

60. Marteau P, Seksik P, Jian R. Probiotics and health: new facts and ideas. Curr Opin Biotechnol 2002; 13:486–489.

61. Kronin V, Hochrein H, Shortman K, et al. Regulation of T cell cytokine production by dendritic cells. Immun Cell Biol 2000; 78:214–223.

62. Christensen HR, Frokiaer H, Pestka JJ. Lactobacilli differentially modulate expression of cytokines and maturation surface markers in murine dendritic cells. J Immunol. 2002; 168:171–178.

63. O'Mahony L, Zong Y, Sharma S, et al. Probiotic bacteria and pathogenic bacteria elicit differential cytokine responses from dendritic cells (abstract). Gastroenterology 2001; 120:1625.

64. Macpherson AJ, Uhr T. Induction of protective IgA by intestinal dendritic cells carrying commensal bacteria. Science 2004; 303:1624–1625.

65. O'Mahony L, Feeney M, MacSharry J, et al. Probiotics, mononuclear cells and epithelial cells: an anti-inflammatory network (abstract). Gastroenterology 2000; 118:680.

66. Borruel N, Carol M, Casellas F, et al. Increased mucosal TNF-α production in Crohn's disease can be downregulated *ex vivo* by probiotic bacteria. Gut 2002; 51:659–664.

67. Borruel N, Casellas F, Antolin M, et al. Effects of non-pathogenic bacteria on cytokine secretion by human intestinal mucosa. Am J Gastroenterol 2003; 98:865–870.

68. Miettinen M, Matikainen S, Vuopio-Varkila J, et al. Lactobacilli and streptococci induce interleukin-12 (IL-12), IL-18, and gamma inteferon production in human peripheral blood mononuclear cells. Infect Immun 1998; 66:6058–6062.

69. Miettinen M, Lehtonen A, Julkunen I, et al. Lactobacilli and streptococci activate NF-kappaB and STAT signalling pathways in human macrophages. J Immunol 2000; 164:3733–3740.

70. Haller D, Serrant P, Granato D, et al. Activation of human NK cells by staphylococci and lactobacilli requires cell contact-dependent costimulation by autologous monocytes. Clin Diagn Lab Immunol 2002; 9:649–657.

71. Haller D, Blum S, Bode C, et al. Activation of human peripheral blood mononuclear cells by non-pathogenic bacteria in vitro: evidence of NK cells as primary targets. Infect Immun 2000; 68:752–759.

72. McCarthy J, O'Mahony L, O'Callaghan L, et al. Double-blind, placebo-controlled trial of two probiotic strains in interleukin 10 knockout mice and mechanistic link with cytokine balance. Gut 2003; 52:975–980.

73. Sheil B, McCarthy J, O'Mahony L, et al. Is the mucosal route of administration essential for probiotic function? Subcutaneous administration is associated with attenuation of murine colitis and arthritis Gut 2004; 53:694–700.

74. Hemmi H, Takeuchi O, Kavai T, et al. A Toll-like receptor recognizes bacterial DNA. Nature 2000; 408:740–745.

75. Wagner H. Toll meets bacterial CpG-DNA. Immunity 2001; 14:499–502.

76. Neish AS, Gerwirtz AT, Zeng H, et al. Prokaryotic regulation of epithelial responses by inhibition of IkappaB-alpha ubiquitination. Science 2000; 289:1560–1563.

77. Kelly D, Campbell JI, King TP, et al. Commensal anaerobic gut bacteria attenuate inflammation by regulating nuclear-cytoplasmic shuttling of PPAR-γ and RelA. Nature Immunol 2004; 5:104–112.

78. Schmitz H, Barmeyer C, Fromm M, et al. Altered tight junction structure contributes to the impaired epithelial barrier function in ulcerative colitis. Gastroenterology 1999; 116:301–309.

79. Isolauri E, Majamaa H, Arvola T, et al. *Lactobacillus casei* strain GG reverses increased intestinal permeability induced by cow milk in suckling rats. Gastroenterology 1993; 105:1643–1650.

80. Madsen K, Cornish A, Soper P, et al. Probiotic bacteria enhance murine and human intestinal epithelial barrier function. Gastroenterology 2001; 121:580–591.

81. Madsen KL, Doyle JS, Jewell LD, et al. *Lactobacillus* sp prevents development of enterocolitis in interleukin-10 deficient mice. Gastroenterology 1999; 116:1107–1114.

82. O'Mahony L, Feeney M, O'Halloran S, et al. Probiotic impact on microbial flora, inflammation, and tumour development in IL-10 knockout mice. Aliment Pharmacol Ther 2001; 15:1219–1225.

83. McCarthy J, O'Mahony L, O'Callaghan L, et al. Double-blind, placebo-controlled trial of two probiotic strains in interleukin 10 knockout mice and mechanistic link with cytokine balance. Gut 2003; 52:975–980.

84. Shultz M, Veltkamp C, Dieleman LA, et al. *Lactobacillus plantarum* 299v in the treatment and prevention of spontaneous colitis in Interleukin-10-deficient mice. Inflamm Bowel Dis 2002; 8:71–80.

85. Mao Y, Nobaek S, Kasravi B, et al. The effects of *Lactobacillus* strains and oat fibre on methotrexate induced enterocolitis in rats. Gastroenterology 1996; 111:334–344.

86. Dieleman LA, Goerres MS, Arends A, et al. *Lactobacillus* GG prevents recurrence of colitis in HLA-B27 transgenic rats after antibiotic treatment. Gastroenterology 2001; 118:A4312.

87. Murphy L, Byrne F, Collins JK, Shanahan F, O'Sullivan GC, Aranda R. Evaluation and characterisation of probiotic therapy in the CD45Rbhi transfer model. Gastroenterology 1999; 116:A3882 (abstract).

88. Rachmilewitz D, Katakura K, Karmeli F, et al. Toll like receptor 9 signaling mediates the anti-inflammatory effects of probiotics in murine experimental colitis. Gastroenterology 2004; 126:520–528.

89. Kruis W, Schutz E, Fric P, Fixa B, Judmaier G, Stolte M. Double blind comparison of an oral *Escherichia coli* preparation and mesalazine in maintaining remission of ulcerative colitis. Aliment Pharmacol Ther 1997; 11:853–858.

90. Rembacken BJ, Snelling AM, Hawkey PM, Axon ATR. Non-pathogenic Escherichia coli versus mesalazine for the treatment of ulcerative colitis: a randomised trial. Lancet 1999; 354:635–639.

91. Venturi A, Gionchetti P, Rizzello F, et al. Impact on the composition of the faecal flora by a new probiotic preparation: preliminary data on maintenance treatment of patients with ulcerative colitis. Aliment Pharmacol Ther 1999; 13:1103–1108.

92. Kruis W, Fric P, Stolte M. The Mutflor Study Group. Maintenance of remission in ulcerative colitis is equally effective with *Escherichia coli* Nissle 1917 and with standard mesalazine. Gastroenterology 2001; 120:A680 (abstract).

93. Guslandi M, Giollo P, Testoni PA. A pilot trial of *Saccharomyces boulardii* in ulcerative colitis. Eur J Gastroenterol Hepatol 2003; 15:697–698.

94. Friedman G, George J. Treatment of refractory 'pouchitis' with probiotic and prebiotic therapy. Gastroenterology 2000; 118:A4167.

95. Gionchetti P, Rizzello F, Venturi A, et al. Oral bacteriotherapy as maintenance treatment in patients with chronic pouchitis: a double-blind, placebo-controlled trial. Gastroenterology 2000; 119:305–309.

96. Gionchetti P, Rizzello F, Helwig U, et al. Prophylaxis of pouchitis onset with probiotic therapy: A double-blind, placebo- controlled trial. Gastroenterology 2003; 124:1202–1209.

97. Mimura T, Rizzello F, Helwig U, et al. Once daily high dose probiotic therapy (VSL#3) for maintaining remission in recurrent or refractory pouchitis. Gut 2004; 53:108–114.

98. Plein K, Holz J. Therapeutic effects of *Saccharomyces boulardii* on mild residual symptoms in a stable phase of Crohn's disease with special respect to chronic diarrhoea-a pilot study. Z Gastroenterol 1993; 31:129–134.

99. Malin M, Suomalainen H, Saxelin M, et al. Promotion of IgA immune response in patients with Crohn's disease by oral bacteriotherapy with *Lactobacillus* GG. Ann Nutr Metab 1996; 40:137–145.

100. Malchow HA. Crohn's disease and Escherichia coli. A new approach in therapy to maintain remission of colonic Crohn's disease? J Clin Gastroenterol 1997; 25:653–658.

101. Gupta P, Andrew H, Kirschner BS, et al. Is Lactobacillus GG helpful in children with Crohn's disease? Results of a preliminary open-label study J Pediatr Gastroenterol Nutr 2000; 31:453–457.

102. Campieri M, Rizzello F, Venturi A, et al. Combination of antibiotic probiotic treatment is efficacious in prophylaxis of post-operative recurrence of Crohn's disease: a randomised controlled study vs mesalamine. Gastroenterology 2000; 118:A4179 (abstract).

103. Guslandi M, Mezzi G, Sorghi M, Testoni PA. Saccharomyces boulardii in the maintenance of Crohn's disease. Dig Dis Sci 2000; 45:1462–1464.

104. McCarthy J, O'Mahony L, Dunne C, et al. An open trial of a novel probiotic as an alternative to steroids in mild/moderately active Crohn's disease (abstract). Gut 2001; 49:2447.

105. Prantrera C, Scribano ML, Falasco G, Andreoli A, Luzi C. Ineffectiveness of probiotics in preventing recurrence after curative resection for Crohn's disease: a randomised controlled trial with *Lactobacillus* GG. Gut 2002; 51:405–409.

12

The Gastrointestinal Microbiota in Cancer

Patricia M. Heavey
School of Life Sciences, Kingston University, Kingston-upon-Thames, U.K.

Ian R. Rowland
Northern Ireland Center for Food and Health, University of Ulster, Coleraine, Northern Ireland, U.K.

Joseph J. Rafter
Department of Medical Nutrition, Novum, Huddinge University Hospital, Karolinska Institutet, Stockholm, Sweden

INTRODUCTION

The microbiota of the human gastrointestinal tract and in particular the large intestine, comprises a large and diverse range of microorganisms, with over 10^{12} bacteria per gram of contents (1). It is therefore not surprising that the activities of this microbial population have a significant impact on the health of the host. The microbiota interacts with its host at both the local (intestinal mucosa) level, and systemically, resulting in a broad range of immunological, physiological, and metabolic effects. From the standpoint of the host, these effects have both beneficial and detrimental outcomes for nutrition, infections, xenobiotic metabolism, toxicity of ingested chemicals, and cancer.

The participation of intestinal bacteria in carcinogenesis continues to be controversial partly due to the lack of agreement on the molecular mechanisms involved in the development of this disease. In normal adult tissues, proliferation, apoptosis, and DNA repair are in equilibrium and this ensures a steady state of healthy cells. In the progression of changes leading from a normal mucosa to carcinoma, at least five to seven major molecular alterations need to occur. Extensive studies on colorectal cancer (CRC) have identified specific genetic changes in various proto-oncogenes, tumor suppressor genes, and DNA mismatch repair genes, as well as alterations in DNA methlyation status and inherited genetic defects. Subsequently, several molecular pathways have been identified which can contribute to the development of CRC. In 1990, Fearon and Vogelstein (2) proposed a genetic pathway of colorectal tumorigenesis, which is now generally accepted as the classical model for the development of CRC. The model postulates that at least five to seven major molecular alterations need to occur for a normal epithelial cell to proceed to carcinoma. This process is now accepted as central to the majority of cancers and has been studied extensively in CRC.

Bacteria have been linked to cancer by two mechanisms: induction of chronic inflammation following bacterial infection and production of toxic bacterial metabolites.

The latter mechanism has a strong link with diet. Carcinogenic agents may be present in the diet or formed in vivo during digestion. Many of these mechanisms involve the metabolic activities of the microbiota normally resident in the human colon. This paper discusses both the detrimental and beneficial consequences of bacterial activity of the gastrointestinal tract focusing on the stomach and large intestine.

THE STOMACH

The pH of the gastric contents of the fasting normal human is usually less than three, which is sufficient to kill most commensal bacteria (3). However, during a meal the gastric acid is buffered, allowing bacteria ingested with food to survive at least until the pH falls, and thus permitting a transient gastric microbiota. However, where gastric acid secretion is impaired, bacteria can survive longer and even proliferate in the elevated pH conditions. Reduced gastric acid secretion (hypochlorhydria) occurs naturally with ageing (4) and is common after gastric surgery. Certain diseases such as pernicious anemia and hypogammaglobulinaemia are associated with achlorhydria, which results in the gastric pH rising to seven and above (4). This allows a diverse microbiota with up to 10^9 organisms per gram to establish, consisting usually of species of salivary bacteria of the genera *Streptococcus, Neisseria, Staphylococcus,* and *Veillonella,* although *Bacteroides, Lactobacillus* and *Escherichia* species are also found (4). Hypochlorhydria is also common in patients with atrophic gastritis associated with chronic *Helicobacter pylori H. pylori* infection.

The presence of a gastric microbiota in hypochlorhydric and achlorhydric individuals has potential toxicological sequelae since it increases the probability of xenobiotic metabolism by the bacteria, particularly since the gastric emptying time of such patients may be up to 5 hours (4). It has been suggested that the increased gastric cancer risk of achlorhydric patients is linked to increased formation of N-nitroso compounds (NOC) by their gastric microbiota (5).

Helicobacter pylori

H. pylori is a Gram-negative bacterium found in the human stomach and plays an important role in the pathogenesis of chronic gastritis and peptic ulcers (6). Additionally, both epidemiological and clinical evidence has indicated that *H. pylori* is associated with an increased risk of gastric carcinoma (7,8) and as such it is the first bacterium to be termed a definitive cause of cancer by the International Agency for Research into Cancer (IARC). The *cag* pathogenicity island appears to play an important role in the aetiology of the disease since, in developed countries, strains of *H. pylori* that carry it are associated with an increased risk of peptic ulcer and adenocarcinoma than strains that are negative for the *cag* island (9).

The precise mechanisms involved in its pathogenesis have yet to be fully elucidated although numerous clues can be derived from in vitro models and animal studies.

The inflammatory effects of *H. pylori* infection have been related to cancer due to increased cell proliferation and production of mutagenic free radicals and NOC (10). In the Mongolian gerbil model of *H. pylori* infection, it has been shown that *H. pylori* inoculation can induce abnormality in gastric mucosal cell proliferation (11).

Infection with *H. pylori* is associated with significant epithelial cell damage as well as an increased level of apoptosis. However, the mechanism for *H. pylori* induced apoptosis in gastric epithelial cells remains uncertain. Apoptosis is a genetically programmed mode of cell death that is regulated by many genes, including oncogenes and oncosuppressor genes, which may be mutated, delayed or abnormally expressed in neoplasia, thus altering tumor cell susceptibility to apoptosis (12). The role of the *p53* tumor supressor gene in apoptosis is currently of particular interest. Genetic abnormalities in this gene have been observed in a wide range of human cancers, and are also closely associated with the transition from adenoma to carcinoma (13). The mutational inactivation of *p53* function allows cells to continue with their cell cycle, meaning damaged or mutated DNA is propagated in the next generation of cells.

Zhang and coworkers (14) examined the effect of *H. pylori* on gastric epithelial cells and the role of *p53* and showed that the organism induced a time and dose dependent inhibition of cell growth and apoptosis over 72 hours. In agreement with other findings (15), at low inoculations of *H. pylori*, cell DNA synthesis was stimulated compared to the controls. They also demonstrated no difference in the induction of gastric cell epithelial cell apoptosis and cell proliferation between cells exposed to *cag*A positive and *cag*A negative strains. In addition, *H. pylori* infection was associated with changes in oncogene and tumor suppressor gene expression as shown by increased *ras p21* expression and *p53* mutation in *H. pylori* positive cases of gastric cancer (16). Cell cycle regulatory proteins have also been identified as critical targets during carcinogenesis. It has been shown that chronic *H. pylori* infection is associated with decreased expression of the cyclin dependent kinase inhibitor (CDI) *p27kip1*. Another CDI, *p16Ink4a* (*p16*) is over-expressed in gastric epithelial cells of *H. pylori* patients and this is associated with an increase in apoptosis (17).

High dose vitamin C has been shown to inhibit *H. pylori* growth and colonization (18) and at physiological concentrations it induced *H. pylori* associated apoptosis and cell cycle arrest in vitro (19). Such effects may account for the observed negative association between dietary vitamin C intake and gastric cancer risk (20) although other mechanisms include the ability of vitamin C to scavenge reactive oxygen species and inhibit NOC formation. Other studies have implicated cigarette smoking and low levels of dietary vitamin C as a contributing factor in those high risk individuals with *H. pylori* infection (21,22).

Overexpression of cyclooxygenase 2 (COX-2) has also been observed in tissues of human gastric cancer. There are two isoforms of COX; COX-1 and COX-2. These are key enzymes that convert arachidonic acid to prostaglandins. COX-1 is expressed in most human tissues, whereas COX-2 is usually undetectable. Overexpression of COX-2 has been implicated in a number of cancers including gastric and colon cancer. It has been shown that COX-2 was overexpressed in 84% of gastric cancer specimens and those specimens with *cag*A positive strain expression had a significantly higher expression of COX-2 than the specimens with *cag*A negative strain expression (23). It has therefore been suggested that the application of COX-2 selective inhibitors may be an effective preventive strategy for gastric cancer and in particular those that would not cause gastrointestinal complications. Both nonsteroidal anti-inflammatory drug (NSAID) use and *H. pylori* infection independently and significantly increase the risk of peptic ulcer and ulcer bleeding. In a meta-analysis of the data it was interpreted that there was synergism for the development of peptic ulcer and ulcer bleeding between *H. pylori* infection and NSAID use (24).

The prevalence of *H. pylori* infection is falling in developing countries and this has been linked to changes in the epidemiology of gastrointestinal diseases, in particular reduced incidence of gastric cancers in western countries (25,26). Improved nutrition, water supplies and reduced family sizes have been associated with reduced *H. pylori* colonization (25). Novel treatment of this infection using probiotics is in the initial stages and results indicate only a slight improvement (27).

THE LARGE INTESTINE

It is becoming increasingly evident that the large and complex bacterial population of the large intestine and their metabolism has an important role in toxicity of ingested chemicals and in cancer (28–31). A number of potential mechanisms have been proposed whereby gut bacteria may impact carcinogenesis. They may have a direct effect through the binding of potential mutagens and thus reduce exposure to the host (32). The normal microbiota present in the gut is known to produce and release toxins, which can bind specific cell surface receptors and affect intracellular signal transduction (33). Bacterial involvement in CRC has been widely studied with most information being derived from animal work and some human studies. Evidence from a wide range of sources supports the view that the colonic microbiota is involved in the etiology of cancer (Table 1).

Gut Bacterial Involvement in Colorectal Cancer

Comparisons of the fecal microbiota of healthy subjects and colon cancer patients have not revealed any consistent patterns, possibly due to the difficulties in culturing and identifying gut organisms. Elevated numbers of *Bacteroides* have been associated with increased colon cancer risk in humans (34,35). Similarly, lecithinase-negative *Clostridium* and *Lactobacillus* were more abundant in colon cancer patients (36) although in another study, some *Lactobacillus* species and *Eubacterium aerofaciens* have been associated with reduced risk (35).

In animals, the presence of the intestinal microbiota has a major impact on colonic tumor formation (37,38). In a study conducted by Reddy and coworkers (38) the rate of tumor formation was much more rapid in conventional than in germ-free rats treated with the tumor initiator 1,2-dimethylhydrazine (DMH). After 20 weeks, 17% of conventional rats had colon carcinomas, whereas there were no tumors (adenomas or carcinomas) in the germ-free animals. At 40 weeks, two out of 18 germ-free rats had developed benign adenomas (although still none had carcinomas), compared to six out of 24 conventional rats with tumors (4 cancers, 2 adenomas); thus the gut microbiota had a tumor-promoting effect when DHM was the tumor initiator.

A high incidence of spontaneous CRC has been demonstrated in the T-cell receptor (TCR) β chain and *p53* double-knockout mice. In one study, 70% of the animals with a conventional microbiota developed adenocarcinomas, whereas adenocarcinoma of the colon did not occur in germ-free TCR $\beta^{-/-}p53^{-/-}$ mice, thus indicating a major role for the intestinal microbiota (39).

Table 1   Evidence That the Colonic Microbiota Is Involved in the Etiology of Colon Cancer

Human feces have been shown to be mutagenic, and genotoxic substances of bacterial origin have been isolated

Intestinal bacteria can produce, from dietary components, substances with genotoxic, carcinogenic, and tumor-promoting activity

Gut bacteria can activate procarcinogens to DNA reactive agents

Germ-free rats fed human diets exhibit lower levels of DNA adducts in tissues than conventional rats

Germ-free rats treated with the carcinogen 1,2-dimethylhydrazine have a lower incidence of colon tumors than similarly treated rats having a normal microbiota

Germ-free T-cell receptor chain and *p53* double-knockout (TCRβ$^{-/-}$ p53$^{-/-}$) mice did not develop adenocarcinoma of the colon at 4 months of age. Adenocarcinomas of the ileocecum and cecum were detected in 70% of the conventional TCRβ$^{-/-}$ p53$^{-/-}$ mice

Streptococcus bovis has been implicated in colonic neoplasia and supplements of this strain of bacteria and antigens extracted from the bacterial cell wall were shown to induce formation of hyperproliferative aberrant colonic crypts and increase the expression of proliferation markers in carcinogen treated rats (40). The effect of individual bacteria on cancer risk varies. Mice mono-associated with *Mitsuokella multiacida*, *Clostridium butyricium* or *Bifidobacterium longum* had a higher incidence of colonic adenoma (68% in each case) as compared to those associated with *Lactobacillus acidophilus* (30%) (41).

Gut Bacterial Metabolism and CRC Risk

The enormous numbers and diversity of the human gut microbiota is reflected in a large and varied metabolic capacity, particularly in relation to xenobiotic biotransformation, carcinogen synthesis and activation. The metabolic activities of the gut microbiota can have wide-ranging implications for the health of the host (42). To date the vast majority of mechanisms whereby bacteria are involved in carcinogenesis involve toxic or protective products of bacterial metabolism. Such metabolic activities include numerous enzymatic reactions and degradation of undigested dietary residues. Diet can substantially modulate these activities by providing a vast array of substrates. A wide range of enzyme activities capable of generating potentially carcinogenic metabolites in the colon are associated with the gut microbiota, including β-glucuronidase, β-glucosidase, nitrate reductase and nitro-reductase. These are usually assayed in fecal suspensions and appear to be present in many bacterial types (43–52).

A major role for the intestinal microbiota has been identified in the metabolism of the bile acids. The primary bile acids, chenodeoxycholic acid and cholic acid, are subject to extensive metabolism by the intestinal microbiota (53), predominantly 7-α-dehydroxylation, which converts cholic to deoxycholic acid (DCA) and chenodeoxycholic acid to lithocholic acid (LCA). These secondary bile acids exert a range of biological and metabolic effects in vitro and in vivo including cell necrosis, hyperplasia, and tumor-promoting activity in the colon, induction of DNA damage and apoptosis (54). It has also been suggested that secondary bile acids influence CRC by selecting for apoptosis-resistant cells or by interacting with various secondary messenger signaling systems.

A number of human observational studies in patients with adenomas or CRC have reported a correlation between fecal bile acid (FBA) concentrations and CRC risk (55,56). Some studies have also suggested that high fecal DCA concentrations and DCA to LCA ratio are associated with increased CRC risk (57). However, not all studies have confirmed this relationship between bile acids and CRC risk (58).

Formation of Protective Agents During Fermentation

Both dietary and endogenous carbohydrate substrates (e.g., starch and non-starch polysaccharides and intestinal mucins) are hydrolyzed by gut bacterial enzymes to produce the short chain fatty acids (SCFAs), acetate, propionate, and butyrate (59). These SCFAs provide an energy source for the intestinal cells and are also thought to confer beneficial effects on the host. SCFAs decrease colonic and fecal pH and this acidic environment is thought to be beneficial to the host (60). Specific oligosaccharides and resistant starch that result in SCFAs, and in particular butyrate (61) may have the potential to decrease CRC risk. This SCFA is of specific interest since it has been shown to induce apoptosis in colon adenoma and colon cell lines. In vitro studies have shown that increased butyrate supply to colon cells induces growth of the gut epithelium whereas reduced butyrate supply causes gut atrophy and functional impairments (62). Sodium butyrate has been observed to induce apoptosis and to alter the resistance of colonic tumor cells to apoptosis (62). However, the majority of these results have come from experiments conducted in vitro and again there have been conflicting views (63).

It follows from the above that modification of the gut microbiota may exert a beneficial effect on the process of carcinogenesis and this opens up the possibility for dietary modification of colon cancer risk. Probiotics and prebiotics, which modify the microbiota by increasing the numbers of lactobacilli and/or bifidobacteria in the colon, have been a particular focus of attention in this regard. In general species of *Bifidobacterium* and *Lactobacillus* have low activities of those enzymes involved in carcinogen formation and metabolism by comparison to other major anaerobes in the gut such as *Bacteroides*, *Eubacteria* and clostridia (44). This suggests that increasing the proportion of lactic acid bacteria (LAB) in the gut could modify, beneficially, the levels of xenobiotic metabolizing enzymes. This manipulation of the gut is discussed in greater detail in other chapters within this book. Overall, experimental and animal research show encouraging effects of several probiotic strains to decrease colon cancer, leading the way to the development of well-designed human intervention trials.

Effects of Gut Microbiota on Gene Expression

To date, there are only a few molecular descriptions of how bacteria in the normal microbiota regulate gene products with presumed positive functions in the intestine or systemically. Dramatic changes in gene expression were noted when germ-free mice were mono-colonized with *Bacteroides thetaiotaomicron*, a component of the normal microbiota of adult mice and humans (64). A number of genes involved in general mechanisms like nutrient uptake, fortification of the intestinal epithelial barrier, postnatal development, and angiogenesis are regulated in response to this commensal microbe. In addition, it is becoming clear that metabolic products, produced by the gut microbiota, can alter gene expression in the colonocyte [e.g., butyrate, produced by bacterial fermentation of dietary fiber, induces p21/Cip1/WAF1 mRNA (important in cell cycle control)] and secondary bile acids, produced from primary bile acids by the gut microbiota, alter AP-1-dependent and *COX-2* gene transcription) (65,66).

SURROGATE MARKERS FOR DIET-RELATED COLON CANCER STUDIES

As discussed above, the gut microbiota has been implicated in the etiology of CRC by a number of studies and these observations form the theoretical basis for the use of several gut microbiota biomarkers (fecal biomarkers) in studies on diet and colon cancer. They are composed of two main categories; those examining the activity of bacterial enzymes or bacterial metabolites and those based on bioassays on fecal water. For a more thorough review of this subject, the reader is referred to Rafter and coworkers (67).

Bacterial Enzymes

A wide range of enzyme activities capable of generating potentially carcinogenic metabolites in the colon are associated with the gut microbiota, including β-glucuronidase β-glucosidase, nitrate- and nitro-reductase. These are usually assayed in fecal suspensions and appear to be present in many bacterial types. Of these enzymes, β-glucuronidase has been the most extensively investigated as a biomarker of CRC risk. It should be noted that these factors are associated with the generation of carcinogens and promoters and do not have a direct link with tumors.

β-Glucuronidase

Many carcinogenic compounds are metabolized in the liver and then conjugated to glucuronic acid before being excreted via the bile into the small intestine. In the colon bacterial β-glucuronidase can hydrolyze the conjugates, releasing the parent compound or its activated, hepatic metabolite.

The activity of β-glucuronidase in the colon can alter the likelihood of tumor induction in animal models of CRC. The use of a β-glucuronidase inhibitor administered in conjunction with the carcinogen azoxymethane (which undergoes activation and conjugation in the liver) significantly reduces the number of tumors formed in the rat colon, indicating that microbiota β-glucuronidase has a role in tumor induction. Metabolic epidemiological studies have shown that populations at high risk of CRC have high levels of fecal β-glucuronidase activity. Furthermore, fecal β-glucuronidase activity in colon cancer patients is significantly higher than in healthy controls.

The activity of β-glucuronidase is influenced by diet. High risk diets for CRC have consistently been shown to increase β-glucuronidase activity relative to low risk diets. Furthermore, various types of fiber decrease the activity of β-glucuronidase in rats.

Although it represents a simple reproducible marker, evidence for a role for β-glucuronidase in human CRC is indirect and is remote from the final endpoint (tumors).

Metabolites

A wide range of metabolites with potential genotoxic, tumor-promoting and anti-carcinogenic activities have been identified in feces.

N-Nitroso Compounds

Nitrate, ingested via diet and drinking water, is reduced by gut bacterial nitrate reductase to its more reactive and toxic reduction product, nitrite. Nitrite reacts with nitrogenous compounds in the body to produce NOC. The reaction can occur chemically in the acidic conditions prevalent in the human stomach and can also be catalyzed at neutral pH by gut bacteria in the colon.

The term NOC covers a wide range of compounds including N-nitrosamines, N-nitrosamides, N-nitrosoguanidines, and N-nitrosoureas, the majority of which are highly carcinogenic, DNA alkylating agents. However, the genotoxic or carcinogenic activity of the NOC produced by the bacterial N-nitrosation process in the large intestine has not yet been established.

Fecal apparent total NOC (ATNC) excretion is increased by red meat consumption. In conjunction with high meat intakes, wheat bran, resistant starch and vegetable consumption had no effect on fecal ATNC excretion or concentration.

Secondary Bile Acids

The primary bile acids, chenodeoxycholic acid and cholic acid, are subject to extensive metabolism, predominantly 7-α-dehydroxylation, by the intestinal microbiota, which converts cholic to DCA and chenodeoxycholic to LCA. These are termed secondary bile acids.

Epidemiological studies indicate that concentrations of secondary bile acids are higher in populations at high risk of CRC and in case control studies 7-α-dehydroxylase activity is higher in cases than controls. In human studies, high fat intake, which correlates with CRC risk, increases FBA concentrations, whereas increased consumption of wheat bran (negatively correlated with CRC risk) reduces FBA concentration.

Short Chain Fatty Acids

The SCFAs acetate, propionate, and butyrate are the principal end-products of carbohydrate fermentation. These are absorbed from the colonic lumen and metabolized by various body tissues. Butyrate is preferentially metabolized by colonocytes.

There is evidence from in vitro studies and animal models (where cecal SCFA concentrations can be measured) that the type of carbohydrate has an important influence on the amount and proportions of SCFA produced, with starch and wheat bran being particularly associated with elevated butyrate production. In human studies, inulin has been shown to enhance excretion of total SCFA in human feces, whereas wheat bran increased absolute or relative proportions of butyrate in feces. Where the butyrate is produced relative to proximal and distal regions of the colon is important and should be a methodological consideration.

Gut bacterial enzymes and fecal metabolites are relatively simple to measure routinely and in general may be of use in assessing effects of diet on modulating exposure of the colon to potential carcinogens, rather than reflecting cancer risk.

Fecal Water Activities

Fecal Water Cytotoxicity

There is considerable evidence that colon tumors are a result of gut luminal factors damaging the mucosa. Furthermore, free reactive and soluble factors are more likely to affect the epithelium than substances bound to the insoluble matrix such as fiber. Therefore, an alternative approach to assaying enzymes or metabolites in feces is to assess toxicological activity of fractions using short-term tests for toxicity, genotoxicity, and mutagenicity. Usually the aqueous phase of the human feces (fecal water) is used, since this will contain most of the free reactive species. For assessment of fecal water cytotoxicity, the effect on proliferation of human colon carcinoma cells in culture is used.

Proliferative zone expansion in the colonic crypts and an increased rate of epithelial proliferation are considered to be an early step in carcinogenesis. Stimulation of proliferative activity in colonic epithelium may in part be mediated via cytotoxic mechanisms, resulting in increased cell loss at the epithelial surface and a compensatory rise in mitotic activity of the crypts. Such considerations led to the development of assays to assess cytotoxic activity in fecal water towards colon cells in vitro. It is thought that bile acids, especially secondary bile acids, make a major contribution to fecal water cytotoxicity. In a comparison of fecal water cytotoxicity in patients at low (no colon adenomas) medium (small colorectal adenomas) and high (large tubular adenomas) risk of CRC, no significant differences between the groups were observed.

Interventions using dietary regimes associated with increased or decreased CRC risk have been shown to modulate appropriately fecal water cytotoxicity. For example, dietary calcium has frequently been shown to reduce the cytotoxicity of fecal water presumably by precipitating soluble bile acids. Fecal water cytotoxicity was higher in subjects on a high fat, low calcium, low fiber diet compared with those on a low fat, high calcium, high fiber regime. In rats, a high red meat consumption increases the cytotoxicity of fecal water. This effect was independent of the fat and bile acid content of the fecal water and may be related to dietary haem.

Fecal Water Genotoxicity

The presence of DNA damaging activity towards human cultured colon cells has been demonstrated in samples of fecal water from healthy human subjects. A wide variation was found ranging from negligible to high activity. The presence of genotoxic activity in fecal water can be considered to reflect exposure of the colonic mucosa to carcinogens.

There is now convincing evidence that CRC is induced by a series of mutational events in a number of critical genes. Sporadic colorectal tumors have been shown to contain mutations and deletions in oncogenes, and tumor suppressor genes such as Apc, K-ras, and *p53*. DNA damage has been detected in biopsies of colon tissue derived from laboratory animals and human subjects. Thus, the presence in the colonic lumen of DNA damaging agents could represent an important risk factor for CRC. There are as yet no reports of validation studies for the endpoint in patients at different risk of CRC.

In healthy subjects, a diet high in fat and meat, but low in dietary fiber (hence considered to be of high CRC risk) was associated with a significantly increased fecal water genotoxicity by comparison to a diet low in fat and meat.

Cytotoxicity and particularly genotoxicity of fecal water have a good mechanistic link with colon carcinogenesis and hence provide potentially valuable, non-invasive methods for assessing CRC risk in human subjects. However, there is a need for more extensive validation of these endpoints.

CONCLUSION

It is becoming increasingly evident that the microbiota of the gastrointestinal tract and in particular that of the large intestine interacts with its host and may exert either harmful or protective effects, thus participating in the etiology of cancer. Gastric adenocarcinoma is the second leading cause of cancer-related deaths in the world and has been associated with the presence of *H. pylori* in the stomach. Several mechanisms of how this bacterium may affect tumorigenesis have been identified as well as dietary and environmental agents, which may confer either protective or detrimental effects. Colon cancer is the fourth most common cancer worldwide and again environmental factors and in particular diet play an important role in this disease. It has been shown that the microbiota of the gut interacts with its host both locally and systemically resulting in a broad range of effects, which may have both beneficial and detrimental outcomes, for nutrition, infections, xenobiotic metabolism, toxicity of ingested chemicals, and cancer. It is important to gain more insight into the pathogenesis of these cancers in order to develop more effective preventive and treatment strategies. The use of pro- and prebiotics may serve to induce beneficial effects on the host. Further research from well-planned intervention trials is required to further our understanding of the role of these agents in human carcinogenesis. Finally, as our understanding of the role of the gut microbiota in health and disease improves, we will be able to develop even better surrogate markers for use in human dietary intervention studies.

REFERENCES

1. Cummings JH, MacFarlane GT. The control and consequences of bacterial fermentation in the human colon. J Appl Bacteriol 1991; 70:443–459.
2. Fearon ER, Vogelstein B. A genetic model for colorectal tumorgenesis. Cell 1990; 61:759–767.
3. Draser BS. The bacterial flora of the intestine. In: Rowland IR, ed. The role of the Gut Flora in Toxicity and Cancer. London: Academic Press, 1988:23–38.
4. Hill MJ. In: Hill MJ, ed. The role of Gut Bacteria in Human Toxicology and Pharmacology. London: Taylor and Francis, 1995.
5. Hill MJ. Gut flora and cancer in humans and laboratory animals. In: Rowland IR, ed. The role of the Gut Flora in Toxicity and Cancer. London: Academic Press, 1988:461–502.
6. Graham DY, Lew GM, Klein PD, et al. Effect of treatment of *Helicobacter pylori* infection on the long-term recurrence of gastric or duodenal ulcer: a randomised, controlled study. Ann Int Med 1992; 116:705–708.
7. Asaka M, Kimura T, Kato M, et al. Possible role of *Helicobacter pylori* infection in early gastric cancer development. Cancer 1994; 73:2691–2694.
8. Kikuchi S, Wada O, Nakajaima T, et al. Serum anti-*Helicobacter pylori* antibody and gastric carcinoma among young adults. Research group on prevention of gastric carcinoma among young adults. Cancer 1995; 75:2789–2793.
9. Atherton JC. H. pylori virulence factors. Br Med Bull 1998; 54:105–120.
10. Moss SF. The carcinogenic effect of *H. pylori* on the gastric epithelial cell. J Physiol Pharmacol 1999; 50:847–856.
11. Yao YL, Xu B, Song YG, et al. Effect of *Helicobacter pylori* infection on gastric mucosal cell proliferation in mongolian gerbils. Di Yi Junyi Daxue Xuebao 2002; 22:348–350.
12. Arends MJ. How do cancer cells die? Apoptosis and its role in neoplastic progression. In: Leake R, Gore M, Ward RH, eds. The Biology of Gynaecological Cancer. Royal College of Obstetricians and Gynaecological Press, London, 1995:73–91.
13. Cho KR, Vogelstein B. Suppressor gene alterations in the colorectal adenoma-carcinoma sequence. J Cell Biochem 1992; 16G:137–141.
14. Zhang ZW, Patchett SE, Farthing MJ. Role of *Helicobacter pylori* and p53 in regulation of gastric epithelial cell cycle phase progression. Dig Dis Sci 2002; 47:987–995.
15. Wagner S, Beil W, Westermann J, et al. Regulation of gastric epithelial cell growth by Helicobacter pylori: offdence for a major role of apoptosis. Gastroenterology 1997; 113:1836–1847.
16. Wang J, Chi DS, Kalin GB, et al. *Helicobacter pylori* infection and oncogene expressions in gastric carcinoma and its precursor lesions. Dig Dis Sci 2002; 47:107–113.
17. Shirin H, Hibshoosh H, Kawabata Y, et al. P16Ink4a is overexpressed in *H. pylori*-associated gastritis and is correlated with increased epithelial apoptosis. Helicobacter 2003; 8:66–71.
18. Correa P, Malcolm G, Schmidt B, et al. Antioxidant micronutrients and gastric cancer. Aliment Pharmacol Ther 1998; 12:73–82.
19. Zhang HM, Wakisaka N, Maeda O, Yamamoto T. Vitamin C inhibits the growth of a bacterial risk factor for gastric carcinoma: *Helicobacter pylori*. Cancer 1997; 80:1897–1903.
20. Zhang ZW, Abdullahi M, Farthing MJ. Effect of physiological concentrations of vitamin C on gastric cancer cells and *Helicobacter pylori*. Gut 2002; 50:165–169.
21. Kaaks R, Tuyns AJ, Haelterman M, Riboli E. Nutrient intake patterns and gastric cancer risk: a case control study in Belgium. Int J Cancer 1998; 78:415–420.
22. You WC, Zhang L, Gail MH, et al. Gastric dysplasia and gastric cancer: *Helicobacter pylori*, serum vitamin C and other risk factors. J Natl Cancer Inst 2000; 92:1607–1612.
23. Guo X, Wang L, Yuan Y. Association between *Helicobacter pylori* cagA strain infection and expression of cyclooxygenase 2 in gastric carcinoma. Chin Med J 2002; 82:868–871. Breuer NF, Dommes P, Jaekel S, Goebell H. Fecal bile acid excretion pattern in colonic cancer patients. Dig Dis Sci 1985; 30:852–859.
24. Huang JQ, Sridhar S, Hunt RH. Role of *Helicobacter pylori* infection, and non-steroidal anti-inflammatory drugs in peptic-ulcer disease: a meta-analysis. Lancet 2002; 359:14–22.

25. Blaser MJ. Helicobacters are indigenous to the human stomach: duodenal ulceration is due to changes in microecology in the modern era. Gut 1998; 43:721–727.

26. Logan RP, Walker MM. ABC of the upper gastrointestinal tract: epidemiology and diagnosis of *Helicobacter pylori* infection. BMJ 2001; 323:920–922.

27. Michetti P, Dorta G, Wiesel PH, et al. Effect of whey-based culture supernatant of *Lactobacillus acidophilus* (johnsonii) La1 on *Helicobacter pylori* infection in humans. Digestion 1999; 60:203–209.

28. Mitsuoka T. Recent trends in research on intestinal flora. Bifidobacteria Microflora 1982; 3:3–24.

29. Draser BS. The bacterial flora of the intestine. In: Rowland IR, ed. The Role of the Gut Flora in Toxicity and Cancer. London: Academic Press, 1998:23–38.

30. Rowland IR, ed. Role of the Gut Flora in Toxicity and Cancer. London: Academic Press, 1988.

31. Rowland IR. Toxicology of the colon: role of the intestinal flora. In: Gibson GR, MacFarlane GT, eds. Human Colonic Bacteria: Role in Nutrition, Physiology and Pathology. Boca Raton: CRC Press, 1995:155–174.

32. Orrhage K, Sillerstrom E, Gustafsson JA, et al. Binding of the mutagenic heterocyclic amines by intestinal and lactic acid bacteria. Mutat Res 1994; 311:239–248.

33. Fassano A. Cellular microbiology: can we learn cell physiology from microorganisms? Am J Physiol 1999; 276:C765–C776.

34. Hill MJ, Draser BS, Hawksworth G, et al. Bacteria and aetiology of cancer of the large bowel. Lancet 1971; i:95–100.

35. Moore WE, Moore LH. Intestinal floras of populations that have a high risk of colon cancer. Appl Environ Microbiol 1995; 61:3202–3207.

36. Kanazawa K, Konishi F, Mitsuoka T, et al. Factors influencing the development of sigmoid colon cancer. Bacteriologic and biochemical studies. Cancer 1996; 77:1701–1706.

37. Reddy BS, Narisawa T, Maronpot R, et al. Animal models for the study of dietary factors and cancer of the large bowel. Cancer Res 1975; 35:3421–3426.

38. Reddy BS, Weisburger JH, Narisawa T, Wynder EL. Colon carcinogenesis in germ-free rats with dimethylhydrazine, and N-nitrosamines in health and gastroduodenal disease. Lancet 1974; ii:550–552.

39. Kado S, Uchida K, Funabashi H, et al. Intestinal microflora are necessary for development of spontaneous adenocarcinoma of the large intestine in T-cell receptor β chain and p53 double-knockout mice. Cancer Res 2001; 61:2395–2398.

40. Ellmerich S, Djouder N, Schöller M, Klein JP. Production of cytokines by monocytes, epithelial and endothelial cells activated by *Streptococcus bovis*. Cytokine 2000; 12:26–31.

41. Horie H, Kanawawa K, Okada M, et al. Effects of intestinal bacteria on the development of colonic neoplasm: an experimental study. Eur J Cancer Prev 1999; 8:237–245.

42. Rowland IR, Gangolli SD. Role of gastro-intestinal microflora in the metabolic and toxicological activities of xenobiotics. In: Ballentyne B, Marrs, TC, Syveron T eds. General and Applied Toxicology 1999; P562–576.

43. Cole CB, Fuller R, Mallett AK, Rowland IR. The influence of the host on expression of intestinal microbial enzyme activities involved in metabolism of foreign compounds. J Appl Bacteriol 1985; 58:549–553.

44. Saito Y, Takano T, Rowland IR. Effects of soybean oligosaccharides on the human gut microflora in *in vitro* culture. Microb Ecol Health Dis 1992; 5:105–110.

45. Tadaka H, Hirooka T, Hiramatsu Y, Yamamoto M. Effects of beta-glucuronidase inhibitor on azoxymethane-induced colonic carcinogenesis in rats. Cancer Res 1982; 42:331–334.

46. Grasten SM, Juntunen KS, Poutanen KS, et al. Rye bread improves bowel function and decreases the concentrations of some compounds that are putative colon cancer risk markers in middle-aged women and men. J Nutr 2000; 130:2215–2221.

47. Kim DH, Jin YH. Intestinal bacterial beta-glucuronidase activity of patients with colon cancer. Arch Pharmacol Res 2001; 24:564–567.

48. Eriyamremu GE, Osagie VE, Alufa OI, et al. Early biochemical events in mice exposed to cycas and fed a nigerian-like diet. Ann Nutr Metab 1995; 39:42–51.

49. Rowland IR, Tanaka R. The effects of transgalactosylated oligosaccharides on gut flora metabolism in rats associated with a human fecal microflora. J Appl Bacteriol 1993; 74:667–674.

50. Mallet AK, Rowland IR. Factors affecting the gut microflora. In: Rowland IR, ed. Role of the Gut Flora in Toxicity and Cancer. London: Academic Press, 1988:347–382.

51. Hughes R, Cross AJ, Pollock JR, Bingham S. Dose-dependent effect of dietary meat on endogenous colonic N-nitrosation. Carcinogenesis 2001; 22:199–202.

52. Silvester KR, Bingham SA, Pollock JR, et al. Effect of meat and resistant starch on fecal excretion of apparent N-nitroso compounds and ammonia from the human large bowel. Nutr Cancer 1997; 29:13–23.

53. MacDonald IA, Sutherland JD, Cohen BI, Mosbach EH. Effect of bile acid oxazoline derivatives on microorganisms participating in 7 alpha-hydroxyl epimenzation of primary bile acids. J Lipid Res 1983; 24:1150–1559.

54. Gill CI, Rowland IR. Diet and cancer: assessing the risk. Br J Nutr 2002; 88:S73–S87.

55. Imray CHE, Radley S, Davis A, et al. Fecal unconjugated bile acids in patients with colorectal cancer or polyps. Gut 1992; 33:1239–1245.

56. Stadler J, Yeung KS, Furrer R, et al. Proliferative activity of rectal mucosa and soluble fecal bile acids in patients with normal colons and in patients with colonic polyps or cancer. Cancer Lett 1988; 38:315–320.

57. Owen RW. Feacal steroids and colorectal carcinogenesis. Scand J Gastroenterol 1997; 222:76–82.

58. de Kok TM, van Maanen JM. Evaluation of fecal mutagenicity and colorectal cancer risk. Mutat Res 2000; 463:53–101.

59. Silvi S, Rumney CJ, Cresci A, Rowland IR. Resistant starch modifies gut microflora and microbial metabolism in human flora-associated rats inoculated with feces from Italian and U.K. donors. J Appl Microbiol 1999; 86:521–530.

60. Scheppach W. Effects of short chain fatty acids on gut morphology and function. Gut 1994; 1:S35–S38.

61. Bird AR, Brown IL, Topping DL. Starches, resistant starches, the gut microflora and human health. Curr Issues Intest Microbiol 2000; 1:25–37.

62. Archer S, Meng S, Shei A, Hodin RA. P21WAFI is required for butyrate-mediated growth inhibition of human colon cancer cells. Proc Natl Acad Sci USA 1998; 95:6791–6796.

63. Ishizuka S, Sonoyama K, Kassai T. Changes in the number and apoptosis of epithelial cells in the colorectum of wheat bran-fed rats soon after administering 1,2-dimethylhydrazine. Biosci Biotechnol Biochem 1997; 61:1337–1341.

64. Xu J, Bjursell MK, Himrod J, et al. A genomic view of the human-*Bacteroides thetaiotaomicron* symbiosis. Science 2003; 299:2074–2076.

65. Smith JG, Yokoyama WH, German BG. Butyric acid from the diet: actions at the level of gene expression. Clin Rev Food Sci 1998; 38:259–297.

66. Hodin RA, Meng S, Archer S. Cellular growth state differentially regulates enterocyte gene expression in butyrate-treated HT-29 cells. Cell Growth Differ 1996; 7:647–653.

67. Rafter J, Govers M, Martel P, et al. PASSCLAIM—Diet-related cancer. European Journal of Nutrition,. 2004; 43:II47–II84.

13

In Vitro Methods to Model the Gastrointestinal Tract

Harri Mäkivuokko and Päivi Nurminen
Danisco Innovation, Kantvik, Finland

INTRODUCTION

The human intestinal microbiota has never been so intensively studied as in this current period. Over the last decade, the use of molecular methods, especially those based on 16S ribosomal RNA, have generated much knowledge on the composition of the intestinal microbiota of especially humans but also animals. The relatively easy accessible fecal sample is the main source of intestinal microbiota used for various analyses. It is uncertain how well fecal samples reflect the composition of the microbiota in the proximal parts of the colon (1,2) but it is certainly very different from the small intestine. In order to study the microbial composition and activity in these sites, one would need in vivo samples from a large number of healthy individuals. Invasive sampling from healthy people is ethically not acceptable. Animal models can be used for invasive sampling (see chapter by Henriksson); however, due to physiological and anatomical differences, animals will have a different microbiota. Therefore, in vitro techniques complement animal studies and offer means to test specific hypotheses in a controlled, replicable manner without using animal models or clinical samplings. With in vitro models, it is possible to simulate the conditions in the human oral cavity, stomach, duodenum, jejunum, ileum, and in the ascending, transverse, and descending sections of the colon.

TYPES OF INTESTINAL SIMULATOR MODELS

In vitro models can be divided into batch cultures, chemostat-type simulators, including semi-continuous and continuous cultures, and non-chemostat-type simulators. All models of the gastrointestinal tract (GIT) have strictly anaerobic conditions in order to simulate the environment that supports the growth of microbiota obtained from the GIT of humans or other mammals. In vitro models can be used sequentially, so that in the simulators of stomach and small intestine the food matrix can be digested using conditions and enzymes representing the physiological conditions in the upper GIT, while the colon simulators continue by simulating the microbial metabolism of the nondigestible residue. The different chemostat- and non-chemostat-type models have major structural differences, but the batch fermentors are generally similarly structured, small-scale bottle fermentors. The chemostat models can be run using inocula in either an in vitro steady-state (the exponential growth of the bacterial has stabilized) achieved with several days of pre-fermentation of the fecal inoculum or after a short (16-24 hours) pre-fermentation.

Batch-Type Simulators

The simplest and most commonly used in vitro method in microbiological studies is the use of batch fermentation with intestinal fluid or fecal slurry to study the effects of different added ingredients. These chemostat are typically anaerobically sealed bottles with fecal, caecal or rumen material and these models simulate only a certain part of the animal's GIT, e.g., mouse cecum or cow's rumen. The transit times of the intestinal fluids through those areas are relatively short and therefore the run-times in batch fermenting simulations range from 2–24 hours (3–7). The accumulation of fermentation products (e.g., SCFAs) can change the conditions in the batch fermentation from the microbially balanced starting point to a more competitive environment for the fermentative microbiota, thus affecting the in vivo relevance in longer simulations. More complex fermentation models with several vessels and fluid transitions between vessels either continuously or semi-continuously avoid this accumulation of metabolites and depletion of nutrients.

Chemostat-Type Simulators

The in vitro colon simulators were introduced for the first time in 1981 (8), and all models functioning today have a lot in common with this model. Rumney and Rowland reviewed the first decade of in vitro simulators in their excellent article (3). Of the models reviewed by Rumney and Rowland, the Reading model introduced by Gibson and co-workers in 1988 (9), revised 1998 by Macfarlane and co-workers (10), is still actively being used and two new interesting models have been described in the literature. Of these, the SHIME (Simulator for Human Intestinal Microbiological Ecosystem) model introduced by Molly et al. in 1993 (11) and the EnteroMix® colon simulator introduced by Mäkivuokko et al.

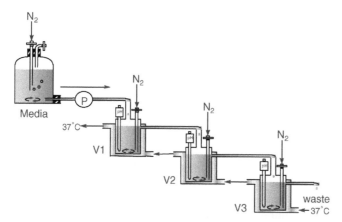

Figure 1 The Reading model. This model represents the human colon in three vessels: V1 proximal, V2 transverse, and V3 distal colon. Media is pumped to system continuously, and at the same time there is a continuous overflow from vessel to vessel. *Source*: From Ref. 9.

Table 1 Colon Simulator Models

	Reading	SHIME	EnteroMix®	TIM 1	TIM 2
Simulation area	Colon	Stomach to colon	Colon	Stomach to ileum	Colon
Vessel volumes	220–320 ml	300–1600 ml	6–15 ml	200 ml	200 ml
pH levels	5.8–6.8	5.0–7.0	5.5–7.0	1.8–6.5	5.8
Running times	14 days to steady state	30 days per cycle	2 days	~1 day	~3 days

Abbreviations: SHIME, Simulator for Human Intestinal Microbiological Ecosystem; TIM, TNO Intestinal Model.

in 2005 (12), together with the Reading model, are structurally chemostat models having 3–6 sequentially attached fermenting vessels with computer controlled fluid transition systems (Fig. 1) and (Table 1). The Reading model and the EnteroMix® model both simulate only the human colon, and a similar artificial simulator media described by Macfarlane et al. (10) is used in them to simulate the fluid entering the colon from the small intestine. The SHIME model simulates the whole human GIT from stomach to colon using artificial SHIME media, which has much in common with the medium described by Macfarlane and co-workers (10). These three models have three different designs in fluid transition. Fluids are either pumped semi-continuously to the subsequent vessels in three-hour intervals (EnteroMix® model), there is a continuous overflow of fluids between vessels (the Reading model), or the model can be a combination of these two types (SHIME).

Reading Simulator

The Reading simulator (Fig. 1) simulates the gut using a 3 stage continuous culture with three glass vessels (220 ml, 320 ml and 320 ml) and different pH in each vessel (5.8, 6.2, and 6.8); mimicking the human proximal, transverse, and distal colon, respectively.

In the beginning of the simulation, each vessel is inoculated with 100 ml of 20% (wt/vol) of human feces. The system is incubated in a batch overnight, after which a continuous pumping of fresh simulator fluid to the first vessel is started. At the same time a continuous overflow from vessel to vessel begins and the system is run for at least 14 days in order achieve a steady-state condition in the vessels. The excess fluid from the third vessel is collected to a waste container. The total retention time of the system can vary, e.g., between 27 and 67 hours (10). The viability of the microbiota is determined by taking samples at regular intervals from the vessels. After the incubation period, the test substance is added to the system mixed in the fresh simulation fluid and the system is then run to new steady state [e.g., for 22 days (9)]. The last phase is the washout period [e.g., for 50 days (9)] with the original simulation fluid to determine how long the changes induced by the test substance can still be measured in the absence of the substrate itself.

SHIME Model

The current SHIME model is a single six-stage system, where the first three glass vessels simulate stomach and small intestine and the subsequent three glass vessels the large intestine (11a). The original SHIME model (Fig. 2) (11) was a single five-stage system without the stomach compartment. Working volumes in these vessels are 300 ml for stomach and small intestine, 1000 ml for ceacum and ascending colon, 1600 ml for

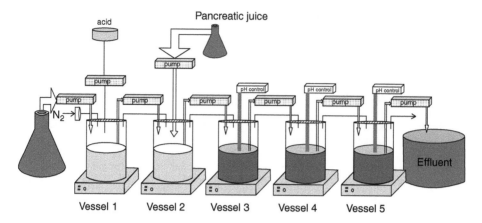

Figure 2 The original SHIME model. Vessels 1–5 in the figure mimic the different compartments of the human GIT: duodenum + jejunum, ileum, caecum + ascending colon, transverse colon and distal colon, respectively. In the revised version of this system, a vessel representing the stomach has been added before vessel 1. First five pumps work semi-continuously, and pumps between vessels, 3–5 and effluent work continuously. *Source*: From Ref. 11.

transverse colon, and 1200 ml for descending colon. pH is controlled in vessels 2, 3, 4, 5, and 6 in the ranges 5.0–6.5, 6.5–7.0, 5.5–6.0, 6.0–6.5 and 6.5–7.0, respectively.

The system is inoculated by introducing 10 ml supernatant of a human western diet suspension per day to the three first vessels for eight successive days. The remaining three vessels 4–6 representing the different compartments of the colon are inoculated with 50 ml of fecal suspension for 10 successive days. The contents of these three vessels are pumped continuously from vessel to vessel and finally to a discard bottle. The transit time of the whole system is 84 hours.

In the beginning of the simulation, 200 ml of fresh SHIME media (11) is added to vessel 1 (stomach) three times per day. Every 2–3 hours, the acidic (pH 2.0) contents of the first vessel is pumped to vessel 2 (duodenum + jejunum) along with 100 ml of pancreatic juice, supplemented with bile, to neutralize the acidity of the gastric effluent. After four hours the contents of vessel 2 is pumped to vessel 3 (ileum).

After eight days of using SHIME media only, the actual test substrate mixed with the SHIME media is introduced to the system. Feeding of the substrate is continued for 12 days, followed by another SHIME media-only period for 8–10 days. This cycle of three periods is repeated for all the studied substrates and samples are taken after each period.

The EnteroMix® Colon Simulator

The EnteroMix® model (Fig. 3) has four parallel units each comprising four glass vessels, allowing four simulations to be run simultaneously using the same fecal inoculum (12). EnteroMix® model vessels 1, 2, 3, and 4 have the smallest working volumes (6, 8, 10, and 12 ml, respectively) of the three models presented here (Table 1). The pH levels in the vessels (5.5, 6.0, 6.5, and 7.0, respectively) are similar to the other models. Because of the small volumes of vessels, a 40 ml inoculum of 25% wt/vol human feces and only 4 g of test substrate is needed for four parallel 48-hour simulations.

The simulation begins by filling the vessels of each of the four units with 0.9 mM anaerobic NaCl (3, 5, 7, and 9 ml to vessels 1, 2, 3, and 4, respectively) and inoculating the

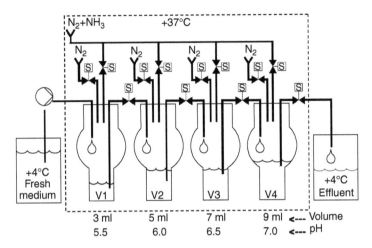

Figure 3 The EnteroMix® model. The figure represents the initial volumes of the system before fresh medium is added to begin the simulation. The vessels V1 to V4 are mimicking different sections of the human colon: caecum+ascending, transverse, descending, and distal colon, respectively. pH controlling and semi-continuous fluid transitions are operated via opening and closing of computer controlled valves (S).

first vessel with 10 ml of fecal inoculum. The inoculum is mixed in the vessel with NaCl and 10 ml of the mixed culture is pumped to the next vessel. This procedure continues through the vessels and finally the excess inoculum is pumped to waste container from the fourth vessel. After three hours of the incubation, 3 ml of fresh simulator media with (three test channels) or without (one control channel) test substance is pumped to the first vessel. The media is fermented in the first vessel for three hours, after which 3 ml of the fermented media is transferred to the second vessel, and 3 ml of fresh media is pumped to the first vessel. This procedure of transferring liquid to the next vessel continues through all the vessels, so that finally after 15 hours, when 3 ml of fermented fluid has been transferred from vessel four to the waste container for the first time, vessels 1, 2, 3, and 4 have respective volumes of 6, 8, 10, and 12 ml of fermenting fluid. The fermentation and three-hourly fluid transfers continue for 48 hours, after which the system is stopped and samples are collected from each vessel.

Other Simulators

In addition to simulate different parts of the GIT, chemostat-type simulators have also been used to simulate the oral cavity, in particular to investigate plaque formation (13); and to simulate the urinary bladder to investigate antibiotic sensitivity of urinary tract infection–causing pathogens (14). These simulators usually consist of a single chemostat.

Non-Chemostat Models

The third type of model is actually comprised of two complementary parts, the TIM (TNO Intestinal Model) systems 1 and 2 introduced by Minekus et al. in 1995 (15) and 1999 (16). The TIM 1 system (Fig. 4) comprises eight sequentially attached glass modules and mimics the stomach and small intestine, while the TIM 2-system consists of four glass modules in a loop mimicking the proximal colon of monogastric animals (Fig. 5). These

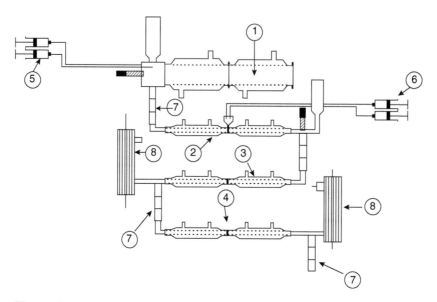

Figure 4 TIM 1 model. The model is mimicking the different sections of the human small intestine: the gastric compartment (1), duodenum (2), jejunum (3) and ileum (4). Gastric (5) and intestinal secretions (6), peristaltic valve pumps (7) and dialysis devices (8) are also included in this simulator. *Source*: From Ref. 17.

dynamic models differ from the three previously presented models in two main aspects: fluid transportation from vessel to vessel is executed via peristaltic valve-pumps and there is a constant absorption of water and fermentation products through dialysis membranes. In both systems the peristaltic movement of the intestinal fluid flowing in a flexible tube in the middle of the modules is achieved by changing the pressure of the 37°C heated water circulating between the module walls and the flexible tube. The peristaltic pressure around the flexible tube is controlled via computer-controlled valves to mimic the gastric emptying times. For the simulation of intestinal absorption TIM 1 has two integrated 5 kDa dialysis membranes, after jejunal and ileal modules, and TIM 2 has one, a hollow-fiber membrane (molecular mass cut-off value 50 kDa) in the lumen of the system. The TIM 1 dialysis membranes allow real-time collection of absorbable metabolites and water that would be absorbable in the human jejunum and ileum. In the tube membrane of TIM 2 circulates dialysis fluid allowing absorption of e.g., water, and short-chain fatty acids. The pH-values are monitored in each compartment.

   In a TIM 1 simulation, a homogenized human meal is introduced into the gastric compartment in pre-set times. From the stomach, the fluid is pumped through the following six compartments. During the simulation, the secretion of enzymes, bile, and pancreatic juice and the pH-controlling of the stomach (a pH gradient from 5.0 to 1.8 in 80 minutes from the beginning) and duodenum (constant pH 6.5) is regulated via computer.

   In a TIM 2 simulation the model is first inoculated with 200 ml of fecal inoculum. Microbiota is allowed to adapt to the conditions for 16 hours, after which the actual simulation is started by adding ileal medium semi-continuously with or without the tested substrate to the system. The pH is constantly maintained constant at 5.8 representing the pH-level in the proximal colon. Samples can be taken both from the lumen of the simulator and from the dialysis liquid during the simulation.

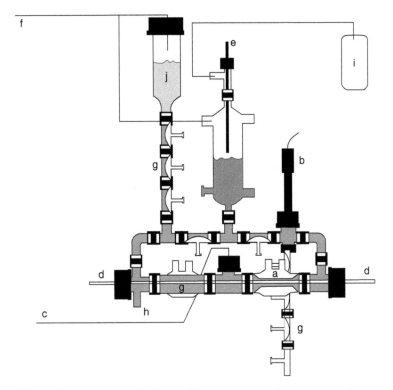

Figure 5 TIM 2 model: The model represents the human proximal colon in one loop-shaped system: peristaltic mixing with flexible walls inside (a), pH electrode (b), alkaline pump (c), dialysis system (d), fluid level sensor (e), nitrogen inlet (f), peristaltic valves (g), sample port (h), gas sampling (i) and ileal medium reservoir. *Source*: From Ref. 18.

Comparison of the Models

The four colon simulation models presented here have structural and functional differences (Table 1), but the solutions used to reproduce the critical conditions that influence the microbiology of the colon are similar in all four models. Firstly the colonic microbiota is simulated in each model using fecal samples from a single donor or several donors in a pooled sample, because more realistic samples of gastrointestinal tract bacteria from the ileum or cecum of humans are very difficult to obtain both ethically and technically. Secondly all the colon simulators use similar growth media that originate from media originally published by Gibson et al. in 1988 (9) mimicking the ileal fluids obtained from sudden-death victims. Thirdly all the colon models have strictly anaerobic conditions, similar pH set-points representing the in vivo situation in the colon of healthy humans (19) and all the functions of these systems are computer-controlled.

The Reading model and the SHIME system are both run until a steady state in microbial growth is reached, while TIM 2 and the EnteroMix® model are run for a pre-determined time (2 or 5 days). The SHIME system is the only one of the above-mentioned four systems having a continuous line from stomach to distal colon, thus enabling the simulation of the whole GI-tract in one run. The simulated ileal fluid coming from TIM 1 can also be used indirectly as growth medium in TIM 2. The EnteroMix® model has the smallest working volumes (Table 1) in the vessels, enabling the simulation of small concentrations of the tested substrate. On the other hand the small volumes do not allow any samplings during the simulation run, which is possible in all the other models, because the volume of microbiota would be too heavily affected in the vessels. The EnteroMix® model is also the only model having parallel channels in the same simulator allowing four parallel simulations to be run at the same time with the same fecal inoculum.

SIMULATING THE RUMEN

Although the simulators described above are mainly aimed at simulating the human GIT, the models can also be used to simulate the GIT of other monogastric animals. However for the simulation of the ruminant GIT different factors have to be taken into consideration; in particular the different functioning of the rumen, retaining and fermenting solid material while liquid phase is allowed to pass on into the GIT.

The anaerobic environment of the rumen is heterogeneous in nature: a large volume of free liquid, a complex solid mass of digesta, and a gas phase. Within this mixture, the diverse microbial population of bacteria, protozoa, and anaerobic fungi can be described as occurring in four different compartments (1) the microbes living free in suspension, (2) the microbes loosely associated with the solid material, (3) the microbes that are trapped in the solid material, and (4) the microbes close to or attached to the rumen wall (20). The complexity is still increased due to the different removal rates of the solid and liquid portions of rumen contents, revealing the dynamic nature of the rumen.

Rumen Simulators

The artificial rumen techniques developed over the past five decades for investigation of rumen physiology as well as evaluation of feed rations, have ranged from batch fermentations to more complicated continuous incubations. In addition, the absorption function of the rumen wall has been included in some designs, in which a semi-permeable membrane is applied for removal of the fermentation end products.

Batch Culture

The most simplistic, in vitro fermentations representing the rumen were performed in different kinds of tubes (21–23). Another way to conduct a static, batch simulation is to use closed glass serum bottles. As an example, in the study of Lopez et al. (24) 0.2 g of diet (ground to pass through 1 mm screen) was weighed into the 120 ml serum bottles and the fermentation process started by dispensing 50 ml of strained, 1:4 (v/v) buffered rumen fluid under CO_2 flushing. The bottles were sealed with butyl rubber stoppers and aluminium caps and incubated in a shaking water bath at +39°C. After 24-hour incubation, total gas production and pH were measured and samples for methane, hydrogen, and short chain fatty acid analysis taken.

The durations of the reported batch fermentations employing rumen microbes have varied from six (25) to 96 hours (26) or even up to 168 hours (27). The buffer systems applied in batch simulations are quite often adopted from by Menke et al. (28), McDougall (29), or Goering and van Soest (30).

Due to the fact that gas production has been used as an indirect measure of digestibility and fermentation kinetics of ruminant feeds, a scaled glass syringe (volume of 100–150 ml) has also been used as a fermentation vessel (28,37). The piston is allowed to move upward without restrain and thus indicates the amount of gas released due to microbial activity. The more sophisticated ways to measure gas production kinetics have been reported, for example the syringe/electronic pressure transducer-equipment (32), which measured and released the accumulated gas. However more automated systems were, both an apparatus which combined electronic pressure transducers and electric micro-valves (33) and the automated pressure evaluation system (APES) (34) where the overpressure was released by use of pressure sensitive switches and solenoid valves.

Semi-Continuous Culture (Rusitec)

The structure of semi-continuous rumen simulation technique Rusitec (Fig. 6), which was described by Czerkawski and Breckenridge (35), provides three of the four microbial compartments mentioned earlier. A Rusitec reaction vessel with capacity of one liter consisted of a Perspex cylinder (254×76) with an inlet at the bottom. The cylinder was sealed by flat Perspex cover provided with a screw flange for easy access. The cover is provided with two outlets, one for sampling and the other for effluent overflow and gas collection. The solids (feed or digesta) were placed in nylon bags (pore size 50–100 μm) inside a perforated container. This "cage" then slid up and down inside the reaction vessel, allowing the effluent to flush the solids. At the bottom of the vessel, the artificial saliva (29) was continuously infused and the excess liquid and the gases are forced out through an overflow by a slight positive

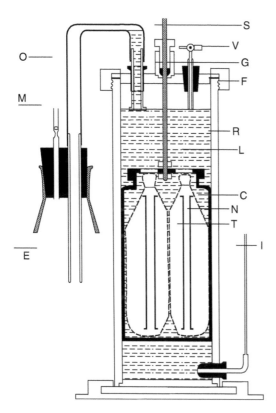

Figure 6   A schematic diagram of semi-continuous Rusitec unit: driving shaft (S), sampling valve (V), gas-tight gland (G), flange (F), main reaction vessel (R), rumen fluid (L), perforated food container (C), nylon gauze bag (N), rigid tube (T), inlet of artificial saliva (I), outlet through overflow (O), line to gas-collection bag (M), vessel for collection of effluent (E). *Source*: From Ref. 35.

pressure in the gas space. The proper fermentation temperature was maintained by incubating the reaction vessel in water bath at 39°C during the experiment.

The fermentation in Rusitec was started by placing solid rumen digesta in one nylon bag and an equal amount of feed to be used in a second nylon bag. The reaction vessel was filled up to overflow with strained diluted rumen contents. After 24 hours the inoculum bag was removed and replaced with a new bag of food. Removal of the oldest bag (48 hours) and adding a new bag was repeated each day. At the beginning of the experiment and during feeding, the gas space was flushed with the mixture of CO_2 and N_2 (5:95 v/v). The removed bag is drained, placed in a plastic bag and the solids washed twice with the artificial saliva. This rumination mimicking process includes gentle pressing of the solids and squeezing out excess liquid, which is combined and returned to the reaction vessel.

The Rusitec technique has been quite widely applied as such. It has been used by a number of authors to study, for example, decreased methanogenesis (36,37) and efficiency of recovery of particle-associated microbes from ruminal digesta (38). In reported Rusitec studies at least up to 16 reaction vessels have been applied simultaneously (39). The running times of sample collection periods have exceeded from five (40) to 36 days (36) after stabilizing the microbial population for 12 hours (39) to 17 days (40).

Continuous Culture

One of the earliest reports of continuous culture apparatus (Fig. 7) is the work of Stewart et al. (41). With the device designed by Quinn (42) the incubation time could exceed more beyond 24 hours because of the pH control system. In these simulation systems the

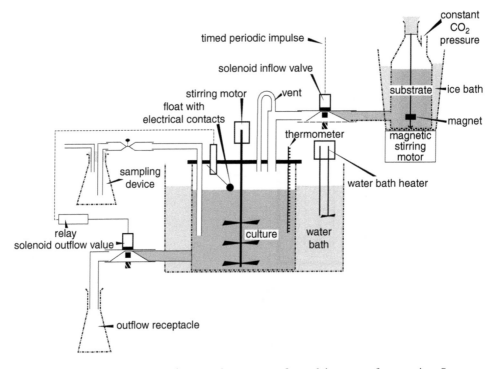

Figure 7   One of the earliest continuous culture systems for studying rumen fermentation. *Source*: From Ref. 41.

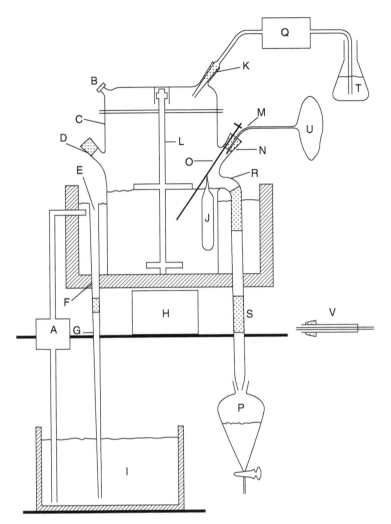

Figure 8 A continuous culture apparatus providing absorption of fermentation products: centrifugal water pump (A), gas-sampling port (B), fermentor (C), feeding port (D), water-drainage pipe (E), Plexiglas reservoir (F), drainage tube (G), magnetic stirrer (H), water bath (I), dialysis sac with cation-exchange resin (J), saliva-inflow ground-glass joint (K), fermentor stirring device (L), gas-outlet tube (M), fermentor port (N), sampling glass tube and resin holder (O), liquid-effluent collection funnel (P), peristaltic pump (Q), effluent outlet (R), effluent rubber tubing (S), saliva-water reservoir (T), gas-collection bladder (U), feed-input apparatus (V). Ports D and N are shown 90° out of phase from their actual position to simplify the drawing. *Source*: From Ref. 44.

water insoluble substrates were continuously delivered to the vessel in the form of a slurry. One of the few devices taking the absorption of fermentation end products into account was developed by Rufener et al. (43) and improved by Slyter et al. (44). The apparatus (Fig. 8) consisted of six independent fermentation chambers (500 ml) with accessories providing anaerobiosis, constant volume, agitation of the fermentation mixture and collection of effluents and gases. For controlling the pH, this system included a dialysis bag containing a mixture of ion-exchange resins, which absorbed the short chain fatty acids. The fermentors were reported to reach the steady state in three to four days of operation. One criterion for this conclusion was the stabilization of protozoal numbers even though their density in the vessels was merely 2% of that found in the inoculum.

The dual flow continuous culture system described by Hoover et al. (45) and modified later by Crawford et al. (46) and Hannah et al. (47) simulates the differential flows of liquids and solids that occur in the rumen. In the design described by Hannah et al. (Fig. 9) (47), the mineral buffer solution (48) supplemented with urea is infused to maintain fixed liquid dilution rate, and solids retention is regulated by adjusting the ratio of the filtered to overflow effluent volumes using a filtering device. Temperature of the vessel is kept constant at +39°C and pH is adjusted by infusion of 5N HCl or 5N NaOH. The vessel is constantly purged with N_2 to preserve anaerobic conditions and mixing of the fermentation broth is performed with magnetic impeller system. The ground and pelleted diet is semi-continuously fed to the vessel in eight equal portions over the 24-hour period by use of an automated feeder.

In typical experiments, durations of stabilization periods have varied from five to seven days followed by three-day effluent sampling period. Fermentation gases are neither collected nor analyzed from this simulation system. Depending on the experiment, systems consisting of four (49) to eight (50) glass vessels with a volume of 1.0 (49) to 1.26 liters (51) have been reported.

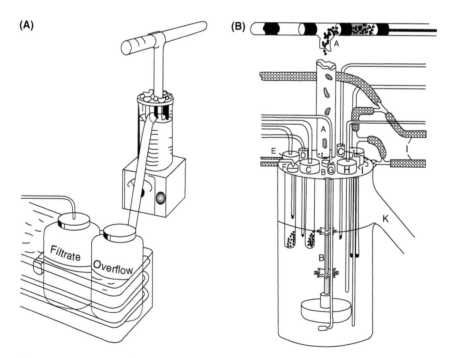

Figure 9 (**A**) General schematic of dual flow continuous culture system. (**B**) Schematic of fermenter flask components. A, Automatic feeding device and feed input port; B, magnetic impeller assembly; C, sodium hydroxide infusion port; D, hydrochloric acid infusion port; E, filters; F, buffer infusion port; G, nitrogen sparger; H, thermocouple assembly; I, coaxial heat exchanger apparatus; J, pH electrode; K, overflow port. *Source*: From Ref. 47.

Possibilities and Limitations of Rumen Simulation Methods

The in vitro environmental conditions (temperature, pH, buffering capacity, osmotic pressure, dry matter content and oxidation-reduction potential) should represent as closely as possible those of the rumen. Irrespective of the technique applied, the quality of the inoculum is one of the most important aspects in rumen simulations. In most studies the rumen fluid is strained through two, sometimes even four layers of cheesecloth. As a result, the inoculum is likely to represent only the microbes occurring in free liquid and a major part of the cellulolytic micro-organisms is lost.

Efforts that can more effectively reproduce the real conditions within the rumen will be very useful. Nevertheless the designs may be too complicated for routine and easy use: particle block up in the outlet filter or daily opening of the fermentor for feeding the microbes prevents the usability. A continuous culture system of two (52) to 21 (53) reaction vessels with running times of three to four weeks is not a very rapid method for analyzing the effects of feed substances on fermentation patterns of rumen microbes. The advantage of a batch simulation over continuous one is not only the possibility to have more replicates but also the flexibility to test a greater number of different treatments simultaneously.

The duration of the fermentation in closed batch culture should be adjusted carefully according to the substrates and cell density to prevent the deprivation and inhibitory effects of accumulating metabolites. As a consequence in either case, the most fastidious bacteria and protozoa are at risk of being lost. A shorter incubation time should be used with substrates that are rapidly fermented. By using actual feed components and compositions, the risk of substrate deprivation during simulations is reduced. For example, Leedle and Hespell (54) have reported the selective effects of single or purified carbohydrates and nitrogen substrates on microbial population. The amount of feed should be not only adequate in relation to the microbial density in vitro, but also in relation to the calculated total digestive nutrient requirement of the host (44).

The lack of substrates or excess of accumulated end products are, more rarely, the reasons for microbial changes in continuous culture systems. Those fermentors, which have a uniform and fast turnover rate for the total contents, quickly lose part or all of the protozoa. Stabilization of the system for several days will lead to selection and survival of those microbes best adapted to that environment. Irrespective of the artificial rumen technique, the longer the simulation is run, the greater the difference that will develop in the microbial populations compared to the original inoculum. However, a stable fermentation that can be maintained long enough to allow microbial adaptation, is considered desirable by continuous culture users (36,55). The use of actual feed components and compositions presumably assists the maintenance of a representative population also in continuous culture systems.

Although some of the artificial rumen techniques are more superior in taking into account the microbial compartments or the different transfer rates of liquids and solids, none of them include the activity of bacteria associated with the rumen wall or the interaction with the host immune system. It is both challenging and difficult to mimic ruminal fermentation and measure the parameters as they actually happen in the rumen. The real long-term effects of a test substance on rumen microbes and animal physiology can be evaluated neither with a short batch simulation nor with continuous culture simulation run for several weeks. Nevertheless, simulation of the rumen in vitro is a valuable technique for evaluating particular feed components and testing new diets before undertaking animal experiments.

CONCLUSION

Despite the advanced techniques used in the simulators described here, they will remain only limited models of the authentic gastrointestinal tract. In particular, the interaction between the microbes and the host is absent including contact with the mucosa and the intestinal immune system. Some of these issues may be addressed by the use of intestinal cell lines, either in the simulator, as a separate loop in the simulator or by using simulator effluent. While the latter would remain approximations of the real situation, they would nevertheless be very valuable for providing further insight into the dynamics and activity of the gastrointestinal microbiota.

REFERENCES

1. Zoetendal EG, von Wright A, Vilpponen-salmela T, Ben-Amor K, Akkermans ADL, de Vos WM. Mucosa-associated bacteria in the human gastrointestinal tract are uniformly distributed along the colon and differ from the community recovered from feces. Appl Environ Microbiol 2002; 68:3401–3407.
2. Ouwehand AC, Salminen S, Arvola T, Ruuska T, Isolauri E. Microbiota composition of the intestinal mucosa: association with fecal microbiota? Microbiol Immunol 2004; 48:497–500.
3. Rumney CJ, Rowland IR. In vivo and in vitro models of the human colonic flora. Crit Rev Food Sci Nutr 1992; 31:299–331.
4. Edwards CA, Eastwood MA. Comparison of the effects of ispaghula and wheat bran on rat caecal and colonic fermentation. Gut 1992; 33:1229–1233.
5. Barry J-L, Hoebler C, Macfarlane GT, et al. Estimation of fermentability of dietary fibre in vitro: a European interlaboratory study. Br J Nutr 1995; 74:303–322.
6. Deprez S, Brezillon C, Rabot S, et al. Polymeric proanthocyanidins are catabolized by human colonic microflora into low-molecular-weight phenolic acids. J Nutr 2000; 130:2733–2738.
7. Oufir LE, Barry JL, Flourie B, et al. Relationships between transit time in man and in vitro fermentation of dietary fiber by fecal bacteria. Eur J Clin Nutr 2000; 54:603–609.
8. Miller TL, Wolin MJ. Fermentation by the human large intestine microbial community in an in vitro semicontinuous culture system. Appl Environ Microbiol 1981; 42:400–407.
9. Gibson GR, Cummings JH, Macfarlane GT. Use of a three-stage continuous culture system to study the effect of mucin on dissimilatory sulfate reduction and methanogenesis by mixed populations of human gut bacteria. Appl Environ Microbiol 1988; 54:2750–2755.
10. Macfarlane GT, Macfarlane S, Gibson GR. Validation of a three-stage compound continuous culture system for investigating the effect of retention time on the ecology and metabolism of bacteria in the human colon. Microb Ecol 1998; 35:180–187.
11. Molly K, Vande WM, Verstraete W. Development of a 5-step multi-chamber reactor as a simulation of the human intestinal microbial ecosystem. Appl Microbiol Biotechnol 1993; 39:254–258.
11a. Alander M, De Smet I, Verstraete W, von Wright A, Mattila-Sandholm T. The effect of probiotic strains on the microbiota of the Simulator of the Human Intestinal Microbial Ecosystem (SHIME). Int J Food Microbiol 1999; 46:71–79.
12. Mäkivuokko H, Nurmi J, Nurminen P, Stowell J, Rautonen N. In vitro effects on polydextrose by colonic bacteria and Caco-2 cell cyclooxygenase gene expression. Nutr Cancer 2005; 52:93–103.
13. Herles S, Olsen S, Afflitto J, Gaffar A. Chemostat flow cell system: an in vitro model for the evaluation of antiplaque agents. J Dent Res 1994; 73:1748–1755.
14. Takahashi S, Sano M, Nishimura M, et al. Bactericidal effect of levofloxacin on strains with equal susceptibility in an in vitro urinary bladder model. Chemotherapy 1998; 44:337–342.

15. Minekus M, Marteau P, Havenaar R, Huis in't Veld HJ. A multicompartmental dynamic computer-controlled model simulating the stomach and small intestine. ATLA 1995; 23:197–209.

16. Minekus M, Smeets-Peeters M, Bernalier A, et al. A computer-controlled system to simulate conditions of the large intestine with peristaltic mixing, water absorption and absorption of fermentation products. Appl Microbiol Biotechnol 1999; 53:108–114.

17. Marteau P, Minekus M, Havenaar R, Huis In't Veld JHJ. Survival of lactic acid bacteria in a dynamic model of the stomach and small intestine: validation and the effects of bile. J Dairy Sci 1997; 80:1033.

18. Krul C, Humblot C, Philippe C, et al. Metabolism of sinigrin (2-propenyl glucosinolate) by the human colonic microflora in a dynamic in vitro large-intestinal model. Carcinogenesis 2002; 23:1011.

19. Macfarlane GT, Gibson GR, Cummings JH. Comparison of fermentation reactions in different regions of the human colon. J Appl Bacteriol 1992; 72:57–64.

20. Czerkawski JW. Compartmentation in the rumen, p. 65–82. An Introduction to Rumen Studies. Oxford OX3 OBW: Pergamon Press Ltd, 1986.

21. Morgavi DP, Newbold CJ, Beever DE, Wallace RJ. Stability and stabilization of potential feed additive enzymes in rumen fluid. Enzyme Microb Technol 2000; 26:171–177.

22. Hatfield RD, Weimer PJ. Degradation characteristics of isolated and in situ cell wall Lucerne pectic polysaccharides by mixed ruminal microbes. J Sci Food Agric 1995; 69:185–196.

23. Susmel P, Spanghero M, Marchetti S, Moscardini S. Trypsin inhibitory activity of raw soya bean after incubation with rumen fluid. J Sci Food Agric 1995; 67:441–445.

24. Lopez S, Valdes C, Newbold CJ, Wallace RJ. Influence of sodium fumarate addition on rumen fermentation in vitro. Br J Nutr 1999; 81:59–64.

25. Gomez JA, Tejido ML, Carro MD. Influence of disodium malate on microbial growth and fermentation in rumen-simulation technique fermenters receiving medium- and high-concentrate diets. Br J Nutr 2005; 93:479–484.

26. Blummel M, Karsli A, Russell JR. Influence of diet on growth yields of rumen micro-organisms in vitro and in vivo: influence on growth yield of variable carbon fluxes to fermentation products. Br J Nutr 2003; 90:625–634.

27. Ranilla MJ, Carro MD, Lopez S, Newbold CJ, Wallace RJ. Influence of nitrogen source on the fermentation of fibre from barley straw and sugarbeet pulp by ruminal micro-organisms in vitro. Br J Nutr 2001; 86:717–724.

28. Menke KH, Raab L, Salewski A, Steingass H, Fritz D, Schneider W. The estimation of the digestibility and metabolizable energy content of ruminant feedingstuffs from the gas production when they are incubated with rumen liquor in vitro. J Agric Sci Camb 1979; 93:217–222.

29. McDougall EO. Studies on ruminant saliva I. The composition and output of sheep's saliva. Biochem J 1948; 43:99.

30. Goering KH, Van Soest PJ. Forage fiber analysis (apparatus, reagents, procedures, and some repplication). Agricultural Handbook. Washington D.C.: Agricultural Research Counsil, U.S.Department of Agriculture, 1970.

31. Wallace RJ, Wallace SJ, McKain N, Nsereko VL, Hartnell GF. Influence of supplementary fibrolytic enzymes on the fermentation of corn and grass silages by mixed ruminal microorganisms in vitro. J Anim Sci 2001; 79:1905–1916.

32. Theodorou MK, Williams BA, Dhanoa MS, McAllan AB, France J. A simple gas production method using a pressure transducer to determine the fermentation kinetics of ruminant feeds. Animal 1994; 48:185–197.

33. Cone JW, van Gelder AH, Visscher GJW, Oudshoorn L. Influence of rumen fluid and substrate concentration on fermentation kinetics measured with a fully automated time related gas production apparatus. Animal 1996; 61:113–128.

34. Davies ZS, Mason D, Brooks AE, Griffith GW, Merry RJ, Theodorou MK. An automated system for measuring gas production from forages inoculated with rumen fluid and its use in determining the effect of enzymes on grass silage. Animal 2000; 83:205–221.

35. Czerkawski JW, Breckenridge G. Design and development of a long-term rumen simulation technique (Rusitec). Br J Nutr 1977; 38:371–384.

36. Wallace RJ, Czerkawski JW, Breckenridge G. Effect of monensin on the fermentation of basal rations in the Rumen Simulation Technique (Rusitec). Br J Nutr 1981; 46:131–148.

37. Dong Y, Bae HD, McAllister TA, Mathison GW, Cheng K-J. Lipid-induced depression of methane production and digestibility in the artificial rumen system (RUSITEC). Canadian J Anim Sci 1997; 77:269–278.

38. Ranilla MJ, Carro MD. Diet and procedures used to detach particle-associated microbes from ruminal digesta influence chemical composition of microbes and estimation of microbial growth in Rusitec fermenters. J Anim Sci 2003; 81:537–544.

39. Wang Y, McAllister TA, Rode LM, et al. Effects of an exogenous enzyme preparation on microbial protein synthesis, enzyme activity and attachment to feed in the Rumen Simulation Technique (Rusitec). Br J Nutr 2001; 85:325–332.

40. Russi JP, Wallace RJ, Newbold CJ. Influence of the pattern of peptide supply on microbial activity in the rumen simulating fermenter (RUSITEC). Br J Nutr 2002; 88:73–80.

41. Stewart DG, Warner RG, Seeley HW. Continuous culture as a method for studying rumen fermentation. Appl Microbiol 1961; 9:150–156.

42. Quinn LY. Continuous culture of ruminal microorganisms in chemically defined medium I. Design of continuous-culture apparatus. Appl Microbiol 1962; 10:580–582.

43. Rufener WH, Nelson WO, Jr., Wolin MJ. Maintenance of the rumen microbial population in continuous culture. Appl Microbiol 1963; 11:196–201.

44. Slyter LL, Nelson WO, Wolin MJ. Modifications of a device for maintenance of the rumen microbial population in continuous culture. Appl Microbiol 1964; 12:374–377.

45. Hoover WH, Crooker BA, Sniffen CJ. Effects of differential solid-liquid removal rates on protozoa numbers in continuous cultures of rumen contents. J Anim Sci 1976; 43:528–534.

46. Crawford RJ, Jr., Shriver BJ, Varga GA, Hoover WH. Buffer requirements for maintenance of pH during fermentation of individual feeds in continuous culture. J Dairy Sci 1983; 66:1881–1890.

47. Hannah SM, Stern MD, Ehle FR. Evaluation of a dual flow continuous culture system for estimating bacterial fermentation in vivo of mixed diets containing various soya bean products. Animal 1986; 16:51–62.

48. Weller RA, Pilgrim AF. Passage of protozoa and volatile fatty acids from the rumen of the sheep and from a continuous in vitro fermentation system. Br J Nutr 1974; 32:341–351.

49. Karunanandaa K, Varga GA. Colonization of rice straw by white-rot fungi (Cyathus stercoreus): effect on ruminal fermentation pattern, nitrogen metabolism, and fiber utilization during continuous culture. Animal 1996; 61:1–16.

50. Mansfield HR, Stern MD, Otterby DE. Effects of beet pulp and animal by-products on milk yield and in vitro fermentation by rumen microorganisms. J Dairy Sci 1994; 77:205–216.

51. Hoover WH, Kincaid CR, Varga GA, Thayne WV, Junkins LL, Jr. Effects of solids and liquid flows on fermentation in continuous cultures IV. pH and dilution rate. J Anim Science 1984; 58:692–699.

52. Fellner V, Sauer FD, Kramer JK. Steady-state rates of linoleic acid biohydrogenation by ruminal bacteria in continuous culture. J Dairy Sci 1995; 78:1815–1823.

53. Newbold CJ, Wallace RJ, Chen XB, McIntosh FM. Different strains of Saccharomyces cerevisiae differ in their effects on ruminal bacterial numbers in vitro and in sheep. J Anim Sci 1995; 73:1811–1818.

54. Leedle JAZ, Hespell RB. Brief incubations of mixed ruminal bacteria: effects of anaerobiosis and sources of nitrogen and carbon. J Dairy Sci 1983; 66:1003–1014.

55. Jayasuriya MCN, Hamilton R, Rogovic B. The use of an artificial rumen to assess low quality fibrous feeds. Biol Wastes 1987; 20:241–250.

14

Animal Models for the Human Gastrointestinal Tract

Anders Henriksson
DSM Food Specialties, Sydney, Australia

INTRODUCTION

Scientific research is continuously generating new ingredients for food and pharmaceutical products. In pace with consumer awareness of healthy products, considerable efforts are made to find new ingredients with beneficial effects on human health. The health benefits of such ingredients need to be assessed in human trials prior to being developed as a product for wider human consumption. Animal trials, conducted prior to human trials, offer a sound filtering system that provide the opportunity to identify those ingredients that are worthy of the relatively costly human studies that may follow. Animal models are important tools used in the study of human gastrointestinal (GI) microbiology. Specifically, animal models are used when considering the effect of food and pharmaceutical ingredients on GI health and disease. These effects include the metabolic and immunological activities of microorganisms that colonize the human gastrointestinal tract (GIT).

This chapter deals with issues related to the use of animal models in studies of human GI microbiota or specific microorganisms of human origin. It discusses similarities and differences between human and animal physiology and microbiota with specific focus on categories of animal models. The following discussion focuses predominantly on rodents and highlights some limitations and opportunities that relate to categories of rodent models such as "germ-free," "human flora associated" and "surgically or chemically modified."

Physiology and Microbiology of the GI Tract

Physiology

The human GIT is the most appropriate environment to conduct studies on the human GI microbiota but for practical reasons animal models are used extensively for these types of studies. The wide range of similarities between the animal and human GIT makes it possible to draw reasonable parallels between these two hosts, however, results from studies on the human microbiota in animals may not entirely reflect processes occurring in the human GIT. The reason for this is that there are also many differences between human and animal gut physiology, diets, and behavior. Rodents are the most extensively used animals in the research of human GI microbiota. The differences between the human and rodent GIT may be important in interpreting any research findings.

When considering the differences between the human and rat GITs, the issue of size is certainly obvious. This difference has impact on transit time of GI contents. Also, the rate of passage can vary between the type of diet, the particle size of digesta and morphological characteristics of the GITs. In rats the transit time is 12–35 hours depending on the type of diet and transit markers used (1,2). In humans, native Africans, consuming a traditional diet, have an average GI transit time of 33 hours, which is approximately half of the transit time that has been observed in Europeans or Africans on a Western diet.

Many more subtle and potentially important morphological and physiological differences exist between the human and rat GITs. An example of this exists in the fact that the adult human appendix, known to be the undeveloped caecum, does not correspond in function to the developed, functioning rodent ceacum. The adult human GIT is roughly divided into three major regions, namely, the stomach, small intestine and the large intestine (colon). In the human fetus the caecum commences deveploment as a conical diversion. As the rest of the intestine grows, caecal growth is arrested and a vermiform appendix remains. In adult humans, the colon, which is haustred throughout its entire length, takes the shape and function of the caecum which is found in many other animals (Fig. 1**B**) (3).

The mouse and rat GIT is divided into four major regions, namely the stomach, small intestine, caecum, and colon (Fig. 1**A**). In contrast to the human stomach, the

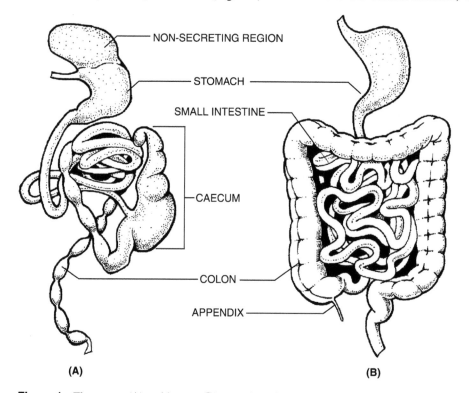

Figure 1   The mouse (**A**) and human (**B**) gastrointestinal tracts.

stomach of rats and mice have a large area of nonsecreting epithelium that expands considerably as the animals are eating. In rodents, microbial fermentation is mainly occurring in the caecum. The colon of these animals is not haustred and is less important for microbial fermentation, compared to the caecum. The large intestine of these animals is important for re-absorption of water and formation of fecal pellets.

Microbiota

The physiological properties of the human GIT, with its many unique features provide a vast number of microbial niches. Host factors such as enzymes, mucins, proteases, bile acids dietary factors and regimes contribute to this diversity. The result is a complex microbial community composed of several hundred microbial species (4) that collectively form the GI microbiota. The mammalian GI microbiota forms dense microbial populations, particularly in the posterior part of the intestine (5). The composition of both human and rodent microbiota has been extensively investigated and discussed in several comprehensive studies and reviews (6–8). The microbial profiles of rodents such as rats and mice are in many ways similar to that of other mammals, including humans (5,9). In such rodents, lactobacilli are present in levels of 10^9 colony forming units (CFU) per gram of feces, (5) whereas in humans, the average levels of fecal lactobacilli are usually 10^4–10^6 CFU per gram of feces (10). As described by Finegold et al. (10), diet has impact on population levels of lactobacillus and other microbial groups in humans. Bifidobacteria may be detected in both human (10) and rodent feces (11), however commercial rodent feed may not support GIT colonization by bifidobacteria as much as some other diets (Fig. 2). This suggests that the type of diet should be considered carefully to ensure that the diet used supports the colonization of important microbial groups. The effect of feed composition is further discussed in the section "Conventional Animals."

There are also behavioral differences between various animal species that may contribute to the resulting GI microbiology of these animals. Rodents are known as coprophages, and unless coprophagy is prevented, it is possible that the GIT of these rodents are continuously re-inoculated with their own fecal microorganisms. This behavior, which could possibly affect the microbial profile, may be inhibited by fitting a tail cup which makes the fecal pellets unavailable to the animals (13). Other techniques have been attempted, such as keeping animals on a grid to allow fecal pellets to fall through and become inaccessible, however coprophagic animals, including rats, usually

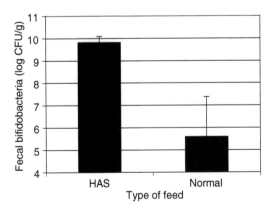

Figure 2 Fecal bifidobacteria of mice (Balb/C) fed high amylomaize starch (HAS) diet, containing 40% starch [AIN 76 (12)] and a commercial rodent feed (Normal). Results presented are the average ± SDV of six animals per group.

collect fecal pellets as they are extruded from anus (14), making such a grid less efficient in preventing coprophagy.

The relative importance of coprophagy, and specifically the rate of microbal re-inoculation, has been investigated in a number of studies. The rat may consume 35–50 percent of the total output of feces, or an even larger proportion if the rat is on a vitamin depleted diet (14). It has been reported that prevention of coprophagy has reduced weight gains in rats and also caused major changes in caecal and fecal lactobacilli, enterococci, and coliforms (15). In another report, prevention of coprophagy made no change in GI microbial profiles, apart from a minor decrease in lactobacilli of the stomach and the lactobacilli of the small intestine (16). A study conducted by Smith (5) indicated that coprophagy has no, or minor effects on gastric microbial populations. These studies, whilst showing dramatically varying conclusions, possibly resulting from varying feed and housing conditions, indicate that coprophagic behavior should remain an important consideration.

The Role of Microbiota on GI Health

The mammalian microbiota has several important functions. It aids in nutrition by degrading complex nutrients and by synthesizing vitamins. It protects against infectious disease, by either preventing invading pathogenic bacteria from establishing in the GIT, or by conditioning of the mucosal immune system. The microbiota may also influence the development of cancer, by modulation of carcinogens, pre-carcinogens or by activation of immunological responses.

Many factors influence the progression and severity of GI infectious disease. Some examples of this are seen in the interaction between various microorganisms and also in their interaction with dietary factors and the host. A pathogen entering the GIT will meet resistance by the microbiota. An invading pathogen is also faced with the host's immune system as well as host factors such as stomach acids, bile acid and enzymes.

The GI microbiota plays an important role in activation of the innate immune system (17–19). Mucosal immune responses are activated as a result of microorganisms interacting with the gut associated lymph tissue (GALT). Interaction of microbes and antigens with GALT leads to a cascade of responses as outlined in the chapter by Moreau. The host mucosal immune system is important in preventing a pathogen from invading the GIT and the translocation of a pathogen to both the mesenteric lymph nodes (MLN) and the internal organs (20–22). The intestinal microbiota and orally administrated probiotics, prebiotics, and other nutrients may also affect the balance of Th1/Th2 cell response, and the production of pro and anti-inflammatory cytokines (23,24). The oral administration of probiotics to rodents may activate macrophages (25) and natural killer (NK) cells (26), in a similar fashion to when they are administered to humans (27). There are a number of described animal models that make research on human GI microbiota possible and bring to light the effects of the human microbiota on nutrition, immunology, and resistance against infections and other diseases (Table 1).

ANIMAL MODELS USED FOR STUDIES
ON THE HUMAN GI MICROBIOTA

Administration Feed and Test Material to the Animal GIT

The effect of specific agents, such as pro and prebiotics or specific chemicals, on the GI microbiota and gut health is monitored after administration of these agents to experimental

Table 1 A Selection of Rodent Models Used in Studies on the Human Microbiota or Specific Microorganisms of Human Origin

Rodent	Strain and/or genotype	Disease induced by the following chemical agent	Type of study	Reference
Mice	C3He/J	N/A	In vivo protein synthesis by probiotics	(28)
Mice	DBA/1	N/A	Effect of probiotics in collagen induced arthritis	(29)
Mice	Balb/C	N/A	Effect of probiotics on phagocytic activity	(30)
Mice	Balb/C	N/A	Probiotics for inhibition of bacterial translocation	(31)
Mice	Balb/C	N/A	Protection against *Salmonella* infection by fecal material from different human donors	(32)
Mice	Balb/C	N/A	Protection against Salmonella infection by specific probiotic strains	(33)
Mice	Balb/C	N/A	Effect of probotics on Th1/Th2 balance	(34)
Mice	Balb/C	N/A	Protection against *Listeria* infection by specific probiotic strains	(35)
Mice	B6C3F1	N/A	Effect of probiotics on Peyers patch lymphocyte populations	(36)
Mice	NMRI	IQ[a], NF[b], AAC[c]	Effect of human intestinal microbiota on mutagenicity and DNA adduct formation	(37)
Mice	IL-2$^{-/-}$[d]	N/A	*Bacteroides vulgatus* for protection against *E. coli* mediated colitis	(38)
Mice	IL-10$^{-/-}$[e]	N/A	Assessment of the role of specific microbes in development of IBD and cancer	(39)
Mice	IL-10$^{-/-}$	N/A	Probiotics for treatment of IBD	(40,41)
Mice	IL-10$^{-/-}$	DSS[f]	Probiotics for treatment of IBD	(42)
Mice	C2H/HN	3-Methylcholanthrene	Probiotics for enhancement of NK cell activity	(26)
Mice	C2H/HN	N/A	Protection against Salmonella infection by specific probiotic strains	(43)
Mice	NIH	N/A	Isolation of probiotic strains that inhibit *Vibrio cholera*	(44)
Mice	Ob/Ob	N/A	Probiotics for treatment of fatty liver disease	(45)
Mice	Balb/C and C3H IL-10$^{-/-}$	DSS	Amelioration of colitis by lysed *E. coli*	(46)
Mice	C57BL/6	Antibiotics	Effect of probotics on Th1/Th2 balance	(24)
Rat	Fisher 344	N/A	Influence of myrosinase on metabolism of glucosinolates	(47)
Rat	Fisher 344	N/A	Effect of dietary fibre on various host enzymes	(48)

(Continued)

Table 1 A Selection of Rodent Models Used in Studies on the Human Microbiota or Specific Microorganisms of Human Origin (*Continued*)

Rodent	Strain and/or genotype	Disease induced by the following chemical agent	Type of study	Reference
Rat	Fisher 344	Azomethane	Effect of carrageenan gel on formation of aberrant crypt foci	(49)
Rat	Sprague-Dawley	Oxalate	Probiotics for reversal of hyperoxaluria	(50)
Rat	Sprague-Dawley	N/A	Probiotics for inhibition of bacterial translocation	(51)
Rat	Sprague-Dawley	DMH[g]	Probiotics for chemically induced cancer	(52)
Rat	Wistar	TNBS/E[h]	Probiotics for improved gut permeability	(53)
Rat	Wistar	N/A	Probiotics for inhibition of bacterial translocation	(54)

[a] 2-Amino-3-methyl-3H:-imidazo [4,5-f] quinoline IQ.
[b] 2-nitroflourene NF.
[c] 2-Amino-alpha-carboline AAC.
[d] Deficient in expression of Interleukin 2 IL–2$^{-/-}$.
[e] Deficient in expression of Interleukin 10 IL–10$^{-/-}$.
[f] Dextran sulfate sodium.
[g] 1-2 dimethylhydrazine DMH.
[h] Trinitrobenzenesulfonic acid/ethanol.
Abbreviations: IBD, inflammatory bowel disease; N/A, not applicable; NK, natural killer cells; Th, T-helper cells.

animals. In this type of studies, the following should be considered: (1) Type of animal feed, (2) Administration of microorganisms (e.g., pathogenic bacteria or probiotics), carcinogens or inflammatory agents, (3) Assessment of animal health and properties of the GI microbiota.

Feed

The effect of specific dietary components are most conveniently assessed after feeding animals a feed containing these compounds. There are several basic feed formulations that may be used for this purpose. The feed which was described by Rickard et al. (12) and modifications thereof (55) are suitable for administration of probiotics. The composition of animal diets may have a significant impact on the composition and activity of the GI microbiota. These effects are further discussed in the section on Human Flora Associated Animals.

Microorganisms

Administration of microorganisms to the GIT of animals can be performed in several ways. Gavage is a method in which microorganisms may be inoculated directly into the stomach of an animal using a gastric probe (33,56). This method allows a known volume containing pathogen, probiotic cultures or complex microbial mixtures to be injected into the stomach. Animals can also be inoculated by administering feed or water containing microorganisms. However, administration through water or feed may not allow for a known inoculum size to be administered at a specific time. In order to avoid this issue, animals can be left for a short time without water or feed to ensure that the contaminated substrate is consumed without delays (43). Alternatively, the microorganisms can be given to animals in a sucrose solution, to improve the rate of consumption (57).

Carcinogens and Inflammatory Agents

Cancer and inflammation may be induced by exposure of animals to specific agents. Administration of carcinogens, pre-carcinogens or pro-inflammatory agents may be introduced orally, by gavage or by feeding animals feeds containing the specific agents. A desired effect may also be induced by intrarectal or systemic inoculation of specific agents.

Conventional Animals

Conventional (CV) animals, being those which have a natural occurring microbiota, may be used to stimulate the human GIT. However, due to differences between the human and animal microbiota, the results from studies in CV animals may be quite different from corresponding studies conducted in humans. CV animal models are useful in studies on orally administrated human microorganisms in vivo, where the activities of the indigenous microbiota are acceptable. CV rodent models have been used for studies of probiotic cultures and particularly their effect on infectious disease. CV animals may also be used to assess the survival of probiotic cultures in vivo. Although the GI conditions of CV animal models are quite different from those of the human, the conditions of the animal GIT are most likely closer to human than what can be simulated in vitro.

There are several examples of studies where CV animals have been used to assess the protective effects of probiotics. These include studies in which *Salmonella* (32,56,58–60), *E. coli* (61,62) and *Listeria* (63,64) have been used as model pathogens. Specific Pathogen Free (SPF) mice have been useful in similar studies where animals were given single or mixed cultures that were considered to be probiotic, before being challenged with *Salmonella*. The progress of infection is determined by monitoring (1) translocation of pathogen to internal organs (31,43), (2) change in animal body weight, and (3) mortality (65) following the challenge. Out of these three general methods, monitoring changes in animal body weight is convenient and relevant in most cases. The virulence of a model pathogen is relevant in this regard, since the virulence will affect the progress of infection. A too virulent strain may induce an unnecessarily severe infection (66). In other cases, human pathogens may not colonize, infect or give a demonstrable effect in an animal model (64,67). If this is the case, then a human pathogen may be replaced with a strain known to be virulent in animals. Examples given so far relate to models used for the monitoring of GI infection and translocation to areas such as MLN and intestinal organs. The protective effect of probiotic cultures can also be monitored by assessing the clearance rate of a specific pathogen from the feces of animals challenged with that pathogen (65). The clearance rate of *Listeria* was measured in the feces of animals that were fed probiotic cultures and meat starter cultures in order to identify specific probiotic cultures that eliminated this particular pathogen (64).

CV animals may also be used as a model system to assess the survival and colonization of probiotic cultures and other microorganisms of human origin. The survival, during passage through the GIT, can be monitored as long as appropriate methods of detection are available. Traditional culturing methods have been important tools used in monitoring the survival of probiotic cultures during GI transit (68–70). Recent years have seen other more efficient detection methods such as molecular probes, which have been developed for the accurate assessments of population sizes of particular probiotic cultures in feces (71). Probes of this type may be used to confirm the identity of particular probiotic strains (72–74) and detect specific strains even at very sparse population levels (75).

GI microbiota is obviously important for the biochemical profile of the GIT. However, simply identifying the survival of microorganisms in feces gives little information about the details of their activity in the GIT. In vivo investigations into the activity of particular microorganisms require methods other than those used for that of detection. Mice or rats may be used to characterize the activity of cultures at specific sites throughout the GIT, something that is very difficult to assess in humans. Traditionally this type of study has been conducted on animals containing microorganisms of interest by describing the biochemical profile of the animal's GI contents. This methodology is adequate in instances where the sum of all microbial and host activities are investigated at the time of sampling. However, it is less appropriate if the activities of specific microorganisms are assessed, where these microorganisms are a part of a complex microbial system. In vivo studies on the activities of specific cultures in a complex ecosystem require different animal models. The development of a lactobacillus free mouse model has provided the opportunity to study the effects of lactobacillus colonization on host physiology, including the effects on fecal bile acids and enzyme activities (76–78). This type of model may be used in studies on the effect of human lactobacillus strains if animals are colonized by strains of human origin.

Germ-Free Animals

Under normal conditions animals are exposed to microorganisms during birth and continue to be exposed to a wide range of microorganisms throughout their lives. These microorganisms form the microbiota, characteristic to CV animals. Hysterectomy at birth, allows the unborn fetus to be transferred from the womb to a sterile chamber. If this process is carried out under sterile conditions, the animal would not be contaminated with microorganisms from the environment. High hygiene standards are required to ensure that animals are maintained and bred under germ-free conditions.

Considerable achievements have been made since the 1970s in investigating the role of the gut microbiota using germ-free animals. Germ-free animals have enabled investigation of animal gut physiology in the absence of the gut microbiota. Studies using germ-free animals have revealed that the gut microbiota is indeed of tremendous importance for the biochemical properties of the GIT, by metabolizing compounds in ingested feed and host factors of mucosal and pancreatic origin. Data from these studies revealed that many physiological and biochemical features of the GIT are indeed the result of microbial gut activity (79,80). The gut of germ-free animals have different physiological and biochemical properties to that of CV animals. The biochemical properties of germ-free and CV animals are often regarded as either germ-free associated characteristics (GAC) or microbiota associated characteristics (MAC). The characteristics of MAC and GAC are described in the chapter by Norin and Midtvedt.

Germ-free animals provide the opportunity to investigate the role of specific microorganisms in the GIT. These microorganisms and their impact on host physiology, can be monitored in an environment that is unaffected by a preexisting microbiota. Ex-germ-free animals have also been used to study the interaction between a controlled composition of microbial species in the GIT (81–84). Germ-free animals have also been useful in research that focuses on the role of the GI microbiota in metabolism of host factors such as mucin and bile acids (85,86). More recently, germ-free technology has been extensively used in research on the effect of specific strains on host immunology (87,88). mucosal physiology and morphology (89–91). Although the absence of a diverse microbiota enables characterization of specific microbes, it cannot be used to characterize their activity in a complex microbial environment. Therefore, animals associated with one or a limited number of strains, may not truly reflect the microbial activity of those that harbor a CV gut ecosystem.

Human Flora–Associated Animals

The establishment of human fecal microbes within animals, provides the opportunity for the study of a microbiota of human origin within these animals. Human flora associated animals (HFA) have proven to be particularly valuable in studies of the metabolic and immunological activities of the human microbiota. Athough HFA animals are valuable for investigations related to the human microbiota, several differences between animal and human physiology may influence colonization by the human microbiota in animal hosts. Such differences may promote host-specific colonization by microorganisms in different animals (92,93). As a result, microbes of human origin may be disadvantaged in the animal GIT, compared to isolates originating from this particular animal species.

HFA animals are created by inoculating germ-free animals with a human fecal homogenate (94). The resulting microbial profile of HFA animals is partly dependent on the differing ability of the various microorganisms in the human fecal sample to colonize the animal GIT. Previous studies have shown that certain microorganisms of human fecal origin were unable to colonize the rodent GIT (95). There may be several reasons for this, such as diets or host factors like transit times and physiological conditions. It has been demonstrated that mice, fed with a commercially available animal feed, may have a reduced, or even undetectable level of bifidobacteria in feces. However, after feeding these mice an alternative diet for several weeks, bifidobacteria could be detected in the mice that were fed sucrose or amylose, with particularly dense populations of bififodobacteria observed in mice that were fed with an amylose rich diet (Fig. 2). This suggests that diet, and specifically dietary ingredients such as certain carbohydrates, are important for the composition of the GI microbota and that previously nondetectable microbial groups may be stimulated to detectable levels. Consideration may be given to the possibility that the growth of microbal populations due to dietary intervention, may be at the expense of less competitive microbial groups.

The colonization of human originated bifidobacteria within germ-free animals is not always successful (95). Hiramaya and co-workers (96) demonstrated that in rodents, the source of fecal material containing bifodobacteria influences the ability of bifodobacteria to colonize the GIT (95). This may be due to the fact that bifodobacteria from different sources possess different natural characteristics. The activity of the human source microbiota that is contained within HFA animals may also be dependent on the cultural origin and dietary habits of the human source (97,98). For instance, fecal material obtained from different human donors has been shown to provide a different degree of effectiveness in protection against *Salmonella* (32). Although this type of model provides a good tool for studying the effects of the human microbiota, it cannot be assumed that the microbial profile of HFA animals is identical to that of the human donor.

HFA rodents are useful in studies of the metabolic activity of the human microbiota. The effects of microbiota on the metabolism of lignans and isoflavones have been investigated in studies using germ-free and HFA rats (98). In similar studies, HFA rats have been used to assess the metabolism of dietary fats (99,100). The usefulness of HFA animals has also been illustrated in studies such as those conducted on the effect of complex carbohydrates on the human microbiota, including the effect of resistant starch (97). Other studies include those relating to the production of short chain fatty acids (85,101) and microbial enzyme activities (102). HFA animals are also valuable for toxicological studies. There are several examples of studies in which HFA animals have been used to assess the effect of the human microbiota on potentially carcinogenic compounds (47,103). Interestingly, both studies indicated that the source of fecal material used to create HFA rats influenced the transformation of pre-carcinogens to carcinogenic componds.

Oozeer and colleagues (28) used a genetically modified *L. casei* strain to assess whether the strain was active throughout the passage of the intestinal tract of HFA mice. This strain was modified by the introduction of genes coding for erythromycin resistance and luciferase. Results from this study indicate that this strain is both metabolically active and able to initiate new protein synthesis during its transit through the GIT. Techniques in transcriptomics and metabolomics are paving the ways for new studies on microbial activity of the gut contents and detailed studies of biochemical properties of host cells lining the GI epithelium.

Surgically Modified Animals

Surgical modification of the GIT gives new opportunities for the study of the GI microbiota. Using surgical procedures, specific parts of the GIT can be removed in order to make modifications to basic physiology. Surgery can also provide the opportunity, by means of cannulation of the GIT, to give repeated post-surgical access to specific sites of the tract. The human GIT lacks some of the areas that may be found in the rodent and porcine GIT, such as the areas of non-secreting epithelium that are found in the stomachs of rodents and pigs. It has been suggested that these areas are the primary sites for *Lactobacillus* colonization within such animals, and that bacterial populations contained at these sites are in fact seeding the intestinal tract with lactobacillus (104,105). If this was to be correct, removal of the non-secreting stomach region could result in a gastric microbial profile that is more in line with that of the human. However, surgical removal of the non-secreting stomach region has no effect on the luminal levels of lactobacillus in either the stomach or in the colon of mice (A. Henriksson, unpublished observations). Therefore, it can be assumed that in mice and possibly other animals, this region is not responsible for the relatively dense lactobacillus populations found in either the intestinal contents or the stomach itself.

The caecum is an important part of the rodent intestine for microbial fermentation. This stands in contrast to the human GIT where the colon is the major site for such fermentation. It has been suggested that this difference is another factor that contributes to various differences between the microbial profiles of rodents and humans (106). However, studies indicate that the microbial biochemical profile of caecectomized mice remains significantly different from that of normal humans or mice (106,107). Most studies on rodents with surgically modified GIT have failed to give microbial profiles that closely resemble that of the human GIT.

Cannulation is performed to provide access to specific sites of the GIT in order to facilitate collection of microbiological samples. Cannulated animals are equipped with a port from which samples can be taken at one or several sites along the GIT. Cannulation has been performed on dogs, pigs, and other larger animals (108–110). This technology has been valuable in assessing the microbial and enzymatic properties of specific areas within the GIT.

Gene Deficient Animals

In recent years, specific mouse strains have been frequently used in studies of colitis. Colitis in mice closely resembles human inflammatory bowel disease (IBD). There are several specific inbred mouse strains that are most useful in this area as they are more likely to develop spontaneous colitis. Strains displaying a disrupted expression of Interleukin (IL)-2, IL-10, and TGF-β have proven to be particularly useful in these studies and have contributed to a broader understanding of the role of the human GI microbiota in IBD. There are a number of different characteristics associated with animals that express irregular cytokine profiles. In mice that are deficient in IL-2, usually when 6–15 weeks old, inflammation occurs in the colon only (111). However, in IL-10 deficient mice, inflammation may also occur in the small intestine as well as the colon (112). TCRα deficient animals have developed inflammation in the caecum, colon, and rectum (113–115), whereas HLA-B27 rats develop inflammation in the colon, duodenum, and caecum (116). These "knock out" models may be used to investigate the effect of the human microbiota, both in terms of the aggravating, as well as alleviating effects on IBD (117).

Immune deficient animals have also been useful in studies relating to the effects of probiotic cultures on colitis. Probiotic cultures investigated in IL-10 deficient mice include *L. salivarius*, *Bifidobacterium lactis* (117). IL-10 knock out mice with colitis have also been used to investigate the effect of a genetically modified (GM) *Lactococcus lactis* that synthesizes IL-10 (42,118).

Chemically Induced Responses

Animals that have been intentionally exposed to specific pro-inflammatory or carcinogenic chemicals have been used in studies on the role of microbiota in the development of both cancer and IBD (Table 1). Cancer, or other malignant abnormalities in the gut mucosa, may be induced by the oral administration of carcinogens. Examples of such carcinogens are 1, 2 dimethylhydrazine (DMH), and N-methyl-N'-nitro-Nitrosogua-din (MNNG). These types of models, which are based on either CV or HFA animals, have also been used to assess the effect of both probiotics and prebiotics on the progression of cancer in its various stages from DNA damage through to differentiation of tissue and formation of tumors.

A study by McIntoch and co-workers (52) investigated the effect of *L. acidophilus* on the incidence of tumor formation as well as the mass of tumors found in animals that had been challenged with DMH. It was demonstrated that the animals that had been given *L. acidophilus* were associated with less tumors than those animals that were given other probiotic cultures. As a result the most effective culture, in terms of protecting animals against cancer, could be isolated out of a range of LAB cultures.

Another way of investigating the effect of microbiota on the formation of cancer is to assess the occurrence of aberrant crypt foci in the intestinal epithelium. In this type of model, increased occurrence of aberrant crypts indicate increased formation of tumors. This model has been used to assess the effect of GI microbiota and specific dietary factors on the development of intestinal cancer (48,119). In similar studies, animals given azoxymethane were used to assess the effect of *L. casei,* of human origin, on the formation of aberrant cells (120). Other studies using 3-methylcholanthrene to induce tumor formation, demonstrated that the same strain delayed the onset of tumor formation. It was suggested that this delay was due to an enhancement of cytotoxicity of NK cells (26). Finally, mucosal carcinogenesis may be assessed by determination of the DNA adduct formation (121). This type of methodology allows assessment of carcinogenesis without visual scoring of aberrant crypts. This method has been successfully used to investigate the effects of human intestinal flora on the mutagenicity of dietary factors by assessing DNA adduct formation (36). Assessment of DNA adduct formation has been used as a tool in investigating the protective effect of potentially probiotic cultures against the formation of cancer (122,123). This type of model provides a cost-effective tool used in studies on the GI microbiota and its role in formation of intestinal cancers.

Although both animal models have been used to demonstrate protection against cancer by probiotic cultures, the difference between how cancer that has developed in the chemically modified animal and how it has developed in the diseased human subject raises questions as to what extent such observations are relevant for the human host. The opportunity to test probiotic cultures in humans that have been intentionally exposed to carcinogens does not exist. However, it is known that some of the probiotic cultures that reduce the incidence of tumor formation in animals have a similar effect on cancer in humans (124,125).

Apoptosis is a mechanism inherent to healthy mucosal cells, which ultimately leads to the death of cancerous cells. The effect of various dietary factors on apoptosis can be assessed in animal models. Several studies have investigated the effect of probiotics and prebiotics on apoptosis. Some of these studies have revealed that prebiotics such as Fructo-Oligosaccharides (FOS) and inulin increase incidence of apoptosis and thereby provide increased protection against the formation of intestinal cancers (126).

A wide range of animal models have been applied to studies on IBD. Naturally occurring animal models have been important tools in studies related to human ulcerative colitis and Crohn's disease. IBD-like symptoms have also been induced chemically. The application of such chemicals may induce ulceration of the intestinal mucosa as well as several immunological responses that are typical to IBD in humans. Simple methods for T-cell induced onset of IBD may be initiated by di-nitro chlorobenzene (DNCP) as described by Glick and Falchuk (127). This method involves both systemic and local application of DNCP. Other chemically induced forms of IBD may be induced by intra rectal inoculation of trinitrobenzene sulphonic acid (TNBS) which is dissolved in alcohol. The latter treatment results in inflammation that lasts for several weeks after exposure to these agents (128). Animals not treated with this agent are normally tolerant to sonicates derived from the heterologus intestine of syngenic littermates (BsH). However, in animals with IBD induced by TNBS, both local and systemic tolerance to BsH is broken (129). Interestingly, this study also demonstrated that tolerance to BsH was abrogated by treatment with IL-10 or antibodies to IL-12.

A study using oral therapy with a probiotic culture had no effect on either the severity of colitis or gut permeability in this TNBS model (53). Similarly, oral therapy with *L. rhamnosus* and a mixture of probiotic cultures has been shown to reduce the extent of colonic damage in TNBS induced colitis (130). However, both *L. rhamnosus* and the culture mixture significantly ameliorated colitis induced by idoacetamide (130). These studies indicate that inflammation induced by a sulfhydryl blocker (e.g., idoacetamide), as described by Rachmilewitz and co-workers (128), may be a better model for assessing the effect of gut microorganisms on colitis.

CONCLUSION

Animal models provide opportunities to investigate the effect of food and pharmaceutical ingredients on GI health and the human microbiota in vivo. A wide range of methods that use animal models have been described, including those based on CV, germ-free, and HFA animals. CV animal models are particularly suitable for studies on the effect of orally dosed probiotic strains, or other microorganisms of human origin, on resistance against infection and aberrant formations in the GI mucosa. Germ-free animals provide opportunities to create HFA animals that are suitable for studies on the effect of the total human microbiota in vivo. HFA animals have been used extensively in studies on the role of the human microbiota in nutrition and metabolism of nutrients. The effect of the microbiota on the immune system can be investigated in chemically modified animals, or specific immune deficient "knock out" models. These models have been used in studies on the effect of the human microbiota and probiotic cultures on the progress of IBD and other diseases that may be caused by a dysfunctional immune system. In addition, chemically modified animals have been used in studies on the effect of probiotic cultures on the development of tumors and other aberrant formations. The usefulness of animals in studies on human microbiota and its effect on GI health has a long standing and clear value. In an age where virtual in vivo simulations are becoming increasingly important, it remains clear that animal models will continue to be highly valuable in research on the functions of the human microbiota and activity of specific microbial strains of human origin.

REFERENCES

1. Luick BR, Penner MH. Nominal response of passage rates to fiber particle size in rats. J Nutr 1991; 121:1940–1947.
2. Sakaguchi E, Itoh H, Uchida S, Horigome T. Comparison of fibre digestion and digesta retention time between rabbits, guinea-pigs, rats and hamsters. Br J Nutr 1987; 58:149–158.
3. Stevens CE. Comparative Physiology of the Vertebrate Digestive System. New York: Cambridge University Press, 1988.
4. Savage DS. Microbial ecology of the gastrointestinal tract. Annu Rev Microbiol 1977; 31:107–133.

5. Smith WH. Observations on the flora of the alimentary tract of animals and factors affecting its composition. J Pathol Bacteriol 1965; 89:95–122.

6. Tannock GW. Normal microbiota of the gastrointestinal tract of rodents. In: Mackie RI, White BA, Isaacson RE, eds. Gastrointestinal Microbiology. New York: Chapman & Hall, 1997 pp. 187–215.

7. Holdeman LV, Good IJ, Moore WEC. Human fecal flora: variation in bacterial composition within individuals and a possible effect on emotional stress. Appl Environ Microbiol 1976; 31:359–375.

8. Moore WEC, Holdeman LV. Human fecal flora: the normal flora of 20 Japanese-Hawaiians. Appl Microbiol 1974; 27:961–979.

9. Drasar BS, Barrow PA. Intestinal Microbiology. Washington: Van Nostrand Reinhold, 1985.

10. Finegold SM, Attebery HR, Sutter VL. Effect of diet on human fecal flora: comparison of Japanese and American diets. Am J Clin Nutr 1974; 27:1456–1469.

11. Mitsuoka T. Taxonomy and ecology of the indigenous intestinal bacteria. Recent Advances in Microbial Ecology. Tokyo: Japan Scientific Societies Press, 1989 pp. 493–498.

12. Rickard KL, Folino M, McIntyre A, Albert V, Muir J, Young GP. Colonic epithelial biology; comparison of resistant starch to soluble and insoluble dietary fibres. Processing Nutr Soc 1994; 18:57.

13. Barnes RH, Fiala G, McGehee B, Brown A. Prevention of coprophagy in the rat. J Nutr 1957; 63:489–528.

14. Barnes RH. Nutritional implications of coprophagy. Nutr Rev 1962; 20:289–291.

15. Fitzgerald RJ, Gustafsson BE, McDaniel EG. Effects of coprophagy prevention on intestinal microflora in rats. J Nutr 1964; 84:155–160.

16. Smith HW, Jones JET. Observations on the alimentary tract and its bacterial flora in healthy and diseased pigs. J Pathol Bacteriol 1963; 86:387–412.

17. Cross ML. Microbes versus microbes: immune signals generated by probiotic lactobacilli and their role in protection against microbial pathogens. FEMS Immunol Med Microbiol 2002; 34:245–253.

18. Schiffrin EJ, Blum S. Interactions between the microbiota and the intestinal mucosa. Eur J Clin Nutr 2002; 56:S60–S64.

19. Borchers AT, Keen CL, Gershwin ME. The influence of yogurt/*Lactobacillus* on the innate and acquired immune response. Clin Rev Allergy Immunol 2002; 22:207–230.

20. Berg RD, Owens WE. Inhibition of translocation of viable *Escherichia coli* from the gastrointestinal tract to the mesenteric lymph nodes and other organs in a gnotobiotic mouse model. Infect Immun 1979; 25:820–827.

21. Berg RD. Inhibition of *Escherichia coli* translocation from the gastrointestinal tract by normal cecal flora in gnotobiotic or antibiotic-decontaminated mice. Infect Immun 1980; 29:1073–1081.

22. Berg RD. Promotion of the translocation of enteric bacteria from the gastrointestinal tracts of mice by oral treatment with penicillin, clindamycin, or metronidazole. Infect Immun 1981; 33:854–861.

23. Perdigon G, Maldonado Galdeano C, Valdez JC, Medici M. Interaction of lactic acid bacteria with the gut immune system. Eur J Clin Nutr 2002; 56:S21–S26.

24. Sudo N, Yu XN, Aiba Y, et al. An oral introduction of intestinal bacteria prevents the development of a long-term Th2-skewed immunological memory induced by neonatal antibiotic treatment in mice. Clin Exp Allergy 2002; 32:1112–1116.

25. Perdigon G, Nader ME, Alvarez MS, Oliver G, Holgado AAPD. Effect of perorally administered lactobacilli on macrophage activation in mice. Infect Immun 1986; 53:404–410.

26. Takagi A, Matsuzaki T, Sato M, Nomoto K, Morotomi M, Yokokura T. Enhancement of natural killer cytotoxicity delayed murine carcinogenesis by a probiotic microorganism. Carcinogenesis 2001; 22:599–605.

27. Gill HS, Rutherfurd KJ, Cross ML. Dietary probiotic supplementation enhances natural killer cell activity in the elderly: an investigation of age-related immunological changes. J Clin Immunol 2001; 21:264–271.

28. Oozeer R, Goupil-Feuillerat N, Alpert CA, et al. *Lactobacillus casei* is able to survive and initiate protein synthesis during its transit in the digestive tract of human flora-associated mice. Appl Environ Microbiol 2002; 68:3570–3574.

29. Kato I, Endo-Tanaka K, Yokokura T. Suppressive effects of the oral administration of *Lactobacillus casei* on the type II collagen-induced arthritis in DBA/1 mice. Life Sci 1998; 63:635–644.

30. Gill HS, Rutherfurd KJ. Viability and dose-response studies on the effects of the immunoenhancing lactic acid bacterium *Lactobacillus rhamnosus* in mice. Br J Nutr 2001; 86:285–289.

31. Gill HS, Shu Q, Lin H, Rutherfurd KJ, Cross ML. Protection against translocating *Salmonella typhimurium* infection in mice by feeding the immuno-enhancing probiotic *Lactobacillus rhamnosus* strain HN001. Med Microbiol Immunol 2001; 190:97–104.

32. Henriksson A, Conway PL. Colonization resistance induced in mice orogastrically dosed with human fecal homogenates from different donors. Microbiol Ecol Health Dis 2001; 13:96–99.

33. Henriksson A, Conway PL. Isolation of the human fecal bifidobacteria which reduce signs of *Salmonella* infection when orogastrically dosed to mice. J Appl Microbiol 2001; 90:223–228.

34. Cross ML, Mortensen RR, Kudsk J, Gill HS. Dietary intake of *Lactobacillus rhamnosus* HN001 enhances production of both Th1 and Th2 cytokines in antigen-primed mice. Med Microbiol Immunol 2002; 191:49–53.

35. Mahoney M, Henriksson A. The effect of processed meat and meat starter cultures on gastrointestinal colonization and virulence of *Listeria monocytogenes* in mice. Int J Food Microbiol 2003; 84:255–261.

36. Pestka JJ, Ha C-L, Warner RW, Lee JH, Ustunol Z. Effects of ingestion of yoghurts containing *Bifidobacterium* and *Lactobacillus* on spleen and peyer's patch lymphocyte populations in the mouse. J Food Prot 2001; 64:392–395.

37. Hirayama K, Baranczewski P, Akerlund JE, Midtvedt T, Moller L, Rafter J. Effects of human intestinal flora on mutagenicity of and DNA adduct formation from food and environmental mutagens. Carcinogenesis 2000; 21:2105–2111.

38. Waidmann M, Bechtold O, Frick JS, et al. *Bacteroides vulgatus* protects against *Escherichia coli*-induced colitis in gnotobiotic interleukin-2-deficient mice. Gastroenterology 2003; 125:162–177.

39. Balish E, Warner T. *Enterococcus fecalis* induces inflammatory bowel disease in interleukin-10 knockout mice. Am J Pathol 2002; 160:2253–2257.

40. Schultz M, Veltkamp C, Dieleman LA, et al. *Lactobacillus plantarum* 299V in the treatment and prevention of spontaneous colitis in interleukin-10-deficient mice. Inflamm Bowel Dis 2002; 8:71–80.

41. McCarthy J, O'Mahony L, O'Callaghan L, et al. Double blind, placebo controlled trial of two probiotic strains in interleukin 10 knockout mice and mechanistic link with cytokine balance. Gut 2003; 52:975–980.

42. Steidler L, Hans W, Schotte L, et al. Treatment of murine colitis by *Lactococcus lactis* secreting interleukin-10, [comment]. Science 2000; 289:1352–1355.

43. Hudault S, Lievin V, Bernet-Camard MF, Servin AL. Antagonistic activity exerted in vitro and *in vivo* by *Lactobacillus casei* (strain GG) against *Salmonella typhimurium* C5 infection. Appl Environ Microbiol 1997; 63:513–518.

44. Silva SH, Vieira EC, Dias RS, Nicoli JR. Antagonism against *Vibrio cholera* by diffusible substances produced by bacterial components of the human fecal microbiota. J Med Microbiol 2001; 50:161–164.

45. Li Z, Yang S, Lin H, et al. Probiotics and antibodies to TNF inhibit inflammatory activity and improve nonalcoholic fatty liver disease, [comment]. Hepatology 2003; 37:343–350.

46. Konrad A, Mahler M, Flogerzi B, et al. Amelioration of murine colitis by feeding a solution of lysed *Escherichia coli*. Scand J Gastroenterol 2003; 38:172–179.

47. Rouzaud G, Rabot S, Ratcliffe B, Duncan AJ. Influence of plant and bacterial myrosinase activity on the metabolic fate of glucosinolates in gnotobiotic rats. Br J Nutr 2003; 90:395–404.

48. Roland N, Nugon-Bandon L, Flinois P, Beaune P. Hepatic and intestinal cytochrome P-450, glutathione-transferase and UDP-glucuronosyl transferase are affected by six types of dietary fiber in rats inoculated with human whole fecal flora. J Nutr 1994; 124:1581–1587.

49. Tache S, Peiffer G, Millet AS, Corpet DE. Carrageenan gel and aberrant crypt foci in the colon of conventional and human flora-associated rats. Nutr Cancer 2000; 37:193–198.

50. Sidhu H, Allison MJ, Chow JM, Clark A, Peck AB. Rapid reversal of hyperoxaluria in a rat model after probiotic administration of *Oxalobacter formigenes*. J Urol 2001; 166:1487–1491.

51. Adawi D, Ahrne S, Molin G. Effects of different probiotic strains of *Lactobacillus* and *Bifidobacterium* on bacterial translocation and liver injury in an acute liver injury model. Int J Food Microbiol 2001; 70:213–220.

52. McIntosh GH, Royle PJ, Playne MJ. A probiotic strain of *L. acidophilus* reduces DMH-induced large intestinal tumors in male Sprague-Dawley rats. Nutr Cancer 1999; 35:153–159.

53. Kennedy RJ, Hoper M, Deodhar K, Kirk SJ, Gardiner KR. Probiotic therapy fails to improve gut permeability in a hapten model of colitis. Scand J Gastroenterol 2000; 35:1266–1271.

54. Eizaguirre I, Urkia NG, Asensio AB, et al. Probiotic supplementation reduces the risk of bacterial translocation in experimental short bowel syndrome. J Pediatr Surg 2002; 37:699–702.

55. Wang X, Brown IL, Evans AJ, Conway PL. The protective effects of high amylose maize (amylomaize) starch granules on the survival of *Bifidobacterium* spp. in the mouse gastrointestinal tract. J Appl Microbiol 1999; 87:631–639.

56. Hudault S, Lievin V, Bernet-Camard MF, Servin AL. Antagonistic activity exerted in vivo and in vitro by *Lactobacillus casei* (strain GG) against *Salmonella typhimurium* C5 infection. Appl Environ Microbiol 1997; 63:513–518.

57. Cohen PS, Rossoll R, Cabelli VJ, Yang S-L, Laux DC. Relationship between the mouse colonizing ability of a human fecal *Escherichia coli* strain and its ability to bind a specific mouse colonic mucous gel protein. Infect Immun 1983; 40:62–69.

58. Paubert-Braquet M, Xiao-Hu G, Gaudichon C, et al. Enhancement of host resistance against *Salmonella typhimurium* in mice fed a diet supplemented with yogurt or milks fermented with warious *Lactobacillus casei* strains. Int J Immunother 1995; XI:153–161.

59. Shu Q, Lin H, Rutherfurd KJ, et al. Dietary *Bifidobacterium lactis* (HN019) ennhances resistance to oral *Salmonella typhimurium* infection in mice. Microbiol Immunol 2000; 44:213–222.

60. Bovee-Oudenhoven IM. Increasing the intestinal resistance of rats to the invasive pathogen *Salmonella enteritidis*: additive effects of dietary lactulose and calcium. Gut 1997; 40:497–504.

61. Shu Q, Gill HS. Immune protection mediated by the probiotic *Lactobacillus rhamnosus* HN001 (DR20) against *Escherichia coli* O157:H7 infection in mice. FEMS Immunol Med Microbiol 2002; 34:59–64.

62. Shu Q, Gill HS. A dietary probiotic (*Bifidobacterium lactis* HN019) reduces the severity of Escherichia coli O157:H7 infection in mice. Med Microbiol Immunol 2001; 189:147–152.

63. DeWaard R, Garssen J, Bokken GCAM, Vos JG. Antagonistic activity of *Lactobacillus casei* strain Shirota against gastrointestinal *Listaria monocytogenes* infection in rats. Int J Food Microbiol 2002; 73:93–100.

64. Mahoney M, Henriksson A. The effect of processed meat and meat starter cultures on gastrointestinal colonization and virulence of *Listeria monocytogenes* in mice. Int J Food Microbiol 2001; 84:255–261.

65. Bernet-Camard MF, Lievin V, Bassart D, Neeser JR, Servin AL, Hudault S. The human *Lactobacillus acidophilus* strain LA1 secrets a nonbacteriocin antibacterial substance(s) active *in vitro* and *in vivo*. Appl Environ Microbiol 1997; 63:2747–2753.

66. Voravuthikunchai SP. Investigation of Caecectomy as a Perturbation of the Mouse Intestinal Ecosystem. Sydney: The University of New South Wales, 1988.

67. Babour AH, Rampling A, Hormaeche CE. Variation in the infectivity of *Listeria monocytogenes* isolates following intregastric inoculation of mice. Infect Immun 2001.

68. Yuki N, Watanabe K, Mike A, et al. Survival of a probiotic, *Lactobacillus casei* strain Shirota, in the gastrointestinal tract: selective isolation from feces and identification using monoclonal antibodies. Int J Food Microbiol 1999; 48:51–57.

69. Saxelin M, Elo S, Salminen S, Vapaatalo H. Dose response colonization of feces after oral administration of *Lactobacillus casei* strain GG. Microb Ecol Health Dis 1991; 4:209–214.

70. Goldin BR, Gorbach SL, Saxelin M, Barakat S, Gualtieri L, Salminen S. Survival of *Lactobacillus* species (strain GG) in human gastrointestinal tract. Dig Dis Sci 1992; 37:121–128.

71. Charteris WP, Kelly PM, Morelli L, Collins JK. Selective detection, enumeration and identification of potentialy probiotic *Lactobacillus* and *Bifidobacterium* species in mixed bacterial populations. Int J Food Microbiol 1997; 35:1–27.

72. Lynch PA, Gilpin BJ, Sinton LW, Savill MG. The detection of *Bifidobacterium adolescentis* by colony hybridization as an indicator of human fecal pollution. J Appl Microbiol 2002; 92:526–533.

73. Lucchini F, Kmet V, Cesena C, Coppi L, Bottazzi V, Morelli L. Specific detection of a probiotic *Lactobacillus* strain in fecal samples by using multiplex PCR. FEMS Microbiol Lett 1998; 158:273–278.

74. Kaneko T, Kurihara H. Digoxigenin-labeled deoxyribonucleic acid probes for the enumeration of bifidobacteria in fecal samples. J Dairy Sci 1997; 80:1254–1259.

75. Su P, Tandianus JE, Park JH, Henriksson A, Dunn NW. A colony hybridization method for *Bifidobacterium lactis* LAFTI B94. Aust Dairy Technol 2003; 58:213.

76. Tannock GW, Crichton C, Welling GW, Koopman JP, Midtvedt T. Reconstitution of the gastrointestinal microflora of *lactobacillus*-free mice. Appl Environ Microbiol 1988; 54:2971–2975.

77. Tannock GW, Dashkevicz MP, Feighner SD. Lactobacilli and bile salt hydrolase in the murine intestinal tract. Appl Environ Microbiol 1989; 55:1848–1851.

78. McConell MA, Tannock GW. Lactobacilli do not influence enzyme activities of duodenal enterocytes of mice. Microb Ecol Health Dis 1993; 6:315–318.

79. Midtvedt T, Gustafsson BE. Microbial convesion of bilirubin to urobilin *in vitro* and *in vivo*. Acta Pathol Microbiol Scand Sect B 1981; 89:57–60.

80. Norin KE, Midtvedt T, Gustafsson BE. Influence of intestinal microflora on the tryptic activity during lactation in rats. Lab Anim Sci 1986; 20:234–237.

81. Corthier C, Dubos F, Ducluzeau R. Prevention of *Clostridium difficile* induced mortality in gnotobiotic mice by *Saccharomyces boulardii*. Can J Microbiol 1986; 32:894–896.

82. Rodrigues ACP, Nardi RM, Bambirra EA, Vieira EC, Nicoli JR. Effects of *Saccharomyces boulardii* against oral infection with *Salmonella typhimurium* and *Shigella flexneri* in conventional and gnotobiotic mice. J Appl Bacteriol 1996; 81:251–256.

83. Wilson KH, Sheagren JN, Freter R, Weatherbee L, Lyerly D. Gnotobiotic models for study of the microbial ecology of *Clostridium difficile* and *Escherchia coli*. J Infect Dis 1986; 153:547–551.

84. Kabir AM, Aiba Y, Takagi A, Kamiya S, Miwa T, Koga Y. Prevention of helicobacter pylori infection by lactobacilli in a gnotobiotic murine model. Gut 1997; 41:49–55.

85. Kleessen B, Hartmann L, Blaut M. Oligofructose and long-chain inulin: influence on the gut microbial ecology of rats associated with a human fecal flora. Br J Nutr 2001; 86:291–300.

86. Chikai T, Nakao H, Uchida K. Deconjugation of bile acids by human intestinal bacteria implanted in germ-free rats. Lipids 1987; 22:669–671.

87. Herías MV, Hessle C, Telemo E, Midtvedt T, Hanson LÅ, Wold AE. Immunomodulatory effects of *Lactobacillus plantarum* colonizing the intestine of gnotobiotic rats. Clin Exp Immunol 1999; 116:283–290.

88. Scharek L, Hartmann L, Heinevetter L, Blaut M. *Bifidobacterium adolescentis* modulates the specific immune response to another human gut bacterium, *Bacteroides thetaiotaomicron*, in gnotobiotic rats. Immunobiology 2000; 202:429–441.

89. Hooper LV, Midtvedt T, Gordon JI. How host-microbial interactions shape the nutrient environment of the mammalian intestine. Annu Rev Nutr 2002; 22:283–307.

90. Husebye E, Hellstrom PM, Sundler F, Chen J, Midtvedt T. Influence of microbial species on small intestinal myoelectric activity and transit in germ-free rats. Am J Physiol Gastrointest Liver Physiol 2001; 280:G368–G380.

91. Kleessen B, Hartmann L, Blaut M. Fructans in the diet cause alterations of intestinal mucosal architecture, released mucins and mucosa-associated bifidobacteria in gnotobiotic rats. Br J Nutr 2003; 89:597–606.

92. Lin JH-C, Savage DC. Host specificity of the colonization of murine gastric epithelium by lactobacilli. FEMS Microbiol Lett 1984; 24:67–71.

93. Conway PL, Henriksson A. Strategies for the isolation and characterization of functional probiotics. Human Health: The Contribution of Microorganisms. London: Springer-Verlag, 1993 pp. 75–94.

94. Mallett AK, Bearne CA, Rowland IR, Farthing MJ, Cole CB, Fuller R. The use of rats associated with a human fecal flora as a model for studying the effects of diet on the human gut microflora. J Appl Bacteriol 1987; 63:39–45.

95. Hirayama K. Ex-germfree mice harboring intestinal microbiota derived from other animal species as an experimental model for ecology and metabolism of intestinal bacteria. Exp Anim 1999; 48:219–227.

96. Hirayama K, Miyaji K, Kawamura S, Itoh K, Takahashi E, Mitsuoka T. Development of intestinal flora of human-flora-associated (HFA) mice in the intestine of their offspring. Exp Anim 1995; 44:219–222.

97. Silvi S, Rumney CJ, Cresci A, Rowland IR. Resistant starch modifies gut microflora and microbial metabolism in human flora-associated rats inoculated with feces from Italian and U.K. donors. J Appl Microbiol 1999; 86:521–530.

98. Bowey E, Adlercreutz H, Rowland I. Metabolism of isoflavones and lignans by the gut microflora: a study in germ-free and human flora associated rats. Food Chem Toxicol 2003; 41:631–636.

99. Ward FW, Coates ME, Cole CB, Fuller R. Effect of dietary fats on endogenous formation of N-nitrosamines from nitrate in germ-free and conventional rats and rats harbouring a human flora. Food Addit Contam 1990; 7:597–604.

100. Kamlage B, Hartmann L, Gruhl B, Blaut M. Linoleic acid conjugation by human intestinal microorganisms is inhibited by glucose and other substrates in vitro and in gnotobiotic rats. J Nutr 2000; 130:2036–2039.

101. Roland N, Nugon-Baudon L, Andrieux C, Szylit O. Comparative study of the fermentative characteristics of inulin and different types of fibre in rats inoculated with a human whole fecal flora. Br J Nutr 1995; 74:239–249.

102. Djouzi Z, Andrieux C. Compared effects of three oligosaccharides on metabolism of intestinal microflora in rats inoculated with a human fecal flora. Br J Nutr 1997; 78:313–324.

103. Lhoste EF, Ouriet V, Bruel S, et al. The human colonic microflora influences the alterations of xenobiotic-metabolizing enzymes by catechins in male F344 rats. Food Chem Toxicol 2003; 41:695–702.

104. Barrow PA, Brooker BE, Fuller BE, Newport MJ. The attachment of bacteria to the gastric epithelium of the pig and its importance in the microbiology of the intestine. J Appl Bacteriol 1980; 48:147–154.

105. Jonsson E, Conway PL. Probiotics for pigs. In: Fuller R, ed. Probiotics, the Scientific Basis. London: Chapman & Hall, 1992:87–110.

106. Voravuthikunchai SP, Lee A. Cecectomy causes long-term reduction of colonization resistance in the mouse gastrointestinal tract. Infect Immun 1987; 55:995–999.

107. Henriksson A, Khaled AKD, Conway PL. The effect of caecectomy on the fecal concentrations of urobilinogen and active trypsin in mice. Microb Ecol Health Dis 1996; 9:61–65.

108. Simmons HA, Ford EJ. Multiple cannulation of the large intestine of the horse. Br Vet J 1988; 144:449–454.

109. Walker JA, Harmon DL, Gross KL, Collings GF. Evaluation of nutrient utilization in the canine using the ileal cannulation technique. J Nutr 1994;2672S–2676S.

110. Hagberg L, Bruce AW, Reid G, Edén CS, Lincoln K, Lidin-Janson G. Colonization of the urinary tract with bacteria from the normal fecal and urethral flora in patients with recurrent urinary tract infections. In: Kass HE, Edén CS, eds. Host-Parasite Interactions in Urinary Tract Infections. Chicago: University of Chicago Press, 1989:194–197.

111. Sadlack B, Merz H, Schorle H, Schimpl A, Feller AC, Horak I. Ulcerative colitis-like disease in mice with a disrupted interleukin-2 gene, [comment]. Cell 1993; 75:253–261.

112. Kûhn RJ, Lohler J, Rennick D, Rajewsky K, Mûller W. Interleukin 10-deficient micedevelop chronic enterocolitis. Cell 1993; 75:263–274.

113. Gaskins HR, Vondrak-Juergens GL, McCracken BA, Woolsey JH. Specific-pathogen-free conditions enhance inflammatory bowel disease in T-cell receptor knockout, but not C3H/HeJBir mice, [comment]. Lab Anim Sci 1997; 47:650–655.

114. Dianda L, Hanby AM, Wright NA, Sebesteny A, Hayday AC, Owen MJ. T cell receptor-alpha beta-deficient mice fail to develop colitis in the absence of a microbial environment. Am J Pathol 1997; 150:91–97.

115. Mombaerts P, Mizoguchi E, Grusby MJ, Glimcher LH, Bhan AK, Tonegawa S. Spontaneous development of inflammatory bowel disease in T cell receptor mutant mice, [comment]. Cell 1993; 75:274–282.

116. Taurog JD, Richardson JA, Croft JT, et al. The germfree state prevents development of gut and joint inflammatory disease in HLA-B27 transgenic rats. J Exp Med 1994; 180:2359–2364.

117. Desreumaux P, Colombel JF. Intestinal flora and Crohn's disease. Ann Pharm Fr 2003; 61:276–281. McCarthy J, O'Mahony L, O'Callaghan L, et al. Double blind, placebo controlled trial of two probiotic strains in interleukin 10 knockout mice and mechanistic link with cytokine balance. Gut 2003; 52:957–980.

118. Steidler L. *In situ* delivery of cytokines by genetically engineered *Lactococcus lactis*. Antonie van Leeuwenhoek 2002; 82:323–331.

119. Hambly RJ, Rumney CJ, Fletcher JM, Rijken PJ, Rowland IR. Effects of high- and low-risk diets on gut microflora-associated biomarkers of colon cancer in human flora-associated rats. Nutr Cancer 1997; 27:250–255.

120. Yamazaki K, Tsunoda A, Sibusawa M, et al. The effect of an oral administration of *Lactobacillus casei* strain shirota on azoxymethane-induced colonic aberrant crypt foci and colon cancer in the rat. Oncol Rep 2000; 7:977–982.

121. Rumney CJ, Rowland IR, Coutts TM, et al. Effects of risk-associated human dietary macrocomponents on processes related to carcinogenesis in human-flora-associated (HFA) rats. Carcinogenesis 1993; 14:79–84.

122. Knasmuller S, Steinkellner H, Hirschl AM, Rabot S, Nobis EC, Kassie F. Impact of bacteria in dairy products and of the intestinal microflora on the genotoxic and carcinogenic effects of heterocyclic aromatic amines. Mutat Res 2001; 480-481:129–138.

123. Horie H, Zeisig M, Hirayama K, Midtvedt T, Moller L, Rafter J. Probiotic mixture decreases DNA adduct formation in colonic epithelium induced by the food mutagen 2-amino-9H-pyrido[2,3-b]indole in a human-flora associated mouse model. Eur J Cancer Prev 2003; 12:101–107.

124. Aso Y, Akaza H, Kotake T, Tsukamoto T, Imai K, Naito S. Preventive effect of a *Lactobacillus casei* preparation on the recurrence of superficial bladder cancer in a double-blind trial. The BLP Study Group. Eur Urol 1995; 27:104–109.

125. Aso Y, Akazan H. Prophylactic effect of a *Lactobacillus casei* preparation on the recurrence of superficial bladder cancer. BLP Study Group. Urol Int 1992; 49:125–129.

126. Hughes R, Rowland IR. Stimulation of apoptosis by two prebiotic chicory fructans in the rat colon. Carcinogenesis 2001; 22:43–47.

127. Glick ME, Falchuk ZM. Dinitrochlorobenzene-induced colitis in the guinea-pig: studies of colonic lamina propria lymphocytes. Gut 1981; 22:120–125.

128. Rachmilewitz D, Simon PL, Schwartz LW, Griswold DE, Fondacaro JD, Wasserman MA. Inflammatory mediators of experimental colitis in rats. Gastroenterology 1989; 97:326–337.

129. Duchmann R, Kaiser I, Hermann E, Mayet W, Ewe K, Meyer zum Buschenfelde KH. Tolerance exists towards resident intestinal flora but is broken in active inflammatory bowel disease (IBD) [comment]. Clin Exp Immunol 1995; 102:448–455.

130. Shibolet O, Karmeli F, Eliakim R, et al. Variable response to probiotics in two models of experimental colitis in rats. Inflamm Bowel Dis 2002; 8:399–406.

15

Born Germ-Free—Microbial Dependent

Elisabeth Norin and Tore Midtvedt
Microbiology and Tumor Biology Center, Karolinska Institutet, Stockholm, Sweden

INTRODUCTION

The essence of research on germ-free life is isolation. Any isolation must be attained mechanically, proven scientifically, and understood philosophically. As early as 1885, Louis Pasteur declared that the concept of a multicellular life free of all demonstrable living microbes could be looked upon as "mission impossible." Germ-free animal research began when Nuttal and Thierfelder in 1895/96 (1) succeeded in keeping a small number of Caesarean-derived guinea pigs alive and germ-free for more than a week. From their work, one can see that the major elements of germ-free research are similar today. They described how to determine the time for partition; developed anesthetic procedures that would not too adversely affect the offspring; and worked out procedures of aseptic Cesarean section and transfer of the offspring from the uterus into a sterile environment and sterilization procedures for food, water, and air, as well as proper methods for testing the sterility of the isolator.

In the decades to follow, several scientists did some work on germ-free multi-cellular organisms, but they all had to work with the first generation. A real breakthrough in germ-free animal research came in 1945, when the second generation of germ-free rats were born at the Lobund Laboratory, Notre Dame, USA. In the following decades, units for germ-free animal research were established in several countries all around the world. Bengt E. Gustafssson's lightweight stainless steel isolators (2) represented a major technical improvement, and so did Trexler's plastic isolators. In the 1980s and 1990s, there has been a temporary decline in germ-free animal research since much resources from bioscience research were allocated to HIV and AIDS. However, in the last 5–7 years, there has been an increased interest in germ-free animals as well as in animals with a specific, known microbiota, i.e., gnotobiotic animals. This increased interest is partly based on progress in molecular methods for studying prokaryot-eukaryote cross-talk in health and disease, partly on the mere fact that investigators, when working with transgenic or knock-out laboratory animals have realized the tremendous influence of the microbiota on the physiological and pathophysiological consequences of the new genetic construct. Therefore, it is easy to forecast that germ-free animals and gnotobiotic technology will be of increasing interest in the years to come. In the following, we will focus on the role of the microbiota on some anatomical structures, physiological, and biochemical functions in the host. Additionally, the immunological impact of the microbiota will briefly be commented on.

TERMINOLOGY

With a slight travesty of the well-known terminology introduced by Claude Bernhard, the mammalian organism itself—a mouse, a rat or a human—can be characterized as a Milieu interieur (MI), a normal microbiota as a Milieu exterieur (ME) and the macroorganism and its microbiota as a Milieu total (MT) (3). In studies on the interplay between MI and ME, two terms, i.e., Microflora Associated Characteristic (MAC) and Germ-free Animal Characteristic (GAC)—have been found to be of considerable value (4). A MAC is defined as the recording of any anatomical structure, physiological, biochemical or immunological function in a macroorganism, which has been influenced by the microbiota. When microorganism(s) influencing the parameter(s) under study are absent—as in germ-free animals, newborns, or in relation to ingestion of antibiotics—the recording of a MAC can be defined as a GAC. Consequently, a germ-free organism is a sum of all GACs, and a normal macroorganism is a sum of MACs. Studies in germ-free animals and healthy newborns have given us the values of GACs, i.e., the MI. When we are investigating conventional organisms—MT—the question *"what have the microbes done,"* can be answered by the equation MT minus MI = ME. A gnotobiotic animal harboring a known microbiota, may present a set-up of some MACs and some GACs, depending on the specific activity of its microbiota.

Over the years, the MAC/GAC concept has been applied in several studies (5). So far, most studies have been related to a phenotypic expression of what the microbes have done. However, the concept is applicable also when studying host-microbe cross-talk on a molecular, genotypic level (6–8). In the following, some major discrepancies between germ-free and conventional animals will be highlighted (Table 1).

GERM-FREE ANIMALS AND DIETARY REQUIREMENTS

Contrary to what is generally believed, germ-free animals require a higher dietary caloric intake than their conventional counterparts. The main reason is very simple. A normal microbiota will break down indigestible dietary substances to compounds that can be absorbed by the host. That is most prominent in ruminants, i.e., the microbiota digest cellulose into short chain fatty acids (SCFAs).

Also contrary to what is generally believed, germ-free animals require a higher intake of nitrogen than their conventional counterparts. The main reason for this is most probably the great loss of non-degraded material from expelled enterocytes that are found in germ-free animals. In conventional animals, the microbiota converts the expelled material into absorbable compounds.

In many germ-free macroorganisms, there might be a demand for an increased dietary intake of some vitamins. Broadly speaking, the gastrointestinal microbiota, placed between the ingesta and the host, may utilize dietary vitamins or produce vitamins themselves.

Among the earliest evidence that the vitamin synthesis is connected to functions by the intestinal microbes was the demonstration that germ-free rats reared without a dietary source of vitamin K developed hemorrhages and hypoprotothrombinemia soon, whereas their conventional controls had normal prothrombin levels and no bleeding tendencies (9).

Table 1 Influences of the Microbiota on Some Intestinal Anatomic, Physiological, and Biochemical Parameters

Parameter	MAC	GAC	Microbes
Anatomical/physiological			
Intestinal wall	Thicker	Thinner	Unknown
Cell kinetics	Fast	Slower	Unknown
Migration motor complexes	Normal	Fewer	Unknown
Production of peptides	Normal	Altered	Unknown
Sensitivity to peptides	Normal	Reduced	Unknown
Caecum size (rodents)	Normal	Enlarged	Partly known
Osmolality	Normal	Reduced	Unknown
Colloid osmotic pressure	Normal	Increased	Unknown
Oxygen tension	Low	High (as in tissue)	Several species
Electropotential Eh, mv	Low (under 100)	High (above 100)	Unknown
Biochemical			
β-aspartylglycine	Absent	Present	Unknown
Bile acid metabolism	Deconjugation	No deconjugation	Many species
	Dehydrogenation	No dehydrogenation	Many species
	Dehydroxylation	No dehydroxylation	Few species
Bilirubin metabolism	Much deconjugation	Little deconjugation	Many species
	Urobilin	No urobilin	One species
Cholesterol	Coprostanol	No coprostanol	Few species
Intestinal gases	Carbon dioxide	Some carbon dioxide	Many species
	Methane	No methane	Few species
	Hydrogene	No hydrogene	Few species
Mucin	Degraded	No degradation	Some species
SCFAs	Large amounts	Far less	Many species
Tryptic activity	Little or absent	High activity	Few species

Abbreviations: MAC, microflora associated characteristic; GAC, germ-free animal characteristic; SCFAs, short chain fatty acids.
Source: From Ref. 8.

Administration of vitamin K_1 restored prothrombin levels to normal values within a few hours, but e.g., vitamin K_3 was less effective. If the germ-free animals were inoculated with an intestinal microbiota from conventional animals, the prothrombin levels were normalized quickly. The vitamin K dependent plasmaprotein factors II, VII, IX, and X are taking part in the blood coagulation cascade. It has been shown that some bacterial strains were effective in reversing vitamin K deficiency (10). It has also been shown, by hindrance of coprophagy in rodents, that the intestinal microbiota supplies the host with parts of the vitamin B complex.

INTESTINAL MICROBIOTA, GROSS ANATOMY, HISTOLOGY, AND MOTILITY

An enlargement of the cecum in the Caesarean-derived guinea pigs was the first anatomical difference observed when the epoch of germ-free research started (1), and similar differences have been observed in all rodents so far investigated. This enlargement might partly be explained by an absence of mucin breakdown in the germ-free animals, partly by a reduced degradation of dietary compounds, such as fiber, and partly by a reduced sensitivity to biogenic amines in germ-free animals (11). Interestingly, it has been shown that a mono-association of germ-free animals with *Clostridium difficile* markedly reduced the cecum size (12).

For years, it was generally accepted that the villi were more slender and uniform in shape and that the crypts were shallower, containing less cells in germ-free animals as compared to their conventional counterparts. Moreover, the lamina propria was supposed to be thinner, and the turn-over rate of epithelial loss was slower. However, most recently it was shown—in germ-free, and conventional rats and mice—that age, gender, and the intestinal compartment actually under study have to be taken into proper consideration before stating significant differences (13–15).

Another striking difference than an enlarged cecum, is a reduction in spontaneous muscular activity in germ-free animals. This may in part be due to a reduced sensitivity to biogenic amines (11), partly also to a reduction in motor migrating complexes (16). Interestingly, it was found that mono-association of germ-free animals with some bacterial species, including a probiotic strain, switches the function from a GAC to an MAC pattern within a few days. Furthermore, the area of endocrine cells in the GI tract is enlarged in germ-free animals (17).

Most recently it has been found that experimental post-surgical intestinal adhesion formation is markedly reduced in germ-free rats (18). After mono-associated with lactobacilli, i.e., a probiotic strain, the animals reacted similar to the germ-free control, whereas they switched to a conventional pattern after being mono-associated with *Escherichia coli*. Obviously, germ-free animals should be used for solving this important question in surgery.

Additionally, germ-free animals may express a compartmentalized reduced osmolarlity in intestinal content, an increased colloid osmotic pressure, a higher oxygen tension, and a higher redox potential than their conventional counterparts. As a consequence of this, strictly anaerobes are often difficult to establish as a monoculture in germ-free animals (this is often a dose-dependency).

BIOCHEMICAL FUNCTIONS AND THE GASTROINTESTINAL MICROBIOTA

Microbial Conversion of Bilirubin to Urobilinogen

Bile pigments, consisting almost exclusively of bilirubin, are the end products of the catabolism of hemoglobin and some other heme-containing enzymes. Bilirubin, taken up by the liver, is conjugated to glucuronate in the liver and excreted with the bile to the intestine, where the bilirubin conjugates are de-conjugated, and transformed to a series of urobilinogens, usually collectively termed *urobilins*. Some intestinal β-glucuronidases are derived from endogenous sources (19), but most of them are of microbial origin (20). The capacity to alter deconjugated bilirubin to urobilins seems to be a rare property among intestinal microorganisms. So far, only one bacterium, a strain of *Clostridium ramosum*, has been found capable of performing this transformation (21,22).

Studies in children as well as adults, in rats and mice, and in pigs and horses show that this is a function normally present in any organism with a normal acting microbiota (8). In infants, this function is established within the first month of life (23). In adults, fecal levels of urobilins are significantly higher in men than in women ($p < 0.05$). Furthermore, in 36 to 50-year-old men the mean level of urobilins is significantly lower than for younger men (< 36 years). In the case of women, the highest fecal values are found in women younger than 35 years of age (24).

Other studies have shown that intake of different antimicrobial drugs used in clinical practice significantly suppressed this MAC (25).

Microbial Bile Acid Metabolism

In all mammals, bile acids are derived from cholesterol in the liver. Cholic acid and chenodeoxycholic acid are common, but many other primary bile acids may be found. The primary bile acids are conjugated, usually with taurine or glycine, sometimes also with sulphate or glucuronate, and excreted into the bile. In the intestinal tract, the conjugated primary bile acids are attacked by microbial enzymes and converted into a variety of metabolites. The so-called secondary bile acids thus formed may then either be excreted with the feces, or reabsorbed, and sometimes further metabolized by hepatic enzymes to so-called tertiary bile acids before re-excretion in the bile. When present in the intestine, the bile acids (primary, secondary or tertiary) are subject to a number of microbial transformations such as deconjugation, desulfatation, deglucuronidation, dehydroxylation, and other oxidation-reduction reactions at the hydroxyl groups (8). In general, the metabolites formed are less water-soluble, less active in forming micelles, and sometimes more toxic to the host.

Over the years, many hypotheses have been brought forward regarding the influence(s) of various bile acids on several host-related signs and symptoms (intestinal motility, cell-turnover, bacterial over-growth, effects similar as pheromones, development of cancer etc). Obviously, further works on these areas are needed.

Microbial Conversion of Cholesterol to Coprostanol

Cholesterol is a component in all mammalian cellular membranes and a precursor of steroid hormones, vitamin D, and bile acids. Pathophysiologically, it is thought to be an important factor in the pathogenesis of atheromatous arterial disease, hypertension, cancer of the large bowel, and other disorders (8). The intestinal cholesterol is derived mainly from two sources—partly from synthesis occurring in the liver and the small intestine and partly from foods of animal origin. The main elimination routes for the plasma cholesterol are biliary excretion of cholesterol into the intestine as well as hepatic conversion of cholesterol to bile acids. The intestinal cholesterol can be absorbed to the entero-hepatic circulation or undergo microbial conversion. The major microbial metabolite is (unabsorbable) coprostanol which is excreted with the feces. The organisms responsible for the conversion are all strictly anaerobic, Gram-positive, nonspore-forming coccoid rods, probably belonging to the genus *Eubacterium*.

By definition, any germ-free organism lacks the intestinal microbial excretion route for cholesterol. From a functional point of view, conversion of cholesterol to coprostanol can be looked upon as a sharp "microbial intestinal knife," influencing the normal entero-hepatic circulation of cholesterol (26). As early as 1959, higher serum cholesterol concentrations were found in germ-free than in conventional rats fed the same diet (27).

Studies in many mammalian species show that this function is present in all animals soon after birth (5). However, data from infants indicate that this function is established—when established—in the second part of their first year. Comparative data from several countries show that one of five healthy adults might be a *"non-excretor"* or *"low-excretor"* of coprostanol. We have hypothesized that a genetically determined receptor determines *whether* an environmental receptor modulation determines *if* a cholesterol converting microbiota will be established. So far, however, the nature of the(se) receptor(s) is still unknown (8).

It has been claimed by some probiotic-producing companies that their microbial products decrease the level of plasma cholesterol, by mechanisms(s) still under discussion. Gnotobiotic animal studies seem very applicable for further mechanistic investigations.

Microbial Degradation of Mucin

Mucin in the GI tract is produced by goblet cells in the mucosa and glandular mucous cells in the submucosa. Mucin consists of a peptide core with oligosaccharide side chains *O*-glycosidically bound, and it has several important physiological and patho-physiological roles. It acts as lubricant, as a barrier and stabilizer for the intestinal microclimate as well as a source of energy for the microbiota. There is growing evidence that the mucin pattern may be a relevant issue to take into account in the pathophysiology of some intestinal diseases, such as ulcerative colitis, Crohn's disease, gastric and duodenal ulceration, and colon adenocarcinoma.

In contrast to conventional rats and healthy adult humans, organisms without any intestinal microbiota excrete large amounts of mucin with their feces (28). The complete degradation of mucin requires various glycosidases and peptidases, and the degradation is a sequential action of several bacterial strains (28,29). However, one *Peptostreptococcus* strain can degrade mucin in vitro and in vivo (30). Additionally, some strains belonging to other species can act upon mucin (31) e.g., *Bifidobacterium* and *Ruminococcus* genera (31) have been isolated and are related to degrading of mucin.

In all mammalian species so far studied, the intestinal microbiota is capable of breaking down mucin (8). In healthy children the function is successively established within the first year of life. It has also been shown that the microbiota might act upon the glycosylation pattern of mucin (32). In fact, alteration in glycosylation was the first observation of a molecular, quorum sensing dependent cross-talk between a host and a single microbial strain present in the GI tract (33). Also regarding this intestinal function, it has been demonstrated that different antibiotics cause disturbance of this microbial function in animals and man (34,35).

Microbial Degradation of Intestinal Enzymes

In the following section, trypsin is used as a model substance for endogenously derived enzymes. It is excreted as a precursor, trypsinogen, from the pancreas, and activated in the small intestine, mainly by brush border enzymes (36). Fecal tryptic activity represents the net sum of processes involving the secretion of trypsinogen, its activation to trypsin, trypsin inactivators, and the presence in the intestine of microbial- and diet-derived compounds and enzymes that inactivate or degrade trypsin and trypsin inactivators. Feces of germ-free rats contain large amounts of tryptic activity, whereas far less is found in their conventional counterparts (37,38). Obviously, intestinal microorganisms are responsible for the inactivation of trypsin, and at least one strain of *Bacteroides distasonis* capable of performing this inactivation, has been described (39).

In most mammals, except man, the intestinal microbiota is breaking down trypsin, yielding fecal tryptic activity to be absent or very low. In man, most adults express tryptic activity in their feces, although the levels are influenced upon by age and gender (24).

Microbial Degradation of β-Aspartylglycine

The biochemical background for the presence of β-aspartylglycine in feces is probably as follows: host-derived intestinal proteolytic enzymes break down some dietary proteins to the β-carboxyl dipeptide β-aspartylglycine. The β-carboxyl dipeptide bonds are then cleaved by proteases derived from microbes (40). This is substantiated by findings in germ-free animals: lambs, piglets, rats, and mice. In feces from germ-free lambs and piglets (Welling, personal communication), and adult germ-free rats and mice (41) β-aspartyl-glycine is always found in germ-free rats and mice, whereas never in samples from their conventional counterparts. Thus, the presence or absence of β-aspartylglycine represents a functional parameter, depending on the presence of dietary precursors, the presence of host-derived proteolytic enzymes, and the presence/absence of microbial derived proteolytic enzymes. Previously it has been shown that the amount of β-aspartylglycine gradually diminishes in feces from ex-germ-free mice, as the number of microbes in their GI microbiota gradually increases (8). This dipeptide has been suggested as an indicator for colonization resistance i.e., a barrier against opportunistic pathogens and other microbes (42). Thus, presence of the dipeptide β-aspartylglycine in feces indicates that the normal intestinal microbial ecosystem is seriously altered.

Microbial Production of Short-Chain Fatty Acids

All Short-Chain Fatty Acids (SCFAs) but acetic acid are microbial anabolic and catabolic products following microbial degradation of many exogenously and endogenously derived compounds in the GI tract of all mammalian species. Endogenous production of acetate may occur in the liver or in the peripheral tissue. However, in intestinal contents nearly all acetate present derives from microbial metabolism. The microbial origin of GI SCFAs has been substantiated by comparative studies in germ-free and conventional animals (43). In conventional animals, the total GI production will partly be influenced by anatomical factors (far more in ruminants than in monogastric animals), partly by dietary habits (more in herbivores than in carnivores). Fecal SCFAs represent the net sum of production, absorption, and possible secretion of SCFAs throughout the GI tract. In short, each mammalian species can be expected to have its own "excretion profile" (8).

The mere fact that so many physiological and clinical roles, ranging from sodium absorption (5,23) to cancer pathogenesis are attributed to SCFAs, make them an extremely interesting parameter. Significant alteration in fecal SCFAs profiles have been found in atopic children (44). Intake of antibiotics (45,46) and dietary changes (47,48) may also cause alterations in fecal SCFAs. Therefore, when studying this parameter in gnotobiotic animals—or in patients—a consequence analysis, as outlined in Table 2, might give an extra incitement for a proper evaluation.

IMMUNOLOGY AND GERM-FREE LIFE

In general, the major difference between germ-free and conventional animals is on a quantitative rather than a qualitative level. This seems to hold true for innate as well as for acquired immunity.

Serum from germ-free animals contains complement in similar amounts as in conventional animals, whereas the levels of specific antibodies are reduced. On a cellular level, polymorphonuclear neutrophiles (PMNs) from germ-free animals are equal to their conventional counterparts with regard to phagocytic capacity (49) and chemotaxis (50), and an apparent reduction in phagocytosis is due to humoral factors, i.e., reduced antibodies (52).

The most striking difference between germ-free and conventional animals is found with regard to the lymphoid immune system. In most—if not all—conventional

Table 2 A Consequence Analysis of One Microbiota Associated Characteristic

Statement: SCFAs are normally produced in high amounts by the intestinal microbiota; they are
  partly absorbed, and partly excreted in feces
Mechanism behind possible consequences
  Biochemical: SCFAs are involved in several metabolic pathways
  Immunological: Uncertain consequences
Place
  Locally
    In the intestinal lumen
    At the mucosa surface
    Within the mucosa cells
  Distant
    In the liver, pancreas, brain etc.
Form
  Direct
    Locally
      Main anions in intestinal content
      Growth promotion of some microbes
      Growth suppression of others
      Growth regulation of mucosa cells
    Distant
      Energy supply to the general metabolism
  Indirect
    Locally
    "Promoted" microbes produce suppressive bacteriocins
    Direct suppression provides niches for other microbes to grow
    Distant
    Metabolic alternations act on production of insulin, etc.
Consequence
  Physiological
    Locally the SCFAs are parts of direct/indirect regulatory
    Mechanisms for water and electrolyte absorption; the net effect is antidiarrheic, involved in
      regulation of carbohydrate metabolism, etc.
  Pathophysiological
    Involved in hepatic coma.
      Probably involved in colonic cancer, ulcerative, and pseudomembraneous colitis, etc.

Abbreviation: SCFA, short chain fatty acid.

mammalian species, there are more lymphoid cells associated with the GI tract than with
the spleen, peripheral lymph nodes, and blood taken together, and gut-associated B cells
account for more than 80% of all B cells in the human body (52). The total daily output of
dimeric IgA is 0.8 g per m of intestine, an amount equivalent to the output of a lactating
mammary gland (52).

    The gut-associated lymphoid tissue (GALT) has to be considered both from the
perspective of its composition and spatial complexity, and an extensive evaluation is
beyond the scope of this review. Interested readers may search in Medline for names such
as Bengt Björkstén, Per Brandtzaeg, John Cebra, and Agnes Wold, among others. In the
future, it is reasonable to assume that germ-free animals will be used in several settings, as
(1) germ-free inbred animals, with and without genetic manipulation of defined
components of their immune systems or their epithelium, (2) isogenic strains of a given
bacterial species expressing defined endogenous or foreign epitopes, and (3) prior or

simultaneous administration of other competing organisms. In all these future experiments, it might be wise to keep in mind that it is a constant *"trialogue"* of interactions between intestinal microbes, epithelium, and GALT. As pointed out elsewhere (6) these interactions are probably dynamic, reciprocal, and combinatorial, making it difficult to separate out a single tune in this cacophony of noise. Utilization of gnotobiotic animals might represent a suitable reductionistic *"noise filter,"* allowing us to study host-microbe cross-talks in greater details. For more information on the role of the intestinal microbiota on the immune system, see the chapter by Moreau elsewhere in this book.

CONCLUSION

For more than a century, germ-free and gnotobiotic animals have been used to investigate the influence of the intestinal microbiota and specific members of the intestinal microbiota on the functioning and health of the host. This has provided much insight into the intricate relation between the host and its microbes. However, as outlined above, much still remains to be studied, and germ-free animals will remain an important tool in the study of the interactions between the intestinal microbiota and the host.

REFERENCES

1. Nuttal GHF, Thierfelder H. Tierisches leben ohne Bakterien im Verdauungskanal. Hoppe Seyler's Zeitsch Physiol Chemie 1896–1897; 33:62–73.
2. Gustafsson BE. Lightweight stainless steel systems for rearing germfree animals. Ann N Y Acad Sci 1959; 78:17–28.
3. Midtvedt T. Influence of antibiotics on biochemical intestinal microflora-associated characteristics in man and animals. In: Gillessen G, Opferkuch W, Peters G, Pulverer G, eds. The Influence of Antibiotics on the Host-Parasite Relationship III. Berlin, Heidelberg: Springer Verlag, 1989:209–215.
4. Midtvedt T, Björneklett A, Carlstedt-Duke B, et al. Germfree research; microflora control and its application to the biomedical sciences. In: Wostmann BS, Pleasants JR, Pollard M, Teah BA, Wagner M, eds. The Influence of Antibiotics Upon Microflora-Associated Characteristics in Man and Animals. New York: Alan R Liss Inc, 1985:241–244.
5. Collinder E, Björnhag G, Cardona M, Norin E, Rehbinder C, Midtvedt T. Gastrointestinal host-microbial interactions in mammals and fish. Comparative studies in man, mice, rats, pigs, horses, cows, elks, reindeers, salmon, and cod. Microb Ecol Health Dis 2003; 15:66–78.
6. Falk GP, Hooper LV, Midtvedt T, Gordon JI. Creating and maintaining the gastrointestinal ecosystem; what we know and need to know from gnotobiology. Microbiol Molecul Biol Rev 1998;1157–1170.
7. Hooper LV, Midtvedt T, Gordon JI. How host-microbial interactions shape the nutrient environment of the mammalian intestine. Ann Rev Nutr 2002; 22:283–307.
8. Midtvedt T. Probiotics, other nutritional factors, and intestinal microflora: microbial functional activities. In: Hansson LÅ, Yolken RH, eds. In: Nestlé Nutrition Workshop Series. Philadelphia: Lippincott-Raven Publishers, 1999:79–96.
9. Gustafsson BE. Vitamin K deficiency in germfree rats. Ann N Y Acad Sci 1959; 78:166–174.
10. Gustafsson BE, Draft FS, McDaniel EG, Smith JC, Fitzgerald RJ. Effects of vitamin K-active compounds and intestinal s in vitamin K deficient germfree rats. J Nutr 1962; 78:461–468.
11. Strandberg K, Sedvall G, Midtvedt T, Gustafsson BE. Effect of some biologically active amines on the cecum wall of germ-free rats. Proc Soc Exp Biol Med 1966; 121:699–702.

12. Gustafsson BE, Midtvedt T, Strandberg K. Effects of microbial contamination on the cecum enlargement of germ-free rats. Scand J Gastroent 1970; 5:309–314.

13. Banasaz M, Alam M, Norin E, Midtvedt T. Gender, age, and microbial status influence upon cell kinetics in a compartmentalised manner. An experimental study in germ-free and conventional rats. Microb Ecol Health Dis 2000; 12:208–218.

14. Banasaz M, Norin E, Midtvedt T. The role of gender, age, and microbial status on cell kinetics in the gastrointestinal tract of mice. An experimental study in germ-free and conventional mice. Microb Ecol Health Dis 2001; 13:135–142.

15. Banasaz M, Åkerlund T, Norin E, Burman L, Midtvedt T. Reduced mitotic activity of intestinal mucosal cells in germ-free rats mono-associated with toxin producing *Clostridium difficile*. Microb Ecol Health Dis 2003; 4:146–152.

16. Husebye E, Hellström PM, Midtvedt T. Intestinal microflora stimulates myoelectric activity of rat wall intestine by promoting cyclic initiation and abnormal propagation of migrating myoelectric complexes. Dig Dis Sci 1992;946–956.

17. Alam M, Midtvedt T. In: Hashimoto K, ed. Microflora and Gastrointestinal Peptides. Shiozawa, Japan: XII IGS Publishing Committee, 1996:409–412.

18. Bothin C, Okada M, Midtvedt T, Perbeck L. The intestinal flora influences adhesion formation around surgical anastomoses. Br J Surg 2001; 88:143–145.

19. Röd TO, Midtvedt T. The origin of intestinal β-glucuronidase in germfree, monocontaminated, and conventional rats. Acta Path Microbiol Scand Sect B 1977; 85:271–276.

20. Gadelle D, Raibaud P, Sacquet E. β-glucuronidase activities of intestinal bacteria determined both in vitro and in vivo in gnotobiotic rats. Appl Environ Microbiol 1985; 49:682–685.

21. Gustafsson BE, Swenander-Lanke L. Bilirubin, and urobilins in germ-free, ex-germ-free, and conventional rats. J Exp Med 1960; 112:975–981.

22. Midtvedt T, Gustafsson BE. Microbial conversion of bilirubin to urobilins in vitro and in vivo. Acta Pathol Microbiol Scand Sect B 1981; 89:57–60.

23. Midtvedt AC, Carlstedt-Duke B, Norin KE, Saxerholt H, Midtvedt T. Development of five metabolic activities associated with the intestinal microflora of healthy infants. J Pediatr Gastroenterol Nutr 1988; 7:559–567.

24. Benno P, Alam M, Collinder E, Norin E, Midtvedt T. Fecal tryptic activity and excretion of urobilinogens in 573 healthy subjects living in Sweden, Norway, and Scotland. Microb Ecol Health Dis 2003; 15:169–175.

25. Saxerholt H, Carlstedt-Duke B, Höverstad T, et al. Influence of antibiotics on the fecal excretion of bile pigments in healthy subjects. J Gastroenterol 1986; 2:991–996.

26. Midtvedt T. Metabolism of endogenous substances. In: Clercq De, ed. Frontiers in Microbiology. From Antibiotics to AIDS. Dordrecht/Boston/Lancaster: Martinus Nijhoff Publishers, 1987:79–87.

27. Danielsson H, Gustafsson BE. On serum-cholesterol levels and neutral fecal sterols in germfree rats. Bile acids and steroids 59. Arch Biochem Biophys 1959; 83:482–485.

28. Gustafsson B, Carlstedt-Duke B. Intestinal water-soluble mucins in germfree, ex-germfree, and conventional animals. Acta Path Microbiol Immunol Scand Sect B 1984; 92:247–252.

29. Hoskins LC, Agustines M, McKee WB, Boulding ET, Kriaris M, Niedermeyer G. Mucin degradation in human colon ecosystems. Isolation and properties of fecal strains that degrade ABH blood group antigens and oligosacharrides from mucin glycoproteins. J Clin Invest 1985; 75:944–953.

30. Carlstedt-Duke B, Midtvedt T, Nord CE, Gustafsson BE. Isolation and characterization of a mucindegrading strain of *Peptostreptococcus* from rat intestinal tract. Acta Path Microbiol Immunol Scand Sec B 1986; 94:292–300.

31. Karlsson KA. Animal glucosphingolipids as membrane attachment sites for bacteria. Ann Rev Biochem 1989; 58:309–350.

32. Freitas M, Axelsson LG, Cayuela C, Midtvedt T, Trugnan G. Microbial-host interactions specifically control the glycosylation pattern in intestinal mouse mucosa. Histochem Cell Biol 2002; 118:149–161.

33. Bry L, Falk PG, Midtvedt T, Gordon JI. A model of host-microbial interactions in an open mammalian ecosystem. Science 1996; 6:1380–1383.

34. Carlstedt-Duke B, Høverstad T, Lingaas E, et al. Influence of antibiotics on intestinal mucin in healthy subjects. Eur J Clin Microbiol 1986; 5:634–638.

35. Carlstedt-Duke B, Alm L, Høverstad T, et al. Influence of clindamycin, administered together with or without lactobacilli, upon intestinal ecology in rats. FEMS Microbiol Ecol 1987; 45:251–259.

36. Norin KE, Midtvedt T, Gustafsson BE. Influence of the intestinal microflora on the tryptic activity during lactation. Lab Anim 1986; 20:234–237.

37. Norin KE. The regulatory and protective role of the normal microflora. In: Grubb R, Midtvedt T, Norin KE, eds. The Normal Microflora and Intestinal Enzymes. New York: Stockton Press, 1988:219–237.

38. Norin KE, Gustafsson BE, Midtvedt T. Strain differences in fecal tryptic activity of germfree and conventional rats. Lab Anim 1986; 20:67–69.

39. Ramare F, Hautefort I, Verhe F, Raibaud P, Iovanna J. Inactivation of tryptic activity by a human-derived strain of *Bacteroides distasonis* in the large intestines of gnotobiotic rats and mice. Appl Environ Microbiol 1996; 62:1434–1436.

40. Welling GW, Helmus G, de Vries-Hospers HG, et al. Germ-free research; microflora control, and its application to the biomedical sciences. In: Wostmann BS, Pleasants JR, Pollard M, Teah BA, Wagner M, eds. Rationale for Use of β-Aspartylglycine as Indicator of Colonization Resistance. New York: Alan R Liss Inc., 1985:155–158.

41. Welling GW, Groen G, Tuinte HM, et al. Biochemical effects on germ-free mice of association with several strains of anaerobic bacteria. J Gen Microbiol 1980; 117:57–63.

42. van der Waaij D, van der Waaij BD. The colonization resistance of the digestive tract in different animal species and in man: a comparative study. Epidemiology and Infections 1990; 105:237–243.

43. Høverstad T, Midtvedt T. Short-chain fatty acids in germ-free mice and rats. J Nutr 1986; 116:1772–1776.

44. Bottcher M, Norin EK, Sandin A, Midtvedt T, Björkstén B. Microflora associated characteristics in faeces from allergig and nonallergic infants. Clin Exp Allergy 2000; 30:1590–1596.

45. Høverstad T, Carlstedt-Duke B, Lingaas E, et al. Influence of ampicillin, clindamycin, and metronidazole on fecal excretion of short-chain fatty acids in healthy subjects. Scand J Gastroenterol 1986; 21:621–626.

46. Høverstad T, Carlstedt-Duke B, Lingaas E, et al. Influence of oral intake of seven different antibiotics on fecal of short-chain fatty acid excretion in healthy subjects. Scand J Gastroent 1986; 21:997–1003.

47. Siigur U, Norin KE, Allgood G, Schlagheck T, Midtvedt T. Effects of olestra on fecal water and short-chain fatty acids. Microb Ecol Health Dis 1996; 9:9–17.

48. Siigur U, Norin KE, Allgood G, Schlagheck T, Midtvedt T. Effect of olestra upon intestinal microecology as reflected by five microflora associated characteristics in man. Microb Ecol Health Dis 1996; 9:297–303.

49. Trippestad A, Midtvedt T. The phagocytic activity of polynuclear leucocytes from germfree and conventional rats. Acta Path Microbiol Scand Sect B 1971; 79:519–522.

50. Trippestad A, Midtvedt T. Chemotaxis of polymorphonuclear leucocytes from germfree rats and generation of chemotactic activity in germfree rat sera. Clin Exp Immunol 1971; 8:639–646.

51. Midtvedt T, Trippestad A. Specificity of opsonic and bactericidal response of gnotobiotic rat sera. Acta Path Microbiol Scand Sect B 1971; 79:291–296.

52. Brandtzaeg P, Halstensen TS, Kett K, et al. Immunobiology and immunopathology of human gut mucosa; humoral immun intraepithelial lymphocytes. Gastroenterology 1989; 97:1562–1584.

16

Modifying the Human Intestinal Microbiota with Prebiotics

Ross Crittenden
The Preventative Health Flagship, Food Science Australia, Werribee, Victoria, Australia

Martin J. Playne
Melbourne Biotechnology, Hampton, Victoria, Australia

INTRODUCTION

The aim of both prebiotic and probiotic functional food ingredients is to improve the health of consumers by selectively altering the composition and/or activity of microbial populations within the gastrointestinal tract. While the probiotic approach endeavors to directly deliver supplemental beneficial bacteria to the gut, prebiotics offer an alternative strategy. Rather than supplying an exogenous source of live bacteria, prebiotics aim to selectively stimulate the proliferation and/or activity of desirable bacterial populations *already resident* in the consumer's intestinal tract.

The prebiotic strategy offers a number of practical and theoretical advantages over modifying the intestinal microbiota using probiotics or antibiotics. This chapter aims to provide an overview of the prebiotic approach, modes of action, and an evaluation of their effectiveness in modulating intestinal microbial populations and providing health benefits to consumers. The production, properties and applications of prebiotics are outlined and likely future developments in prebiotics are discussed. However, before exploring the concept of modifying the intestinal microbiota using prebiotics, it is perhaps pertinent to first reflect briefly on why we might want to alter the composition and activity of the intestinal microbiota in the first place.

WHY MODIFY THE INTESTINAL MICROBIOTA?

Far from being inconsequential to our lives, the bacteria residing within our gastrointestinal tracts are highly important to our health and well-being. They provide us with a barrier to infection by intestinal pathogens (1), much of the metabolic fuel for our colonic epithelial cells (2), and contribute to normal immune development and function (3,4). Intestinal bacteria have also been implicated in the etiology of some chronic diseases of the gut such as inflammatory bowel disease (IBD) (5,6). As we age, changes occur in the composition of the intestinal microbiota that may contribute to an increased level of undesirable microbial metabolic activity and subsequent degenerative diseases of the intestinal tract (7,8).

Modifying the composition of the intestinal microbiota to restore or maintain a beneficial population of micro-organisms would appear to be a reasonable approach in cases where a deleterious or sub-optimal population of micro-organisms has colonized the gut. The difficulty facing intestinal microbiologists is trying to determine what constitutes a "normal," healthy intestinal microbiota. A switch in recent years from culture-based, phenotypic examination of microbial ecosystems to the application of culture-independent, molecular techniques has helped speed progress. It has also provided new insights into the great diversity of bacteria within the human intestinal tract. Historical estimates based on culture methods did recognize the complexity of the ecosystem, placing the number of bacterial species within the gastrointestinal microbiota at around 400, dominated by perhaps 30–40 (9). However, it is now believed to be far richer, with the number of identified taxa expected to eventually exceed 1000 (10).

It is clear that we are only at the very beginning of understanding the role of individual bacterial populations in health and disease and their interactions with each other, the host, and the diet. Addressing these fundamental questions is an essential prerequisite to targeted disease intervention strategies involving modification of the intestinal microbiota. While acknowledging that the science of manipulating the intestinal microbiota to achieve improved health is still very much in its infancy, progress is being made, and strategies that may lead to tangible health benefits in specific populations are emerging.

THE PREBIOTIC STRATEGY TO MODIFYING THE INTESTINAL MICROBIOTA

For a variety of reasons, the two bacterial genera most often advocated as beneficial organisms with which to augment the intestinal microbiota are lactobacilli and bifidobacteria, both of which are common members of the human intestinal microbiota (11,12). These bacteria are numerically common, non-pathogenic, non-putrefactive, non-toxigenic, saccharolytic organisms that appear from available knowledge to provide little opportunity for deleterious activity in the intestinal tract. As such, they are reasonable candidates to target in terms of restoring a favorable balance of intestinal species.

While the probiotic strategy aims to supplement the intestinal microbiota via the ingestion of live bacteria, the prebiotic strategy aims to stimulate the proliferation and/or activity of beneficial microbial populations already resident in the intestine. The characteristics shared by all successful prebiotics is that they remain largely undigested during passage through the stomach and small intestine and selectively stimulate only beneficial populations of bacteria in the colon. That is not to say that prebiotics cannot be theoretically designed to target bacteria within the stomach and small intestine, but rather those currently developed tend to target bifidobacteria, which predominantly reside in the colon. Importantly, prebiotics should not stimulate the proliferation or pathogenicity of potentially deleterious micro-organisms within the intestinal microbiota. To date, most prebiotics have been non-digestible carbohydrates, particularly oligosaccharides. Since the prebiotics identified to date promote the proliferation of bifidobacteria in particular, they are often referred to as bifidogenic factors or bifidus factors. Historically, lactobacilli and bifidobacteria have been targeted as beneficial organisms with which to augment the intestinal tract. However, as discussed later in this chapter, the manipulation more broadly of the metabolic activity of the microbiota is of increasing interest for improving intestinal health (13).

A number of largely prophylactic health targets have been proposed for prebiotics that, as might be expected, overlap considerably with the targets of probiotic interventions. The mechanisms of action remain largely theoretical, but rational hypotheses have been developed as our understanding of the intestinal microbiota has advanced. Proposed benefits in the gut include protection against enteric infections, increased mineral absorption, immunomodulation, trophic and anti-neoplastic effects of short chain fatty acids (SCFA), fecal bulking, and reduced toxigenic microbial metabolism (Figs. 1–4).

A BRIEF HISTORY OF THE DEVELOPMENT OF BIFIDUS FACTORS AND PREBIOTICS

Bifidogenic or bifidus factors were recognized as early as 1954 with Gyorgy et al. (14,15) describing such components in milk and colostrum, including a range of amino sugars and non-glycosylated casein peptides. Glycoproteins from whey were also shown to have bifidogenic potential (16) along with lactoferrin (17,18). Bifidogenic effects have been reported for pantethine from carrot extracts (19,20) and for 2-amino-3-carboxy-1,4-naphthoquinone (ACNQ), a compound isolated from *Propionibacterium freudenreichii* (21,22).

Interest in bifidogenic compounds accelerated with the identification of non-digestible oligosaccharides (NDOs) in human milk as major factors responsible for maintaining an intestinal microbiota numerically dominated by bifidobacteria in breast-feeding infants. In contrast, infants fed cow's milk-based formula developed a mixed microbiota, including higher levels of potentially deleterious organisms (23,24). Human

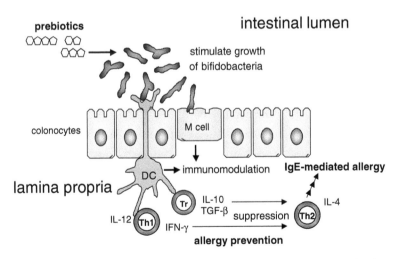

Figure 1 Proposed mechanisms of immunomodulation by prebiotics for the prevention of IgE-mediated food allergies that are mediated by a skewing of the immune response at the T helper (Th) cell level towards a Th2 response. Prebiotics stimulate the growth of bifidobacteria that are sampled by the gut-associated lymphoid tissue via M-cells or dendritic cells (DC). The commensal bacteria drive a counterbalancing Th1 response producing interferon-γ (IFN-γ), and/or a tolerogenic response by regulatory T-cells (Tr) producing the anti-inflammatory cytokines interleukin-10 (IL-10) and transforming growth factor-β (TGF-β) that quell the allergenic Th2 response.

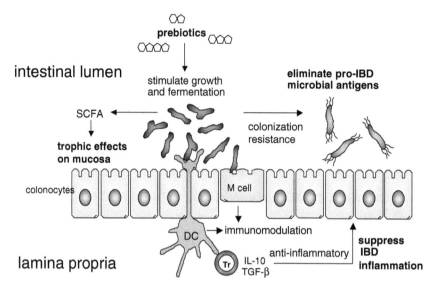

Figure 2 Proposed mechanisms by which prebiotics may ameliorate inflammatory bowel disease (IBD). *Abbreviations*: DC, dendritic cell; IL, interleukin; SCFA, short chain fatty acids; Tr, regulatory T cell; TGF, transforming growth factor.

milk oligosaccharides (HMOs) (discussed later in this chapter) were then, and remain today, too complex to be synthesized commercially. However, other NDOs were shown to replicate the bifidogenic effect of milk oligosaccharides. The Japanese research community in particular studied the ability to modify the intestinal microbiota using lactulose and fructo- and galacto-oligosaccharides. Although often lacking rigorous design, early studies (25–30) at least provided the impetus for later, randomized controlled studies that have demonstrated the notion that some NDOs selectively promote the proliferation of bifidobacteria in the intestinal tract.

Concurrently in the late 1980s and early 1990s, interest was rising in the use of probiotics to modify the intestinal microbial balance. The term "prebiotic" was coined by

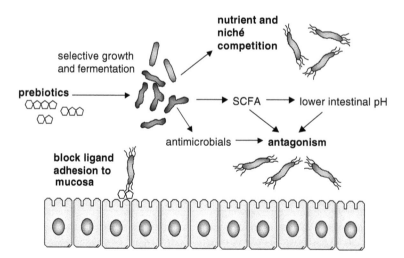

Figure 3 Proposed mechanisms by which prebiotics may enhance colonization resistance against bacterial pathogens in the gastrointestinal tract. *Abbreviation*: SCFA, short chain fatty acids.

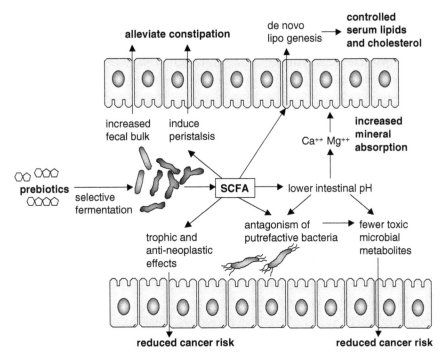

Figure 4 Proposed mechanisms by which the selective fermentation of prebiotics and subsequent production of short chain fatty acids (SCFA) improve bowel habit, increase dietary mineral absorption, and may reduce the risk of colon cancer.

Gibson and Roberfroid in 1995 (31) and effectively linked these two concepts for promoting beneficial populations of intestinal bacteria. Gibson and Roberfroid (31) broadened the narrow bifidogenic target to include the specific stimulation of any potentially beneficial microbial genera. There is an obvious potential for synergy between prebiotic and probiotic ingredients, and hence, foods containing both prebiotic and probiotics ingredients were termed "synbiotics."

CURRENTLY AVAILABLE PREBIOTIC CARBOHYDRATES

The prebiotics most commonly used as functional food ingredients are non-digestible oligosaccharides (NDOs), of which a variety of types are commercially available (32). Most of these NDOs are natural components of many common foods including honey, milk, and various fruits and vegetables (32–34). Commercially, they are produced as food ingredients by four main processes:

1. Extraction and purification from plants, e.g., soybean oligosaccharides and inulin from chicory
2. Controlled enzymatic degradation of polysaccharides, e.g., xylo-oligosaccharides, isomalto-oligosaccharides, and some fructo-oligosaccharides
3. Enzymatic synthesis from disaccharides, e.g., some fructo-oligosaccharides, galacto-oligosaccharides and lactosucrose (32,33)
4. Chemical isomerization, e.g., lactulose.

In nearly all cases, the commercial oligosaccharide products contain a range of oligosaccharide structures of differing molecular weights and often with a variety of glycosidic linkages between sugar moieties. To date, the largest number of reported studies and the most consistent evidence accumulated for prebiotic effects have been for fructo-oligosaccharides and the polyfructan inulin (34–39). Good evidence from human studies also exists for the prebiotic activities of galacto-oligosaccharides (40–43) and lactulose (44–47). Boehm and Stahl (48) have summarized 28 of the human studies conducted on the physiological effects of galacto-oligosaccharides and fructans (fructo-oligosaccharides and inulin). Most of these studies were between one and three weeks in duration. Commercial food-grade oligosaccharide was fed at between 8 and 15 g/day in most experiments. Higher levels (40 g/day) were fed when inulin was used. They list 14 trials on galacto-oligosaccharides involving 298 adults and 27 infants, and another 14 with fructans involving 238 adults and 34 infants. In nearly all cases, only healthy volunteers were tested.

A number of other NDOs, to which less rigorous study has been so far applied, have at least indications of prebiotic potential. These include lactosucrose (49–52), gluco- (53), xylo- (54,55), isomalto- (56–59), and soybean oligosaccharides (60–63). Additionally, bifidogenic effects have been reported for lactitol (45), polydextrose (64) and glucono-δ-lactone (65) in small human feeding studies.

Evidence that some dietary fibers, such as resistant starches (66–72), arabinoxylan (73,74) and plant gums (75) have prebiotic potential is accumulating, but to date remains limited largely to in vitro and animal studies. These large carbohydrates may have some advantages in the intestinal tract over rapidly fermented oligosaccharides. They minimize rapid gas formation and osmotic effects in the gut, which can lead to intestinal discomfort, flatulence and diarrhea at high doses of NDOs (typically above 15–20 g per day). Additionally, they persist as substrates for saccharolytic fermentation more distally in the colon where carbohydrate limitation is believed to promote toxigenic microbial reactions leading to an increased risk of colorectal cancer (76–79).

The molecular structure of the prebiotic can be expected to determine its physiological effects as well as which microbial species are able to utilize it as a carbon and energy source in the bowel. However, it appears that despite the diversity in molecular sizes, sugar compositions, and structural linkages within the range of prebiotic carbohydrates, it is the bifidobacteria that are almost universally observed to respond. Some established and emerging prebiotics, including lactulose (46), galacto-oligosaccharides (40,80,81) and resistant starches (69,71) have been sporadically reported to stimulate intestinal *Lactobacillus* populations. Indeed, some lactobacilli have been shown to possess the metabolic machinery to use fructo-oligosaccharides (82,83). Despite this, bifidobacteria remain the major beneficiaries of these substrates in the gut. Given the benefits attributed to probiotic lactobacilli, the development of novel prebiotics directly targeting *Lactobacillus* species remains an opportunity. The rise in these beneficial bacterial populations during prebiotic feeding has often been shown to be accompanied by concomitant reductions in the numbers of putrefactive organisms such as clostridia and *Bacteroides* spp. and Enterobacteriaceae (31,44–46,60,84), possibly due to antagonism by SCFA production, acidification of the colonic environment, or direct antagonism (Figs. 3–4).

MODIFYING THE INTESTINAL BIFIDOBACTERIUM POPULATION

The composition of the human intestinal microbiota changes naturally with age, and prebiotic strategies need to be targeted to reflect the desired outcome for specific demographics. This section describes how prebiotics might provide benefits for specific human populations in relation to the characteristics of their own particular intestinal microbiota, and outlines some of the evidence for health effects accumulated so far. A brief summary of the main physiological effects of prebiotics is listed in Table 1.

Infants

Bifidobacteria colonize the human intestinal tract during or soon after birth and in breast-fed infants they eventually dominate the microbiota (85). The numerical dominance of bifidobacteria is induced by bifidogenic components in breast milk, including oligosaccharides (85,86). Indeed, human milk oligosaccharides (HMOs) are the original prebiotics. The concentration of oligosaccharides found in human milk (5 to 10 g/L) is about 100 times that found in cow's milk (0.03 to 0.06 g/L). HMOs are complex with more than 130 identified structures (87). Each individual oligosaccharide is based on a variable

Table 1 Summary of Physiological Effects of Prebiotics

Level of substantiation	Effect	Comments
Strong	Increase intestinal numbers of bifidobacteria	Magnitude of bifidogenic effect is inversely proportional to the size of the initial intestinal *Bifidobacterium* population. Best evidence for FOS, inulin, lactulose, and GOS, with emerging evidence for a range of NDOs and some dietary fibers.
	Improved bowel habit	Improved frequency of defecation and stool consistency demonstrated with many prebiotics. Lactulose has a long history of pharmaceutical use as a laxative.
	Alleviate hepatic encephalopathy	Lactulose has a long history of use as a pharmaceutical.
	Increase calcium absorption	Positive results in animal studies and now more consistently in human trials. Prebiotics appear to enhance colonic Ca^{++} uptake. Indications that larger prebiotics with sustained colonic fermentation may be the more effective.
Moderate	Control of serum lipid levels	Consistently positive results in animal studies, but mixed results in human trials. Mechanism appears to be control of de novo lipogenesis via SCFA.
	Prevention of colorectal cancer	Demonstrations of anti-cancer effects in rodent models for a range of prebiotics. Reduced intestinal genotoxicity in human studies.
	Improved colonization resistance	Lactulose effective against chronic *Salmonella* infection. Some evidence from animal studies for other prebiotics against intestinal and systemic infections. Possible deleterious effects of rapid acidification on gut mucosa require investigation.
Weak	Immune modulation	Limited evidence from animal studies for anti-allergy effects. Suggestions of immunomodulation from antiviral effects and enhanced immune responses to vaccination.

Abbreviations: NDOs, nondigestible oligosaccharides; SCFA, short chain fatty acids; FOS, fructo-oligosaccharides; GOS, galacto-oligosaccharides.

combination of glucose, galactose, sialic acid, fucose and/or N-acetylglucosamine, with varied sizes and linkages accounting for the considerable variety (88).

In contrast to breast-fed infants, infants fed cow's milk-based formulae develop a more mixed intestinal microbiota, with lower counts of bifidobacteria and higher counts of clostridia and enterococci (89). Formula-fed infants have also been observed to have higher fecal ammonia and other potentially harmful bacterial products (90,91). The bifidogenic effect of HMOs can be emulated using FOS and GOS (40,41,92). However, there is increasing evidence for roles of HMO outside their bifidogenic impact in the gut. These include blocking adhesion of pathogens to the intestinal mucosa (93–95) and roles in developing cognition (96). Hence, N-acetylneuraminic acid derivatives or sialyl-lactose are also commonly added to infant milk formulae. The complexity of HMOs has thwarted attempts to synthesize their full range of structures commercially, although specific oligosaccharides have been synthesized using chemical and biotechnological approaches (97–100). There is a ready market in infant milk formulas for oligosaccharides that more closely replicate all of the properties of HMOs and research to synthesize them will no doubt continue.

Effects on Immune System Maturation

There is a growing recognition of the importance of the intestinal microbiota to the healthy maturation of the host's immune system, including appropriate programming of oral tolerance to dietary antigens (101). Differences have been observed between the intestinal bacterial populations of healthy infants and those suffering from atopic eczema. These included differences within the genus *Bifidobacterium*, which are found in lower numbers in the feces of allergic infants (102–108) and with a more adult-like species composition dominated by *Bif. adolescentis* (109,110) rather than the usual species associated with the infant intestine such as *Bif. bifidum, Bif. breve*, and *Bif. longum* (= *Bif. infantis*) (111,112).

Recent indications that probiotics may reduce the severity of atopic eczema in infants (113,114) has led to interest in understanding if similar effects can be achieved with prebiotics. The proposed mechanisms are outlined in Figure 1 and involve stimulation of Th1 cells and/or regulatory T cells. Nagura et al. (115) tested the ability of raffinose consumption to re-balance a Th2-biased immune response in a controlled study using an engineered murine model of IgE-mediated allergy to ovalbumin. Feeding a relatively high dose of raffinose stimulated a counterbalancing Th1-type immune response, reduced Th2 cell activity and suppressed the synthesis of serum IgE to ovalbumin in response to long-term allergen challenge. Using a similar model, Yoshida et al. (116) recently reported similar positive results for bifidogenic alginate-oligosaccharides, indicating that prebiotics may be able to replicate the benefits seen for probiotics in allergy prevention.

Adults

The proportion of bifidobacteria in the colonic microbiota drops following weaning and the introduction of solid food. In adults, they account for 1–5% of the total bacteria in feces. Although they form a slightly higher proportion of total bacteria in the caecum (117–121), the total numbers of bifidobacteria per gram of intestinal contents increases approximately 100-fold with passage from the caecum to the colon. In the feces of healthy adults bifidobacteria are found in numbers generally in the order of 10^8–10^{10} cells per gram. (10,122–125). While these figures represent the typical *Bifidobacterium* cell density, a proportion of healthy adults harbor considerably lower numbers of *Bifidobacteria* in their gut (by several orders of magnitude) without any discernable adverse effects (125–128).

It is yet to be determined how the total number of bifidobacteria within a stable microbiota influences the long-term health of the human host. In individuals with naturally low levels of bifidobacteria, other micro-organisms with similar functionalities may occupy a similar niche and fulfill a similar role in the intestinal tract.

It is clear from the number of human feeding studies reported to date that consumption of prebiotics can increase the numbers of bifidobacteria in the colon of adults. For NDOs, consumption of typically 10–15 g/day can induce 10- to 100-fold increases in *Bifidobacterium* numbers (129,130). However, a range of factors may influence the size of any increase in *Bifidobacterium* numbers, the most important being the initial size of the population within the intestinal tract. In comparing different trials conducted using fructo-oligosaccharides, Rao (130) observed that the size of the bifidogenic response was inversely proportional to the size of the initial *Bifidobacterium* population rather than showing a strong dose response. In individuals colonized with an already large population of bifidobacteria, prebiotic consumption appears not to increase the total *Bifidobacterium* population size further.

Bif. adolescentis, *Bif.catenulatum/pseudocatenulatum*, *Bif. bifidum*, and *Bif. longum* are the most frequently reported *Bifidobacterium* species in the intestines of adults, with considerable variation between individuals (121,125,126,131,132). To date, no clear rationale for promoting one species of *Bifidobacterium* over others has emerged. Indeed, it may be quite difficult to achieve major shifts within the population dynamic of bifidobacteria at the species level even if this was desirable. In one study to investigate this, feeding 8 g/day of galacto-oligosaccharides to healthy adult volunteers did not result in marked changes in the composition of their intestinal bifidobacterial populations at the species level (133–135). Similarly, despite observing increases in total *Bifidobacterium* numbers, Harmsen et al. (136) also saw no changes in the species composition of bifidobacteria in a study where adult volunteers were fed 9 g/day of inulin. The species composition within intestinal bifidobacteria has been shown to remain fairly stable over many months in adults (10,121,125,137,138) suggesting that day-to-day fluctuations in diet have little impact on the species dynamic.

Even if they do not significantly alter the bacterial population dynamics in all individuals, prebiotics may still be effective in providing benefits to the consumer if they beneficially modulate the metabolic activity of the microbiota. Hypothetical examples might be increased production of SCFA or vitamins that benefit the health of the colonic epithelium, or synthesis of antagonistic metabolites that augment colonization resistance against pathogens. Tannock et al. (139) used molecular techniques to investigate both phylogenetic (DNA-DGGE) and metabolic (RNA-DGGE) changes in the intestinal microbiota induced by galacto- or fructo-oligosaccharides. While no discernable changes were observed in bacterial communities using DNA-DGGE (nor increases in total *Bifidobacterium* numbers by traditional culturing), RNA-DGGE analysis revealed that the prebiotics increased the activity of some bacterial groups including bifidobacteria. A current research need is to identify metabolic activities of the microbiota that affect the health of the host (positively or negatively) and to demonstrate that these can be specifically modulated with prebiotics in situ.

Prebiotics in the Treatment of Inflammatory Bowel Disease

A genetic predisposition to develop an over-zealous inflammatory immune response to components of the intestinal microbiota has been implicated in the etiology of IBD (140). Elimination of specific bacterial antigens, immunomodulation, and trophic effects of SCFA on the intestinal epithelium have all been proposed as mechanisms by which prebiotics could alleviate IBD (Fig. 2). The size of the intestinal *Bifidobacterium* population has been shown to be relatively small (141,142) in subjects afflicted with IBD, although cause and effect links between disease and a diminished intestinal *Bifidobacterium* population remain to be established. Interventions with prebiotics have shown some benefit in ameliorating inflammation in both animal and human feeding trials. Using differing rodent models of IBD, a number of research groups have demonstrated amelioration of inflammation using prebiotic interventions. These include studies with lactulose (143), inulin (140) and fructo-oligosaccharides (144). In contrast, Holma et al. (145) observed no reduction in inflammation by intervention with galacto-oligosaccharides despite an increase in *Bifidobacterium* numbers.

In addition to NDOs, larger polysaccharides with prebiotic potential have also been shown to have promise in the treatment of IBD. Resistant starch was demonstrated to ameliorate IBD in rodent models of disease (146,147), and in one study (147) out-performed a diet with an equivalent dose of fructo-oligosaccharides. Additionally, an arabinoxylan-rich germinated barley product has been reported to have benefits in the treatment of active IBD. This ingredient was shown to induce the proliferation of bifidobacteria in the human intestine (148), consistent with other in vitro and animal studies of the fermentation of arabinoxylans by intestinal bacteria (73,74,149). In rodent models of IBD, and in two small, non-blinded human studies of subjects with ulcerative colitis, consumption of the germinated barley product reduced inflammation (150–153).

These results suggest that prebiotics have at least some potential to benefit IBD sufferers. However, convincing evidence of a consistent clinical benefit in the treatment of IBD remains to be demonstrated in large, randomized, double-blind, placebo-controlled trials.

Elderly

The proposed benefits of prebiotics for the elderly have been based on early studies using culture methods that showed *Bifidobacterium* levels substantially decreased as a proportion of the total fecal microbiota in elderly Japanese, while the numbers of putrefactive bacteria such as clostridia increased (154). These findings have only recently been re-addressed using modern bacteriological and molecular techniques, with mixed results. While a study from the United Kingdom (155,156) supported the earlier observations of a drop in *Bifidobacterium* numbers, other studies of elderly Italians and Dutch did not show any reduction in the size of the *Bifidobacterium* population (157,158). Still, prebiotics may be of benefit in elderly subjects with a low level of bifidobacteria and high levels of deleterious bacteria. One such group is elderly people with *Clostridium difficile*-associated diarrhea (CDAC) who have been shown to have a diminished bifidobacterial population (159). Prebiotic intervention may eventually prove to be beneficial in the prevention of such conditions in the elderly.

Prebiotics are also hypothesized to have potential to provide protection for degenerative diseases in the elderly such as colon cancer and osteoporosis, and experimental evidence for benefits in these conditions are discussed in a later section of this chapter.

SYNBIOTICS

Products containing both prebiotic and probiotic ingredients are termed "synbiotics" due to the obvious potential for synergy between these ingredients. Although the prebiotic may not necessarily be utilized by the included probiotic bacteria in a synbiotic food, attempts have been made to maximize potential synergies by using complementary prebiotics that may aid the colonization and in situ functionality of the included probiotic strains. In a study in pigs, Brown et al. (160) showed that the inclusion of resistant starch or oligosaccharides in a synbiotic combination with a probiotic *Bifidobacterium* resulted in significantly higher numbers of bifidobacteria in the intestinal tract than with feeding of the probiotic alone. Continuing prebiotic feeding after the cessation of probiotic feeding also significantly extended the intestinal persistence of the probiotic.

In terms of potential health benefits, synbiotic combinations have shown enhanced impact over feeding solely probiotics or prebiotics in rodent models investigating anti-cancer effects (161–164) and colonization resistance (165). To increase the specificity of synbiotics for the added probiotic strains, bifidobacteria have themselves been exploited as a source of enzymes to synthesize NDOs (166–168) including synthesis of galacto-oligosaccharides in yoghurt during fermentation (166).

MECHANISMS OF THE BIFIDOGENIC EFFECT

The mechanism(s) by which prebiotics promote the relatively specific proliferation of bifidobacteria remain speculative. It is probably due to the efficient utilization of these carbohydrates as carbon and energy sources by bifidobacteria relative to other intestinal bacteria, and their tolerance to the SCFA and acidification of the microenvironment resulting from fermentation. Additionally, many bifidobacteria adhere to large granular substrates such as resistant starch and these may provide a site for colonization as well as a substrate (13,169). The ability of bifidobacteria to use a wide variety of oligosaccharides and other complex carbohydrates reflects their evolution in the hind-gut of humans and animals where the ability to metabolize a diverse range of food and host-derived complex carbohydrates and glycoproteins provides a competitive advantage. Analysis of the *Bif. longum* genome has revealed a large number of proteins specialized for the catabolism of carbohydrates (170).

Interestingly, while many bifidobacteria grow well when cultured with prebiotic oligosaccharides as their sole carbon and energy source, they often do not grow when supplied only with the monosaccharides from which these oligosaccharides are composed (74,171,172). This physiology may be another consequence of their evolution in an environment with a limited availability of simple sugars. It suggests that bifidobacteria lack transport mechanisms for many monosaccharides and import prebiotic oligosaccharides before hydrolyzing and metabolizing them. This presumably minimizes the availability of released simple sugars for cross-feeding by other intestinal bacteria and may be another factor contributing to the specific bifidogenic effect of NDOs.

ADVANTAGES OF THE PREBIOTIC STRATEGY

While both antibiotics (see chapter by Sullivan and Nord) and probiotics (see chapter by Khedkar and Ouwehand) can modify the intestinal microecology, the prebiotic strategy offers a number of advantages over these two approaches.

Advantages over Antibiotics

Eliminating pathogenic groups with antibiotics is an obvious approach to beneficially modifying the intestinal microbiota. However, perturbation of indigenous microbial ecosystems caused by the collateral damage to desirable populations can lead to potentially serious side effects. These include antibiotic-associated diarrhea and pseudomembranous colitis involving overgrowth of *Clostridium difficile* as well as oral or vaginal candidiasis (173–175). Prebiotics and probiotics can ameliorate the potential of opportunistic infections caused by disturbances to the microbiota by restoring populations of beneficial bacteria (176–179). No long-term side effects have been reported for either prebiotic or probiotic ingredients, enabling their safe long-term use in prophylactic strategies to minimize disease. In contrast, long-term use of antibiotics may elicit a range of side-effects including liver damage, hypersensitivity, sensitivity to sunlight, and increasing the risk of developing antibiotic-resistant bacterial strains (180,181). This latter risk is particularly serious, and applies also to the sub-therapeutic use of antibiotics in intensive livestock farming in order to minimize infections and maximize yields, particularly for poultry and pork. Alternatives to antibiotics are urgently sought, and there has been considerable interest in the use of both prebiotics and probiotics in animal feeds to aid production. Although they have shown some promise (182,183), further research is needed into their application within an overall management strategy in order to match the performance of antibiotics.

Advantages over Probiotics

Storage Stability

With the exception of some mechanisms of immunomodulation, the theoretical basis for many of the anticipated probiotic effects of bifidobacteria rely on the bacteria being viable in the intestinal tract. Currently, probiotics are limited by their stability largely to fresh food products such as fermented dairy products and juices, and nutraceutical products where they are formulated as dried powders. In contrast, prebiotics are stable, can be heat-processed, and can therefore be incorporated into a wider range of processed foods and beverages with longer shelf lives than probiotics.

Host-Microbiota Compatibility

It is clear that selected probiotic bifidobacteria do survive transit through the stomach and small intestine and can be recovered in feces. However, in most cases, ingested probiotic strains persist only transiently in the intestine (134,184–188). An introduced probiotic strain must compete with an already established microbiota. The application of molecular techniques to profile the complex microbial communities has revealed that each person has a unique intestinal microbiota at the community, genus, and species level (137–139). This has been demonstrated in the case of bifidobacteria using PCR-DGGE analysis of *Bifidobacterium* species in feces, where each individual has their own particular combination of species (121,125). This uniqueness appears to extend to the strain level too, with molecular fingerprinting techniques showing that each person generally harbors multiple and unique *Bifidobacterium* strains (138,189–191). This host-microbiota stability and individuality suggests that certain host-microbiota compatibilities exist, and using prebiotics that augment an individual's own bacteria may prove more successful than introducing an exogenous strain for some applications.

The importance of host species/probiotic species specificity remains a contentious question. It is often recommended that probiotics be selected from bacteria indigenous to the intestinal tract of the targeted host species (192). However, the predominant probiotic *Bifidobacterium* species currently used in human probiotics is *Bif. animalis* (=*Bif. lactis*) (11), which is not an autochthonous member of the human intestinal microbiota. This species is taxonomically distant from human intestinal species (193), but is used because of its superior technological stability compared with human intestinal isolates. The prebiotic strategy overcomes any potential host/probiotic strain compatibility issues by targeting those strains already resident in the intestinal tract of an individual.

Inhibition of Pathogen Adhesion

One mechanism by which oligosaccharides may provide protection against infection by pathogenic micro-organisms has been hypothesized to be that of blocking adhesion to intestinal mucosa by acting as soluble receptor analogues (Fig. 3) (194–196). Microbial virulence factors, such as fimbriae and other membrane-based adhesins, control mucosal attachment and colonization of tissues. The recognition domains of fimbriae are similar to lectins that bind to carbohydrate epitopes on membrane glycocojugates of epithelial cells. Kunz and Rudloff (197) have listed the receptor specificities of glyco- and lactose-derived oligosaccharides and various pathogenic bacteria and viruses. Carbohydrate-mediated cell interactions affect cell-cell interactions, as well as bacterium, viral and toxin interactions with epithelial cells. The specificity of attachment provides potential for control of gastro-intestinal infections through the use of specific oligosaccharide structures.

Stimulation of Fermentative Activity in the Gut

In addition to modifying population dynamics, prebiotics also modify the activity of the microbiota by providing a source of readily fermentable carbohydrate. Indeed, it may be this dietary fiber-like characteristic of modifying the fermentative activity of the existing microbiota that is the important factor in providing a number of health benefits to consumers (Figs. 2–4). Proposed health effects of prebiotics that are speculated to be largely contingent on modifications to metabolic activity of the microbiota include reductions in risk factors for colon cancer, increased mineral absorption, improved lipid metabolism, and increased resistance to intestinal pathogens.

Reduced Risk Factors for Colon Cancer. The intestinal microbiota has a number of biochemical activities relevant to colon cancer risk that relate to the composition and activity of different bacterial populations. Hence, prebiotics may have a role in reducing risk factors for colon cancer. Since they supply a source of fermentable carbohydrate to the colon, dietary fiber-like anti-carcinogenic effects have been proposed for prebiotics (Fig. 4). Proposed mechanisms include supplying the colonic epithelium with SCFA (particularly butyrate); suppression of microbial protein metabolism, bile acid conversion and other mutagenic and toxigenic bacterial reactions; and immunomodulation. Butyrate production in the distal colon is suspected to be beneficial in preventing the development of colorectal cancers (198–200). While *Lactobacillus* and *Bifidobacterium* probiotics do not produce butyrate as major fermentation end products, prebiotics can stimulate butyrate production by the colonic microbiota, which provides a potential advantage of this approach (37,201). To date, the capacity of prebiotics to significantly contribute to a reduced incidence of colorectal cancer remains unproven. However, the results of preliminary human and animal experiments have provided sufficient encouragement to maintain the impetus for continued research into the protective effects of prebiotics.

Numerous studies in humans and animals have shown that consumption of prebiotics can produce an improved colonic environment in terms of reducing the levels of mutagenic enzyme activities (e.g., β-glucuronidase and azoreductase) and bacterial metabolites (e.g., secondary bile acids, phenols and indoles) that are purportedly associated with colon cancer risk. Examples include studies with lactulose (44,45,202), galacto-oligosaccharides (203), resistant starch (69,204–206) and lactosucrose (51). However, not all prebiotic feeding studies have shown improvements in these parameters (46,47,66,207), and in any case, the quantitative importance of these markers to eventual cancer development remains to be established.

A growing number of studies report protection by prebiotics against the development of pre-neoplastic lesions and/or tumors in rodent models of colon carcinogenesis. Again, these have used a variety of prebiotics including fructo-oligosaccharides and inulin (summarized by Pool-Zobel et al. (37)), lactulose (161,208) and resistant starch (209,210). Dose effects have been observed (37), but in general, very high doses of NDOs have been used in the animal studies. An important question that is beginning to be addressed is the significance of the sustainability of fermentation provided by different prebiotics during passage through the colon on their effectiveness in preventing colon cancer. The distal colon and rectum are the major sites of disease in humans, but SCFA produced by bacterial fermentation in the colon are rapidly absorbed by the colonic mucosa near the site of their production. Hence, prebiotics that can supply a persistent source of fermentable carbohydrate that sustains SCFA synthesis through to the distal colon may prove to be the most effective. Indeed, studies with different molecular sized fructan prebiotics have reported increased protection with the larger, more slowly fermented prebiotics (37).

Improving Mineral Absorption. As seen for dietary fibres, a number of prebiotics have been shown to increase mineral absorption in animal models (211–214). The precise mechanisms of prebiotic-mediated improvements in mineral uptake remain unclear, but fermentative activities of the microbiota including SCFA production and reductions in luminal pH are believed to be involved (Fig. 4) (213). Calcium and magnesium are the main minerals for which uptake is improved. Under normal circumstances dietary calcium is predominately absorbed in the small intestine with little calcium absorbed in the colon (215). However, prebiotic fermentation is believed to extend calcium uptake into the colon (34). In rats, increased calcium uptake has led to improved bone mineralization for animals fed galacto-oligosaccharides (216), lactulose (217) and fructo-oligosaccharides (218).

Although two human studies have shown little impact on mineral uptake (219,220), a number have reported beneficial effects on calcium (221–225) and magnesium absorption (226) using fructo-oligosaccharides, inulin and galacto-oligosaccharides. Differences in results have been attributed to differences in study designs and treatment populations (212,225). Griffin et al. (225) saw no effect with short chain fructo-oligosaccharides in a population of pubertal girls, but a significant increase in the calcium absorption and balance was observed when the girls consumed a mixture of fructo-oligosaccharides and inulin, perhaps reflecting a more sustained colonic fermentation. Overall, results so far are encouraging of a role for prebiotics in improving calcium uptake. Further research is warranted to investigate links between long-term prebiotic consumption and improved bone density in humans at risk of developing osteoporosis.

Effects on Serum Lipids and Cholesterol. A role for prebiotics in controlling hyperlipidemia has been proposed and a relatively large number of animal and human studies have focused on the effects of oligosaccharide and inulin intake on lipid metabolism. These include eight human trials summarized by van Loo et al. (34), and more recent trials (227–231). The mechanism by which lowering of serum lipids and cholesterol may occur has been speculated to be regulation of host de novo lipogenesis via SCFA absorbed from the gut (Fig. 4) (232). While convincing positive effects on lowering serum triglycerols and cholesterol have often been reported in animal studies (233) the results from human studies have tended to be contradictory, although no deleterious effects have been reported (232). The trials conducted to date indicate that while there is certainly potential for prebiotics to control serum lipids, more research is needed to identify the most appropriate target populations, the impact of background diet, and the mechanisms of action.

Improving Colonization Resistance in the Gut. The ability of prebiotics to improve colonization resistance and prevent bacterial infections from the gut has been only scantly explored, but results so far indicate a potential application for lactulose and NDOs in this capacity. Lactulose has the most accumulated evidence. Özaslan et al. (234) observed lower caecal overgrowth and translocation of *Escherichia coli* in rats with obstructive jaundice when they were fed lactulose, while Bovee-Oudenhoven et al. (235) reported that consumption of lactulose increased colonization resistance against the invasive pathogen *Salmonella enteritidis* in rats. Indeed, lactulose consumption at high doses (up to 60 g per day) is effective in eliminating salmonella from the intestinal tract of chronic human carriers and is used as a pharmaceutical for this purpose in some countries (236). The mode of action is speculated to be acidification of the gut that prevents growth of this acid-sensitive pathogen.

The anti-infective effects of fructo-oligosaccharides and inulin have been examined in mice challenged with the enteric pathogen *Candida albicans* and with systemic infections of *Salmonella* and *Listeria monocytogenes* (237). Prebiotic feeding significantly reduced intestinal colonization by *Candida* and the mortality of the mice with the systemic infections, the latter effect hypothesized as being due to gut microbiota-induced immunomodulation. However, two randomized, blinded, and controlled trials in which Peruvian infants living in environments with a high burden of gastrointestinal and other infections were fed oligofructose failed to show any significant benefit in terms of preventing diarrhea or the use of health care resources (238), although a high level of breast feeding amongst these infants may have limited the opportunity for effect. Prebiotic intervention may prove effective in rapidly restoring colonization resistance and preventing infections in cases where the intestinal microbiota has been perturbed.

Other Physiological and Technological Benefits of Prebiotics

In addition to the effects elicited by prebiotics discussed thus far, prebiotics have a number of other functional properties that make them attractive pharmaceuticals and food ingredients. Through their action in fecal bulking and water retention in the bowel, prebiotics are effective in relieving constipation and maintaining normal stool frequency (34). Additionally, by stimulating bacterial protein synthesis and reducing production of ammonia by the microbiota, lactulose is effective in the treatment of hepatic encephalopathy (236). NDOs are sweet and can be used as low-cariogenic and low-calorific sugar substitutes, while polysaccharides such as inulin are used as fat replacers. Their indigestibility and subsequent impact on glucose and insulin responses also make them suitable for diabetics. In terms of food technology, NDOs supply a number of valuable physicochemical functionalities. They can be used to increase viscosity, reduce Malliard reactions, alter water retention, depress freezing points, and suppress crystal formation. Hence, they are used commercially in a wide variety of foods and beverages.

DISADVANTAGES OF THE PREBIOTIC APPROACH

While there are many advantages of the prebiotic approach, the use of this strategy to modify the intestinal microbiota is not without its disadvantages. First among these is the potential for intestinal side-effects if excessive doses of prebiotic oligosaccharides are consumed (discussed in more detail in the following section). Secondly, there are instances where probiotics may be more applicable to restoring colonization resistance in the gut. One example is during episodes of diarrhea when mucosal damage may lead to reduced capacity for sugar digestion. Ingestion of prebiotic oligosaccharides under these conditions may exacerbate symptoms associated with sugar malabsorption even at usually tolerable doses. Thirdly, there may be mechanisms, such as immunomodulation, where the introduction of an exogenous probiotic strain could theoretically provide a superior stimulus. Finally, some effects of probiotics are known to be strain specific and prebiotics cannot at this stage emulate that specificity.

SAFE DOSAGE LEVELS

Safety of use must always be a dominant issue in the development of new food products. Fortunately, it is well established that lactulose, short-chain oligosaccharides, inulin, resistant starch and dietary fiber are not toxic, even in high doses. Non-digestible carbohydrates are consumed as part of the normal daily diet, as they are natural components of most plants (239). Estimates of resulting intakes of fructo-oligosaccharides and inulin are between 1 and 10 g/day from normal diets in Europe and the United States of America (239,240). It is likely that intakes of around 8 g/day by adults are normal. Thus, any recommended dosages of non-digestible carbohydrates will be additional to the natural basal dose consumed. Recommended effective doses of prebiotic oligosaccharides in adults usually range from 10 to 15 g/day. With the shorter chain oligosaccharides, such as fructo- and galacto-oligosaccharides, intakes exceeding 15 g/day in adults can lead to flatulence, abdominal discomfort and cramping (241–243). With adaptation, larger doses of up to of 25–30 g/day can be tolerated with few ill effects. Excessive consumption of lactulose and NDOs can result in diarrhea due to osmotic water retention in the colon, with the offending dose depending on the weight of the individual, rate of consumption (single dose or frequent smaller doses spread over the day), and the composition and activity of the intestinal microbiota.

A possible side-effect from the consumption of rapidly fermented, acidogenic prebiotic sugars was recently identified by Dutch researchers (244,245). While investigating the effects of lactulose and fructo-oligosaccharides on the translocation of *Salmonella* in rats, the researchers noted that feeding the prebiotics left the animals more susceptible to pathogen translocation from the gut. Intestinal acidification was observed due to the rapid prebiotic fermentation, and while this inhibited the acid-sensitive pathogen in the intestinal lumen, it possibly also damaged the mucosa leading to an impaired barrier effect. Further research is needed to investigate possible negative impacts of high doses of rapidly fermented sugars on the intestinal mucosa.

CONCLUSION AND FUTURE DIRECTIONS

There is little doubt from the volume of accumulated evidence from human and animal studies that prebiotics can modify the dynamics of the colonic microbiota. Bifidobacteria are the dominant group of bacteria stimulated by all prebiotics developed so far. That such a range of diverse carbohydrate structures can promote the selective proliferation of bifidobacteria is testament to the remarkable metabolic agility of these organisms. The magnitude of the bifidogenic effect is largely affected by the size of the intestinal *Bifidobacterium* population, and little impact on *Bifidobacterium* numbers is observed in individuals who already harbor high numbers of these bacteria.

Beyond Bifidobacteria

Although traditional microbiology culture methods have enabled some assessment of the selectivity of prebiotics, new molecular techniques that enable analysis of non-cultivable bacteria are starting to be applied in studies investigating the impact of prebiotics on the colonic microbiota. Almost certainly, other bacterial populations that are affected by the intake of current prebiotics will emerge. While evidence to date supports the beneficial role of bifidobacteria and lactobacilli in the intestinal tract (11,12), they are but two of a multitude of bacterial genera within the intestinal microbiota that potentially confer benefits to the host. As we gradually shed light on the activities of newly identified intestinal bacteria and their interactions with the host in health and disease new beneficial and detrimental organisms will be undoubtedly be identified. The challenge will be to find or design selective prebiotics to modulate populations and activities of these particular organisms.

Phylogenetic vs. Physiological Modulation of the Microbiota

It should be emphasized that altering the microbial population dynamic is only one aspect of prebiotic action. While stimulating the proliferation of particular groups of bacteria might be important for some health effects (e.g., immunomodulation), this may be secondary to specifically altering the metabolic activity of the microbiota for other effects (e.g., anti-cancer). Marked differences between the phylogenetic and physiologic effects of prebiotics on particular groups of organisms have been observed (139). Because of its trophic and anti-neoplasic effects on the colonic epithelium, stimulating specific populations of butyrigenic bacteria in the colon may well be the next important target for prebiotics. In situ measurement of specific bacterial activities remains problematic, but advances in functional genomics may provide a new avenue to explore the interactions between prebiotics, the intestinal microbiota and the host in health and disease.

Blurring the Distinctions Between Prebiotics, Dietary Fibers, and Other Fermentable Dietary Carbohydrates in the Colon

The greatest volume of research and evidence for prebiotic effects has been accrued for fructo-oligosaccharides and inulin, but there is accumulating evidence of prebiotic actions by a number of non-digestible carbohydrates. Lactulose and galacto-oligosaccharides have strong claims to be classified as prebiotics, while there is promising evidence for prebiotic activity by isomalto-, xylo-, and soybean-oligosaccharides. There is growing interest in the impact of dietary fibers on the composition as well as the activity of the intestinal microbiota, and resistant starches and arabinoxylans in particular warrant further study for bifidogenic and other prebiotic effects.

It has been hypothesized that synergies might exist between NDOs that stimulate a bifidogenic response and SCFA production in the proximal colon and larger polysaccharides that sustain a source of fermentable carbohydrate through to the distal colon. ORAFTI (Belgium) market a prebiotic (Synergy 1) that includes both short chain fructo-oligosaccharides and the longer chain fructan inulin and have reported synergistic effects in this combination for a range of physiological effects (246). Similarly, complementary effects have been noted for FOS/inulin and resistant starches (72,247). Development of synergistic prebiotic combinations to optimize the composition and activity of the microbiota throughout the length of the intestinal tract, or to target specific intestinal regions (e.g., for treatment of IBD) is set to provide continuing avenues for future research.

Effects on Human Health

A growing understanding of the intestinal microbiota and its contribution to health and disease has enabled rational hypotheses to be developed for prebiotic interventions targeted to specific human populations. Testing of these hypotheses is still mostly centered at the animal model or pilot human trial stage. Prebiotic oligosaccharides are already used in some infant formulas and efforts to replicate the activities of HMOs are likely to continue. Although the effects of prebiotics overlap somewhat with probiotics, the prebiotic strategy does provide some potential advantages. Despite these physical and potentially physiological advantages, research into the clinical effects of prebiotics still lags that devoted to probiotics.

There is good evidence that prebiotics can relieve constipation and control hepatic encephalopathy, and lactulose is currently used pharmaceutically for these purposes. Additionally, a number of other health targets proposed for prebiotics have accumulating evidence of benefits. The most promising targets have been discussed in this chapter and include increasing calcium uptake, boosting colonization resistance against intestinal pathogens, and ameliorating IBD. Evidence for these benefits is still largely preliminary, but is sufficiently encouraging to warrant continuing investigation. While research efforts have naturally focused on the health benefits of prebiotics, and to date few reports of deleterious effects have surfaced, further quantification of the potential risks of prebiotics at different doses, in combination with different diets, and for different demographics, both healthy and diseased should be conducted. It is also important that prebiotics be trialed in the context of total diets, since other dietary components, for example the presence of dietary fibers that influence intestinal transit rates, can be expected to affect the clinical outcomes.

Recent years have seen marked progress in our understanding of the microecology of the gastrointestinal tract. However, we are still only at the very beginning of developing an appreciation of the functional relationships between the microbiota and the host, in health and disease. A more profound understanding of what constitutes a "healthy" intestinal microbiota composition, and which microbial groups and activities are involved in health and disease, is a prerequisite to the future development of prebiotics with specifically targeted health effects. The challenge remains to demonstrate clinically relevant benefits to health by prebiotic interventions in well-designed and controlled human trials.

REFERENCES

1. Bourlioux P, Koletzko B, Guarner F, Braesco V. The intestine and its microflora are partners for the protection of the host: report on the Danone Symposium " The Intelligent Intestine," held in Paris, June 14, 2002. Am J Clin Nutr 2003; 78:675–683.

2. Topping DL, Clifton PM. Short-chain fatty acids and human colonic function: roles of resistant starch and nonstarch polysaccharides. Physiol Rev 2001; 81:1031–1064.
3. Blum S, Schiffrin EJ. Intestinal Microflora and Homeostasis of the Mucosal Immune Response: Implications for Probiotic Bacteria? Curr Issues Intest Microbiol 2003; 4:53–60.
4. Tlaskalova-Hogenova H, Stepankova R, Hudcovic T, et al. Commensal bacteria (normal microflora), mucosal immunity and chronic inflammatory and autoimmune diseases. Immunol Lett 2004; 93:97–108.
5. Cummings JH, Macfarlane GT, Macfarlane S. Intestinal bacteria and ulcerative colitis. Curr Issues Intest Microbiol 2003; 4:9–20.
6. Marteau P, Seksik P, Shanahan F. Manipulation of the bacterial flora in inflammatory bowel disease. Best practice res:. Clin Gastroenterol 2003; 17:47–61.
7. Saunier K, Doré J. Gastrointestinal tract and the elderly: functional foods, gut microflora and healthy ageing. Dig Liver Dis 2002; 34:S19–S24.
8. Guarner F, Malagelada JR. Gut flora in health and disease. Lancet 2003; 361:512–519.
9. Tannock GW. Molecular assessment of intestinal microflora. Am J Clin Nutr 2001; 73:410S–414S.
10. Vaughan EE, de Vries MC, Zoetendal EG, Ben-Amor K, Akkermans ACL, de Vos WM. The intestinal LABs. Antonie van Leeuwenhoek 2002; 82:341–352.
11. Crittenden R. An update on probiotic bifidobacteria. In: Salminen S, von Wright A, Ouwerhand A, eds. Lactic Acid Bacteria: Microbiological and Functional Aspects. New York: Marcel Dekker, 2004:125–157.
12. Salminen S, Gorbach S, Lee Y-K, Benno Y. Human studies on probiotics: what is scientifically proven today? In: Salminen S, von Wright A, Ouwerhand A, eds. Lactic Acid Bacteria: Microbiological and Functional Aspects. New York: Marcel Dekker, 2004:515–530.
13. Topping DL, Fukushima M, Bird AR. Resistant starch as a prebiotic and synbiotic: state of the art. Proc Nutr Soc 2003; 62:171–176.
14. Gyorgy P, Norris RF, Rose CS. Bifidus factor I. A variant of *Lactobacillus bifidus* requiring a special growth factor. Arch Biochem Biophys 1954; 48:193–201.
15. Gyorgy P, Kuhn R, Rose CS, Zilliken F. Bifidus factor II. Its occurrence in milk from different species and in other natural products. Arch Biochem Biophys 1954; 48:202–208.
16. Rasic JL, Kurmann JA. Bifidobacteria and their role. Boston, MA, U.S.A.: Birkhauser Verlag, 1983.
17. Modler HW. Bifidogenic factors—sources, metabolism and applications. Int Dairy J 1994; 4:383–407.
18. Saito H, Miyakawa H, Ishibashi N, Tamura Y, Hayasawa H, Shimamura S. Effect of iron-free and metal-bound forms of lactoferrin on the growth of bifidobacteria *E.coli* and *S.aureus*. Biosci Microflora 1996; 15:1–7.
19. Samejima K, Yoshioka M, Tamura Z. Bifidus factors in carrot I. Chem Pharm Bull (Tokyo) 1971; 19:166–177.
20. Yoshioka M, Tamura Z. Bifidus factors in carrot II. Chem Pharm Bull (Tokyo) 1971; 19:178–185.
21. Kaneko T. A novel bifidogenic growth stimulator produced by *Propionibacterium freudenreichii*. Biosci Microflora 1999; 18:73–80.
22. Mori H, Sato Y, Taketomo N, et al. Isolation and structural identification of bifidogenic growth stimulator produced by *Propionibacterium freudenreichii*. J Dairy Sci 1997; 80:1959–1964.
23. Benno K, Sawada T. The intestinal microflora of infants: composition of fecal flora in breast-fed and bottle-fed infants. Microbiol Immunol 1984; 28:975–986.
24. Balmer SE, Wharton BA. Diet and fecal flora in the newborn: breast milk and infant formula. Arch Dis Child 1989; 64:1672–1677.
25. Petuely F. Bifidusflora bei flaschenkindern durch bifidogene substanzen (bifidusfaktor). Z Kinderheilkd 1957; 79:174–179 (In German).

26. Ruttloff H, Taufel A, Krause W, Haenel H, Taufel K. Die intestinal-enzymatische spaltung von galakto-oligosacchariden im darm von tier und menschen mit besonderer berucksichtigung von Lactobacillus bifidus. II. Mitt. zum intestinalen verhalten der lactulose. Nahrung 1967; 11:39–46 (In German).

27. Yazawa K, Imai K, Tamura Z. Oligosaccharides and polysaccharides specifically utilizable by bifidobacteria. Chem Pharm Bull (Tokyo) 1978; 26:3306–3311.

28. Minami Y, Yazawa K, Tamura Z, Tanaka T, Yamamoto T. Selectivity of utilisation of galactosyl-oligosaccharides by bifidobacteria. Chem Pharm Bull 1983; 31:1688–1691.

29. Hidaka H, Eida T, Takizawa T, Tokunaga T, Tashiro Y. Effects of fructooligosaccharides on intestinal flora and human health. Bifidobacteria Microflora 1986; 5:37–50.

30. Mitsuoka T, Hidaka H, Eida T. Effect of fructooligosaccharides on intestinal microflora. Nahrung 1987; 31:427–436.

31. Gibson GR, Roberfroid MB. Dietary modulation of the human colonic microbiota—Introducing the concept of prebiotics. J Nutr 1995; 125:1401–1412.

32. Playne MJ, Crittenden R. Commercially available oligosaccharides. Bull Int Dairy Fed 1996; 313:10–22.

33. Crittenden RG, Playne MJ. Production, properties and applications of food-grade oligosaccharides. Trends Food Sci Technol 1996; 7:353–361.

34. van Loo J, Cummings J, Delzenne N, et al. Functional food properties of non-digestible oligosaccharides: a consensus report from the ENDO project (DGXII AIRII-CT94-1095). Br J Nutr 1999; 81:121–132.

35. Kolida S, Tuohy K, Gibson GR. Prebiotic effects of inulin and oligofructose. Br J Nutr 2002; 87:S193–S197.

36. Taper HS, Roberfroid MB. Inulin/oligofructose and anticancer therapy. Br J Nutr 2002; 87:S283–S286.

37. Pool-Zobel B, van Loo J, Rowland I, Roberfroid MB. Experimental evidences on the potential of prebiotic fructans to reduce the risk of colon cancer. Br J Nutr 2002; 87:S273–S281.

38. Buddington KK, Donahoo JB, Buddington RK. Dietary oligofructose and inulin protect mice from enteric and systemic pathogens and tumor inducers. J Nutr 2002; 132:472–477.

39. Harmsen HJM, Raangs GC, Franks AH, Wildeboer-Veloo ACM, Welling GW. The effect of the prebiotic inulin and the probiotic Bifidobacterium longum on the fecal microflora of healthy volunteers measured by FISH and DGGE. Microb Ecol Health Dis 2002; 14:211–219.

40. Ben XM, Zhou XY, Zhao WH, et al. Supplementation of milk formula with galacto-oligosaccharides improves intestinal micro-flora and fermentation in term infants. Chinese Med J 2004; 117:927–931.

41. Boehm G, Lidestri M, Casetta P, et al. Supplementation of a bovine milk formula with an oligosaccharide mixture increases counts of fecal bifidobacteria in preterm infants. Arch Dis Childhood 2002; 86:F178–F181.

42. Bouhnik Y, Flourie B, D'Agay-Abensour L, et al. Administration of transgalacto-oligosaccharides increases fecal bifidobacteria and modifies colonic fermentation metabolism in healthy humans. J Nutr 1997; 127:444–448.

43. Ito M, Deguchi Y, Matsumoto K, Kimura M, Onodera N, Yajima T. Influence of galacto-oligosaccharides on human fecal microflora. J Nutr Sci Vitaminol 1993; 39:635–640.

44. Terada A, Hara H, Kataoka M, Mitsuoka T. Effect of lactulose on the composition and metabolic activity of the human fecal flora. Microbial Ecol Health Dis 1992; 5:43–50.

45. Ballongue J, Schumann C, Quignon P. Effects of lactulose and lactitol on colonic microflora and enzymatic activity. Scand J Gastroenterol 1997; 32:41–44.

46. Tuohy KM, Ziemer CJ, Klinder A, Knobel Y, Pool-Zobel BL, Gibson GR. A human volunteer study to determine the prebiotic effects of lactulose powder on human colonic microbiota. Microb Ecol Health Dis 2002; 14:165–173.

47. Bouhnik Y, Attar A, Joly FA, Riottot M, Dyard F, Flourie B. Lactulose ingestion increases fecal bifidobacterial counts: a randomised double-blind study in healthy humans. Eur J Clin Nutr 2004; 58:462–466.

48. Boehm G, Stahl B. In: Mattila-Sandholm T, Saarela M, eds. Oligosaccharides. In: Functional Dairy Products. Cambridge: Woodhead Publishing/CRC Press, 2003:203–243.

49. Fujita K, Hara K, Sakai S, et al. Effects of 4-β-D-galactosylsucrose (lactosucrose) on intestinal flora and its digestibility in humans. J Jpn Soc Starch Sci 1991; 38:249–255.

50. Yoneyama M, Mandai T, Aga H, Fujii K, Sakai S, Katayama Y. Effects of 4-β-D-galactosylsucrose (lactosucrose) intake on intestinal flora in healthy humans. J Jpn Soc Nutr Food Sci 1992; 45:101–107.

51. Hara H, Li S-T, Sasaki M, et al. Effective dose of lactosucrose on fecal flora and fecal metabolites of humans. Bifidobacteria Microflora 1994; 13:51–63.

52. Ohkusa T, Ozaki Y, Sato C, Mikuni K, Ikeda H. Long-term ingestion of lactosucrose increases bifidobacterium sp. in human fecal flora. Digestion 1995; 56:415–420.

53. Djouzi Z, Andrieux C, Pelenc V, et al. Degradation and fermentation of alpha-gluco-oligosaccharides by bacterial strains from human colon: in vitro and in vivo studies in gnotobiotic rats. J Appl Bacteriol 1995; 79:117–127.

54. Okazaki M, Fujikawa S, Matsumoto N. Effect of xylo-oligosaccharide on the growth of bifidobacteria. Bifidobacteria Microflora 1990; 9:77–86.

55. Campbell JM, Jr., Fahey GC, Wolf BW. Selected indigestible oligosaccharides affect large bowel mass, cecal and fecal short-chain fatty acids, pH and microflora in rats. J Nutr 1997; 127:130–136.

56. Kohmoto T, Fukui F, Takaku H, Machida Y, Arai M, Mitsuoka T. Effect of isomalto-oligosaccharides on human facal flora. Bifidobacteria Microflora 1988; 7:61–69.

57. Kohmoto T, Fukui F, Takaku H, Mitsuoka T. Dose-response test of isomaltooligosaccharides for increasing fecal bifidobacteria. Agric Biol Chem 1991; 55:2157–2159.

58. Kaneko T, Kohmoto T, Kikuchi H, et al. Effects of isomaltoologosaccharides intake on defecation and intestinal environment in healthy volunteers. J Home Econ Jpn 1993; 44:245–254 (in Japanese).

59. Kaneko T, Kohmoto T, Kikuchi H, Shiota M, Iino H, Mitsuoka T. Effects of isomaltooligosaccharides with different degrees of polymerisation on human fecal bifidobacteria. Biosci Biotech Biochem 1994; 58:2288–2290.

60. Benno Y, Endo K, Shiragani N, Sayama K, Mitsuoka T. Effects of raffinose intake on human fecal microflora. Bifidobacteria Microflora 1987; 6:59–63.

61. Hara T, Ikeda N, Hatsumi K, Watabe J, Iino H, Mitsuoka T. Effects of small amount ingestion of soybean oligosaccharides on bowel habits and fecal flora of volunteers. Jpn J Nutr 1997; 55:79–84.

62. Hayakawa K, Mitzutani J, Wada K, Masai T, Yoshihara I, Mitsuoka T. Effects of soybean oligosaccharides on human fecal microflora. Microb Ecol Health Dis 1990; 3:293–303.

63. Wada K, Watabe J, Mizutani J, Tomoda M, Suzuki H, Saitoh Y. Effects of soybean oligosaccharides in a beverage on human fecal flora and metabolites. J Agric Chem Soc Jpn 1992; 66:127–135.

64. Jie Z, Bang-Yao L, Ming-Jie X, et al. Studies on the effects of polydextrose intake on physiologic functions in Chinese people. Am J Clin Nutr 2000; 72:1503–1509.

65. Asano T, Yuasa K, Kunugita K, Teraja T, Mitsuoka T. Effects of gluconic acid on human fecal bacteria. Microbial Ecol Health Dis 1994; 7:247–256.

66. Kleessen B, Stoof G, Proll J, Schmiedl D, Noack J, Blaut M. Feeding resistant starch affects fecal and cecal microflora and short-chain fatty acids in rats. J Anim Sci 1997; 75:2453–2462.

67. Brown IL, Wang X, Topping DL, Playne MJ, Conway PL. High amylose maize starch as a versatile prebiotic for use with probiotic bacteria. Food Aust 1998; 50:603–610.

68. Wang X, Brown IL, Evans AJ, Conway PL. The protective effects of high amylose maize (amylomaize) starch granules on the survival of *Bifidobacterium* spp. in the mouse intestinal tract. J Appl Microbiol 1999; 87:631–639.

69. Silvi S, Rumney CJ, Cresci A, Rowland IR. Resistant starch modifies gut microflora and microbial metabolism in human flora-associated rats inoculated with feces from Italian and U.K. donors. J Appl Microbiol 1999; 86:521–530.

70. Bielecka M, Biedrzycka E, Majkowska A, Juskiewicz J, Wroblewska M. Effect of non-digestible oligosaccharides on gut microecosystem in rats. Food Res Int 2002; 35:139–144.

71. Wang X, Brown IL, Khaled D, Mahoney MC, Evans AJ, Conway PL. Manipulation of colonic bacteria and volatile fatty acid production by dietary high amylose maize (amylomaize) starch granules. J Appl Microbiol 2002; 93:390–397.

72. Le Blay G, Michel C, Blottiere H, Cherbut C. Raw potato starch and short-chain fructo-oligosaccharides affect the composition and metabolic activity of rat intestinal microbiota differently depending on the caecocolonic segment involved. J Appl Microbiol 2003; 94:312–320.

73. van Laere KM, Hartemink R, Bosveld M, Schols HA, Voragen AG. Fermentation of plant cell wall derived polysaccharides and their corresponding oligosaccharides by intestinal bacteria. J Agric Food Chem 2000; 48:1644–1652.

74. Crittenden R, Karppinen S, Ojanen S, et al. In vitro fermentation of cereal dietary fiber carbohydrates by probiotic and intestinal bacteria. J Sci Food Agric 2002; 82:1–9.

75. Okubo T, Ishihara N, Takahashi H, et al. Effects of partially hydrolyzed guar gum intake on human intestinal microflora and its metabolism. Biosci Biotech Biochem 1994; 58:364–1369.

76. Muir JG, Yeow EG, Keogh J, et al. Combining wheat bran with resistant starch has more beneficial effects on fecal indexes than does wheat bran alone. Am J Clin Nutr 2004; 79:1020–1028.

77. Champ MM. Physiological aspects of resistant starch and in vivo measurements. J AOAC Int 2004; 87:749–755.

78. Andoh A, Tsujikawa T, Fujiyama Y. Role of dietary fiber and short-chain fatty acids in the colon. Curr Pharm Design 2003; 9:347–358.

79. Muir JG, Yeow EGW. Importance of combining indigestible carbohydrate with protein sources in the diet: implications for reducing colorectal cancer risk. Proc Nutr Soc Aust 2000; 24:196–204.

80. Smiricky-Tjardes MR, Grieshop CM, Flickinger EA, Bauer LL, Fahey GC, Jr. Dietary galactooligosaccharides affect ileal and total-tract nutrient digestibility, ileal and fecal bacterial concentrations, and ileal fermentative characteristics of growing pigs. J Animal Sci 2003; 81:2535–2545.

81. Ito M, Deguchi Y, Miyamori A, et al. Effects of administration of galacto-oligosaccharides on the human fecal microflora, stool weight and abdominal sensation. Microb Ecol Health Dis 1990; 3:285–292.

82. Barrangou R, Altermann E, Hutkins R, Cano R, Klaenhammer TR. Functional and comparative genomic analyses of an operon involved in fructooligosaccharide utilization by Lactobacillus acidophilus. Proc Nat Acad Sci USA 2003; 100:8957–8962.

83. Kaplan H, Hutkins RW. Metabolism of fructooligosaccharides by Lactobacillus paracasei 1195. Appl Environ Microbiol 2003; 69:2217–2222.

84. Roberfroid MB, Van Loo JAE, Gibson GR. The bifidogenic nature of chicory inulin and its hydrolysis products. J Nutr 1998; 128:11–19.

85. Harmsen HJM, Wildeboer-Veloo ACM, Raangs GC, et al. Analysis of intestinal flora development in breast-fed and formula-fed infants by using molecular identification and detection methods. J Pediatr Gastroenterol Nutr 2000; 30:61–67.

86. Mountzouris KC, McCartney AL, Gibson GR. Intestinal microflora of human infants and current trends for its nutritional modulation. Br J Nutr 2002; 87:405–420.

87. Brand-Miller J, McVeagh P. Human milk oligosaccharides: 130 reasons to breast-feed. Br J Nutr 1999; 82:333–335.

88. Kunz C, Rudloff S, Baier W, Klein N, Strobel S. Oligosaccharides in human milk: structural, functional, and metabolic aspects. Ann Rev Nutr 2000; 20:699–722.

89. Adlerberth I. Establishment of a normal intestinal microflora in the newborn infant. In: Hansen LA, Yolken RH, eds. Probiotics,Other Nutritional Factors and Intestinal Microflora (Nestle Nutrition Workshop Series, Vol 42, Nestec Ltd, Vevey). Philadelphia: Lippincott-Raven, 1999:63–78.

90. Heavey PM, Savage SA, Parrett A, Cecchini C, Edwards CA, Rowland IR. Protein-degradation products and bacterial enzyme activities in feces of breast-fed and formula-fed infants. Br J Nutr 2003; 89:509–515.

91. Edwards CA, Parrett AM. Intestinal flora during the first months of life: new perspectives. Br J Nutr 2002; 88:S11–S18.

92. Moro G, Minoli I, Mosca M, et al. Dosage-related bifidogenic effects of galacto- and fructooligosaccharides in formula-fed term infants. J Pediatr Gastroenterol Nutr 2002; 34:291–295.

93. Newburg DS. Human milk glycoconjugates that inhibit pathogens. Curr Med Chem 1999; 6:117–127.

94. Martin-Sosa S, Martin MJ, Hueso P. The sialylated fraction of milk oligosaccharides is partially responsible for binding to enterotoxigenic and uropathogenic *Escherichia coli* human strains. J Nutr 2002; 132:3067–3072.

95. Morrow AL, Ruiz-Palacios GM, Altaye M, et al. Human milk oligosaccharides are associated with protection against diarrhea in breast-fed infants. J Pediatr 2004; 145:297–303.

96. Wang B, Brand-Miller J. The role and potential of sialic acid in human nutrition. Eur J Clin Nutr 2003; 57:1351–1369.

97. Vincent SJ, Faber EJ, Neeser JR, Stingele F, Kamerling JP. Structure and properties of the exopolysaccharide produced by *Streptococcus macedonicus* Sc136. Glycobiology 2001; 11:131–139.

98. Rencurosi A, Poletti L, Guerrini M, Russo G, Lay L. Human milk oligosaccharides: an enzymatic protection step simplifies the synthesis of $3'$- and $6'$-O-sialyllactose and their analogues. Carbohydr Res 2002; 337:473–483.

99. La Ferla B, Prosperi D, Lay L, Russo G, Panza L. Synthesis of building blocks of human milk oligosaccharides. Fucosylated derivatives of the lacto- and neolacto-series. Carbohydr Res 2002; 337:1333–1342.

100. Dumon C, Samain E, Priem B. Assessment of the two *Helicobacter pylori* alpha-1,3-fucosyltransferase ortholog genes for the large-scale synthesis of LewisX human milk oligosaccharides by metabolically engineered *Escherichia coli*. Biotechnol Prog 2004; 20:412–419.

101. Ouwehand A, Isolauri E, Salminen S. The role of the intestinal microflora for the development of the immune system in early childhood. Eur J Nutr 2002; 41:I32–I37.

102. Björkstén B, Naaber P, Sepp E, Mikelsaar M. The intestinal microflora in allergic Estonian and Swedish 2-year old children. Clin Exp Allergy 1999; 29:342–346.

103. Björkstén B, Sepp E, Julge K, Voor T, Mikelsaar M. Allergy development and the intestinal microflora during the first year of life. J Allergy Clin Immunol 2001; 108:516–520.

104. Böttcher M, Sandin A, Norin E, Midtvedt T, Björkstén B. Microflora associated characteristics in feces from allergic and non-allergic children. Clin Exp Allergy 2000; 30:590–596.

105. Grönlund MM, Arvilommi H, Kero P, Lehtonen OP, Isolauri E. Importance of intestinal colonization in the maturation of humoral immunity in early infancy: a prospective follow up study of healthy infants aged 0-6 months. Arch Dis Child 2000; 83:F186–F192.

106. Kalliomäki M, Kirjavainen P, Eerola E, Kero P, Salminen S, Isolauri E. Distinct patterns of neonatal gut microflora in infants in whom atopy was and was not developing. J Allergy Clin Immunol 2001; 107:129–134.

107. Kirjavainen PV, Apostolou E, Arvola T, Salminen SJ, Gibson GR, Isolauri E. Characterizing the composition of intestinal microflora as a prospective treatment target in infant allergic disease. FEMS Immunol Med Micrbiol 2001; 32:1–7.

108. Kirjavainen PV, Arvola T, Salminen SJ, Isolauri E. Aberrant composition of gut microbiota of allergic infants: a target of bifidobacterial therapy at weaning? Gut 2002; 51:51–55.

109. He F, Ouwehand AC, Isolauri E, Hashimoto H, Benno Y, Salminen S. Comparison of mucosal adhesion and species identification of bifidobacteria isolated from healthy and allergic infants. FEMS Immunol Med Microbiol 2001; 30:43–47.

110. Ouwehand AC, Isolauri E, He F, Hashimoto H, Benno Y, Salminen S. Differences in *Bifidobacterium* flora composition in allergic and healthy infants. J Allergy Clin Immunol 2001; 108:144–145.

111. Mackie RI, Sghir A, Gaskins HR. Developmental microbial ecology of the neonatal gastrointestinal tract. Am J Clin Nutr 1999; 69:1035S–1045S.

112. Matsuki T, Watanabe K, Tanaka R, Fukuda M, Oyaizu H. Distribution of bifidobacterial species in human intestinal microflora examined with 16S rRNA-gene-targeted species-specific primers. Appl Environ Microbiol 1999; 65:4506–4512.

113. Kalliomäki M, Salminen S, Arvilommi H, Kero P, Koskinen P, Isolauri E. Probiotics in primary prevention of atopic disease: a randomised placebo-controlled trial. Lancet 2001; 357:1076–1079.

114. Kalliomäki M, Salminen S, Poussa T, Arvilommi H, Isolauri E E. Probiotics and prevention of atopic disease: 4-year follow-up of a randomised placebo-controlled trial. Lancet 2003; 361:1869–1871.

115. Nagura T, Hachimura S, Hashiguchi M, et al. Suppressive effect of dietary raffinose on T-helper 2 cell-mediated immunity. Br J Nutr 2002; 88:421–427.

116. Yoshida T, Hirano A, Wada H, Takahashi K, Hattori M. Alginic acid oligosaccharide suppresses Th2 development and IgE production by inducing IL-12 production. Int Arch Allergy Immunol 2004; 133:239–247.

117. Marteau P, Pochart P, Dore J, Bera-Maillet C, Bernalier A, Corthier G. Comparative study of bacterial groups within the human cecal and fecal microbiota. Appl Environ Microbiol 2001; 67:4939–4942.

118. Franks AH, Harmsen HJ, Raangs GC, Jansen F, Schut GW. Variations of bacterial populations in human feces measured by fluorescent in situ hybridization with group-specific 16S rRNA-targeted oligonucleotide probes. Appl Environ Microbiol 1998; 64:3336–3345.

119. Sghir A, Gramet G, Suau A, Rochet V, Pochart P, Dore J. Quantification of bacterial groups within human fecal flora by oligonucleotide probe hybridization. Appl Environ Microbiol 2000; 66:2263–2266.

120. Harmsen HJ, Wildeboer-Veloo AC, Grijpstra J, Knol J, Degener JE, Welling GW. Development of 16S rRNA-based probes for the *Coriobacterium* group and the *Atopobium* cluster and their application for enumeration of *Coriobacteriaceae* in human feces from volunteers of different age groups. Appl Environ Microbiol 2000; 66:4523–4527.

121. Requena T, Burton J, Matsuki T, et al. Identification, detection, and enumeration of human *Bifidobacterium* species by PCR targeting the transaldolase gene. Appl Environ Microbiol 2002; 68:2420–2427.

122. Tannock GW. Analysis of the intestinal microflora using molecular methods. Eur J Clin Nutr 2002; 56:S44–S49.

123. Mangin I, Bouhnik Y, Bisetti N, Decaris B. Molecular monitoring of human intestinal *Bifidobacterium* strain diversity. Res Mincrobiol 1999; 150:343–350.

124. Tannock GW. The bifidobacterial and *Lactobacillus* microflora of humans. Clin Rev Allergy Immunol 2002; 22:231–253.

125. Satokari RM, Vaughan EE, Akkermans ADL, Saarela M, de Vos WM. Bifidobacterial diversity in human feces detected by genus-specific PCR and denaturing gradient gel electrophoresis. Appl Environ Microbiol 2001; 67:504–513.

126. Matsuki T, Watanabe K, Tanaka R, Fukuda M, Oyaizu H. Distribution of bifidobacterial species in human intestinal microflora examined with 16S rRNA-gene-targeted species-specific primers. Appl Environ Microbiol 1999; 65:4506–4512.

127. Hayashi H, Sakamoto M, Benno Y. Phylogenetic analysis of the human gut microbiota using 16S rDNA clone libraries and strictly anaerobic culture-based methods. Microbiol Immunol 2002; 46:535–548.

128. Matsuki T, Watanabe K, Fujimoto J, et al. Development of 16S rRNA-gene-targeted group-specific primers for the detection and identification of predominant bacteria in human feces. Appl Environ Micrbiol 2002; 68:5445–5451.

129. Crittenden RG. Prebiotics. In: Tannock GW, ed. Probitoics: A Critical Review. Wymondham U.K.: Horizon Scientific Press, 1999:141–156.

130. Rao AV. Dose-response effects of inulin and oligofructose on intestinal bifidogenesis effects. J Nutr 1999; 129:1442S–1445S.

131. Saito Y, Hamanaka Y, Saito K, Takizawa S, Benno Y. Stability of species composition of fecal bifidobacteria in human subjects during fermented milk administration. Curr Microbiol 2002; 44:368–373.

132. Ventura M, Elli M, Reniero R, Zink R. Molecular microbial analysis of *Bifidobacterium* isolates from different environments by the species-specific amplified ribosomal DNA restriction analysis (ARDRA). FEMS Microbiol Ecol 2001; 36:113–121.

133. Alander M, Mättö J, Kneifel W, et al. Effect of galacto-oligosaccharide supplementation on human fecal microflora and on survival and persistence of *Bifidobacterium lactis* Bb-12 in the gastrointestinal tract. Int Dairy J 2001; 11:817–825.

134. Satokari RM, Vaughan EE, Akkermans ADL, Saarela M, de Vos WM. Polymerase chain reaction and denaturing gradient gel electrophoresis monitoring of fecal Bifidobacterium populations in a prebiotic and probiotic feeding trial. Syst Appl Microbiol 2001; 24:227–231.

135. Malinen E, Mättö J, Salmitie M, Alander M, Saarela M, Palva A. PCR-ELISA—II: Analysis of *Bifidobacterium* populations in human fecal samples from a consumption trial with *Bifidobacterium lactis* Bb-12 and a galacto-oligosaccharide preparation. Syst Appl Microbiol 2002; 25:249–258.

136. Harmsen HJM, Raangs GC, Franks AH, Wildeboer-Veloo ACM, Welling GW. The effect of the prebiotic inulin and the probiotic *Bifidobacterium longum* on the fecal microflora of healthy volunteers measured by FISH and DGGE. Microb Ecol Health Dis 2002; 14:211–219.

137. Zoetendal EG, Akkermans AD, de Vos WM. Temperature gradient gel electrophoresis analysis of 16S rRNA from human fecal samples reveals stable and host-specific communities of active bacteria. Appl Environ Microbiol 1998; 64:3854–3859.

138. Tannock GW. A fresh look at feces. Microbiol Aust 2003; 24:34–35.

139. Tannock GW, Munro K, Bibiloni R, et al. Impact of consumption of oligosaccharide-containing biscuits on the fecal microbiota of humans. Appl Environ Microbiol 2004; 70:2129–2136.

140. Schultz M, Timmer A, Herfarth HH, Sartor RB, Vanderhoof JA, Rath HC. *Lactobacillus* GG in inducing and maintaining remission of Crohn's disease. BMC Gastroenterol 2004; 15:5.

141. Linskens RK, Huijsdens XW, Savelkoul PHM, Vandenbroucke-Grauls CMJE, Meuwissen SMG. The bacterial flora in inflammatory bowel disease: Current insights in pathogenesis and the influence of antibiotics and probiotics. Scand J Gastroenterol 2001; 36:S29–S40.

142. Favier C, Neut C, Mizon C, Cortot A, Colombel JF, Mizon J. Fecal beta-D-galactosidase production and *Bifidobacteria* are decreased in Crohn's disease. Dig Dis Sci 1997; 42:817–822.

143. Rumi G, Tsubouchi R, Okayama M, Kato S, Mozsik G, Takeuchi K. Protective effect of lactulose on dextran sulfate sodium-induced colonic inflammation in rats. Dig Dis Sci 2004; 49:1466–1472.

144. Cherbut C, Michel C, Lecannu G. The prebiotic characteristics of fructooligosaccharides are necessary for reduction of TNBS-induced colitis in rats. J Nutr 2003; 133:21–27.

145. Holma R, Juvonen P, Asmawi MZ, Vapaatalo H, Korpela R. Galacto-oligosaccharides stimulate the growth of bifidobacteria but fail to attenuate inflammation in experimental colitis in rats. Scand J Gastroenterol 2002; 37:1042–1047.

146. Jacobasch G, Schmiedl D, Kruschewski M, Schmehl K. Dietary resistant starch and chronic inflammatory bowel diseases. Int J Colorectal Dis 1999; 14:201–211.

147. Moreau NM, Martin LJ, Toquet CS, et al. Restoration of the integrity of rat caeco-colonic mucosa by resistant starch, but not by fructo-oligosaccharides, in dextran sulfate sodium-induced experimental colitis. Br J Nutr 2003; 90:75–85.

148. Kanauchi Y, Fujiyama K, Mitsuyama Y, et al. Increased growth of *Bifidobacterium* and *Eubacterium* by germinated barley foodstuff, accompanied by enhanced butyrate production in healthy volunteers. Int J Mol Med 1999; 3:175–179.

149. Oikarinen S, Heinonen S, Karppinen S, et al. Plasma enterolactone or intestinal Bifidobacterium levels do not explain adenoma formation in multiple intestinal neoplasia (Min) mice fed with two different types of rye-bran fractions. Br J Nutr 2003; 90:119–125.

150. Bamba T, Kanauchi O, Andoh A, Fujiyama Y. A new prebiotic from germinated barley for nutraceutical treatment of ulcerative colitis. J Gastroenterol Hepatol 2002; 17:818–824.

151. Fukuda M, Kanauchi O, Araki Y, et al. Prebiotic treatment of experimental colitis with germinated barley foodstuff: a comparison with probiotic or antibiotic treatment. Int J Mol Med 2002; 9:65–70.

152. Kanauchi K, Mitsuyama Y, Araki A, Andoh A. Modification of intestinal flora in the treatment of inflammatory bowel disease. Curr Pharm Des 2003; 9:333–346.

153. Kanauchi T, Suga M, Tochihara T, et al. Treatment of ulcerative colitis by feeding with germinated barley foodstuff: first report of a multicenter open control trial. J Gastroenterol 2002; 37:67–72.

154. Mitsuoka T. Recent trends in research on intestinal flora. Bifidobacteria Microflora 1982; 1:3–24.

155. Hopkins MJ, Sharp R, Macfarlane GT. Age and disease related changes in intestinal bacterial populations assessed by cell culture, 16S rRNA abundance, and community cellular fatty acid profiles. Gut 2001; 48:198–205.

156. Hopkins MJ, Sharp R, Macfarlane GT. Variation in human intestinal microbiota with age. Dig Liver Dis 2002; 34:S12–S18.

157. Canzi E, Casiraghi MC, Zanchi R, et al. Yogurt in the diet of the elderly: a preliminary investigation into its effect on the gut ecosystem and lipid metabolism. Lait 2002; 82:713–723.

158. Silvi S, Verdenelli MC, Orpianesi C, Cresci A. EU project Crownalife: functional foods, gut microflora and healthy ageing - Isolation and identification of *Lactobacillus* and *Bifidobacterium* strains from fecal samples of elderly subjects for a possible probiotic use in functional foods. J Food Eng 2003; 56:195–200.

159. Hopkins MJ, MacFarlane GT. Changes in predominant bacterial populations in human feces with age and with *Clostridium difficile* infection. J Med Micrbiol 2002; 51:448–454.

160. Brown I, Warhurst M, Arcot J, et al. Fecal numbers of bifidobacteria are higher in pigs fed *Bifidobacterium longum* with a high amylose cornstarch than with a low amylose cornstarch. J Nutr 1997; 127:1822–1827.

161. Challa A, Rao DR, Chawan CB, Shackelford L. *Bifidobacterium longum* and lactulose suppress azoxymethane-induced colonic aberrant crypt foci in rats. Carcinogenesis 1997; 18:517–521.

162. Rowland IR, Rumney CJ, Coutts JT, Lievense LC. Effect of *Bifidobacterium longum* and inulin on gut bacterial metabolism and carcinogen-induced aberrant crypt foci in rats. Carcinogenesis 1998; 19:281–285.

163. Gallaher DD, Khil J. The effect of synbiotics on colon carcinogenesis in rats. J Nutr 1999; 129:1483S–1487S.

164. Femia AP, Luceri C, Dolara P, et al. Antitumorigenic activity of the prebiotic inulin enriched with oligofructose in combination with the probiotics *Lactobacillus rhamnosus* and *Bifidobacterium lactis* on azoxymethane-induced colon carcinogenesis in rats Carcinogenesis 2002; 23:1953–1960.

165. Asahara T, Nomoto K, Shimizu K, Watanuki M, Tanaka R. Increased resistance of mice to *Salmonella enterica* serovar Typhimurium infection by synbiotic administration of Bifidobacteria and transgalactosylated oligosaccharides. J Appl Microbiol 2001; 91:985–996.

166. Lamoureux L, Roy D, Gauthier SF. Production of oligosaccharides in yogurt containing bifidobacteria and yogurt cultures. J Dairy Sci 2002; 85:1058–1069.

167. Hung MN, Lee BH. Purification and characterization of a recombinant beta-galactosidase with transgalactosylation activity from *Bifidobacterium infantis* HL96. Appl Micrbiol Biotechnol 2002; 58:439–445.

168. Jorgensen F, Hansen OC, Stougaard P. High-efficiency synthesis of oligosaccharides with a truncated beta-galactosidase from *Bifidobacterium bifidum*. Appl Microbiol Biotechnol 2001; 57:647–652.

169. Crittenden R, Laitila A, Forssell P, et al. Adhesion of bifidobacteria to granular starch and its implications in probiotic technologies. Appl Environ Microbiol 2001; 67:3469–3475.

170. Schell MA, Karmirantzou M, Snel B, et al. The genome sequence of *Bifidobacterium longum* reflects its adaptation to the human gastrointestinal tract. Proc Natl Acad Sci USA 2002; 99:14422–14427.

171. Hopkins MJ, Cummings JH, Macfarlane GT. Inter-species differences in maximum specific growth rates and cell yields of bifidobacteria cultured on oligosaccharides and other simple carbohydrate sources. J Appl Microbiol 1998; 85:381–386.

172. Crittenden RG, Morris LF, Harvey ML, Tran LT, Mitchell HL, Playne MJ. Selection of a *Bifidobacterium* strain to complement resistant starch in a synbiotic yoghurt. J Appl Microbiol 2001; 90:268–278.

173. Beaugerie L, Petit JC. Microbial-gut interactions in health and disease. Antibiotic-associated. Best Pract Res Clin Gastroenterol 2004; 18:337–352.

174. Sherman RG, Prusinski L, Ravenel MC, Joralmon RA. Oral candidosis. Quintessence Int 2002; 33:521–532.

175. Dan M, Kaneti N, Levin D, Poch F, Samra Z. Vaginitis in a gynecologic practice in Israel: causes and risk factors. Isr Med Assoc J 2003; 5:629–632.

176. Plummer S, Weaver MA, Harris JC, Dee P, Hunter J. *Clostridium difficile* pilot study: effects of probiotic supplementation on the incidence of *C. difficile*. Int Microbiol 2004; 7:59–62.

177. Gill HS. Probiotics to enhance anti-infective defences in the gastrointestinal tract. Best Pract Res Clin Gastroenterol 2003; 17:755–773.

178. Marteau P, Seksik P, Jian R. Probiotics and intestinal health effects: a clinical perspective. Br J Nutr 2002; 88:S51–S57.

179. Kanamori Y, Hashizume K, Kitano Y, et al. Anaerobic dominant flora was reconstructed by synbiotics in an infant with MRSA enteritis. Pediatr Int 2003; 45:359–362.

180. Stern RS. Photocarcinogenicity of drugs. Toxicol Lett 1998; 102-103:389–392.

181. Dancer SJ. The problem with cephalosporins. J Antimicrob Chemother 2001; 48:463–478.

182. Patterson JA, Burkholder KM. Application of prebiotics and probiotics in poultry production. Poultry Sci 2003; 82:627–631.

183. Flickinger EA, van Loo J, Jr., Fahey GC. Nutritional responses to the presence of inulin and oligofructose in the diets of domesticated animals: a review. Crit Rev Food Sci Nutr 2003; 43:19–60.

184. Mattila-Sandholm T, Blum S, Collins JK, et al. Probiotics: towards demonstrating efficacy. Trends Food Sci Technol 1999; 10:393–399.

185. von Wright A, Vilpponen-Salmela T, Llopis MP, et al. The survival and colonic adhesion of *Bifidobacterium infantis* in patients with ulcerative colitis. Int Dairy J 2002; 12:197–200.

186. Shimakawa Y, Matsubara S, Yuki N, Ikeda M, Ishikawa F. Evaluation of *Bifidobacterium breve* strain Yakult-fermented soymilk as a probiotic food. Int J Food Microbiol 2003; 81:131–136.

187. Fujiwara S, Seto Y, Kimura A, Hashiba H. Intestinal transit of an orally administered streptomycin-rifampicin-resistant variant of *Bifidobacterium longum* SBT2928: its long-term survival and effect on the intestinal microflora and metabolism. J Appl Micrbiol 2001; 90:43–52.

188. Brigidi P, Swennen E, Vitali B, Rossi M, Matteuzzi D. PCR detection of *Bifidobacterium* strains and *Streptococcus thermophilus* in feces of human subjects after oral bacteriotherapy and yogurt consumption. Int J Food Microbiol 2003; 81:203–209.

189. McCartney AL, Wenzhi W, Tannock GW. Molecular analysis of the composition of the bifidobacterial and *Lactobacillus* microflora of humans. Appl Environ Microbiol 1996; 62:4608–4613.

190. Kimura K, McCartney AL, McConnell MA, Tannock GW. Analysis of fecal populations of bifidobacteria and lactobacilli and investigation of the immunological responses of their human hosts to the predominant strains. Appl Environ Microbiol 1997; 63:3394–3398.

191. Mangin I, Bouhnik Y, Bisetti N, Decaris B. Molecular monitoring of human intestinal *Bifidobacterium* strain diversity. Res Mincrobiol 1999; 150:343–350.

192. Dunne C, O'Mahony L, Murphy L, et al. In vitro selection criteria for probiotic bacteria of human origin: correlation with in vivo findings. Am J Clin Nutr 2001; 73:386S–392S.

193. Germond JE, Mamin O, Mollet B. Species specific identification of nine human *Bifidobacterium* spp. in feces. Syst Appl Micrbiol 2002; 25:536–543.

194. Zopf D, Roth S. Oligosaccharide anti-infective agents. Lancet 1996; 347:1017–1021.

195. Kunz C. Microbial receptor analogs in human milk: structural and functional aspects. In: Hansen LA, Yolken RH, eds. Probiotics, Other Nutritional Factors and Intestinal Microflora. (Nestle Nutrition Workshop Series, Vol. 42, Nestec Ltd, Vevey). Philadelphia: Lippincott-Raven, 1999:157–173.

196. Kunz C, Rudloff S. Health benefits of milk-derived carbohydrates. Bull Int Dairy Fed 2002; 375:72–79.

197. Kunz C, Rudloff S. Biological functions of oligosaccharides in human milk. Acta Paediatr 1993; 82:903–912.

198. Orchel A, Dzierzewicz Z, Parfiniewicz B, Weglarz L, Wilczok T. Butyrate-induced differentiation of colon cancer cells is PKC and JNK dependent. Dig Dis Sci 2005; 50:490–498.

199. Miller SJ. Cellular and physiological effects of short-chain fatty acids. Mini Rev Med Chem 2004; 4:839–845.

200. Lupton JR. Microbial degradation products influence colon cancer risk: the butyrate controversy. J Nutr 2004; 134:479–482.

201. Young GP, Le Leu RK. Resistant starch and colorectal neoplasia. J AOAC Int 2004; 87:775–786.

202. De Preter V, Geboes K, Verbrugghe K, et al. The in vivo use of the stable isotope-labelled biomarkers lactose-[N-15]ureide and [H-2(4)]tyrosine to assess the effects of pro- and prebiotics on the intestinal flora of healthy human volunteers. Br J Nutr 2004; 92:439–446.

203. Rowland IR, Tanaka R. The effects of transgalactosylated oligosaccharides on gut flora metabolism in rats associated with a human fecal microflora. J Appl Bacteriol 1993; 74:667–674.

204. Phillips J, Muir JG, Birkett A, et al. Effect of resistant starch on fecal bulk and fermentation-dependent events in humans. Am J Clin Nutr 1995; 62:121–130.

205. Hylla S, Gostner A, Dusel G, et al. Effects of resistant starch on the colon in healthy volunteers: possible implications for cancer prevention. Am J Clin Nutr 1998; 67:136–142.

206. Grubben MJ, van den Braak CC, Essenberg M, et al. Effect of resistant starch on potential biomarkers for colonic cancer risk in patients with colonic adenomas: a controlled trial. Dig Dis Sci 2001; 46:750–756.

207. Bouhnik Y, Flourié B, Ouarne F, et al. Effects of prolonged ingestion of fructo-oligosaccharides (FOS) on colonic bifidobacteria, fecal enzymes and bile acids in humans. Gastroenterol 1994; 106:A598.

208. Rowland IR, Bearne CA, Fischer R, Pool-Zobel BL. The effect of lactulose on DNA damage induced by DMH in the colon of human flora-associated rats. Nutr Cancer 1996; 26:37–47.

209. Perrin P, Pierre F, Patry Y, et al. Only fibers promoting a stable butyrate producing colonic ecosystem decrease the rate of aberrant crypt foci in rats. Gut 2001; 48:53–61.

210. Nakanishi S, Kataoka K, Kuwahara T, Ohnishi Y. Effects of high amylose maize starch and *Clostridium butyricum* on metabolism in colonic microbiota and formation of azoxymethane-induced aberrant crypt foci in the rat colon. Microbiol Immunol 2003; 47:951–958.

211. Scholz-Ahrens KE, Schaafsma G, van den Heuvel EG, Schrezenmeir J. Effects of prebiotics on mineral metabolism. Am J Clin Nutr 2001; 73:459S–464S.

212. Scholz-Ahrens KE, Schrezenmeir J. Inulin, oligofructose and mineral metabolism—experimental data and mechanism. Br J Nutr 2002; 87:S179–S186.

213. Morohashi T. The effect on bone of stimulated intestinal mineral absorption following fructooligosaccharide consumption in rats. Biosci Microflora 2002; 21:21–25.

214. Sakuma K. Molecular mechanism of the effect of fructooligosaccharides on calcium absorption. Biosci Microflora 2002; 21:13–20.

215. Hillman LS, Tack E, Covell DG, Vieira NE, Yergey AL. Measurement of true calcium absorption in premature infants using intravenous ^{46}Ca and oral ^{44}Ca. Pediatr Res 1988; 23:589–594.

216. Chonan K, Matsumoto M, Watanuki. Effects of galactooligosaccharides on calcium absorption and preventing bone loss in ovariectomized rats. Biosci Biotechnol Biochem 1995; 59:236–239.

217. Mizota T. Lactulose as a growth- promoting factor for *Bifidobacterium* and its physiological aspects. Bull Int Dairy Fed 1996; 313:43–48.

218. Ohta A, Osakabe N, Yamada K, Saito Y, Hidaka H. Effects of fructooligosaccharideson Ca, Mg and P absorption in rats. J Jap Soc Nutr Food Sci 1993; 46:123–129.

219. van den Heuvel EG, Schaafsma G, Muys T, van Dokkum W. Nondigestible oligosaccharides do not interfere with calcium and nonheme-iron absorption in young, healthy men. Am J Clin Nutr 1998; 67:445–451.

220. Teuri U, Korpela R, Saxelin M, Montonen L, Salminen S. Increased fecal frequency and gastrointestinal symptoms following ingestion of galacto-oligosaccharide-containing yogurt. J Nutr Sci Vitaminol 1998; 44:465–471.

221. Coudray C, Bellanger J, Castiglia-Delavaud C, Remesy C, Vermorel M, Rayssignuier Y. Effect of soluble or partly soluble dietary fibers supplementation on absorption and balance of calcium, magnesium, iron and zinc in healthy young men. Eur J Clin Nutr 1997; 51:375–380.

222. van den Heuvel EG, Muijs T, van Dokkum W, Schaafsma G. Lactulose stimulates calcium absorption in postmenopausal women. J Bone Miner Res 1999; 14:1211–1216.

223. van den Heuvel EG, Muys T, van Dokkum W, Schaafsma G. Oligofructose stimulates calcium absorption in adolescents. Am J Clin Nutr 1999; 69:544–548.

224. van den Heuvel EG, Schoterman MHC, Muijs T. Transgalactooligosaccharides stimulate calcium absorption in postmenopausal women. J Nutr 2000; 130:2938–2942.

225. Griffin IJ, Davila PM, Abrams SA. Non-digestible oligosaccharides and calcium absorption in girls with adequate calcium intakes. Br J Nutr 2002; 87:S187–S191.

226. Coudray C, Demigne C, Rayssiguier Y. Effects of dietary fibers on magnesium absorption in animals and humans. J Nutr 2003; 133:1–4.

227. van Dokkum W, Wezendonk B, Srikumar TS, van den Heuvel EG. Effect of nondigestible oligosaccharides on large-bowel functions, blood lipid concentrations and glucose absorption in young healthy male subjects. Eur J Clin Nutr 1999; 53:1–7.

228. Jackson KG, Taylor GR, Clohessy AM, Williams CM. The effect of the daily intake of inulin on fasting lipid, insulin and glucose concentrations in middle-aged men and women. Br J Nutr 1999; 82:23–30.

229. Kruse HP, Kleessen B, Blaut M. Effects of inulin on fecal bifidobacteria in human subjects. Br J Nutr 1999; 82:375–382.

230. Brighenti F, Casiraghi MC, Canzi F, Ferrari A. Effect of consumption of a ready-to-eat breakfast cereal containing inulin on the intestinal milieu and blood lipids in healthy male volunteers. Eur J Clin Nutr 1999; 53:726–733.

231. Alles MS, de Roos MN, Bakx JC, van de Lisdonk E, Zock PL, Hautvast JGAC. Consumption of fructo-oligosaccharides does not favourably affect blood glucose and serum lipids in non-insulin dependent diabetic patients. Am J Clin Nutr 1999; 69:64–69.

232. Williams CM, Jackson KG. Inulin and oligofructose: effects on lipid metabolism from human studies. Br J Nutr 2002; 87:S261–S264.

233. Delzenne NM, Daubioul C, Neyrinck A, Lasa M, Taper HS. Inulin and oligofructose modulate lipid metabolism in animals: review of biochemical events and future prospects. Br J Nutr 2002; 87:S255–S259.

234. Özaslan C, Türkçapar AG, Kesenci M, et al. Effect of Lactulose on Bacterial Translocation. Eur J Surg 1997; 163:463–467.

235. Bovee-Oudenhoven IMJ, Termont DMSL, Heidt PJ, Van der Meer R. Increasing the intestinal resistance of rats to the invasive pathogen *Salmonella enteritidis*: additive effects of dietary lactulose and calcium. Gut 1997; 40:497–504.

236. Schumann C. Medical, nutritional and technological properties of lactulose. An update. Eur J Nutr 2002; 41:17–25.

237. Buddington KK, Donahoo JB, Buddington RK. Dietary oligofructose and inulin protect mice from enteric and systemic pathogens and tumor inducers. J Nutr 2002; 132:472–477.

238. Duggan C, Penny ME, Hibberd P, et al. Oligofructose-supplemented infant cereal: 2 randomized, blinded, community-based trials in Peruvian infants. Am J Clin Nutr 2003; 77:937–942.

239. Moshfegh AJ, Friday JE, Goldman JP, Chug Ahuja JK. Presence of inulin and oligofructose in the diets of Americans. J Nutr 1999; 129:1407S–1411S.

240. van Loo J, Coussement P, De Leenheer L, Hoebregs H, Smits G. On the presence of inulin and oligofructose as natural ingredients in the Western diet. Crit Rev Food Sci Nutr 1995; 35:525–552.

241. Smith PB. Safety of short-chain fructo-oligosaccharides and GRAS affirmation by the U.S.FDA. Biosci Microflora 2002; 21:27–29.

242. Deguchi Y, Matsumoto K, Ito A, Watanuki M. Effects of β 1-4 galacto-oligosaccharides administration on defaecation of healthy volunteers with a tendency to constipation. Jap J Nutr 1997; 55:13–22.

243. Saavedra JM, Tschernia A. Human studies with probiotics and prebiotics: clinical implications. Br J Nutr 2002; 87:S241–S246.

244. Bovee-Oudenhoven IMJ, ten Bruggencate SJM, Lettink-Wissink MLG, van der Meer R. Dietary fructo-oligosaccharides and lactulose inhibit intestinal colonization but stimulate translocation of *Salmonella* in rats. Gut 2003; 52:1572–1578.

245. ten Bruggencate SJ, Bovee-Oudenhoven IM, Lettink-Wissink ML, van der Meer R. Dietary fructo-oligosaccharides dose-dependently increase translocation of *Salmonella* in rats. J Nutr 2003; 133:2313–2318.

246. van Loo JA. Prebiotics promote good health: the basis, the potential, and the emerging evidence. J Clin Gastroenterol 2004; 38:S70–S75.

247. Younes H, Coudray C, Bellanger J, Demigne C, Rayssiguier Y, Remesy C. Effects of two fermentable carbohydrates (inulin and resistant starch) and their combination on calcium and magnesium balance in rats. Br J Nutr 2001; 86:479–485.

17

Modifying the Gastrointestinal Microbiota with Probiotics

Chandraprakash D. Khedkar
Department of Dairy Microbiology and Biotechnology (Maharashtra Animal and Fishery Sciences University, Nagpur), College of Dairy Technology, Warud (Pusad), India

Arthur C. Ouwehand
Danisco Innovation, Kantvik, and Functional Foods Forum, University of Turku, Turku, Finland

INTRODUCTION

The origin of fermentations involving the production of lactic acid are lost in the ancient times, but it is not difficult to imagine how nomadic communities gradually acquired the art of preserving their meager supplies of milk by storing them in animal skins or crude earthenware pots. Initially, the intention could well have been simply to keep the milk cool through the evaporation of whey from the porous surface, but the chance transformation of the raw milk into a refreshing, slightly viscous foodstuff would soon have been recognized as a desirable innovation resulting in yogurt-like products.

At the beginning of last century, Eli Metchnikoff proposed the, now classic, theory that the apparent longevity of Bulgarian tribesmen was a direct result of their lifelong consumption of yogurt-like fermented milk products, probably mostly fermented by lactobacilli (1). This inspired an interest in the nutritional and therapeutic characteristics of these products. The validity of these hypotheses was debated for many years but one undeniable effect of his work was a marked increase in the popularity of yogurt throughout Europe. At about the same time, Henri Tissier suggested that bifidobacteria could be administered to children with diarrhea to help restore their gut microbiota balance (2).

Fermented milk products like yogurt and other products containing beneficial or "probiotic" cultures, such as lactobacilli, bifidobacteria, lactococci, and propionibacteria are currently among the best-known examples of functional foods in many countries around the world. These products are associated with a range of health claims, some more documented then others, including alleviation of symptoms of lactose intolerance (3), treatment of diarrhea (4), cancer risk reduction (5) and restoration of gastrointestinal (6) and urogenital microbiota (7), and constipation (8). Milk is an ideal food system to act as a carrier of these versatile bacteria to the human gastrointestinal tract (GIT) and support their viability. From these beginnings, the probiotic concept has progressed considerably and is now the focus of much research attention worldwide. Significant advances have been made in the selection and characterization of specific cultures and substantiation of health claims relating to their consumption. Subsequently, the area of probiotics has advanced from anecdotal reports, with scientific evidence now accumulating to back up health claim properties of specific strains. Nowadays the majority of scientific and commercial attention is concentrated on probiotic microorganisms like *Lactobacillus* and *Bifidobacterium*, with the result that an expanding range of probiotic dairy products containing these species are now available to the consumer.

This paper will critically examine the health claims and evidence for beneficial effects of probiotic organisms in relation to modifying the gastrointestinal microflora and its functioning.

PROBIOTICS

Definition

The term probiotic is derived from Greek, meaning "for life" and originated to describe substances produced by one microorganism which stimulate the growth of others (9). The Food and Agriculture Organization of the United Nations (FAO) and the World Health Organization (WHO) have stated that there is adequate scientific evidence to indicate that there is potential for probiotic foods to provide health benefits and that specific strains are safe for human consumption (10). An expert panel commissioned by FAO and WHO defined probiotics as "Live microorganisms which when administered in adequate amounts confer a health benefit on the host." This definition will be used in the current chapter instead of the term biotherapeutic agents (11) which is sometimes used as well to indicate probiotics.

Probiotic Microorganisms

Lactobacillus and *Bifidobacterium* are the principal bacterial genera central to both probiotic and prebiotic approaches to dietary modulation of the intestinal microflora. In

Table 1 Commonly Used Probiotic Microorganisms

Lactobacillus	*Bifido-bacterium*	*Lacto-coccus lactis subsp*	*Strepto-coccus*	*Entero-coccus*	*Saccharomyces*
acidophilus, brevis, delbruekii, fermentum, gasseri, johnsonii, lactis, paracasei, plantarum, rhamnosus, reuteri	adolescentis, animalis/ lactis,[a] bifidum, breve, infantis, lactis, longum, thermophilum	cremoris, lactis	thermophilus	faecium	cerevisiae (boulardii)[b]

[a] The current taxonomic status of *B. animalis* and *B. lactis* is unclear.
[b] *Saccharomyces boulardii* is likely to be identical to *Saccharomyces cerevisiae*.
Source: From Refs. 12–16.

addition, there are many different microorganisms currently used as probiotics. A list of microbes commonly used as probiotics is given in Table 1. Some of these organisms have been studied much more extensively than others. It is therefore important that probiotics are referred to by their strain designation as well as their species. Although other members of the same species share most characteristics, different probiotic strains may differ in some essential properties (17).

Desirable Characteristics of Probiotic Microorganisms

Many desirable characteristics have been proposed by various researchers for probiotic lactobacilli and bifidobacteria (and other microbes) to be used as dietary adjuncts for gastrointestinal and related health benefits. These organisms should have the ability to survive in sufficient numbers, the acidity of the gastric juices and to pass in a viable state to the small intestinal region (18–23). Ability of these organisms to proliferate and/or colonize the gut is also an important desirable, although appears not so common, property. In practice the desired properties of these microorganisms are dependent on the host for which probiotic administration is intended, the anatomical site within the host toward which the probiotic is directed (most often the GIT) and the desired effect at that site are the principal focus of probiotic applications (19,21,23).

A general set of desirable properties of probiotic microorganisms, regardless of the intended host or site of application is presented in Table 2. In vitro tests based on these selection criteria, although not a definite means of strain selection, may provide useful initial information. In addition, well-characterized, and validated model systems such as the TNO Intestinal Models (TIM-I and II) and the Simulator of the Human Intestinal Microbial Ecosystem (SHIME), which aim to mimic complex physiological and physicochemical in vivo reactions, may also be of value in strain selection (for a description of intestinal models, see the chapter by Mäkivuokko and Nurminen). Several tests for gastric passage and gastric digestion of the candidature organisms, as well as pH resistance and ability to pass through the stomach are presented in Table 3. Such types of tests are less expensive than human or animal trials and do not have the associated ethical drawbacks (29). However, ultimate proof of probiotic effects requires validation in well-designed, randomized, double-blind, placebo controlled, statistically sound clinical trials (30).

Administration of a large number of these organisms will increase the number of surviving microbes, but various strains of these organisms may differ in acid tolerance and survival. However, as the transit time of fermented milk products through the stomach is shorter than many other foods (31), and fermented dairy products provide a buffer towards gastric juice (32), this has been shown to lead to the appearance in high numbers of the administered strains in the feces (33).

PROPOSED HEALTH BENEFITS OF PROBIOTICS

The health benefits of probiotics can be direct or indirect through modulation of the composition and/or activity of the endogenous microbiota or of the immune system. Many health claims have been made concerning probiotics, especially concerning their potential to prevent or help cure gastrointestinal and related ailments. These include improved lactose digestion and other direct enzymatic effects, prevention, and curative treatment of gastroenteritis, antibiotic-associated diarrhea, traveler's diarrhea, constipation, intestinal

Table 2 Desirable Properties of Probiotics

Probiotic characteristics	Technological/Functional properties
Stability: bile salts and gastric acidity	Survival in human gastrointestinal tract
Adherence: ability to adhere to the intestinal mucosa	Immune cell modulation and competitive inhibition of the pathogenic organisms
Transient colonization	Growth and multiplication in the human gastrointestinal tract
Safety	Well-documented clinical safety, organism must be accurately identified to strain level before recommending its use. It should be non-toxic, non-pathogenic, non-allergenic, non-mutagenic, non-carcinogenic and have no transferable antibiotic resistance
Antagonism: against pathogenic and putre-factive organisms	Prevention of pathogen colonization through competition for nutrients and binding sites and through production of antimicrobial substances
Proven health effects	Clinically documented and validated therapeutic effects. Dose-response data for minimum effective dosage of the probiotic organism in different formulations
Stability: stability/viabil-ity during processing and storage	All of the aforementioned desirable characteristics should be maintained during processing and storage of these products organism should be genetically stable, no plasmid transfer
Technological suitability	Culture should be suitable for production of acceptable quality finished products with desirable viable counts

infections and to suppress colonization of the gut by pathogenic organisms colonized in gut, irritable bowel syndrome (IBS) and various conditions of diarrhea, hypocholesterolaemea, urogenital tract infection, atopic diseases, skin diseases, gastrointestinal well-being, inflammatory bowel disease (IBD) and colon cancer (16,31,34,35).

Table 3 Experiments Demonstrating Resistance Tests for Survival of an Organism in the Upper Digestive Tract for Selected Probiotic Strains

Resistance test method	Organisms tested	Reference
Gastric digestion in vivo (mixture of HCl + pepsin + rennet)	*Lb. acidophilus* (survival)	(24)
	Propionibacterium freudenreichii (survival without loss of vitality)	(25)
	Yogurt, buttermilk and sour milk cultures (survival with different digestion times)	
pH	*Bif. bifidum* (4 strains) 2 hours at pH 2.4 and 6.5 (strong action at pH 2)	(26)
Human gastric juices conditions of the stomach: cultured milk mixed with gastric juice (70:30)	*Lb. acidophilus* (survival) yogurt and sour milk cultures (addition of gastric juices with pH 3.48–6.75, no bacteriocidal or bacteriostatic effect observed)	(27)
Artificial gastric juices (at pH 3.0 incubation by 37°C)	*Lb. acidophilus* and *Lb. plantarum* survive 3 hours; *Lb. bulgaricus* less resistant, survives only for 1 hour	(28)

These microorganisms possess various immunological functions viz., mitogenic activity (36), adjuvant activity (21), macrophage activation (37), enhancement of antibody production (38), induction of interferon-γ production (39) and antitumor effects (40), amongst others. It has further been indicated by a number of studies that both the cell wall and cytoplasm of specific probiotic bacteria induced mitogenic responses of spleen cells (37,41,42).

Health benefits of probiotic organisms related that may impact on the gut microbiota are summarized in the following paragraphs.

Use of Probiotics to Combat Gastrointestinal Infections

Probiotics have been shown to be useful in the treatment of a variety of gastrointestinal disorders, and the details are presented in Table 4. A number of these disorders have a significant inflammatory component in the small and/or large intestine and there is a growing body of research to suggest that probiotic bacteria may be useful particularly in many of these pediatric gastrointestinal conditions. Specific strains of *Lactobacillus rhamnosus, Lb. reuteri, Lb. plantarum, Bifidobacterium lactis*, and *Saccharomyces cerevisiae (boulardii)* have all been extensively studied. Probiotics can reduce the duration and severity of rotaviral enteritis, as well as decrease the risk of antibiotic-associated diarrhea in children and *Clostridium difficile* diarrhea in adults. Prevention of viral diarrhea in day-care centers as well as traveler's diarrhea has been demonstrated with some probiotics, although not all are equally effective (67). Small bowel bacterial overgrowth conditions may respond to probiotic use. How the probiotic bacteria counteract the inflammatory process by enhancing the degradation of external antigens, reducing the secretion of inflammatory mediators and maintaining the healthy gut microbiota by exclusion of pathogens is schematically shown in Figure 1.

Possible Mode of Action of Probiotics in Reducing the Duration of Diarrhea

Several potential mechanisms have been proposed for how probiotics reduce the duration of rotavirus diarrhea, but none have been proven and each theory has its limitations. The first is competitive blockage of receptor sites (69) in which probiotics bind to receptors, thereby preventing adhesion and invasion of the virus. This concept might be plausible if there was evidence for specific receptor competition. In most cases, by the time a probiotic is ingested, the patient will already have had diarrhea for possibly 12 hours. By this time, the virus has infected mature enterocytes in the mid- and upper region of the small intestinal villi. The virus and/or its enterotoxin, NSP4, will then have disturbed fluid and electrolyte transport, thereby lowering fluid and glucose absorption. The toxin could have then potentially activated secretory reflexes, causing loss of fluids from secretory epithelia, resulting in diarrhea (70). At best, subsequent competitive exclusion of viruses would only be effective for attachment of progeny, and it is not known whether such inhibition would reduce diarrhea. If probiotic organisms somehow competed with the toxin or peptides released from villous endocrine cells, it is feasible that the cascade that leads to diarrhea could be prevented.

The second potential mechanism may be that the immune response is enhanced by probiotics, leading to the observed clinical effect (45). This is supported by the protective effect which local immunoglobulin A (IgA) antibodies appear to confer against rotavirus (71). However, a problem with this theory is given that diarrhea appears to cease within 1 to 3 days in patients who would otherwise suffer for 4 to 6 days; the probiotics would need

Table 4 Examples of the Effects of Probiotics on Microbial Infections

Disorder	Subject	Probiotics	Effect	Reference
Infantile diarrhea	Human	*Lactobacillus* GG	Reduced duration of diarrhea etc.	(43–47)
	Human	*Lb. reuteri*	Reduced duration of diarrhea	(49)
	Human	*Bif. Bifidum* + *Str. thermophilus*	Prevention of diarrhea	(23)
	Human	*Bif. breve*	Prevention of diarrhea	(51)
Antibiotic-associated diarrhea	Human	*Bif. longum*	Decreased course of erythromycin-induced diarrhea	(52)
	Human	*Lactobacillus* GG	Decreased course of erythromycin-induced diarrhea, and other side effects of erythromycin	(53)
	Human	*Str. faecium*	Decreased diarrhea associated with anti-tubercular drugs administered for pulmonary TB	(54)
	Human	*Sc. boulardii*	Reduce incidence of diarrhea	(55) (56) (56a)
Relapsing *C. difficile* colitis	Human	*Lactobacillus* GG	Improves/terminates colitis	(57) (58)
	Human	*Lactobacillus* GG	Eradicated associated diarrhea	(59) (60)
Travelers' diarrhea	Human	*Lb. acidophilus* + *Bif. bifidum*	Decrease frequency, not duration of diarrhea	(61) (61a)
	Human	*Lactobacillus* GG		(62)
Foodborne pathogen exclusion	Male BALB/c Mice	*Lb. casei Shirota*	Increased resistance to lethal infection with Salmonella, *E. coli*, and *L. monocytogenes*	(62a)
	Male rat		Increased resistance to salmonellosis infection	(63)
	Mice	*Bif. lactis* HN019	Increased survival of Salmonella infection	(49a)
	Mice	*Lb. rhamnosus* HN001	Increased survival of *E. coli* O157:H7 infection	(50)
	In vitro	Yogurt bacteria	Inhibit growth of Salmonella	(64,65)
	Human	*Lb. acidophilus* + *Lactobacillus* GG	Decreased shigellosis-associated diarrhea	(66)

Abbreviations: Bif, Bifidobacterium; C, Clostridium; E, Escherichia; L, Listeria; Lb, Lactobacillus; Sc, Saccharomyces; Str, Streptococcus.

to trigger the antibody response rapidly so that it interfered with further viral activity. Animal studies do indicate that secretory IgA can be triggered by probiotic ingestion (72), but the rate was not determined, nor was the influence on cessation of fluid loss across the secretory cell membranes. Modification of the cytokine profile to one that enhances anti-inflammatory cytokines (73) or attenuation of the virus' and/or toxin's effect on the enteric nervous system might provide rapid cessation of epithelial secretion and diarrhea. Alternatively, stimulation of T cells to produce gamma interferon, leading to potential inhibition of chloride secretion, might also inhibit diarrhea. One aspect of the immunity theory that needs to be clarified is why lactobacilli, which we assume are present in the child intestine, appear unable to prevent infection; yet those administered orally thereafter help to clear the diarrhea.

A third mechanism could involve a signal(s) from probiotics to the host that down-regulates the secretory and motility defenses designed to remove perceived noxious substances. Glycosylated intestinal mucins inhibit rotaviruses (74), and MUC2 and MUC3 mRNA expression is increased in response to probiotics signaling, protecting cells against pathogenic bacterial adhesion (13). However, direct host cell signaling between probiotic organism and secretory cells has not yet been investigated. Attachment of the virus causes cytokine prostaglandin and nitric oxide to be released from the enterocytes, both of which could affect motility. The possibility exists that lactobacilli could alter this release (75). The intestinal host defense mechanisms comprise complex systems involving the innate and adaptive immune responses, and protective effects of the indigenous microbiota. The commensal microorganisms colonizing the intestinal mucosa provide a barrier effect against pathogens by using a variety of mechanisms, such as occupation of habitats, competition for nutrients, and production of antimicrobials. It is also established that the probiotic organisms can modulate the homeostasis of the host's defense mechanisms, both innate and adaptive immune functions (4).

A final theory is that the probiotics produce substances that inactivate the viral particles. This has been shown in vitro (76), with supernatants from *Lactobacillus rhamnosus* GR-l and *L. fermentum* RC-14 inactivating 10^9 particles of the double-stranded DNA adenovirus and the negative-stranded RNA vesicular stomatitis virus within 10 minutes. The effect was likely due to acid, but more specific antiviral properties have not been ruled out. Whether or not viral inactivation can inhibit diarrhea remains to be confirmed.

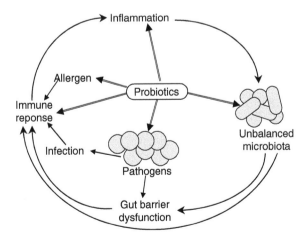

Figure 1   Schematic representation of the possible ways by which probiotics may counteract ($\Rightarrow$) the intestinal inflammatory process. *Source*: From Ref. 68.

More detailed investigation is needed to understand how probiotic strains reduce the duration of diarrhea in conjunction with rehydration therapy. Such studies could lead to a better understanding of the dynamics within the intestinal microbiota that is being disrupted and depleted by rapid fecal loss. In doing so, new intervention therapies should be generated to quickly and effectively trigger the cessation of not only rotavirus infections but also other gastrointestinal infections that debilitate patients for 2 to 3 days.

The possible mode of action for diarrhea and other gastrointestinal diseases, such as IBS and IBD, are the subject of intense investigation in many labs, using genomics, knockout mice models, etc., (77,78).

MODIFYING INTESTINAL MICROBIOTA COMPOSITION THROUGH INTAKE OF PROBIOTICS

In the human GIT, variability exists in bacterial numbers and composition between the stomach, small intestine and colon. The total bacterial count in gastric contents is usually below 10^3 per gram contents with numbers in the small intestine ranging from about 10^4 per ml of contents to about 10^6–10^7 at the terminal ileum (79). In comparison to other regions of the GIT, the human large intestine is a complex, heavily populated and diverse microbial ecosystem. Bacterial numbers in the human large intestine are in the range of 10^{11}–10^{12} for every gram of the gut contents (80). The colonic microbiota is capable of responding to anatomical and physicochemical variations that are present. The right or proximal colon is characterized by a relatively high substrate availability (due to dietary input), a pH of around 5.5–6.0 (from acids produced during microbial fermentation) and a more rapid transit than the distal region. The left or distal area of colon has a lower concentration of available substrate, in particular carbohydrates, the pH is approximately 6.5–7.0, and the flow of the digesta is slower. The proximal region tends to be a more saccharolytic environment than the distal gut, the latter having higher bacterial proteolysis. Several hundred different species of bacteria are known to be present in the large intestine (see also the chapter by Ben Amor and Vaughan). Gram-negative rods belonging to the *Bacteroides fragilis* group are the numerically predominant culturable bacteria in the colon. The other main groups consist of different (Gram-positive) rods and cocci, such as bifidobacteria, clostridia, peptococci, streptococci, eubacteria, lactobacilli, peptostreptococci, ruminococci, enterococci, coliforms, methanogens, dissimilattory sulfate-reducing bacteria, and acetogens. The microbiota includes saccharolytic organisms, proteolytic species and bacteria that can metabolize gases. Despite the huge diversity of bacteria present in the large gut (estimated over 1000 species), it is certain that the vast majority has hitherto not been identified or cultured [(81), see also the chapter by Ben Amor and Vaughan].

Increasing Numbers of Beneficial Microbes

One of the properties thought to be important for the health benefits of consumed probiotic organisms is their ability to adhere to the intestinal mucosa. As such they can resist peristalsis and occupy a habitat at the expense of potentially harmful organisms. The probiotic applications to the human gut are already widespread, and evidence is mounting that these organisms have a beneficial effect on the host. It is now well established that the probiotic organisms can transiently establish themselves in the GIT and inhibit the adhesion and growth of enteropathogens. Table 5 delineates the effect of feeding selected probiotic preparations on the human gut microbiota.

Suppressing Numbers of Potentially Harmful Microbes

The artificial manipulation of the human intestinal microbiota by consumption of large numbers of probiotic microorganisms may lead to the presence of large numbers of lactic acid-producing microorganisms in the small intestine. Any available sugars will be quickly fermented to various organic acids and/or ethanol. This leads to a change in the environment where the production of various low-molecular toxic metabolites and antigenic macromolecules by various intestinal, potentially pathogenic microbes and the effects of endotoxins may be strongly reduced (Table 5). The intestinal growth of all other types of nonintestinal pathogens is strongly inhibited by abundant probiotic fermentation in the small intestine. Reduction of viral infectivity was attributed to ethanol or acid-mediated denaturation of viral envelope proteins. In addition to organic acids, bacteriocins, such as e.g., Lactacin F (88), and some unidentified compounds synthesized by probiotic organisms

Table 5   Effect of Feeding Selected Probiotic Preparations on Human Gut Microbiota

Type of probiotic organisms	Effect on gut microbiota	Reference
Lb. rhamnosus GG	Attachment of probiotic organism to CaCo-2 intestinal cell line and in vivo to human colonic mucosa	(82)
Lb. rhamnosus GG	Increased the number of fecal bifidobacteria and lactobacilli Concomitant decrease in clostridia counts	(83)
Lb. plantarum (VTTE-79098)	Reduction in enterobacteriaceae counts of 4 log cycles, Clostridia 1 log cycle, and slight decreases in enterococci counts in a SHIME reactor	(82)
Lb. paracasei ssp. paracasei (VTTE-94506)		
Lb. paracasei ssp. paracasei (VTTE-94510)		
L. rhamnosus (VTTE-94510)		
Bifidobacterium sp. (VTTE-94508)		
L. casei Shirota	Balancing of intestinal microbiota	(84)
Bif. bifidum	Balancing of intestinal microbiota	(84a)
Lb. acidophilus-LBKV3	Highly significant increases in fecal lactobacilli, bifidobacteria, propionibacteria and lacto-cocci counts and concomitant decreases in coliforms, clostridia, staphylococci and enterococci in tribal kids of 2–5 years	(85)
Lb. acidophilus-LBKV3 supplimented with *Propionibacterium freundenrichii ssp. Shermanii*	Increases in vivo antimicrobial activity of the microflora against putrefactive organisms in the gut of tribal kids of 2–5 years	(86)
Bif. lactis HN019	Increase in fecal lactobacilli and bifidobacteria	(87)

Abbreviation: SHIME, Simulator for Human Intestinal Microbiological Ecosystem.

may confer an additional growth-inhibiting effect (89). However, it is still uncertain whether such substances are produced in situ in the intestine and are effective.

MODIFYING THE MICROBIAL METABOLIC ACTIVITY

Due to its numbers and taxonomic diversity, the intestinal microbiota has an enormous metabolic potential. The microbiota's metabolic activity is comparable to that of the liver, our metabolically most active organ. This metabolism has a pronounced influence on the health and well being of the host, as described in more detail in the chapter by Goldin. Probiotics have been shown to be able to change the metabolic activity of the intestinal microbiota. In part, this may relate to a direct change in its composition, but it may also relate to a change in metabolism of some members of the microbiota in response to a shift in the intestinal environment. The main metabolic markers that are potentially influenced by probiotics are the production of short chain fatty acids (SCFA) and fecal enzyme activity.

Short Chain Fatty Acid Production

Principal end products of bacterial fermentation in the colon are SCFA, i.e., acetate, propionate, and butyrate. Other fermentation products include ethanol, lactate, succinate, formate, valerate, and caproate. Branched chain fatty acids such as isobutyrate, 2-methyl-butyrate, and isovalerate may also be formed from the fermentation of amino acids.

Short Chain Fatty Acids

The production of SCFA by the intestinal microbiota serves to salvage energy from the digesta that would otherwise be lost for the host (90). Butyrate provides an important energy source for the intestinal epithelium. Propionate is metabolized in the liver where it possibly serves as a precursor for gluconeogenesis. Acetate is mainly taken up by muscle tissue but is also used by adipocytes for lipogenesis. Lactate is also metabolized by muscle tissue. However, despite the fact that enterocytes only slowly absorb lactate, it is usually found only at low concentrations in the digesta as it is used to a large extent by members of the intestinal microbiota (91) and only accumulates in disease (92).

Probiotics and Short Chain Fatty Acids

Probiotics will, when they are metabolically active, produce organic acids in the intestine; these will mainly be lactate and acetate. Furthermore, the metabolic activity will influence the metabolism of other microbes present in the intestine, through competition for nutrients and through the production of metabolites. It is, however, not really known to what extent probiotics are metabolically active in the human intestine, in particular in the colon, and whether probiotics produce antimicrobials such as bacteriocins in situ. Studies in mice, colonized with a human microbiota, do however indicate metabolic activity (93).

Assessment of the data presented in Table 6 indicates that most probiotics tested do not affect the composition of the fecal short chain fatty acid composition. This may be explained by the lack of metabolic activity of the probiotics in the colon, but it is more likely to reflect the efficient absorption of fatty acids by the colon (2). Therefore, to assess the influence of probiotics, and for that matter also prebiotics, on the availability of SCFA,

Table 6 Influence of Probiotics on Fecal Short Chain Fatty Acids (SCFA) and Fecal Enzyme Activity in Humans, Selected References

Probiotic	Dose (CFU/day)	Duration	Subjects	SCFA change	Fecal enzyme activity change	Reference
B. lactis Bb-12	2.8×10^{10}	6 hours	Ileostomists	No change	–	(94)
S. cerevisiae boulardii	1 g	6 days	Healthy adults	No change	–	(95)
S. cerevisiae boulardii	1 g	6 days	Patients with total enteral nutrition	Increase	–	(95)
Yogurt + *L. acidophilus* 145 + *B. longum* 913	3×10^8 *L. acidophilus* 3×10^7 *B. longum*	6 months	Healthy adults	No change	–	(13)
Kefir			Healthy adults	Increase, though not different from control (milk)	–	(96)
L. plantarum 299v		4 weeks	Healthy adults	No change	No change	(97)
L. casei Shirota	3×10^{11}	4 weeks	Healthy adults	Decrease	Decrease	(98)
L. rhamnosus HN019	1.6×10^9	6 months	Healthy adults	No change	No change	(99)
L. casei DN-114 001	1.3×10^{10}	1 month	Healthy infants	No change	Decrease	(100)
L. gasseri SBT2055	$10^9, 10^{10}, 10^{11}$	41 days	Healthy adults	No change	Decrease/ no change	(100a)
L. gasseri ADH	2×10^{10}	11 days	Healthy elderly Elderly with atrophic gastritis	–	Decrease	(101)
L. rhamnosus GG	1.4×10^{10}	4 weeks	Healthy adults	–	No change	(86)
L. rhamnosus GG	2×10^{10}	2 weeks	Healthy elderly	–	Decrease	(102)
L. rhamnosus GG	$1-2 \times 10^{10}$	2 weeks	Healthy adults	–	Decrease	(103)
L. rhamnosus GG	4×10^{10}	4 weeks	Healthy adults	–	Decrease	(103a)

(Continued)

Table 6 Influence of Probiotics on Fecal Short Chain Fatty Acids (SCFA) and Fecal Enzyme Activity in Humans, Selected References (*Continued*)

Probiotic	Dose (CFU/day)	Duration	Subjects	SCFA change	Fecal enzyme activity change	Reference
L. rhamnosus LC-705 *P. freudenreichii* JS	$1-2\times10^{10}$ *L. rhamnosus* $2-4\times10^{10}$ *P. freudenreichii*	4 weeks	Healthy elderly	–	Decrease	(104)
L. reuterii	7.2×10^{8}	4 weeks	Healthy elderly	–	No change	(104)
B. longum	1.3×10^{10}	3–6 weeks	Healthy adults	–	Decrease	(105)
L. acidophilus NCFM	4×10^{10}	4 weeks	Healthy adults	–	Decrease	(106)
VSL#3 (bifidobacteria + lactobacilli + streptococci)	9×10^{11}	20 days	Irritable bowel syndrome patients	–	Decrease/ increase	(107)

sampling should preferably take place in the proximal colon where substrates are more abundant and the microbes more active.

Fecal Enzyme Activity

Fecal Enzymes

One of the detrimental effects the human intestinal microbiota may have on host health is the production of tumor promoters, mutagens, and carcinogens from undigested dietary substrates and endogenous residues. Bacterial enzymes involved in the formation of such substances are β-glucoronidase, azoreductase, nitroreductase, and nitrate reductase (108); see also the chapters by Rafter and Rowland, and Goldin. A reduction in the activity of these enzymes can be expected to lead to a reduced exposure to carcinogenic substances. Animal models have suggested this also leads to a reduced incidence in colorectal cancers (106). However, it is not clear if this also holds true for humans.

Probiotics and Fecal Enzyme Activity

Most of the probiotics (listed in Table 6) tend to induce a reduction in fecal enzyme activity. This appears to be therefore one of the more general and reproducible properties of probiotics. However, since fecal enzyme activity is not a definite biomarker for cancer risk, one should be cautious when drawing conclusions and extrapolating from animal experiments to humans. As with SCFA production, the mechanism behind this is probably competition for nutrients and production of inhibitory metabolites.

CONCLUSION

The area of modulation of gastrointestinal microbiota through intake of probiotics seems to hold much promise for the prophylactic management and/or treatment of gut disorders, as mediated by pathogens. The growing realization by consumers that our food profoundly influences our health has fueled the introduction of food products with health claims such as probiotics into the market. It seems that the use of probiotics in general clinical practice is not far away, given that products such as VSL#3, containing a mixture of lactic acid bacteria probiotics, are already being used. However, it is relevant to note that studies on particular strains may not necessarily be extrapolated to all probiotic microorganisms. Molecular tools will continue to be used to understand and manipulate probiotic bacteria with a view to produce vaccines and new and improved products. The critical step in wider application will be to make products available that are safe and clinically proven in a specific formulation easily accessible to physicians and consumers. Systematically randomized, double-blind and placebo-controlled studies including large numbers of human volunteers are needed to advance the scientific knowledge of probiotics and gastrointestinal microbiota. Technological advances like protective coating(s), micro-encapsulation, or addition of prebiotic compounds that can serve as growth factors for probiotic organisms will improve the survival of strains in the gut of consumers. It is necessary to clearly understand the functionality of these organisms in the intestinal ecosystem.

REFERENCES

1. Metchnikoff E. The Prolongation of Life. New York: G. Putnam's Sons, 1908.
2. Tissier H. Traitement des infections intestinales par la méthode de la flore bactérienne de l'intestin. Crit Rev Soc Biol 1906; 60:359–361.
3. Kim HS, Gilliland SE. *Lactobacillus acidophilus* as a dietary adjunct for milk to aid lactose digestion in humans. J Dairy Sci 1983; 66:959–966.
4. van Niel CW, Freudtner C, Garrison MM. Christakis, D.A. *Lactobacillus therapy* for acute infectious diarrhoea in children: a meta-analysis. Pediatrics 2002; 109:678–684.
5. Ohashi Y, Nakai S, Tsukamoto T, et al. Habitual intake of lactic acid bacteria and risk reduction of bladder cancer. Urol Int 2002; 68:273–280.
6. Orrhage K, Sjöstedt S, Nord CE. Effect of supplements with lactic acid bacteria and oligofructose on the intestinal microflora during administration of cefpodoxime proxetil. J Antimicrob Chemother 2000; 46:603–611.
7. Gardiner GE, Heinemann C, Bruce AW, Beuerman D, Reid G. Persistence of *Lactobacillus fermentrum* RC-14 and *Lactobacillus rhamnosus* GR-1 but not L. rhamnosus GG in the human vagina as demonstrated by randomly amplified polymorphic DNA. Clin Diag Lab Immunol 2002; 9:92–96.
8. Marteau P, Cuillerier E, Meance S, et al. *Bifidobacterium animalis* strain DN-173 010 shortens the colonic transit time in healthy women: a double-blind, randomized, controlled study. Aliment Pharmacol Ther 2002; 16:587–593.
9. Lilly DM, Stillwell RH. Therapeutic properties of substances produced by lactic acid bacteria. Science 1965; 147:747–748.
10. Food and Agriculture Organization of United Nations and World Health Organization 2001. Regulatory and clinical aspects of dairy probiotics. Food and Agriculture

Organization of United Nations and World Health Organization Expert Consultant Report. Food and Agriculture Organization of United Nations and World Health Organization Working Group Report (Online).

11. Food and Agriculture Organization of United Nations and World Health Organization 2002. Guidelines for the evaluation of probiotics in food. Food and Agriculture Organization of United Nations and World Health Organization Working Group Report (Online).

12. Dunne C, O'Mahony L, Murphy L. In vitro selection criteria for probiotic bacteria of human origin:correlation with in vivo findings. Am J Clin Nutr 2001; 73:386S–392S.

13. Mack DR, Michail S, Wei S, McDougall L, Hollingsworth MA. Probiotics inhibit enteropathogenic *E. coli* adherence in vitro by inducing intestinal mucin gene expression. Am J Physiol 1999; 276:G941–G950.

14. Mattila-Sandholm T, Salminen S. Up-to-date on probiotics in Europe. Gastroenterol Intern 1998; 1:8–12.

15. Ventura M, Zink R. Rapid identification, differentiation, and proposed new taxonomic classification of *Bifidobacterium lactis*. Appl Environ Microbiol 2002; 68:6429–6434.

16. Gardiner GE, Heinemann C, Baroja ML, et al. Oral administration of the probiotic combination *Lactobacillus rhamnosus* GR-1 and *L. fermentum* RC-14 for human intestinal applications. Int Dairy J 2002; 12:191–196.

17. Ouwehand AC, Vesterlund S. Health aspects of probiotics. IDrugs 2003; 6:573–580.

18. Collins JK, Thornton G, O'Sullivan G. Selection of probiotic strains for human applications. Int Dairy J 1998; 8:487–490.

19. Gardiner GE, Ross RP, Kelly PM, Stanton C, Collins JK, Fitzgerald G. Microbiology of therapeutic milks. In: Robinson RK, ed. Dairy Microbiology Handbook: The Microbiology of Milk and Milk Products. New York: Wiley Interscience Publishers, 2002:431–468.

20. Havenaar R, Huis in't Veld JHJ. In: Wood BJB, ed. Lactic Acid Bacteria in Health and Disease. London: Elsevier Applied Science Publishers, 1992:151–170.

21. Kohwi Y, Hashimoto Y, Tamura Z. Antitumor and immunological adjuvant effect of *Bifidobacterium adolescentis* M 101–104. Biosci Biotech Biochem 1982; 57:2127–2132.

22. Lee YK, Salminen S. Probiotics in gastrointestinal disorders. Trends Food Sci Technol 1995; 6:241–245.

23. Saavedra JM, Bauman NA, Oung I, Perman JA, Yolken RH. Feeding of *Bifidobacterium bifidum* and *Streptococcus thermophilus* to infants in hospital for prevention of diarrhoea and shedding of rotavirus. Lancet 1994; 344:1046–1049.

24. Breslaw ES, Kleyn DH. Factors affecting growth of Bifidobacteria. Food Sci 1973; 38:1016–1021.

25. Mantere-Alhonen S. Fermented milks and human health. Meijeritieteellinen Aikakauskirja 1983; 41:217–237.

26. Pattersson L, Graf W, Sewelin U. Symposium of the swedish nutrition foundation. XV. In: Hallgren Bo, ed. Nutrition and the Intestinal Flora. Stockholm, Sweden: Almquist & Wiksell International, 1983.

27. Lindwall S, Fonden R. Fermented Milks and Health, FIL/IDF Doc. 1984; 179:21–33

28. Yakult Honsha Co. Ltd. 1971. The summary of Reports Yakult. Yakult Honsha Co. Ltd. 1-1-19, Higshi Shinbashi, Minato-ku, Tokyo, Japan and Intestinal flora of microorganisms and health. Prospectus.

29. Huis in't Veld JHJ, Shortt C. Gut flora and health- past, present and future. In: Leeds RA, Rowland IR, eds. International Congress and Symposium Series No. 219. London: Royal Society of Medicine Press, 1996:27–39.

30. Reid G, Jass J, Sebulsky MT, McCormick JK. Potential uses of probiotics in clinical practice. Clin Microbiol Rev 2003; 16:658–672.

31. Fooks LJ, Gibson GR. Probiotics as modulators of the gut flora. Br J Nutr 2002; 88:S39–S49.

32. Conway PL, Gorbach SL, Goldin BR. Survival of lactic acid bacteria in the human stomach and adhesion to intestinal cells. J Dairy Sci 1987; 70:1–12.

33. Alm L, Leijenmark CE, Persson AK, Midtvedt T. The regulatory and protective role of the normal microflora. In: WennerGren International Symposium Series. 1988;52: 293–297.
34. Mountzouris KC, McCartney AL, Gibson GR. Intestinal microflora of human and infants and current trends for its nutritional modulation. Br J Nutr 2002; 87:405–420.
35. Salminen SJ, Gueimonde M, Isolauri E. Probiotics that modify disease risk. J Nutr 2005; 135:1294–1298.
36. Kado-oka Y, Fujiwara S, Hirota T. Effects of bifidobacteria cells on mitogenic response of splenocytes and several functions of phagocytes. Milchwissenschaft 1991; 46:626–630.
37. Sekine K, Ohta J, Onishi M, et al. Analysis of antitumor properties of effector cells stimulated with a cell wall preparation (WPG) of *Bifidobacterium infantis*. Biol Pharm Bull 1995; 18:148–152.
38. Yasui H, Nagaoka N, Mike A, Hayakawa K, Ohwaki M. Detection of *Bifidobacterium* strains that induce large quantities of IgA. Microb Ecol Health Dis 1992; 5:155–162.
39. Kishi A, Uno K, Matsubara Y, Okuda C, Kishida T. Effect of the oral administration of *Lactobacillus brevis* subsp. coagulans on interferon-α producing capacity in humans. J Am Coll Nutr 1996; 15:408–412.
40. Sekine K, Kawashima T, Hashimoto Y. Comparison of the TNF-α levels induced by human-derived *Bifidobacterium longum* and rat-derived *Bifidobacterium animalis* in mouse peritoneal cells. Bifidobact Microflora 1994; 13:79–89.
41. Hosono A, Lee J, Ametani A, et al. Characterization of a water-soluble polysaccharide fraction with immunopotentiating activity from *Bifidobacterium adolescentis* M 101-4. Biosci Biotech Biochem 1997; 61:312–316.
42. Tone-Shimokawa Y, Toida T, Kawashima T. Isolation and structural analysis of polysaccharide containing galactofuranose from the cell walls of *Bifidobacterium infantis*. J Bacteriol 1996; 178:317–320.
43. Isolauri E, Juntunen M, Rautanen T, Sillanaukee P, Koivula T. A human *Lactobacillus* strain (*Lactobacillus* GG) promotes recovery from acute diarrhoea in children. Paediatrics 1991; 88:90–97.
44. Isolauri E, Kaila M, Mykkanen H, Ling WH, Salminen S. Oral bacteriotherapy for gastroenteritis. Digest Dis Sci 1994; 39:2595–2600.
45. Kalia M, Isolauri E, Soppi E, Virtanen E, Laine S, Arvilommi H. Enhancement of the circulating antibody secreting cell response in human diarrhoea by human *Lactobacillus* strain. Paed Res 1992; 32:141–144.
46. Majaama H, Isolauri E, Saxelin M, Vesikari T. Lactic acid bacteria in the treatment of acute rotavirus gastroenteritis. J Paed Gastroenterol Nutr 1995; 20:333–339.
47. Raza S, Graham SM, Allen SJ, Sultana S, Cuevas L, Hart CA. *Lactobacillus* GG promotes recovery form acute non-bloody diarrhoea in Pakistan. Pediatr Infect Dis J 1995; 14:107–111.
48. Salminen S, Ouwehand AG, Benno Y, Lee YK. Probiotics: how should they be defined? Trends Food Sci Technol 1999; 10:107–110.
49. Shornikova AV, Casas IA, Isolauri E, Mykkanan H, Vesikari T. *Lactobacillus reuteri* as a therapeutic agent in acute diarrhoea in young children. J Pediatr Gastroenterol Nutr 1997; 24:399–404.
49a. Shu Q, Hai L, Rutherfurd KJ, Fenwick SG, Gopal PK, Gill HS. Dietary Bifidobacterium lactis (HN019) enhances resistance to oral Salmonella typhimurium in mice. Microbiol Immunol 2000; 44:213–222.
50. Shu Q, Gill HS. Imune protection mediated by the probiotic Lactobacillus rhamnosus HN001 (DR10™) against Escherichia coli O157:H7 infection in mice. FEMS Immunol Med Microbiol 2002; 34:59–64.
51. Hotta M, Sato SIN. Clinical effects of Bifidobacterium preparation on pediatric intractable diarrhoea. Keio J Med 1987; 36:298–314.
52. Colombel JF, Corot A, Neut C, Romond C. Yogurt with *Bifidobacterium longum* reduces erythromycin-induced gastro-intestinal effects. Lancet 1987; 2:43–44.

53. Sittonen S, Vapaatalo H, Salminen S, et al. Effect of *Lactobacillus* GG yogurt in prevention of antibiotic associated diarrhoea. Ann Med 1990; 22:57–59.

54. Borgia M, Sepe N, Brancato V, Borgia RA. A controlled clinical study on *Streptococcus faecium* preparation for the prevention of side reactions during long-term antibiotic therapy. Curr Therapeutic Res 1982; 31:265–271.

55. Surawicz CM, Elmer LW, Speelman P, McFarland LV, Chinn J, vanBelle G. Prevention of antibiotic-associated diarrhoea by *Saccharomyces boulardii*: a prospective study. Gastroenterology 1989; 96:981–988.

56. Macfarlane GT, Cummings JH. Probiotics and prebiotics: can regulating the activities of intestinal bacteria benefit health. BMJ 1999; 318:999–1003.

56a. Elmer GW, Surawicz C, McFarland L. Biotherapeutic agents: a neglected modality for the treatment and prevention of selected intestinal and vaginal infections. J. Am. Med. Asso. 1996; 275:870–876.

57. Gorbach SL, Chang TW, Goldin B. Successful treatment of relapsing *Clostridium difficile* Colitis with *Lactobacillus* GG. Lancet 1987; 26:1519–1522.

58. Bartlett JG, Chang TW, Gurwith M. Antibiotic-associated pseudomembraneous colitis due to toxin-producing clostridia. N Engl J Med 1982; 57:141–145.

59. Bennet RG, Laughon B, Lindsay J. 1990. *Lactobacillus* GG treatment of *Clostridium difficile* infection in nursing home patients. Abstract, 3rd International Conference on Nosocomial Infections. Atlanta.

60. Biller JA, Katz AJ, Flores AF, Buie TM, Gorbach SL. Treatment of recurrent *Clostridium difficile* colitis with *Lactobacillus* GG. J Pediatr Gastroenterol Nutr 1995; 21:224–226.

61. Black FT, Andersen PL, Orskov F, Gaarslev K, Laulund S. Prophylactic efficacy of lactobacilli on traveller's diarrhoea. Travellers Med 1989; 8:1750–1753.

61a. Oksanen PJ, Salminen S, Saxelin M, Hamalmein P, Arja IV, Leena MI, Seppo N, Oksanen T, Posti I, Salminen E, Siitonen S, Struckey H, Topilla A, Vapaatalo H. Prevention of traveller's diarrhoea by *Lactobacillus* GG. Ann Med 1990; 22:53–56.

62. Hilton E, Kolakowski P, Smith M, Singer C. Efficacy of *Lactobacillus* GG as a diarrhoea preventive measure. J Travellers Med 1996; 4:3–7.

62a. Nomoto K, Nagaoka M, Yokokura T, Mutai M. Augmentation of resistance of mice to bacterial infection by a polysaccharide peptidoglycan complex (PSPG) extracted from *Lactobacillus casei*. Biotherapy 1989; 1:169–171.

63. Hitchins AD, Wells P, McDonough FE, Wong NP. Amelioration of the adverse effect of gastrointestinal challenge with *Salmonella enteriditis* on weaning rats by a yogurt diet. Am J Clin Nutr 1986; 41:91–100.

64. Bovee-Oudenhoven I, Termont D, Dekker R, Van der Meer R. Calcium in milk and Fermentation by yogurt bacteria increase the resistance of rats to *Salmonella* infection. Gut 1996; 38:59–65.

65. Brassart D, Neeser JR, Michetti P, Sevin A. The selection of dairy bacterial strains with probiotic properties based on their adhesion to human intestinal epithelial cells. Les Bacteries *Lactiques/Lact ic Acid Bacteria*. Adria Normandie: Presses Unversitaires de Caen, 1995 pp. 201–202.

66. Sepp E, Tamm E, Torm S, Lutsar I, Mikelsaar M, Salminem S. Impact of a *Lactobacillus* probiotics on the faecal microflora in children with shigellosis. Microecol Ther 1995; 23:74–80.

67. Majaama H, Isolauri E, Saxelin M, Vesikari T. Lactic acid bacteria in the treatment of acute rotavirus gastroenteritis. J Pediatr Gastroenterol Nutr 1995; 20:333–338.

68. Isolauri E, Kirjavainen PV, Salminen S. Probiotics: a role in the treatment of intestinal infection and inflammation? Gut 2002; 50:iii54–iii59.

69. Bernet MF, Brassart D, Neeser JR, Servin AL. *Lactobacillus acidophilus* LA 1 binds to cultured human intestinal cell lines and inhibits cell attachment and cell invasion by enterovirulent bacteria. Gut 1994; 35:483–489.

70. Lundgren O, Svensson L. Pathogenesis of rotavirus diarrhea. Microbes Infect 2001; 3:1145–1156.

71. Ward LA, Rosen BI, Yuan L, Saito LJ. Pathogenesis of an attenuated and a virulent strain of group A human rotavirus in neonatal gnotobiotic pigs. J Gen Virol 1996; 77:1431–1441.

72. Reid G, Charbonneau D, Gonzalez S, Gardiner G, Erb J, Bruce AW. Ability of *Lactobacillus* GR-l and RC-14 to stimulate host defences and reduce gut translocation and infectivity of *Salmonella typhlmurium*. Nutraceut Food 2002; 7:168–173.

73. Christensen HR, Frokiaer H, Pestka JJ. Lactobacilli differentially modulate expression of cytokines and maturation surface markers in murine dendritic cells. J Immunol 2002; 168:171–178.

74. Yolken RH, Ojeh C, Khatri IA, Sajjan U, Forstner JF. Intestinal mucins inhibit rotavirus replication in an oligosaccharide-dependent manner. J Infect Dis 1994; 169:1002–1006.

75. Xu J, Verstraete W. Evaluation of nitric oxide production by lactobacilli. Appl Microbiol Biotechnol 2001; 56:504–507.

76. Cadieux P, Burton J, Gardiner G, et al. *Lactobacillus* strains and vaginal ecology. JAMA 2002; 287:1940–1941.

77. Rachmilewitz D, Katakura K, Karmeli F, et al. Toll-like receptor 9 signaling mediates the anti-inflammatory effects of probiotics in murine experimental colitis. Gastroenterology 2004; 126:520–528.

78. Mohamadzadeh M, Olson S, Kalina WV, et al. Lactobacilli activate human dendritic cells that skew T cells toward T helper 1 polarization. Proc Natl Acad Sci USA 2005; 102:2880–2885.

79. Gorbach SL, Nahas L, Lerner PI. Studies on intestinal microflora I: Effects of diet, age and periodic sampling on numbers of faecal microorganisms in man. Gastroenterology 1967; 53:845–855.

80. Cummings JH, Macfarlane GT. A review: the control and consequences of bacterial fermentation in the human colon. J Appl Bacteriol 1991; 70:443–459.

81. Blaut M, Collins MD, Welling GW, Dore J, van Loo J, de Vos W. Molecular biological methods for studying the gut microbiota: the EU human gut flora project. Br J Nutr May 2002; 87:S203–S211.

82. Alander M, De Smet I, Nollet L, Verstraete W, Mattila-Sandholm T. The effect of probiotic strains on the microbiota of the simulator of the human intestinal microbial ecosystem (SHIME). Int J Food Microbiol 1999; 46:71–79.

83. Benno Y, Hosono M, Hashimoto H, Kojima T, Yamazaki K, Iino H. Effects of *Lactobacillus* GG yoghurt on human intestinal microecology in Japanese subjects. Nutr Today 1996; 31:9S–12S.

84. Aso Y, Akazan H. Prophylactic effect of *Lactobacillus casei* preparation on the recurrence of superficial bladder cancer. Urol Int 1992; 49:125–129.

84a. Marteau P, Flourie B, Pochart P. Effect of chronic ingestion of a fermented dairy product containing *lactobacillus* and *Bifidobacterium bifidum* on metabolic activities of the colonic flora in humans. Am J Clin Nutr 1990; 52:685–688.

85. Khedkar CD, Patil MR, Gyananath G, et al. Studies on implantationability of a probiotic culture of *Lactobacillus acidophilus* in gastrointestinal tract of tribal children. In Proceedings of International Seminar and Workshop on Fermented Foods, health status and social well-being, held at Anand(India) 2003:62–63.

86. Khedkar CD, Gyananath G, Patil MR, Bajad DN, Khojare AS, Sarode AR. 2006. Studies on effect of feeding probiotic *Lactobacillus* supplemented with *Propionibacterium* culture on gastrointestinal microflora of tribal children. Lait (In press).

87. Gopla PK, Prasad J, Gill HS. Effects of consumption of *Bifidobacterium lactis* DR10 and galactooligosaccharides on the microecology of the gastrointestinal tract in human subjects. Nutr Res 2001; 23:1313–1318.

88. Altermann E, Russell WM, Azcarate-Peril MA, et al. Complete genome sequence of the probiotic lactic acid bacterium *Lactobacillus acidophilus* NCFM. Proc Natl Acad Sci USA 2005; 102:3906–3912.

89. Bongaerts GPA, Severijnen RSVM. The beneficial, antimicrobial effect of probiotics. Med Hypotheses 2001; 56:174–177.

90. Hooper LV, Midtvedt T, Gordon JI. How host-microbial interactions shape the nutrient environment of the mammalian intestine. Annu Rev Nutr 2002; 22:283–307.

91. van Nuenen MHMC, Meyer PD, Venema K. The effect of various inulins and *Clostridium difficile* on the metabolic activity of the human colonic microbiota in vitro. Microb Ecol Health Dis 2003; 15:137–144.

92. Tsukahara T, Iwasaki Y, Nakayama K, Ushida K. Microscopic structure of the large intestinal mucosa in piglets during an antibiotic associated diarrhea. J Vet Med Sci 2003; 65:301–306.

93. Oozeer R, Mater DD, Goupil-Feuillerat N, Corthier G. Initiation of protein synthesis by a labeled derivative of the *Lactobacillus casei* DN-114 001 strain during transit from the stomach to the cecum in mice harboring human microbiota. Appl Environ Microbiol 2004; 70:6992–6997.

94. Hove H, Nordgaard-andersen I, Bröbech Mortensen P. Effect of lactic acid bacteria on the intestinal production of lactate and short-chain fatty acids, and the absorption of lactose. Am J Clin Nutr 1994; 59:74–79.

95. Girard-Pipau F, Pompei A, Schneider S, et al. Intestinal microflora, short chain and cellular fatty acids, influence of a probiotic *Saccharomyces boulardii*. Microb Ecol Health Dis 2002; 14:220–227.

96. St-Onge M-P, Farnworth ER, Savard T, Chabot D, Mafu A, Jones PJH. Kefir consumption does not alter plasma lipid levels or cholesterol fractional synthesis rates relatiev to milk in hyperlipidemic men: a randomized controlled trial. BMC Complement. Altern Med 2002; 2:1–789.

97. Goossens D, Jonkers D, Russel M, Stobberingh E, van den Bogaard A, Stockbrügger R. The effect of *Lactobacillus plantarum* 299v on the bacterial composition and metabolic activity in faeces of healthy volunteers: a placebo-controlled study on the onset and duration of effects. Aliment Pharmacol Ther 2003; 18:495–505.

98. Spanhaak S, Havenaar R, Schaafsma G. The effect of consumption of milk fermented by *Lactobacillus casei* strain Shirota on the intestinal microflora and immune parameters in humans. Eur J Clin Nutr 1998; 52:899–907.

99. Tannock GW, Munro K, Harmsen HJM, Welling GW, Smart J, Gopal PK. Analysis of the fecal microflora of human subjects consuming a probiotic product containing *Lactobacillus rhamnosus* DR20. Appl Environ Microbiol 2000; 66:2578–2588.

100. Guerin-Danan C, Chabanet C, Pedone C, et al. Milk fermented with yogurt cultures and *Lactobacillus casei* compared with yogurt and gelled milk: influence on intestinal microflora in healthy infants. Am J Clin Nutr 1998; 67:111–117.

100a. Fujiwara S, Seto Y, Kimura A, Hashiba H. Establishment of orally administered *Lactobacillus* gasseri SBT 2055 SR in the gastrointestinal tract of humans and its influence on intestinal microflora and metabolism. J Appl Microbiol 2001; 90:343–352.

101. Pedrosa MC, Golner BB, Goldin BR, Barakat S, Dallal GE, Russell RM. Survival of yoghurt-containing organisms and *Lactobacillus gasseri* (ADH) and their effect on bacterial enzyme activity in the gastrointestinal tract of healthy and hypochlorhydric elderly subjects. Am J Clin Nutr 1995; 61:353–359.

102. Ling WH, Hänninen O, Mykkänen H, Heikura M, Salminen S, von Wright A. Colonization and fecal enzyme activities after oral *Lactobacillus* GG administration in elderly nursing home residents. Ann Nutr Metab 1992; 36:162–166.

103. Hosoda M, He F, Kojima T, Hashimoto H, Iino H. Effects of fermented- milk with *Lactobacillus rhamnosus* GG strain administration on defecation, putrefactive metabolites and fecal microflora of healthy volunteers. J Nutr Food 1998; 1:1–9.

103a. Goldin BR, Gorbach SL, Saxelin M, Barakat S, Gualtieri L, Salminen S. Survival of *Lactobacillus* species (strain GG) in human gastrointestinal tract. Dig Dis Sci 1992; 37:121–128.

104. Ouwehand AC, Lagström H, Suomalainen T, Salminen S. The effect of probiotics on constipation, faecal azoreductase activity and faecal mucins. Ann Nutr Metab 2002; 46:159–162.

105. Tomoda T, Nakano Y, Kageyama T. Effect of yogurt and yogurt supplemented with *Bifidobacterium* and/or lactulose in healthy persons: a comparative study. Bifidobact. Microflora 1991; 10:123–130.

106. Goldin BR, Gorbach SL. The relationship between diet and rat faecal bacterial enzymes implicated in colon cancer. J Natl Cancer Inst 1976; 57:371–375.

107. Brigidi P, Vitali B, Swennen E, Bazzocchi G, Matteuzzi D. Effects of probiotic administration upon the composition and enzymatic activity of human faecal microbiotia in patients with irritable bowel syndrome or functional diarrhea. Res Microbiol 2001; 152:735–741.

108. Hughes R, Rowland IR. Metabolic activities of the gut microflora in relation to cancer. Microb Ecol Health Dis 2000; 11:179–185.

18
Modifying the Intestinal Microbiota with Antibiotics

Åsa Sullivan and Carl Erik Nord
Department of Laboratory Medicine, Karolinska University Hospital, Karolinska Institutet, Stockholm, Sweden

INTRODUCTION

The human host and the microorganisms colonizing skin and mucous surfaces constitute dynamic biological communities or ecosystems. The composition of the microbiota is relatively stable but fluctuations occur intra-individually over time, and there are also large inter-individual differences. The specific microbiota at each ecological habitat is referred to as the normal microbiota. The numerically and the most diverse human normal microbiota is found in the gastrointestinal tract, and some of the species are potential pathogens that may cause disease under certain circumstances (1). One of the functions of the gastrointestinal microbiota is to act as a barrier against overgrowth of such organisms and also to prevent colonization of pathogenic bacteria from the environment. This phenomenon is termed, "colonization resistance" (2). Treatment with antimicrobial agents disturbs the ecological balance between the host and the normal microbiota and overgrowth of yeasts and *Clostridium difficile*, or of intrinsically or acquired resistant microorganisms, may occur. Horizontal spread of resistance genes by conjugation or transformation to other microbial species can take place. The gastrointestinal normal microbiota plays an important role in this development (3).

Orally administered antimicrobial drugs that are incompletely absorbed or excreted via bile or transluminally, frequently give rise to a reduced colonization resistance. Other factors of importance are the antimicrobial spectrum of the agents, the dose as well as the pharmacokinetic properties of the agents. The outcome of antimicrobial treatment with respect to disturbances in the intestinal microbiota may further vary between individuals. Apart from different anatomical and physiological qualities of the host, the ability of some microorganisms to produce substances that inactivate antimicrobial agents and binding of agents to intestinal material renders the prediction of the effects difficult (4). However, the ecological impact is of great importance in the clinical situation and guidance may be acquired by knowledge of the results from studies on the ecological influence of antimicrobial agents on the normal microbiota.

ANTIMICROBIAL AGENTS THAT INHIBIT THE SYNTHESIS OF THE BACTERIAL CELL WALL—β-LACTAM ANTIBIOTICS

Penicillins

The effect of penicillins on the gastrointestinal microbiota is summarized in Table 1.

Phenoxymethylpenicillin

Phenoxymethylpenicillin has been shown to induce minor variations in numbers of aerobic and anaerobic gastrointestinal microorganisms in healthy adults (5,6) and in infants treated for upper or lower respiratory tract infections or otitis media (7). Penicillin that reaches the gastrointestinal tract is destroyed by beta-lactamase produced by the microorganisms. Despite the low concentration of the agent in feces, generally under the detection level, occasional new colonization with Gram-negative aerobic rods has been observed during administration.

Ampicillin

Ampicillin has a broader antimicrobial spectrum than phenoxymethylpenicillin and is active also against species of Gram-negative microorganisms. The effect on the normal gastrointestinal microbiota is moderate with suppressed numbers of enterococci, streptococci, corynebacteria and enterobacteria. Minor effects on anaerobic species have also been observed in one study. Overgrowth of resistant aerobic Gram-negative rods is common and occasionally also of *Candida* species (8–10). The disturbances are increasing with increased doses.

Ampicillin/Sulbactam

The impact of ampicillin/sulbactam on the intestinal microbiota has been studied in patients undergoing colorectal surgery (11,12). From an ecological point of view it should be expected that it would be less favorable to combine ampicillin with a beta-lactamase inhibitor like sulbactam since the antimicrobial spectrum increases. The effect in particular on the aerobic microbiota has been shown to be mild while the number of anaerobic microorganisms was suppressed. With higher doses, overgrowth of yeasts has been observed and occasionally also overgrowth of *Pseudomonas fluorescens*.

Amoxicillin

Amoxicillin is an agent closely related to ampicillin and with a similar spectrum. Amoxicillin is acid-stable and is therefore better adsorbed. Overgrowth and emergence of amoxicillin-resistant enterobacteria have been the main outcome in studies on the effects on the normal gastrointestinal microbiota both in patients (13,14,16,19), in healthy

Table 1 Impact of Penicillins on the Intestinal Microbiota

Agent	Dose mg/day	Days of administration	Number of subjects	Impact on			Emergence of resistance			Overgrowth of		Concentration range mg/kg	Reference
				Aerobic G+ cocci	Enterobacteria	Anaerobic bacteria	Enterococci	Enterobacteria	Bacteroides	C. difficile	Candida		
Phenoxymethylpenicillin	1000×2	10	10	−	−	−	−	+	−	−	−	<d	(5)
	800×2	7	10	−	−	−	−	−	−	−	−	<d	(6)
Ampicillin	50 mg/kg	3–6	9	−	−	−	−	−	−	−	−	n.e.	(7)
	500×3	7	10	−	−	−	−	+	−	−	−	n.e.	(8)
	1000–3000	5	10	++	++	+	−	+	−	−	−	n.e.	(9)
Ampicillin/sulbactam	500×3	5	10	+	+	−	−	+	−	−	+	n.e.	(10)
	2000/1000×3	2	10	+	+	+	−	−	−	−	+	1.7–27.6	(11)
	500/500×3	2	21	−	−	+	−	−	−	−	−	0.1–21.6	(12)
Amoxicillin	1000×2	14	14	−	+	+	−	+	−	−	+	n.e.	(13)
	2000	15	8	+	+	+	−	−	−	−	+	n.e.	(14)
	500×3	7	6	−	+	−	−	+	−	−	−	n.e.	(15)
	500×3	7	40	−	+	−	−	+	−	−	−	n.e.	(16)
	500×3	7	10	−	+	−	−	+	−	−	−	<d	(17)
	500×3	7	10	+	+	−	−	+	−	−	−	<d	(18)
	250×3	5	10	−	+	+	−	+	−	−	−	n.e.	(10)
	250×3	7	38	−	+	−	−	+	−	(+)	+	n.e.	(19)
Amoxicillin/clavulanic acid	40 mg/kg	3–6	9	−	+	−	−	−	−	−	−	n.e.	(7)
	875/125	7	12	+	+	−	−	−	−	+	−	<d	(20)
	500/250×3	3	6	−	−	−	−	−	−	−	−	n.e.	(21)
	500/125×3	7	6	−	−	−	−	+	−	−	−	n.e.	(15)

(Continued)

Table 1 Impact of Penicillins on the Intestinal Microbiota (*Continued*)

Agent	Dose mg/day	Days of administration	Number of subjects	Impact on — Aerobic G+ cocci	Impact on — Enterobacteria	Impact on — Anaerobic bacteria	Emergence of resistance — Enterococci	Emergence of resistance — Enterobacteria	Emergence of resistance — Bacteroides	Overgrowth of — C. difficile	Overgrowth of — Candida	Concentration range mg/kg	Reference
	375×3[a]	5	4	+	+	−	−	+	−	−	−	n.e.	(22)
	187.5×3[a]	5	4	+	+	−	−	+	−	−	−	n.e.	(22)
	250/125×3	7	6	+	−	−	−	+	−	−	−	n.e.	(23)
	27.5 mg/kg×4	10–11	11	−	−	−	−	+	−	−	−	n.e.	(24)
Bacampicillin	1600	15	8	−	−	−	−	−	−	−	−	n.e.	(14)
	400×3	7	12	−	−	+	−	−	−	−	−	<d	(25)
Pivampicillin	700×4	3	10	−	+	−	−	+	−	−	+	n.e.	(26)
Talampicillin	250×3	5	10	−	+	−	−	+	−	−	−	n.e.	(10)
Azlocillin	5000×3	7–8	6	+	+	+	−	+	−	−	−	n.e.	(27)
Piperazillin	4000×3	2	20	+	+	+	−	−	−	+	−	<d–101.2	(28)
Piperazillin/ tazobactam	4000/ 500×3	4–8	20	+	+	−	−	−	−	−	−	1.2–276/ <d–22.2	(29)
Pivmecillinam	600×4	7	10	+	++	+	−	−	−	−	−	n.e.	(30)
	400×3	7	5	+	+	+	−	−	−	−	−	n.e.	(30)
	400×2	7	15	−	+	−	−	−	−	−	−	<d–15.6	(31)
Ticarcillin/ clavulanic acid	5000/200 ×3	7	10	+	+	−	−	−	−	−	−	<d	(32)

[a] Ratio amoxicillin/clavulanic acid 2/1.

Abbreviations: +, mild to moderate effect, increase or decrease 2–4log10 cfu/g feces; ++, strong impact >4log10 cfu/g feces; −, no significant changes; d, the detection limit; n.e., not examined.

volunteers (10,15,17,18) and in infants (7). In patients, in contrast to in healthy persons, amoxicillin has also a suppressive effect on the anaerobic microbiota (13,14,19).

Amoxicillin/Clavulanic Acid

Administration of amoxicillin/clavulanic acid has been shown to induce increased numbers of amoxocillin-resistant enterobacteria both in healthy adults (15,20–23) and in child patients (24). In some of the mentioned studies, there were also disturbances in the numbers of aerobic cocci, mainly observed as increased numbers of enterococci.

Bacampicillin, Pivampicillin, and Talampicillin

Bacampicillin, pivampicillin and talampicillin are esters of ampicillin that are better absorbed than ampicillin, and thereby a more favorable ecological effect on the intestinal microbiota is expected.

No major changes in the intestinal microbiota have been observed during long-term treatment of patients with bacampicillin (14) or in connection with shorter administration to healthy volunteers (25). However, the anaerobic microbiota was affected in some of the subjects in the latter study. Subjects receiving bacampicillin in tablets had an undisturbed intestinal microbiota in contrast to subjects receiving bacampicillin in syrup.

Pivampicillin and talampicillin have been shown to give rise to increased numbers of enterobacteria in healthy volunteers (10,26) and increased numbers of *Candida* species have been observed in a few subjects during administration with pivampicillin.

Azlocillin

The impact of azlocillin on the intestinal microbiota has been studied in connection with treatment of patients suffering from skin and soft tissue infections (27). Suppressed numbers of both aerobic and anaerobic species were observed and overgrowth of resistant enterobacteria occurred in some patients.

Piperacillin and Piperacillin/Tazobactam

Piperazillin is excreted in bile leading to high fecal concentrations. Short-term administration to patients undergoing colorectal surgery has resulted in marked effects on both the aerobic and anaerobic intestinal microbiota (28). Addition of tazobactam to piperazillin in treatment of patients reduced the ecological disturbances in the anaerobic microbiota while the aerobic microbiota was still suppressed (29).

Pivmecillinam

Pivmecillinam has a spectrum including in particular Gram-negative aerobic rods and the main impact during administration to healthy volunteers has been seen as reduced numbers of gastrointestinal *Escherichia coli* (30,31). More pronounced changes have been observed to occur at higher doses with decreasing numbers of anaerobic species like lactobacilli and bacteroides and increasing numbers of enterococci (30).

Ticarcillin/Clavulanic Acid

The effect of ticarcillin/clavulanate on the gastrointestinal microbiota has been evaluated in healthy subjects. Only minor disturbances were detected, such as decreased numbers of enterobacteria and a concomitant increase of aerobic cocci (32).

Parenterally Administered Cephalosporins

The spectra of cephalosporins are broader than that of penicillins. Several cephalosporins are excreted biliary and a strong ecological impact can be expected. Enterococci are intrinsically resistant to cephalosporins and their numbers usually increase during administration.

The impact of parenterally administered cephalosporins on the gastrointestinal normal microbiota is summarized in Table 2.

Cefazolin

The impact of intravenously administered cefazolin on the gastrointestinal microbiota has been studied in patients at an intensive care unit (33) and in patients undergoing gastrectomy (34). Overgrowth of resistant *Pseudomonas* species was detected in the first study while increasing numbers of enterococci, reduced numbers of streptococci and also suppressed numbers of some anaerobic species were observed in the second study.

Cefbuperazone

Changes in the intestinal microbiota in connection with short-term administration of cefbuperazone have been studied in patients undergoing colorectal surgery (35). The agent suppressed the aerobic cocci, enterobacteria as well as the anaerobic microbiota.

Cefepime

A selective reduction of the numbers of *E. coli* has been observed during administration of cefepime in healthy volunteers (36).

Cefmenoxime

Significantly decreased numbers of enterobacteria, bifidobacteria and lactobacilli have been observed in connection with parenteral administration of cefmenoxime in healthy subjects. Furthermore, there was a concomitant increase in numbers of clostridia and *Candida* species (37).

Cefoperazone

Cefoperazone is mainly excreted in bile giving rise to high fecal concentrations and thereby major changes in the intestinal microbiota can be expected. The impact of cefoperazone on the fecal microbiota has been evaluated in adult patients (38) and in sick children (39,40). The Gram-negative aerobic rods as well as numbers of staphylococci and streptococci were markedly suppressed in all studies. Overgrowth of resistant enterobacteria, enterococci and *Candida* species were observed and anaerobic species were also suppressed.

Table 2 Impact of Parenterally Administered Cephalosporins on the Intestinal Microbiota

Agent	Dose mg/day	Days of administration	Number of subjects	Impact on			Emergence of resistance			Overgrowth of		Concentration range mg/kg	Reference
				Aerobic G+ cocci	Enterobacteria	Anaerobic bacteria	Enterococci	Enterobacteria	Bacteroides	C. difficile	Candida		
Cefazolin	60–80 mg/kg	4–11	5	−	+	n.e.	−	+	−	n.e.	−	n.e.	(33)
Cefbuperazone	1000×3	4	8	+	−	+	−	−	−	−	−	n.e.	(34)
	1000×2	1	10	+	+	+	−	−	−	−	−	0.8–27.0	(35)
Cefepime	1000×2	8	8	−	+	−	−	−	−	−	−	n.e.	(36)
Cefmenoxime	4000	3	15	−	+	−	−	−	−	−	+	n.e.	(37)
Cefoperazone	2000×2	7–14	28	+	++	+	−	−	−	+	−	n.e.	(38)
	100 mg/kg	4–7	16	+	++	+	−	−	−	−	+	n.e.	(39)
	a	a	5	+	++	n.e.	−	+	n.e.	n.e.	+	>d	(40)
Cefotaxime	a	a	6	−	+	n.e.	−	−	n.e.	n.e.	−	<d	(40)
	100 mg/kg	a	26	+	+	−	−	+	−	−	−	n.e.	(41)
	60–80 mg/kg	4–11	11	−	−	n.e.	−	+	n.e.	n.e.	−	n.e.	(33)
Cefotiam	6000	3	15	−	+	−	−	+	−	−	+	n.e.	(37)
Cefoxitin	2000×4	2	20	+	+	+	−	+	−	−	−	1.5–35.5	(42)
	6000–12000	8–23	6	+	+	+	−	+	−	+	−	<d–32.0	(43)
Cefozopran	500×3	4	8	+	+	+	−	−	−	−	−	n.e.	(34)
Cefpirome	2000×2	7.5	10	−	+	−	−	−	−	−	−	n.e.	(44)
Ceftazidime	4000	1	8	−	+	−	−	−	−	−	−	n.e.	(37)
Ceftizoxime	4000	1	8	−	++	−	−	+	−	−	−	n.e.	(37)
Ceftriaxone	1500×2	7–13	12	++	++	+	−	−	−	+	+	n.e.	(45)

(Continued)

Table 2 Impact of Parenterally Administered Cephalosporins on the Intestinal Microbiota (*Continued*)

Agent	Dose mg/day	Days of administration	Number of subjects	Impact on			Emergence of resistance			Overgrowth of		Concentration range mg/kg	Reference
				Aerobic G+ cocci	Enterobacteria	Anaerobic bacteria	Enterococci	Enterobacteria	Bacteroides	C. difficile	Candida		
	2000	1	10	−	++	−	−	−	−	−	−	n.e.	(46)
	1000	5	11	+	++	n.e.	−	−	n.e.	n.e.	+	<d–1600	(47)
	1000	5	10	n.e.	n.e.	+	n.e.	n.e.	−	+	n.e.	n.e.	(48)
	1000	10	12	+	++	+	−	−	−	−	+	n.e.	(49)
	a	a	9	+	++	n.e.	−	+	n.e.	n.e.	+	n.e.	(40)
Ceftriaxone/ loracarbef	1000/ 400×2	2+8	12	+	+	+	−	−	−	+	+	n.e.	(49)
Flomaxef	1000×3	4	8	+	−	+	−	−	−	+	−	n.e.	(34)
Moxalactam	2000×3	1	10	+	+	+	−	−	−	−	−	0.2–23.0	(50)
	2000	1	10	+	+	+	−	−	−	−	−	0.2–23.0	(50)

a No data available.

Abbreviations: +, mild to moderate effect, increase or decrease 2–4log10 cfu/g feces; ++, strong impact >4log10 cfu/g feces; −, no significant changes; d, the detection limit; n.e., not examined.

Cefotaxime

Cefotaxime is excreted in bile to a lesser extent than cefoperazone and the effects on the intestinal microbiota are usually more moderate. The numbers of enterobacteria are suppressed and overgrowth of *Pseudomonas* species and occasionally of enterococci have been observed (33,40,41).

Cefotiam

Cefotiam has been shown to decrease the numbers of intestinal enterobacteria and lactobacilli and to increase the numbers of *Pseudomonas* and *Candida* species (37).

Cefoxitin

Pronounced changes in the gastrointestinal microbiota have been shown to occur after cefoxitin prophylaxis of patients undergoing colorectal surgery (42) and in hospitalized male patients (43). In both studies the major changes observed were decreased numbers of enterobacteria and Gram-negative anaerobic species, while there was a proliferation of resistant enterococci and enterobacteria. Growth of *C. difficile* was found in 5 of 6 hospitalized patients (43).

Cefozopran

In patients receiving prophylactic antimicrobial treatment after gastrointestinal surgery, cefozopran induced decreased numbers of enterobacteria, streptococci, *Veillonella* and *Lactobacillus* species and overgrowth of enterococci (34).

Cefpirome

Administration of cefpirome to healthy male volunteers suppressed the numbers of *E. coli* below the detection limit (44). No other major changes were observed.

Ceftazidime and Ceftizoxime

The impact of a single dose of ceftazidime or ceftizoxime on the intestinal microbiota has been investigated in healthy volunteers (37). Ceftazidime significantly reduced the numbers of enterobacteria and lactobacilli. The number of enterobacteria was suppressed also by administration of ceftizoxime, and resistant enterobacteria like *Citrobacter* and *Proteus* species proliferated.

Ceftriaxone and Ceftriaxone/Loracarbef

Ceftriaxone is, as well as cefoperazone, to a large extent excreted biliary and the agent induced marked changes in the intestinal microbiota (40,45–49). Ceftriaxone has been shown to give rise to elimination or strong suppression of the numbers of Gram-negative aerobic rods, reduced numbers of streptococci and staphylococci and also to reduced numbers of anaerobic microorganisms. Overgrowth of species resistant to ceftriaxone like enterococci and *Candida* species is common.

The ecological effect of ceftriaxone has been compared with a step-down therapy of ceftriaxone followed by loracarbef in patients with community-acquired pneumonia (49).

Both the aerobic and the anaerobic microbiota were affected in a similar way as with ceftriaxone only, although the reduction of enterobacteria occurred to a lesser extent.

Flomoxef or Moxalactam

Changes in intestinal microbiota have been investigated after administration of flomoxef to patients undergoing gastrectomy (34). The effect on the aerobic microbiota was mainly detected as decreased numbers of streptococci and overgrowth of enterococci. Anaerobic Gram-positive rods and cocci as well as Gram-negative cocci were suppressed.

In an earlier study, the effect of a single dose of moxalactam was compared with a three-dose prophylaxis (50). In both groups of patients there was a reduction in the numbers of enterobacteria and streptococci while enterococci proliferated. Several anaerobic species decreased significantly in connection with the administration.

Perorally Administered Cephalosporins

Studies on the effects of perorally administered cephalosporins are summarized in Table 3.

Cefaclor

Alterations in the intestinal microbiota during administration of cefaclor have been studied in patients (19) and in healthy volunteers (51,52). In the microbiota of patients there were reduced numbers of both aerobic and anaerobic Gram-positive cocci. Enterococci, enterobacteria and *Bacteroides* species increased and there were also increased numbers of *Candida albicans*. In healthy subjects only minor changes occurred in the anaerobic microbiota.

Cefadroxil

Reduced numbers of intestinal viridans streptococci have been observed during administration of cefadroxil in adult healthy subjects (5). In infants being treated for infections, disturbances were restricted to the anaerobic microbiota with reduced numbers of bifidobacteria and bacteroides (7).

Cefetamet/Pivoxil

Cefetamet has a broad spectrum of activity against both aerobic Gram-positive and Gram-negative microorganisms. The modification on the intestinal microbiota during treatment of patients has, however, been shown to be slight and nonsignificant (53).

Cefexime

The ecological effects on the intestinal microbiota of cefixime have been investigated in healthy volunteers (51,54) and in patients with exacerbation of chronic bronchitis (53). In all three studies, disturbances were observed in the aerobic microbiota as reduced numbers of enterobacteria and increased numbers of enterococci. Growth of *C. difficile* was common in all studies while the impact on the anaerobic microbiota varied between the studies, from reduced numbers of clostridia to reductions of several species including bacteroides.

Table 3 Impact of Perorally Administered Cephalosporins on the Intestinal Microbiota

Agent	Dose mg/day	Days of adminis-tration	Number of subjects	Impact on			Emergence of resistance			Overgrowth of		Concen-tration range mg/kg	Reference
				Aerobic G+ cocci	Enter-obac-teria	Anaero-bic bac-teria	Entero-cocci	Entero-bacteria	Bacter-oides	C. difficile	Candida		
Cefaclor	250×3	14	6	−	−	−	−	−	−	+	−	n.e.	(51)
	250×3	7	39	+	+	−	−	+	−	−	+	n.e.	(16)
	250×3	7	10	−	−	−	−	−	−	−	−	<d	(52)
Cefadroxil	500×2	10	10	+	−	−	−	+	−	−	−	<d	(5)
	30 mg/kg	3–6	5	−	−	+	−	−	−	−	−	n.e.	(7)
Cefetamet/pivoxil	500×2	10	8	−	−	−	−	−	−	−	−	<d–38.8	(53)
Cefexime	400×4	10	8	+	+	−	−	−	−	+	−	n.e.	(53)
	400	14	6	+	+	+	−	−	−	+	−	n.e.	(51)
	200×2	7	10	+	+	++	−	−	−	+	−	<d–912	(54)
Cefpodoxime/proxetil	200×2	7	10	+	+	++	−	−	−	+	+	<d–700	(19)
	100×2	7	10	+	+	−	−	−	−	+	−	n.e.	(55)
Cefprozil	500×2	8	8	−	−	−	−	−	−	+	−	n.e.	(56)
Ceftibuten	400	10	14	+	+	+	−	−	−	−	+	<d–3.2	(57)
Cefuroxime/axetil	600×3	3	6	+	+	−	−	−	−	−	−	n.e.	(21)
	250×2	10	8	−	−	−	−	−	−	+	−	n.e.	(53)
	250×2	10	10	+	+	−	−	−	−	+	−	<d–1.35	(58)
	250×2	7	20	+	+	−	−	+	−	−	−	n.e.	(59)
	250×2	5	10	+	+	+	−	+	−	−	+	<d–152	(60)
Cephradine	1000×2	7	6	−	−	−	−	−	−	−	−	n.e.	(23)
Loracarbef	200×2	7	40	−	−	−	−	−	−	−	−	n.e.	(15)
	200×2	7	20	−	−	−	−	−	−	−	−	0.27(0.32)[a]	(61)
	30 mg/kg	3–6	6	+	+	+	−	−	−	−	−	n.e.	(7)

[a] Mean (standard deviation).

Abbreviations: +, mild to moderate effect, increase or decrease 2–4log10 cfu/g feces; ++, strong impact >4log10 cfu/g feces; −, no significant changes; d, the detection limit; n.e., not examined.

Cefpodoxime Proxetil

A marked decrease in numbers of aerobic intestinal microorganisms with disappearance of *E. coli* has been seen during administration of cefpodoxime proxetil in healthy volunteers (18,55). The anaerobic microbiota was also affected and after treatment overgrowth of enterococci, candida and *C. difficile* occurred.

Cefprozil

The ecological impact of cefprozil was determined in a double-blind placebo-controlled study (56). Analysis of the fecal microbiota revealed mainly a moderate decrease in enterobacteria and a few subjects became colonized with *C. difficile*.

Ceftibuten

Ceftibuten administration has been shown to partly affect the aerobic intestinal microbiota (57). The numbers of *E. coli* was significantly reduced while there was an overgrowth of enterococci. Four subjects became colonized with yeasts, mainly *C. albicans*. The anaerobic microbiota was disturbed to a lesser degree. However, six volunteers were colonized by *C. difficile*.

Cefuroxime/Axetil

The effect of cefuroxime/axetil on the gastrointestinal microbiota has been evaluated in patients (53) and in healthy subjects (21,53,58–60). Ecological disturbances have mainly been observed as decreased numbers of enterobacteria, overgrowth of enterococci and in varied changes in the anaerobic microbiota. In several studies, colonization with *Candida* species and *C. difficile* has been observed. Fecal concentrations of cefuroxime/axetil, when measured, have generally been rather low. In one study though, four subjects had very high amounts of the agent in feces and thereby also more pronounced disturbances in the microbiota (60).

Cephradine

Elimination of staphylococci has been shown to be the major significant change in the microbiota occurring during administration of cephradine in healthy volunteers (23).

Loracarbef

No major ecological disturbances in the intestinal microbiota have been detected in connection with administration of loracarbef as treatment for acute bronchitis (16) or in healthy volunteers (61). In patients, new aerobic Gram-negative species were detected during the investigation period. However, all strains were susceptible to loracarbef. Loracarbef therapy caused increasing levels of enterococci in infants but had no significant effect on the anaerobic microbiota (7).

Monobactams

The ecological impact of monobactams on the gastrointestinal microbiota is shown in Table 4.

Table 4 Impact of Monobactam, Carbapenems, and Glycopeptides on the Intestinal Microbiota

Agent	Dose mg/day	Days of administration	Number of subjects	Impact on			Emergence of resistance			Overgrowth of		Concentration range mg/kg	Reference
				Aerobic G+ cocci	Enterobacteria	Anaerobic bacteria	Enterococci	Enterobacteria	Bacteroides	C. difficile	Candida		
Aztreonam	2000×3	7–9	9	−	++	+	−	−	−	−	−	n.e.	(62)
	1000×3	7–9	9	−	++	−	−	−	−	−	−	n.e.	(62)
	1000×3	2	20	+	+	−	−	−	−	−	−	<d–21.4	(63)
	500×3	5	10	+	++	−	−	−	−	−	+	<d–>1000	(64)
	100×3	5	10	−	++	−	−	−	−	−	−	<d–100	(64)
	20×3	5	10	−	+	−	−	−	−	−	−	<d–30	(64)
Imipenem/ cilastin	1000/ 1000×4	2	10	+	+	+	−	−	−	−	−	0.7–11.3	(65)
	500/ 500×4	2	10	+	+	+	−	−	−	−	−	<0.1–5	(65)
	500/ 500×4	6–11	10	−	+	−	−	−	−	−	−	n.e.	(66)
Meropenem	500×3	7	10	+	+	+	−	−	−	−	−	<d	(67)
Lenapenem	500×2	4.5	6	+	−	−	−	−	−	−	−	n.e.	(68)
Vancomycin	125×4	21	6	+	−	−	(+)	−	−	−	−	d–>1800	(69)
	125×4	10	2×10	+	−	+	(+)	+	−	−	−	15–1400	(70)
	125×4	7	10	+	+	+	(+)	+	−	−	−	520(197)[a]	(59)
Teicoplanin	200×2	21	6	+	++	−	+	−	−	−	−	d–<1800	(69)
	100×2	21	10	+	+	−	+	−	−	−	−	d–>1800	(69)

[a] Mean (Standard Deviation).

Abbreviations: +, mild to moderate effect; increase or decrease 2–4log10 cfu/g feces; + +, strong impact >4log10 cfu/g feces; −, no significant changes; d, the detection limit; n.e., not examined.

Aztreonam

The effects of aztreonam on the intestinal microbiota have been studied in patient groups (62,63) and in healthy volunteers (64). The dominating effects of aztreonam on aerobic species have been observed as a marked decrease in numbers of enterobacteria. Emergence of aztreonam-resistant enterococci and reduced numbers of anaerobic microbiota occurred in connection with higher dosing (64).

Carbapenems

The effect of carbapenems on the fecal normal microbiota is shown in Table 4.

Imipenem

The effects of parenteral imipenem/cilastin therapy have been evaluated after prophylactic treatment of patients undergoing colorectal surgery (65) and in hospitalized patients with serious infections (66). In the first study, aerobic Gram-positive cocci, enterobacteria as well as several anaerobic species were significantly suppressed. The major effect in the latter study was observed as decreased numbers of enterobacteria.

Meropenem

The gastrointestinal microbiota has been studied in connection with administration of meropenem to healthy male volunteers (67). No measurable concentrations of meropenem were found in feces but disturbances were seen both in the aerobic and anaerobic microbiota. The numbers of streptococci and enterobacteria decreased while enterococci increased. Clostridia, Gram-negative anaerobic cocci and *Bacteroides* species were also suppressed.

Lenapenem

In a study where lenapenem was given to healthy male volunteers (68), the antimicrobial agent did not influence the total numbers of aerobic or anaerobic bacteria but streptococci and *Veillonella* species were suppressed in numbers.

OTHER AGENTS WITH INHIBITORY EFFECT ON THE SYNTHESIS OF THE CELL WALL—GLYCOPEPTIDES

Glycopeptides are poorly absorbed and reach very high fecal concentrations and major disturbances are expected in the gastrointestinal microbiota.

A summary of the ecological impact of glycopeptides on the intestinal microbiota is shown in Table 4.

Vancomycin

Perorally administered vancomycin has been given to healthy subjects and the effects on the intestinal microbiota have been analyzed (59,69,70). In the aerobic microbiota the total numbers of enterococci and staphylococci have been seen to decrease while resistant

Gram-negative rods and enterococci emerged. Dramatic increase of other naturally resistant species like pediococci and lactobacilli has also been observed. Suppressed numbers of bacteroides and *Bifidobacterium* species were seen in two of the studies (59,70).

Teicoplanin

The ecological impact of teicoplanin has been evaluated in two dosing regimes in healthy volunteers (69). Highly glycopeptide-resistant *Pediococcus* species, enterococci and lactobacilli increased during the administration. After the high-dose regimen treatment the numbers of staphylococci decreased while enterobacteria increased.

ANTIMICROBIAL AGENTS INTERFERING WITH THE SYNTHESIS OF PROTEINS

The impact of macrolides, azalide, ketolide, lincosamide and streptogramin on the gastrointestinal microbiota is shown in Table 5 and the impact of tetracyclines, aminoglycosides, nitrofurantoin and oxazolidone in Table 6.

Macrolides

Clarithromycin

The ecological impact of clarithromycin on the gastrointestinal microbiota has been investigated in several studies on healthy volunteers (71–74). In the aerobic microbiota the numbers of *E. coli* have been observed to decrease significantly while there has been a concomitant overgrowth of other aerobic Gram-negative species. The degree of the disturbances in numbers of enterobacteria has varied depending on the dosing regimen. In the study where the lowest dose was applied, there was a suppression also of the number of streptococci (74). In the anaerobic microbiota decreased numbers have been detected mainly of bifidobacteria, lactobacilli, clostridia and *Bacteroides* species.

Dirithromycin

The influence of dirithromycin on the normal human intestinal microbiota has been evaluated in healthy persons (75). The major route of elimination of the agent is fecal, and high fecal concentrations were demonstrated with apparent disturbances in both the aerobic and anaerobic microbiota. The numbers of *E. coli* decreased, streptococci and staphylococci increased and there was overgrowth of dirithromycin-resistant enterobacteria. Anaerobic Gram-positive cocci, bifidobacteria, eubacteria and *Bacteroides* decreased while clostridia and lactobacilli increased during the treatment period.

Erythromycin

Marked disturbances have been observed in the intestinal microbiota during oral administration of erythromycin in healthy adults (74,76) and in infants (7). The aerobic Gram-positive cocci were reduced in numbers and there were marked reductions in the

Table 5 Impact of Macrolides, Azalide, Ketolide, Lincosamide, and Streptogramin on the Intestinal Microbiota

Agent	Dose mg/day	Days of administration	Number of subjects	Impact on			Emergence of resistance			Overgrowth of		Concentration range mg/kg	Reference
				Aerobic G+ cocci	Entero-bacteria	Anaerobic bacteria	Entero-cocci	Entero-bacteria	Bacteroides	C. difficile	Candida		
Clarithromycin	500×2	10	10	−	++	+	−	+	+	−	−	<d–513	(71)
	500×2	7	12	−	++	+	−	+	−	−	−	128(58)[a]	(72)
	500×2	7	6	−	++	+	−	+	−	−	−	n.e.	(73)
	250×2	7	10	+	+	+	−	+	−	−	−	<d–243	(74)
Dirithromycin	500	7	20	+	+	+	−	+	−	−	−	<d–45	(75)
Erythromycin	1000×2	7	10	+	++	++	−	+	−	−	+	<d–1412	(74)
	500×2	7	10	+	++	++	−	+	−	−	+	<d–1120	(76)
	40 mg/kg	3–6	12	+	+	+	−	+	−	−	−	n.e.	(7)
Roxithromycin	150×2	5	6	−	+	−	−	−	−	−	−	<d–240	(77)
Azithromycin	500	3	6	−	+	−	−	−	−	−	−	n.e.	(73)
Telithromycin	800	10	10	+	+	−	−	+	+	−	−	<d–1330	(71)
Clindamycin	600×3	3	15	+	−	++	−	−	−	−	−	2.1–460	(78)
	150×4	7	10	+	+	++	−	+	−	+	−	<d–200	(79)
	150×4	7	10	+	+	++	−	+	−	+	−	n.e.	(80)
	150×4	7	12	+	+	++	−	+	+	+	−	<d–452	(81)
Quinupristin/ dalfopristin	7.5 mg/kg ×2	5	20	+	+	+	+	−	+	−	−	291(184)/ 42(22)[a]	(82)

[a] Mean(Standard Deviation).

Abbreviations: +, mild to moderate effect; increase or decrease 2–4log10 cfu/g feces; + +, strong impact >4log10 cfu/g feces; −, no significant changes; d, the detection limit; n.e., not examined.

Table 6 Impact of Tetracyclines, Aminoglycosides, Nitrofurantoin, Oxazolidinone, Folic Acid Antagonists, and Nitroimidazoles on the Intestinal Microbiota

Agent	Dose mg/day	Days of administration	Number of subjects	Impact on			Emergence of resistance			Overgrowth of		Concentration range mg/kg	Reference
				Aerobic G+ cocci	Entero bacteria	Anaerobic bacteria	Enterococci	Enterobacteria	Bacteroides	C. difficile	Candida		
Tetracycline	250×4	8–10	15	−	−	−	−	+	−	n.e.	+	n.e.	(83)
Doxycycline	200+ 100×1[a]	8–10	15	−	−	−	−	+	−	n.e.	+	n.e.	(83)
	100	7	10	+	+	−	−	+	+	−	−	<d–56.7	(84)
Tobramycin	100×3	4	8	+	+	−[b]	−	−	n.e.	n.e.	−	<d–100	(85)
	100×2+ 300	4	8	+	+	+[b]	−	−	n.e.	n.e.	−	<d–100	(85)
Nitrofurantoin	100	30	7	−	−	n.e.	−	−	n.e.	n.e.	−	n.e.	(86)
Linezolid	600×2	7	12	+	+	++	−	−	+	−	−	7.1(2.6)[c]	(20)
Co-trimoxazole	40/200	30	7	−	++	n.e.	−	−	n.e.	n.e.	−	n.e.	(86)
	6/30 mg/kg	3–6	8	−	−	+	−	−	−	−	−	n.e.	(7)
Metronidazole	400×3	5–7	10	−	−	−	−	−	−	−	−	<d	(87)

(Continued)

Table 6 Impact of Tetracyclines, Aminoglycosides, Nitrofurantoin, Oxazolidinone, Folic Acid Antagonists, and Nitroimidazoles on the Intestinal Microbiota (*Continued*)

| | | | | Impact on | | | | Emergence of resistance | | | | Overgrowth of | | | Concen- | |
| | | | | | | | | | | | | | | | tration range | |
Agent	Dose mg/day	Days of adminis- tration	Number of subjects	Aerobic G + cocci	Entero bac- teria	Anaero- bic bac- teria	Entero- cocci	Entero- bacteria	Bac- ter- oides	C. difficile	Can- dida	mg/kg	Reference
Tinidazole	800+ 400×2[a]	2	20	+	–	+	–	–	–	–	–	<d–4.8	(88)
Metronida- zole/ amoxycil- lin	150×2 400/1000 ×2	7 7	10 14	– +	– –	– –	– +	– +	– –	– –	– +	<d <d(amoxy)	(89) (90)
Metronida- zole/ clarithromy- cin	400/250 ×2 400/250 ×2	7 7	16 51	+ –	+ +	+ +	+ –	+ +	+ –	– –	– +	88–261 (clarithro) n.e.	(90) (91)

[a] Initial dose 200 mg or 800 mg respectively.
[b] Measured indirectly by β-aspargylglycine concentrations.
[c] Mean(Standard Deviation).

Abbreviations: +, mild to moderate effect, increase or decrease 2–4log10 cfu/g feces; + +, strong impact >4log10 cfu/g feces; –, no significant changes; d = the detection limit; n.e., not examined.

numbers of enterobacteria while new species of resistant Gram-negative rods proliferated. Several subjects became colonized with yeasts. Anaerobic species like bifidobacteria, lactobacilli, clostridia and *Bacteroides* were also suppressed to a varying degree.

Roxithromycin

The consequences of oral treatment with roxithromycin on the intestinal microbiota are more limited than the effects of erythromycin in healthy volunteers (77). The fecal concentrations were also lower and the changes were restricted to a decrease in total counts of Enterobacteriaceae.

Azalide

Azithromycin

The ecological effect of azithromycin has been compared with the effect of clarithromycin in healthy volunteers (73). The main impact of azithromycin was detected as decreased numbers of bacterial species in the family Enterobacteriaceae.

Ketolide

Telithromycin

Moderate disturbances in the gastrointestinal microbiota have been recorded during administration of telithromycin to healthy subjects (71). The numbers of *E. coli* were significantly reduced and overgrowth of staphylococci and resistant enterobacteria was observed. In the anaerobic microbiota there were reduced numbers of lactobacilli and bifidobacteria. A selection of highly resistant *Bacteroides* isolates was also recorded during and after treatment.

Lincosamide

Clindamycin

The ecological impact of clindamycin on the fecal microbiota has been studied after intravenous clindamycin prophylaxis in patients undergoing colorectal surgery (78) and after oral administration in healthy subjects (79–81). Clindamycin is excreted in the bile and high fecal concentrations have been detected with marked disturbances, in particular in the anaerobic microbiota. Enterococci are not susceptible to clindamycin and their numbers have usually increased and so have the numbers of clindamycin-resistant enterobacteria. Anaerobic Gram-positive cocci and rods and anaerobic Gram-negative rods have been markedly suppressed or eliminated during treatment. Emergence of clindamycin-resistant *Bacteroides* species has been detected in one of the studies and colonization with *C. difficile* was common.

Streptogramin

Quinupristin/Dalfopristin

In healthy volunteers treated with quinupristin/dalfopristin (RP59500), the impact on the fecal microbiota has been investigated (82). The numbers of enterococci and *Enterobacteriaceae* increased significantly and anaerobic non-sporulating and Gram-negative bacteria decreased. The total numbers of quinupristin/dalfopristin-resistant and also erythromycin-resistant anaerobes and enterococci increased significantly. The observed modifications disappeared within 12 weeks after the administration.

Tetracyclines

Tetracycline

The ecological effect of tetracycline hydrochloride on the gastrointestinal microbiota has been examined in healthy volunteers (83). Tetracycline had no major effect on the total numbers of intestinal microorganisms although a few subjects acquired new strains of *C. albicans*. However, the major finding was the emergence of resistant *E. coli* strains in 10 of 15 subjects.

Doxycycline

The effect of doxycycline has been evaluated in two studies in healthy subjects (83,84). The results are partly consistent in that new resistant strains were detected during treatment. Acquisition of *C. albicans* occurred in subjects in the first mentioned study and new strains of Enterobacteriaceae in the latter. In this study, the aerobic microbiota was also suppressed while the anaerobic microbiota was not influenced. However, the number of fusobacteria was reduced and a marked emergence of resistance was also observed in anaerobic microorganisms (84).

Aminoglycosides

Tobramycin

Two dosing regimens of tobramycin have been compared for the selective decontamination effect of the digestive tract in healthy volunteers (85). Both regimens markedly suppressed the number of aerobic Gram-negative rods while the higher dose also had an effect on the anaerobic microbiota as evidenced by low concentrations of beta-aspartylglycine.

Nitrofurantoin

Nitrofurantoin has been used for prophylaxis in women with recurrent urinary tract infections (86). The effect on the fecal microbiota was examined semi-quantitatively. The agent had no effect on the numbers of enterococci or enterobacteria and no resistant strains or overgrowth of strains was detected.

Oxazolidinone

Linezolid

Linezolid is a relatively new synthetic antimicrobial agent that has been evaluated for the effects on the gastrointestinal microbiota in healthy male subjects (20). There was a statistically significant reduction of enterococci whereas the numbers of resistant *Klebsiella* strains increased. The agent also exerted changes in the anaerobic microbiota with decreased numbers of lactobacilli, bifidobacteria, clostridia and strains of *Bacteroides*. The minimum inhibitory concentrations (MIC) values of *Bacteroides fragilis* strains increased during administration and returned to pre-treatment levels on day 35.

AGENTS BLOCKING THE METABOLISM OF FOLIC ACID

The impact of folic acid antagonists is summarized in Table 6.

Co-trimoxazole

The ecological effects of co-trimoxazole on the intestinal microbiota have been evaluated in a scheme for prophylaxis in women suffering from recurrent urinary tract infections and in infants being treated for various infections (7,86). In women there was a marked decrease in numbers of Enterobacteriaceae and resistant *E. coli* strains were detected in samples of one woman. In infants, lactobacilli and bifidobacteria were nearly absent but no other significant changes were observed.

ANTIMICROBIAL AGENTS THAT INTERFERE WITH THE SYNTHESIS OF DNA

The ecological impact of nitroimidazoles and combinations of metronidazole and penicillin or macrolide is shown in Table 6 and the impact of quinolones is shown in Table 7.

Nitroimidazoles

Metronidazole

Only minor alterations of the aerobic and anaerobic gastrointestinal microbiota have been shown to occur during metronidazole treatment of patients with different infections (87).

Tinidazole

Parenterally administered tinidazole has been used in order to prevent postoperative infections after abdominal surgery (88). Analyses of the intestinal microbiota revealed that the treatment induced proliferation of the numbers of enterococci and staphylococci. Anaerobic Gram-positive cocci, fusobacteria and bacteroides were also significantly affected during and immediately after the administration period. In connection with oral

Table 7 Impact of Quinolones on the Intestinal Microbiota

Agent	Dose mg/day	Days of adminis- tration	Number of subjects	Impact on			Emergence of resistance			Overgrowth of		Concen- tration range mg/kg	Reference
				Aerobic G+ cocci	Entero- bac- teria	Anaero- bic bac- teria	Entero- cocci	Entero- bacteria	Bacter- oides	C. difficile	Can- dida		
Ciprofloxa- cin	750×2 + 400×2	2	21	+	++	+	-	-	-	-	-	<d-858	(92)
	500×2	5	12	+	++	+	-	-	-	-	-	n.e.	(93)
	500×2	7	12	+	++	-	-	-	-	-	-	<d-2200	(94)
	500×2	5	14	+	+	+	-	-	-	-	-	n.e.	(95)
	500×2	mean 42	15	-	++	+	-	+	-	-	-	n.e.	(96)
	400×2	7	12	-	++	-	-	-	-	-	-	n.e.	(97)
	400×2	4	8	-	++	n.e.	-	+	n.e.	n.e.	-	315-714	(98)
	750	1	10	-	+	-	-	-	-	-	-	<d-3700	(99)
	500	5	6	-	++	-	-	-	-	-	-	n.e.	(100)
	250×2[b]	7(+28)	15	+	++	-	-	-	-	-	-	n.e.	(101)
	500×1	5-10	7	-	++	-	-	-	-	-	-	n.e.	(102)
	250×2	5-10	7	-	++	-	-	-	-	-	-	n.e.	(102)
	250×2	3	7	-	++	-	-	+	+	-	-	n.e.	(103)
	50×4	6	12	+	++	-	-	-	-	-	-	n.e.	(104)
	20	14	5	-	++	n.e.	-	-	n.e.	n.e.	-	n.e.	(105)
Enoxacin	400×2	7	10	-	++	-	+	-	-	-	-	1.1-9.3	(106)
Garenoxacin	100-1200	14	30	+	+	++	+	+	+	-	+	36-263	(107)

Drug	Dose	Days	No.							Value	Ref.
Gatifloxacin	400	10	18	+	++	—	—	—	—	n.e.	(108)
Gemifloxa-cin	320	7	10	+	+	—	—	—	—	58–121	(109)
	320	1	12	—	+	—	—	—	—	<1–194	(110)
Levofloxa-cin	200×3	7	6	+	++	—	+	—	—	<d–163	(111)
	500	7	10	+	+	—	—	—	—	94(57)[a]	(112)
Lomefloxa-cin	400	7	10	—	++	—	—	—	—	<d–203	(113)
Moxifloxa-cin	400	7	12	+	++	—	—	—	—	66(25)[a]	(72)
Norfloxacin	400×2	8	10	—	++	—	—	—	—	n.e.	(114)
	400×2	5	6	+	++	—	—	—	—	n.e.	(115)
	200×2	5	6	—	++	—	—	—	—	n.e.	(115)
	400×2	5	10	+	++	—	—	—	—	n.e.	(116)
	200×2	5	10	—	++	—	—	—	—	n.e.	(116)
	100×2	5	10	—	++	—	—	—	—	n.e.	(116)
	400×2	7	10	—	++	—	—	—	—	n.e.	(117)
	200	7	10	—	++	—	—	—	—	n.e.	(117)
	200×2	7	10	—	++	—	—	—	—	303–1906	(118)
	400	1	12	—	++	—	—	—	—	13–1030	(119)
	200	30	7	—	++	n.e.	—	—	n.e.	n.e.	(86)
Ofloxacin	400	7	10	+	++	—	—	—	—	78(34)[a]	(112)
	200×2	5	5	+	++	—	—	—	+	n.e.	(120)
	400	1	24	+	++	+	—	—	—	n.e.	(121)
Pefloxacin	400×2	10	6	+	++	n.e.	—	—	+	<d–231	(122)

(Continued)

Table 7 Impact of Quinolones on the Intestinal Microbiota (*Continued*)

Agent	Dose mg/day	Days of administration	Number of subjects	Impact on Aerobic G+ cocci	Impact on Enterobacteria	Impact on Anaerobic bacteria	Emergence of resistance Enterococci	Emergence of resistance Enterobacteria	Emergence of resistance Bacteroides	Overgrowth of C. difficile	Overgrowth of Candida	Concentration range mg/kg	Reference
Rufloxacin	400×2	7	15	−	++	−	−	−	−	−	−	n.e.	(104)
	400	1	12	−	++	+	−	−	−	−	−	26–305	(119)
	200	11–35	32	++	++	+	+	−	−	−	−	n.e.	(123)
Sitafloxacin	100×3	7	6	++	++	+	−	−	+	+	+	62(30)[a]	(124)
Sparfloxacin	400+200[c]	8	8	+	++	−	−	−	−	−	−	476(240)[a]	(125)
Trovafloxa-cin	200	10	12	+	++	−	−	+	−	−	−	n.e.	(126)
	200	1	12	−	+	+	−	+	−	−	−	<1–120	(110)

[a] Mean (Standard Deviation).

[b] 250 mg×2 for 7 days, 250 mg for 7 days and 125 mg for 21 days.

[c] Initial dose 400 mg.

Abbreviations: +, mild to moderate effect, increase or decrease 2–4log10 cfu/g feces; ++, strong impact >4log10 cfu/g feces; −, no significant changes; d, the detection limit; n.e., not examined.

administration of tinidazole to healthy subjects, no significant changes have been detected in the gastrointestinal microbiota (89).

Metronidazole in Combination with Amoxycillin

In patients with *Helicobacter pylori* infection treated with omeprazole, metronidazole and amoxycillin, the alterations in the intestinal microbiota have been evaluated (90). Marked ecological disturbances were seen. The numbers of enterococci, enterobacteria, other than *E. coli*, and peptostreptococci increased significantly. Several patients became colonized with *Klebsiella* and *Citrobacter* species as well as with yeasts.

Metronidazole in Combination with Clarithromycin

The influence of *H. pylori* treatment with omeprazole, metronidazole and clarithromycin on the intestinal microbiota has been examined in two groups of patients (90,91). In the first mentioned study, it was found that the numbers of bifidobacteria, clostridia and species of *Bacteroides* were significantly decreased during treatment whereas the numbers of enterococci increased. Strains of enterococci, Enterobacteriaceae and *Bacteroides* spp. had significantly increased MIC values during the administration. In the second study the microbiota was compared to that of healthy subjects. Before treatment, patients were characterized by high concentrations of lactobacilli. Immediately after treatment there was an increased colonization with yeasts and enterobacteria, other than *E. coli*, while the growth of lactobacilli, clostridia and bacteroides decreased. Four weeks after the start of the study the microbiota of patients was similar to that in healthy subjects.

Quinolones

The ecological impact of quinolone administration on the fecal microbiota is described in Table 7.

Ciprofloxacin

The ecological consequences of ciprofloxacin have been evaluated in patients in connection with colorectal surgery (92), in patients with acute leukaemia in remission (96), in prevention of bacterial infections in cirrhosis (101,102) and in treatment of travelers' diarrhea (103). A number of studies have also been performed on healthy volunteers (93–95,97–100,104,105). Ciprofloxacin is excreted in feces in extremely high concentrations and has an activity mainly against Gram-negative aerobic rods. Marked suppression or elimination of enterobacteria has also been shown to occur, both in patients and in the healthy subjects examined. The extension of disturbances has varied depending on the doses. Minor alterations of numbers of Gram-positive aerobic cocci, mainly enterococci, have further been observed and in some studies minor alterations were detected also in the anaerobic microbiota. Ciprofloxacin-resistant species of *Pseudomonas* and *Acinetobacter* have been detected during treatment of patients with acute leukaemia (96) and in healthy volunteers who were given ciprofloxacin intravenously (98). Furthermore, 4 of 7 ciprofloxacin-treated patients with travelers' diarrhea acquired multi-resistant *E. coli* and in 4 subjects increased MIC values of ciprofloxacin for *Bacteroides* spp. were detected (103).

Enoxacin

The effect of enoxacin on the colonic microbiota in human volunteers has been examined (106). Enterobacteria were almost completely suppressed during administration of the drug whereas other aerobic and anaerobic species were not significantly affected.

Garenoxacin

The ecological effect of garenoxacin has been evaluated in healthy individuals receiving oral doses ranging between 100 and 1200 mg daily (107). Higher doses resulted in marked effects on the intestinal microbiota; the strongest effect was noticed in reduced numbers of *Bacteroides* species. Fecal concentrations of garenoxacin also increased with higher doses as well as the selection of resistant strains, mainly enterococci and enterobacteria. In comparison, the *Bacteroides* species strains were less susceptible to the quinolone agent.

Gatifloxacin

Gatifloxacin has been given to healthy subjects in order to study the impact on the normal intestinal microbiota (108). Gatifloxacin possesses a broad spectrum of antimicrobial activity and the administration resulted in not only elimination or strong suppression of *E. coli* strains but also in decreased numbers of enterococci and increased numbers of staphylococci. The numbers of clostridia and fusobacteria decreased significantly in the anaerobic microbiota.

Gemifloxacin

Gemifloxacin is another agent with a broad spectrum of antimicrobial activity. It is active both against Gram-positive and Gram-negative bacteria. The ecological impact of the agent has been investigated in a placebo-controlled study in healthy volunteers (109) and in a randomized cross-over study where the effect of a single dose was investigated in healthy subjects (110). In the first mentioned study, the effect of gemifloxacin was shown to be selective with reduced numbers mainly of enterococci, streptococci and enterobacteria. The single dose caused a pronounced reduction in the numbers of *E. coli* and to a lesser extent also of enterococci and *Bacteroides* species. New quinolone-resistant isolates of Gram-negative aerobes appeared in some subjects.

Levofloxacin

Levofloxacin has been shown to cause a selective reduction in the normal microbiota of healthy subjects, mainly directed towards Gram-negative aerobic rods (111,112). The numbers of enterococci were reduced to a lesser extent. Increased MIC values against strains of *Bacteroides* was detected in one study (111).

Lomefloxacin

Almost a complete eradication of Gram-negative aerobic rods have been shown to occur in the intestinal microbiota of volunteers during administration of lomefloxacin (113). Aerobic Gram-positive and anaerobic microorganisms were virtually unaffected.

Moxifloxacin

The ecological impact of moxifloxacin has been evaluated in healthy subjects (72). The administration caused significant decreases of enterococci and enterobacteria while no other major changes were observed.

Norfloxacin

A number of studies have investigated the ecological effects of norfloxacin on the normal intestinal microbiota (86,114–119). All studies have been performed in healthy subjects and the results have been consistent. Elimination or strong suppression of enterobacteria has been observed and slight reductions of enterococci have been detected in connection with the highest dosing regimens. Only minor fluctuations of other species have been seen.

Ofloxacin

The potential of ofloxacin to disturb the intestinal microbiota has been studied in healthy volunteers (112,120) as well as in patients undergoing gastric surgery (121). In both volunteers and in patients the numbers of enterobacteria were strongly suppressed or eliminated and the numbers of enterococci were significantly reduced. In patients the numbers of lactobacilli, bifidobacteria, eubacteria and species of *Veillonella* and *Bacteroides* were also affected.

Pefloxacin

The influence of pefloxacin on the gastrointestinal microbiota with regard to colonization resistance has been evaluated in two studies on healthy volunteers (104,122). Gram-negative aerobic rods were eliminated during treatment while the numbers of enterococci were slightly suppressed. In one of the studies a significant increase of yeasts was detected in half of the subjects (122).

Rufloxacin

The impact of rufloxacin on intestinal microbiota has been studied in healthy male volunteers after a single dose (119) and in connection with prophylactic treatment of patients with cancer (123). The single dose significantly reduced the numbers of Enterobacteriaceae. This was also observed in patients but the number of *Bacteroides* species was affected as well, however to a lesser extent. The MIC values of rufloxacin for enterococci increased significantly during the second week of treatment.

Sitafloxacin

Sitafloxacin has been shown to markedly suppress both the aerobic and anaerobic intestinal microbiota in healthy persons (124). Most anaerobic microorganisms as well as the aerobic Gram-negative rods were eliminated on the third day of administration until one day after the discontinuation of the drug.

Sparfloxacin

Administration of sparfloxacin to male volunteers has been shown to have a strong impact on *E. coli* and to moderately reduce the numbers of enterococci (125).

Trovafloxacin

The ecological impact of trovafloxacin has been evaluated in connection with multiple (126) and single doses (110) administered to healthy males. The numbers of Enterobacteriaceae were suppressed in both studies, after long-term use below the detection limit. A single dose also resulted in decreased counts of *B. fragilis* group species in some subjects.

CONCLUSION

Antibiotics have a profound place in modern medicine and are indispensable in the treatment of infectious diseases. However, their antimicrobial properties may also affect members of the intestinal microbiota and thereby alter its composition and activity. This may lead to unwanted side effects. It is therefore important to select the appropriate antibiotic and dose that will cause the eradication of the infectious agent but will minimally affect the intestinal microbiota.

REFERENCES

1. Guarner F, Malagelada JR. Gut flora in health and disease. Lancet 2003; 361:512–519 discussion 63–65.
2. Vollaard EJ, Clasener HA. Colonization resistance. Antimicrob Agents Chemother 1994; 38:409–414.
3. Hoiby N. Ecological antibiotic policy. J Antimicrob Chemother 2000; 46:59–62.
4. van der Waaij D, Nord CE. Development and persistence of multi-resistance to antibiotics in bacteria; an analysis and a new approach to this urgent problem. Int J Antimicrob Agents 2000; 16:191–197.
5. Adamsson I, Edlund C, Sjostedt S, Nord CE. Comparative effects of cefadroxil and phenoxymethylpenicillin on the normal oropharyngeal and intestinal microflora. Infection 1997; 25:154–158.
6. Heimdahl A, Kager L, Nord CE. Changes in the oropharyngeal and colon microflora in relation to antimicrobial concentrations in saliva and faeces. Scand J Infect Dis Suppl 1985; 44:52–58.
7. Bennet R, Eriksson M, Nord CE. The fecal microflora of 1-3-month-old infants during treatment with eight oral antibiotics. Infection 2002; 30:158–160.
8. Black F, Einarsson K, Lidbeck A, Orrhage K, Nord CE. Effect of lactic acid producing bacteria on the human intestinal microflora during ampicillin treatment. Scand J Infect Dis 1991; 23:247–254.
9. Knothe H, Wiedemann B. Die Wirkung von Ampicillin auf die Darmflora des gesunden Menschen. Zentralbl Bakteriol Hyg I Abt 1965; 197:234–243.
10. Leigh DA. Pharmacology and toxological studies with amoxicillin, talampicillin and ampicillin and a clinical trial of parenteral amoxycillin in serious hospital infections. Drugs Exp Clin 1979; 5:129–139.
11. Kager L, Malmborg AS, Sjostedt S, Nord CE. Concentrations of ampicillin plus sulbactam in serum and intestinal mucosa and effects on the colonic microflora in patients undergoing colorectal surgery. Eur J Clin Microbiol 1983; 2:559–563.
12. Kager L, Liljeqvist L, Malmborg AS, Nord CE, Pieper R. Effects of ampicillin plus sulbactam on bowel flora in patients undergoing colorectal surgery. Antimicrob Agents Chemother 1982; 22:208–212.

13. Stark CA, Adamsson I, Edlund C, et al. Effects of omeprazole and amoxicillin on the human oral and gastrointestinal microflora in patients with *Helicobacter pylori* infection. J Antimicrob Chemother 1996; 38:927–939.

14. Gipponi M, Sciutto C, Accornero L, et al. Assessing modifications of the intestinal bacterial flora in patients on long-term oral treatment with bacampicillin or amoxicillin: a random study. Chemioterapia 1985; 4:214–217.

15. Mittermayer HW. The effect of amoxycillin and amoxycillin plus clavulanic acid on human bowel microflora. In: Croydon EAP, Michel MF, eds. Augmentin. Amsterdam: Elsevier Science, 1983.

16. Floor M, van Akkeren F, Rozenberg-Arska M, et al. Effect of loracarbef and amoxicillin on the oropharyngeal and intestinal microflora of patients with bronchitis. Scand J Infect Dis 1994; 26:191–197.

17. Edlund C, Stark C, Nord CE. The relationship between an increase in beta-lactamase activity after oral administration of three new cephalosporins and protection against intestinal ecological disturbances. J Antimicrob Chemother 1994; 34:127–138.

18. Brismar B, Edlund C, Nord CE. Impact of cefpodoxime proxetil and amoxicillin on the normal oral and intestinal microflora. Eur J Clin Microbiol Infect Dis 1993; 12:714–719.

19. Christensson B, Nilsson-Ehle I, Ljungberg B, et al. Swedish Study Group. A randomized multicenter trial to compare the influence of cefaclor and amoxycillin on the colonization resistance of the digestive tract in patients with lower respiratory tract infection. Infection 1991; 19:208–215.

20. Lode H, Von der Hoh N, Ziege S, Borner K, Nord CE. Ecological effects of linezolid versus amoxicillin/clavulanic acid on the normal intestinal microflora. Scand J Infect Dis 2001; 33:899–903.

21. Wise R, Bennett SA, Dent J. The pharmacokinetics of orally absorbed cefuroxime compared with amoxicillin/clavulanic acid. J Antimicrob Chemother 1984; 13:603–610.

22. Motohiro T, Tanaka K, Koga T, et al. Effect of BRL 25000 (clavulanic acid-amoxicillin) on bacterial flora in human feces. Jpn J Antibiot 1985; 38:441–480.

23. Brumfitt W, Franklin I, Grady D, Hamilton-Miller JM. Effect of amoxicillin-clavulanate and cephradine on the fecal flora of healthy volunteers not exposed to a hospital environment. Antimicrob Agents Chemother 1986; 30:335–337.

24. Lambert-Zechovsky N, Bingen E, Proux MC, Aujard Y, Mathieu H. Effect of amoxycillin combined with clavulanic acid on the fecal flora of children. Pathol Biol 1984; 32:436–442.

25. Heimdahl A, Nord CE, Weilander K. Effect of bacampicillin on human mouth, throat and colon flora. Infection 1979; 7:S446–S451.

26. Knothe H, Lembke U. The effect of ampicillin and pivampicillin on the intestinal flora of man. Zentralbl Bakteriol [Orig A] 1973; 223:324–332.

27. Nord CE, Bergan T, Aase S. Impact of azlocillin on the colon microflora. Scand J Infect Dis 1986; 18:163–166.

28. Kager L, Malmborg AS, Nord CE, Sjostedt S. The effect of piperacillin prophylaxis on the colonic microflora in patients undergoing colorectal surgery. Infection 1983; 11:251–254.

29. Nord CE, Brismar B, Kasholm-Tengve B, Tunevall G. Effect of piperacillin/tazobactam treatment on human bowel microflora. J Antimicrob Chemother 1993; 31:61–65.

30. Knothe H. The influence of pivmecillinam on the human gut flora. Arzneimittelforschung Forschung (Drug Reg) 1976; 26:427–431.

31. Sullivan A, Edlund C, Svenungsson B, Emtestam L, Nord CE. Effect of perorally administered pivmecillinam on the normal oropharyngeal, intestinal and skin microflora. J Chemother 2001; 13:299–308.

32. Nord CE, Bergan T, Thorsteinsson SB. Impact of ticarcillin/clavulanate on the intestinal microflora. J Antimicrob Chemother 1989; 24:221–226.

33. Vogel F, Knothe H. Changes in aerobic fecal bacterial flora of severely ill patients during antibiotic treatment. Klin Wochenschr 1985; 63:1174–1179.

34. Takesue Y, Yokoyama T, Akagi S, et al. Changes in the intestinal flora after the administration of prophylactic antibiotics to patients undergoing a gastrectomy. Surg Today 2002; 32:581–586.

35. Kager L, Brismar B, Malmborg AS, Nord CE. Impact of cefbuperazone on the colonic microflora in patients undergoing colorectal surgery. Drugs Exp Clin Res 1986; 12:983–986.

36. Bacher K, Schaeffer M, Lode H, Nord CE, Borner K, Koeppe P. Multiple dose pharmacokinetics, safety, and effects on fecal microflora, of cefepime in healthy volunteers. J Antimicrob Chemother 1992; 30:365–375.

37. Knothe H, Dette GA, Shah PM. Impact of injectable cephalosporins on the gastrointestinal microflora: observations in healthy volunteers and hospitalized patients. Infection 1985; 13:S129–S133.

38. Alestig K, Carlberg H, Nord CE, Trollfors B. Effect of cefoperazone on fecal flora. J Antimicrob Chemother 1983; 12:163–167.

39. Lambert-Zechovsky N, Bingen E, Proux MC, Aujard Y, Mathieu H. Effects of cefoperazone on children's fecal flora. Pathol boil 1984; 32:439–442.

40. Guggenbichler JP, Kofler J. Influence of third-generation cephalosporins on aerobic intestinal flora. J Antimicrob Chemother 1984; 14:67–70.

41. Lambert-Zechovsky N, Bingen E, Aujard Y, Mathieu H. Impact of cefotaxime on the fecal flora in children. Infection 1985; 13:S140–S144.

42. Kager L, Ljungdahl I, Malmborg AS, Nord CE, Pieper R, Dahlgren P. Antibiotic prophylaxis with cefoxitin in colorectal surgery: effect on the colon microflora and septic complications–a clinical model for prediction of the benefit and risks in using a new antibiotic in prophylaxis. Ann Surg 1981; 193:277–282.

43. Mulligan ME, Citron D, Gabay E, Kirby BD, George WL, Finegold SM. Alterations in human fecal flora, including ingrowth of *Clostridium difficile*, related to cefoxitin therapy. Antimicrob Agents Chemother 1984; 26:343–346.

44. Knothe H, Schafer V, Sammann A, Badian M, Shah PM. Influence of cefpirome on pharyngeal and fecal flora after single and multiple intravenous administrations of cefpirome to healthy volunteers. J Antimicrob Chemother 1992; 29:81–86.

45. Nilsson-Ehle I, Nord CE, Ursing B. Ceftriaxone: pharmacokinetics and effect on the intestinal microflora in patients with acute bacterial infections. Scand J Infect Dis 1985; 17:77–82.

46. Cavallaro V, Catania V, Bonaccorso R, et al. Effect of a broad-spectrum cephalosporin on the oral and intestinal microflora in patients undergoing colorectal surgery. J Chemother 1992; 4:82–87.

47. de Vries-Hospers HG, Tonk RH, van der Waaij D. Effect of intramuscular ceftriaxone on aerobic oral and fecal flora of 11 healthy volunteers. Scand J Infect Dis 1991; 23:625–633.

48. Welling GW, Meijer-Severs GJ, Helmus G, et al. The effect of ceftriaxone on the anaerobic bacterial flora and the bacterial enzymatic activity in the intestinal tract. Infection 1991; 19:313–316.

49. Vogel F, Ochs HR, Wettich K, et al. Effect of step-down therapy of ceftriaxone plus loracarbef versus parenteral therapy of ceftriaxone on the intestinal microflora in patients with community-acquired pneumonia. Clin Microbiol Infect 2001; 7:376–379.

50. Kager L, Malmborg AS, Nord CE, Sjostedt S. Impact of single dose as compared to three dose prophylaxis with latamoxef (moxalactam) on the colonic microflora in patients undergoing colorectal surgery. J Antimicrob Chemother 1984; 14:171–177.

51. Finegold SM, Ingram-Drake L, Gee R, et al. Bowel flora changes in humans receiving cefixime (CL 284,635) or cefaclor. Antimicrob Agents Chemother 1987; 31:443–446.

52. Nord CE, Heimdahl A, Lundberg C, Marklund G. Impact of cefaclor on the normal human oropharyngeal and intestinal microflora. Scand J Infect Dis 1987; 19:681–685.

53. Novelli A, Mazzei T, Fallani S, Dei R, Cassetta MI, Conti S. Betalactam therapy and intestinal flora. J Chemother 1995; 7:25–31.

54. Nord CE, Movin G, Stalberg D. Impact of cefixime on the normal intestinal microflora. Scand J Infect Dis 1988; 20:547–552.

55. Orrhage K, Sjostedt S, Nord CE. Effect of supplements with lactic acid bacteria and oligofructose on the intestinal microflora during administration of cefpodoxime proxetil. J Antimicrob Chemother 2000; 46:603–612.

56. Lode H, Muller C, Borner K, Nord CE, Koeppe P. Multiple-dose pharmacokinetics of cefprozil and its impact on intestinal flora of volunteers. Antimicrob Agents Chemother 1992; 36:144–149.

57. Brismar B, Edlund C, Nord CE. Effect of ceftibuten on the normal intestinal microflora. Infection 1993; 21:373–375.

58. Edlund C, Brismar B, Sakamoto H, Nord CE. Impact of cefuroxime-axetil on the normal intestinal microflora. Microb Ecol Health Dis 1993; 6:185–189.

59. Edlund C, Barkholt L, Olsson-Liljequist B, Nord CE. Effect of vancomycin on intestinal flora of patients who previously received antimicrobial therapy. Clin Infect Dis 1997; 25:729–732.

60. Leigh DA, Walsh B, Leung A, Tait S, Peatey K, Hancock P. The effect of cefuroxime axetil on the fecal flora of healthy volunteers. J Antimicrob Chemother 1990; 26:261–268.

61. Nord CE, Grahnen A, Eckernas SA. Effect of loracarbef on the normal oropharyngeal and intestinal microflora. Scand J Infect Dis 1991; 23:255–260.

62. Jones PG, Bodey GP, Swabb EA, Rosenbaum B. Effect of aztreonam on throat and stool flora of cancer patients. Antimicrob Agents Chemother 1984; 26:941–943.

63. Kager L, Brismar B, Malmborg AS, Nord CE. Effect of aztreonam on the colon microflora in patients undergoing colorectal surgery. Infection 1985; 13:111–114.

64. de Vries-Hospers HG, Welling GW, Swabb EA, van der Waaij D. Selective decontamination of the digestive tract with aztreonam: a study of 10 healthy volunteers. J Infect Dis 1984; 150:636–642.

65. Kager L, Brismar B, Malmborg AS, Nord CE. Imipenem concentrations in colorectal surgery and impact on the colonic microflora. Antimicrob Agents Chemother 1989; 33:204–208.

66. Nord CE, Kager L, Philipson A, Stiernstedt G. Impact of imipenem/cilastatin therapy on fecal flora. Eur J Clin Microbiol 1984; 3:475–477.

67. Bergan T, Nord CE, Thorsteinsson SB. Effect of meropenem on the intestinal microflora. Eur J Clin Microb Infect Dis 1991; 10:524–527.

68. Nakashima M, Uematsu T, Kosuge K, Nakagawa S, Hata S, Sanada M. Pharmacokinetics and safety of BO-2727, a new injectable 1-beta-methyl carbapenem antibiotic, and its effect on the fecal microflora in healthy male volunteers. J Antimicrob Chemother 1994; 33:987–998.

69. Van der Auwera P, Pensart N, Korten V, Murray BE, Leclercq R. Influence of oral glycopeptides on the fecal flora of human volunteers: selection of highly glycopeptide-resistant enterococci. J Infect Dis 1996; 173:1129–1136.

70. Lund B, Edlund C, Barkholt L, Nord CE, Tvede M, Poulsen RL. Impact on human intestinal microflora of an *Enterococcus faecium* probiotic and vancomycin. Scand J Infect Dis 2000; 32:627–632.

71. Edlund C, Alvan G, Barkholt L, Vacheron F, Nord CE. Pharmacokinetics and comparative effects of telithromycin (HMR 3647) and clarithromycin on the oropharyngeal and intestinal microflora. J Antimicrob Chemother 2000; 46:741–749.

72. Edlund C, Beyer G, Hiemer-Bau M, Ziege S, Lode H, Nord CE. Comparative effects of moxifloxacin and clarithromycin on the normal intestinal microflora. Scand J Infect Dis 2000; 32:81–85.

73. Matute AJ, Schurink CA, Krijnen RM, Florijn A, Rozenberg-Arska M, Hoepelman IM. Double-blind, placebo-controlled study comparing the effect of azithromycin with clarithromycin on oropharyngeal and bowel microflora in volunteers. Eur J Clin Microbiol Infect Dis 2002; 21:427–431.

74. Brismar B, Edlund C, Nord CE. Comparative effects of clarithromycin and erythromycin on the normal intestinal microflora. Scand J Infect Dis 1991; 23:635–642.

75. Eckernas SA, Grahnen A, Nord CE. Impact of dirithromycin on the normal oral and intestinal microflora. Eur J Clin Microbiol Infect Dis 1991; 10:688–692.

76. Heimdahl A, Nord CE. Influence of erythromycin on the normal human flora and colonization of the oral cavity, throat and colon. Scand J Infect Dis 1982; 14:49–56.

77. Pecquet S, Chachaty E, Tancrede C, Andremont A. Effects of roxithromycin on fecal bacteria in human volunteers and resistance to colonization in gnotobiotic mice. Antimicrob Agents Chemother 1991; 35:548–552.

78. Kager L, Liljeqvist L, Malmborg AS, Nord CE. Effect of clindamycin prophylaxis on the colonic microflora in patients undergoing colorectal surgery. Antimicrob Agents Chemother 1981; 20:736–740.

79. Heimdahl A, Nord CE. Effect of erythromycin and clindamycin on the indigenous human anaerobic flora and new colonization of the gastrointestinal tract. Eur J Clin Microbiol 1982; 1:38–48.

80. Orrhage K, Brismar B, Nord CE. Effect of supplements with *Bifidobacterium longum* and *Lactobacillus acidophilus* on the intestinal microbiota during administration of clindamycin. Microb Ecol Health Dis 1994; 7:17–25.

81. Sullivan A, Barkholt L, Nord CE. *Lactobacillus acidophilus*. *Bifidobacterium lactis* and *Lactobacillus* F19 prevent antibiotic-associated ecological disturbances of *Bacteroides fragilis* in the intestine. J Antimicrob Chemother 2003; 52:308–311.

82. Scanvic-Hameg A, Chachaty E, Rey J, et al. Impact of quinupristin/dalfopristin (RP59500) on the fecal microflora in healthy volunteers. J Antimicrob Chemother 2002; 49:135–139.

83. Bartlett JG, Bustetter LA, Gorbach SL, Onderdonk AB. Comparative effect of tetracycline and doxycycline on the occurrence of resistant *Escherichia coli* in the fecal flora. Antimicrob Agents Chemother 1975; 7:55–57.

84. Heimdahl A, Nord CE. Influence of doxycycline on the normal human flora and colonization of the oral cavity and colon. Scand J Infect Dis 1983; 15:293–302.

85. Mulder JG, Wiersma WE, Welling GW, van der Waaij D. Low dose oral tobramycin treatment for selective decontamination of the digestive tract: a study in human volunteers. J Antimicrob Chemother 1984; 13:495–504.

86. Mavromanolakis E, Maraki S, Samonis G, Tselentis Y, Cranidis A. Effect of norfloxacin, trimethoprim-sulfamethoxazole and nitrofurantoin on fecal flora of women with recurrent urinary tract infections. J Chemother 1997; 9:203–207.

87. Nord CE. Ecological impact of narrow spectrum antimicrobial agents compared to broad spectrum agents on the human intestinal microflora. In: Nord CE, Heidt PJ, Rusch VC,

van der Waaij D, eds. Old Herborn University Monograph: Consequences of Antimicrobial Therapy for the Composition of the Microflora of the Digestive Tract. Herborn: Institute for Microbiology and Biochemistry, 1993:8–19.

88. Kager L, Ljungdahl I, Malmborg AS, Nord CE. Effect of tinidazole prophylaxis on the normal microflora in patients undergoing colorectal surgery. Scand J Infect Dis Suppl 1981; 26:84–91.

89. Heimdahl A, Nord CE, Okuda K. Effect of tinidazole on the oral, throat, and colon microflora of man. Med Microbiol Immunol (Berl) 1980; 168:1–10.

90. Adamsson I, Nord CE, Lundquist P, Sjostedt S, Edlund C. Comparative effects of omeprazole, amoxycillin plus metronidazole versus omeprazole, clarithromycin plus metronidazole on the oral, gastric and intestinal microflora in Helicobacter pylori-infected patients. J Antimicrob Chemother 1999; 44:629–640.

91. Buhling A, Radun D, Muller WA, Malfertheiner P. Influence of anti-Helicobacter triple-therapy with metronidazole, omeprazole and clarithromycin on intestinal microflora. Aliment Pharmacol Ther 2001; 15:1445–1452.

92. Brismar B, Edlund C, Malmborg AS, Nord CE. Ciprofloxacin concentrations and impact of the colon microflora in patients undergoing colorectal surgery. Antimicrob Agents Chemother 1990; 34:481–483.

93. Bergan T, Delin C, Johansen S, Kolstad IM, Nord CE, Thorsteinsson SB. Pharmacokinetics of ciprofloxacin and effect of repeated dosage on salivary and fecal microflora. Antimicrob Agents Chemother 1986; 29:298–302.

94. Brumfitt W, Franklin I, Grady D, Hamilton-Miller JM, Iliffe A. Changes in the pharmacokinetics of ciprofloxacin and fecal flora during administration of a 7-day course to human volunteers. Antimicrob Agents Chemother 1984; 26:757–761.

95. Ljungberg B, Nilsson-Ehle I, Edlund C, Nord CE. Influence of ciprofloxacin on the colonic microflora in young and elderly volunteers: no impact of the altered drug absorption. Scand J Infect Dis 1990; 22:205–208.

96. Rozenberg-Arska M, Dekker AW, Verhoef J. Ciprofloxacin for selective decontamination of the alimentary tract in patients with acute leukemia during remission induction treatment: the effect on fecal flora. J Infect Dis 1985; 152:104–107.

97. Enzensberger R, Shah PM, Knothe H. Impact of oral ciprofloxacin on the fecal flora of healthy volunteers. Infection 1985; 13:273–275.

98. Krueger WA, Ruckdeschel G, Unertl K. Influence of intravenously administered ciprofloxacin on aerobic intestinal microflora and fecal drug levels when administered simultaneously with sucralfate. Antimicrob Agents Chemother 1997; 41:1725–1730.

99. Pecquet S, Ravoire S, Andremont A. Fecal excretion of ciprofloxacin after a single oral dose and its effect on fecal bacteria in healthy volunteers. J Antimicrob Chemother 1990; 26:125–129.

100. Holt HA, Lewis DA, White LO, Bastable SY, Reeves DS. Effect of oral ciprofloxacin on the fecal flora of healthy volunteers. Eur J Clin Microbiol 1986; 5:201–205.

101. Borzio M, Salerno F, Saudelli M, Galvagno D, Piantoni L, Fragiacomo L. Efficacy of oral ciprofloxacin as selective intestinal decontaminant in cirrhosis. Ital J Gastroenterol Hepatol 1997; 29:262–266.

102. Esposito S, Barba D, Galante D, Gaeta GB, Laghezza O. Intestinal microflora changes induced by ciprofloxacin and treatment of portal-systemic encephalopathy (PSE). Drugs Exp Clin Res 1987; 13:641–646.

103. Wistrom J, Gentry LO, Palmgren AC, et al. Ecological effects of short-term ciprofloxacin treatment of travellers' diarrhoea. J Antimicrob Chemother 1992; 30:693–706.

104. Van Saene JJ, Van Saene HK, Geitz JN, Tarko-Smit NJ, Lerk CF. Quinolones and colonization resistance in human volunteers. Pharm Weekbl Sci 1986; 8:67–71.

105. van de Leur JJ, Vollaard EJ, Janssen AJ, Dofferhoff AS. Influence of low dose ciprofloxacin on microbial colonization of the digestive tract in healthy volunteers during normal and during impaired colonization resistance. Scand J Infect Dis 1997; 29:297–300.

106. Edlund C, Lidbeck A, Kager L, Nord CE. Effect of enoxacin on colonic microflora of healthy volunteers. Eur J Clin Microbiol 1987; 6:298–300.

107. Nord CE, Gajjar DA, Grasela DM. Ecological impact of the des-F(6)-quinolone, BMS-284756, on the normal intestinal microflora. Clin Microbiol Infect 2002; 8:229–239.

108. Edlund C, Nord CE. Ecological effect of gatifloxacin on the normal human intestinal microflora. J Chemother 1999; 11:50–53.

109. Barker PJ, Sheehan R, Teillol-Foo M, Palmgren AC, Nord CE. Impact of gemifloxacin on the normal human intestinal microflora. J Chemother 2001; 13:47–51.

110. Garcia-Calvo G, Molleja A, Gimenez MJ, et al. Effects of single oral doses of gemifloxacin (320 milligrams) versus trovafloxacin (200 milligrams) on fecal flora in healthy volunteers. Antimicrob Agents Chemother 2001; 45:608–611.

111. Inagaki Y, Nakaya R, Chida T, Hashimoto S. The effect of levofloxacin, an optically-active isomer of ofloxacin, on fecal microflora in human volunteers. Jpn J Antibiot 1992; 45:241–252.

112. Edlund C, Sjostedt S, Nord CE. Comparative effects of levofloxacin and ofloxacin on the normal oral and intestinal microflora. Scand J Infect Dis 1997; 29:383–386.

113. Edlund C, Brismar B, Nord CE. Effect of lomefloxacin on the normal oral and intestinal microflora. Eur J Clin Microbiol Infect Dis 1990; 9:35–39.

114. Leigh DA, Emmanuel FXS, Tighe C, Hancock P, Boddy S, Pharmacokinetic studies on norfloxacin in healthy volunteers and effect on the fecal flora. 14th International Congress of Chemotherapy, Kyoto, Japan, 1985.

115. Pecquet S, Andremont A, Tancrede C. Selective antimicrobial modulation of the intestinal tract by norfloxacin in human volunteers and in gnotobiotic mice associated with a human fecal flora. Antimicrob Agents Chemother 1986; 29:1047–1052.

116. De Vries-Hospers HG, Welling GW, Van der Waaij D. Norfloxacin for selective decontamination: a study in human volunteers. Prog Clin Biol Res 1985; 181:259–262.

117. Meckenstock R, Haralambie E, Linzenmeier G, Wendt F. Die Beeinflussung der Darmflora dürch norfloxacin bei gesunden Menschen. Z Antimikr Antineoplast Chemother 1985; 1:27–34.

118. Edlund C, Bergan T, Josefsson K, Solberg R, Nord CE. Effect of norfloxacin on human oropharyngeal and colonic microflora and multiple-dose pharmacokinetics. Scand J Infect Dis 1987; 19:113–121.

119. Marco F, Gimenez MJ, Jimenez de Anta MT, Marcos MA, Salva P, Aguilar L. Comparison of rufloxacin and norfloxacin effects on fecal flora. J Antimicrob Chemother 1995; 35:895–901.

120. Pecquet S, Andremont A, Tancrede C. Effect of oral ofloxacin on fecal bacteria in human volunteers. Antimicrob Agents Chemother 1987; 31:124–125.

121. Edlund C, Kager L, Malmborg AS, Sjostedt S, Nord CE. Effect of ofloxacin on oral and gastrointestinal microflora in patients undergoing gastric surgery. Eur J Clin Microbiol Infect Dis 1988; 7:135–143.

122. Vollaard EJ, Clasener HA, Janssen AJ. Influence of pefloxacin on microbial colonization resistance in healthy volunteers. Eur J Clin Microbiol Infect Dis 1992; 11:257–260.

123. D'Antonio D, Pizzigallo E, Lacone A, et al. The impact of rufloxacin given as prophylaxis to patients with cancer on their oral and fecal microflora. J Antimicrob Chemother 1996; 38:839–847.

124. Inagaki Y, Yamamoto N, Chida T, Okamura N, Tanaka M. The effect of DU-6859a, a new potent fluoroquinolone, on fecal microflora in human volunteers. Jpn J Antibiot 1995; 48:368–379.

125. Ritz M, Lode H, Fassbender M, Borner K, Koeppe P, Nord CE. Multiple-dose pharmacokinetics of sparfloxacin and its influence on fecal flora. Antimicrob Agents Chemother 1994; 38:455–459.
126. van Nispen CH, Hoepelman AI, Rozenberg-Arska M, Verhoef J, Purkins L, Willavize SA. A double-blind, placebo-controlled, parallel group study of oral trovafloxacin on bowel microflora in healthy male volunteers. Am J Surg 1998; 176:27S–31S.

19

The Intestinal Microbiota of Pets: Dogs and Cats

Minna Rinkinen
Department of Clinical Veterinary Sciences, Faculty of Veterinary Medicine, University of Helsinki, Helsinki, Finland

INTRODUCTION

The knowledge of canine and feline intestinal microbiota is relatively scarce and based mainly on data from laboratory animals, on responses to dietary interventions, or on animals suffering from chronic intestinal disorders believed to be of bacterial nature. Most of the studies are performed on quite low numbers of animals that were often sacrificed and samples of intestinal material collected post-mortem (1,2).

As obtaining fecal samples is much more feasible than sampling the contents of upper intestinal tract, most of the papers have focused on fecal microbiota, which may not be considered to represent the whole intestinal microecology. In addition, observations based on the cultivation of luminal contents may not reflect the microbiota adhered to mucosa.

Most of the bacterial studies have been performed with traditional cultivation and characterization methods, which may have biased the identification and taxonomy of microbiota. In humans, it is estimated that only 40% of intestinal bacteria are culturable (3); a similar outcome can be expected also in dogs and cats. In addition, the bacterial taxonomy and nomenclature have changed during time, so bacteria identified in earlier studies may currently be re-classified under a different name. For a more in-depth description on the analysis of the intestinal microbiota, see the chapter by Ben-Amor and Vaughan in this book.

Proximal small intestine harbors total bacteria of 10^{6-8} CFU/ml of luminal content. The number of intestinal bacteria increases distally, reaching up to 10^{14} CFU/g in feces. In the small intestine aerobic and facultative aerobic bacteria outnumber anaerobic bacteria (4). When moving aborally in the gut, anaerobic bacteria start to dominate and finally gain numbers as high as 10^{10} of CFU anaerobic bacteria/g fecal material (5).

DEVELOPMENT OF INTESTINAL MICROBIOTA IN DOGS AND CATS

Although there is paucity of research data concerning the development of intestinal microbiota of dogs and cats, it can be considered to follow a similar pattern as known for other mammals. Intestinal colonization is a gradual process starting immediately after birth. In newborn puppies and kittens the alimentary canal is sterile but is quickly inhabited by bacteria from birth canal and environment. The dam usually licks the newborn thoroughly thus transferring its own indigenous bacteria to her offspring. Within 24 hours the numbers of bacteria in various parts of the gastrointestinal tract of a newborn puppy are similar to those of an adult dog (2).

The indigenous intestinal microbiota is considered an integral part of the host defense mechanisms. It forms a barrier against pathogen colonization and also influences the host's immunological, biochemical, and physiological features (6).

Once the microbiota has become established, it is relatively stable. Oral antibiotics may have a marked effect on the homeostasis of intestinal microbiota. However, these changes will be re-established relatively soon (7–9). Disturbances in the gut microbiota may result in diarrhea, malabsorption, and chronic intestinal inflammation (10). Acute diarrhea may be fatal as pathogens may invade the host's tissues resulting in bacteremia and sepsis.

Ageing has documented effects on the constitution of intestinal microbiota in dogs. Numbers of bifidobacteria and peptostreptococci diminish with ageing whereas *Clostridium perfringens* and streptococci are more prevalent in the large bowel of elderly dogs (1).

CANINE AND FELINE GASTROINTESTINAL MICROBIOTA

Gram-Positive Intestinal Bacteria

Amongst Gram-positive bacteria residing in the gut, lactic acid bacteria (LAB) make up the largest and most important part of the intestinal microbiota. Although they have a significant protective function in the gut, the present knowledge of canine and feline Gram-positive intestinal microbiota is scant.

Most of the canine LAB belong to the genera *Streptococcus* and *Lactobacillus*. In a recent study, *Streptococcus alactolyticus* was found to be a predominant culturable LAB in jejunal and fecal samples of four beagle dogs. In addition, *Lactobacillus animalis, L. reuteri, L. murinus, L. ruminus* and *S. bovis* are reported to harbor in the gut (11,12).

The presence of bifidobacteria in canine GI tract is controversial. Many papers report absence of bifidobacteria in the canine fecal samples (11,13), whereas others described bifidobacteria as a substantial part of canine fecal microbiota (14–17). Willard and co-workers isolated fecal bifidobacteria from dogs inconstantly and independent on the diet. It was concluded that bifidobacteria may be only sporadically present in the feces of healthy dogs (18).

In healthy cats, the total number of duodenal microbiota is reported to range from 10^5 to 10^9 cfu/ml, most of the bacteria being anaerobic (10,19). The most common anaerobic isolates belonged to groups *Bacteroides, Clostridium, Eubacteria* and *Fusobacteria*, whereas *Pasteurella* spp were the most prevailing aerobic bacteria in feline proximal small intestine. In addition, *Acinetobacter* spp, *Pseudomonas* spp and *Lactobacillus* spp were detected in the duodenal samples of healthy cats (10,19). Lactobacilli were also isolated from feline fecal samples (20).

Intestinal Pathogenic Bacteria

Bacteria are seldom the sole pathogenic factor in canine and feline gastrointestinal disturbances. Some of the pathogens have been linked to clinical disease, but these pathogenic organisms are frequently isolated also in healthy individuals (21–26).

Escherichia Coli

Escherichia coli is a normal intestinal inhabitant in warm-blooded animals, including cats and dogs, although its clinical significance as canine and feline enteropathogen is not very well documented. Colonization is believed to take place within the first days of a newborn animal. Certain strains of *E. coli* may act as intestinal pathogens causing gastrointestinal infections. Enteropathogenic *E. coli* and enterotoxigenic *E. coli* are known to associate with canine diarrhea, especially in young dogs (27–30). However, these strains have been isolated from non–diarrheic animals, too (28,30,31).

Enterohemorrhagic *E. coli* (EHEC) has been isolated occasionally from dogs. Most of these reports are from dogs living in contact with cattle. EHEC has never been documented in cats (24).

Clostridia

Clostridium perfringens

Clostridium perfringens is an anaerobic, spore-forming bacillus associated with acute and chronic diarrhea in dogs and cats. However, the role of *C. perfringens* as an intestinal pathogen is questionable, as it commonly harbors in the intestinal tract of healthy dogs, too (23,32). *C. perfringens* produces toxins, which are classified in five toxigenic types (A–E). *C. perfringens* enterotoxin (CPE) is the best characterized virulence factor and coregulated with sporulation. All *C. perfringens* types can produce CPE, but type A strains are most frequently involved. CPE has been reported to cause nosocomial diarrhea, severe hemorrhagic enteritis, and acute and chronic large bowel diarrhea in dogs (33). On the other hand, CPE is also found in feces of non-diarrheic animals (23,32), although a significant association was present with diarrhea and detection of CPE (23).

One study reports *C. perfringens* carrying ß2 toxin gene (*cpb2*) isolated from diarrheic dogs, suggesting ß2 toxin alone or together with CPE may play a role in canine clostridial diarrhea (34).

Clostridium difficile

C. difficile is associated with diarrhea in dogs, although it has been frequently isolated from dogs with no signs of diarrhea (23,35). *C. difficile*–related diarrhea in humans is principally associated with hospitalization and use of antimicrobials. In dogs, no significant association was found in the prevalence of *C. difficile* along with hospitalization and antibiotic administration, but increased carriage rate was observed in non-hospitalized dogs receiving antibiotics (23).

Salmonella

Both healthy and diarrheic dogs and cats may carry *Salmonella*. Prevalence in healthy dogs is reported to be between 1% and 38% (24,36). Furthermore, *Salmonella* isolation rates in dogs with clinical enteritis is reported low (21,25,37).

The prevalence of *Salmonella* in canine fecal isolates examined has reduced during the past decades. This most likely reflects the change in feeding of dogs, as commercial pet foods have replaced raw meat and offal (36). Feeding bones and raw food diet yielded a 30% *Salmonella* isolation rate in stool samples of dogs consuming this type of diet. Feeding raw chicken and meat to dogs may therefore be a risk for potential transfer of *Salmonella* to humans, too (38,39).

Salmonella is regarded relatively rare in cats, isolation prevalence varying between 0.8% and 18%; in most reports it is approximately 1%. Also cats may be asymptomatic carriers (22,24,40). An outbreak of *Salmonella enterica* serovar Typhimurium in cats was reported in Sweden, where salmonellosis was probably transmitted from wild infected birds hunted by the cats (41).

Campylobacters

Campylobacters are regarded as important zoonotic pathogens. Most of the human infections are food- or water-borne, but infections from pets may also be of concern, especially with immunocompromised people (42–44). Campylobacters have been associated with acute and chronic diarrhea in dogs and cats (43). However, as they are frequently isolated from both healthy and diarrheic animals, it is suggested they are not primary pathogens but more likely opportunistic microbes producing clinical signs in predisposing conditions, such as poor nutrition or housing, or high animal density (45,46). Young dogs seem to be more prone to carry campylobacters, carriage rate being up to 75% of dogs less than 12 months old, whereas the isolation rate in adult dogs was only 32.7% (47,48).

Campylobacter shedding correlates clearly with diarrhea in young dogs, but for dogs older than 12 months there was no evident correlation with shedding and clinical disease. In cats, no significant association was found between campylobacteriosis and diarrhea in any age group (49,50).

In cats and dogs, *C. helveticus*, *C. jejuni*, and *C. upsaliensis* are most prevalent *Campylobacter* strains. *C. helveticus* has been isolated in healthy cats and dogs (47,51,52). One study reported *C. helveticus* to inhabit 21.7% of the cats examined, being the most prevalent *Campylobacter* species isolated (47). In addition, *C. coli*, and *C. lari* have been isolated to lesser extent (43,45,48,50,53–55). However, the traditional phenotypic identification methods have been criticized for being unreliable when identifying thermophilic campylobacters (56). The clinical relevance of these campylobacters is unclear.

Campylobacter upsaliensis

C. upsaliensis is a catalase-negative thermotolerant campylobacter recognized as an emerging human pathogen. In humans it is associated with gastroenteritis and bacteremia (57). It was first isolated from canine feces (54) and some years later also from feline feces (58). It has been reported to be the most prevalent campylobacter in dogs (47,50,56) and cats (50,56). Thus, it is of interest whether household pets may comprise a reservoir for this zoonotic pathogen although human and canine strains are reported to be genotypically distinct (51).

C. upsaliensis has been isolated from feces of both diarrheic and healthy dogs and cats. It is documented to infect puppies at approximately six weeks of age without causing a clinical disease when puppies were raised separately in a breeding kennel, presumably in acceptable conditions. Poor sanitation and high animal density are marked risk factors, increasing the carriage rate of *C. upsaliensis* up to 2.6-fold. These findings support the opportunistic nature of this organism as a canine and feline pathogen (51,59).

Helicobacters

Helicobacter spp. are Gram-negative, microaerophilic curved or spiral-shaped motile bacteria. Many gastric *Helicobacter*-like organisms (GHLO) are frequently found in cats and dogs. Virtually all dogs can be expected to harbor gastric GHLO (60,61), although most of the dogs are asymptomatic. Additionally, the clinical signs in dogs suffering from gastritis may persist despite the eradication of helicobacters. Therefore the role of GHLO as an etiological factor in canine gastritis is currently unclear (62,63).

In dogs, *H. felis*, *H. bizzozeronii*, *H. salomonis*, *"Flexispira rappini,"* *H. bilis*, and *"H. heilmannii"* have been reported to inhabit the gastric mucosa. The human pathogen *H. pylori* has not yet been isolated in canine gastric biopsies. However, a recent paper reports presumably non-cultivable *H. pylori*, or a closely related *Helicobacter* in two dogs, results based on its 16S rRNA sequence (64). Unlike dogs, cats have been documented to acquire *H. pylori*, although very infrequently. Feline *H. pylori* infection has been suggested to be an anthroponosis, i.e., cats are infected by humans carrying *H. pylori* (63,65–67).

In addition to GHLOs, dogs and cats are reported to have also enteric helicobacters. *H. canis* has been isolated from diarrheic cats and dogs (68,69), and *H. marmotae* from cat feces (70).

MODIFYING THE INTESTINAL MICROBIOTA: PRE- AND PROBIOTICS

First documented studies of dietary manipulation of canine and feline intestinal microbiota date back to the beginning of the twentieth century (71).

Today, there is growing interest in modifying their gut microbiota towards what is considered a healthy composition, i.e., increase in LAB and bifidobacteria, and decrease in potential pathogenic bacteria (72). Many commercial pet foods now contain prebiotics (e.g., fructo-oligosaccharides, FOS). In addition, probiotics are also marketed for dogs and cats.

Prebiotics

Prebiotics are reported to have a variable impact on canine fecal and intestinal microbiota. Supplementing dogs' food with FOS and mannanoligosaccharides increased ileal lactobacilli and fecal lactobacilli and bifidobacteria concentrations (73). Feeding short chain FOS to dogs increased the total number of fecal anaerobes and lowered the number of *Clostridium perfringens* (17,74). Similar outcome was achieved with arabinogalactan supplementation (15). On the other hand, no significant differences were noticed in the denaturing gradient gel electrophoresis analysis of fecal bacterial profiles when dogs were fed a diet containing 10% fiber (16), and another study revealed no significant effect of FOS supplementation on canine fecal *Clostridium* spp (18).

FOS supplementation increased fecal lactobacilli and decreased numbers of *E. coli* in healthy cats, but did not alter the duodenal microbiota (75,76). This supports the notion that, as FOS are nondigestible fibers fermented in the proximal gut in humans (mainly in the large intestine) (77), also in cats FOS have only a minimal effect on the microbes residing in the proximal part of GI tract. In a study of eight cats, feeding lactosucrose increased fecal lactobacilli and bifidobacteria counts significantly, while numbers of clostridia and *Enterobacteriaceace* decreased significantly (78).

Probiotics

Currently, there are no commerically available probiotics fulfilling the species specificity criterion applied to probiotics as stated by Saarela and co-workers (79). Despite that, probiotics are utilized in pet animals in the hope to create beneficial alterations in the intestinal microbiota.

Enterococcus faecium SF68 has been documented to enhance specific immuno-logical responses in young dogs (80) and *E. faecalis* FK-23 stimulated non-specific immune functions in healthy adult dogs (81). *E. faecium* is also reported to have an effect on canine enteropathogens. It significantly decreased the canine in vitro mucus adhesion of *C. perfringens* (82). This finding was supported also in vivo (83). On the other hand, *E. faecium* increased both the in vitro adhesion and fecal shedding of campylobacters (82,83). Pasupathy and co-workers (84) evaluated the effect of *Lactobacillus acidophilus* on the digestibility of food and growth of puppies. They concluded that *Lactobacillus* supplementation has a favorable effect during the active growth period, although differences between the study group and control group were not significant.

CONCLUSION

In the recent years the interest in canine and feline gastrointestinal microbiota has increased, resulting in a fair amount of documented information. However, the current knowledge of canine and feline gastrointestinal microbiota is still rather scarce. The growing interest in pre- and probiotics together with the novel microbiological methods has already made a scientific contribution to the field of small animal intestinal microbiology. With this trend likely to continue in the future, our knowledge of the canine and feline gastrointestinal microbiota and the factors related to its regulation will expand.

REFERENCES

1. Benno Y, Nakao H, Uchida K, Mitsuoka T. Impact of the advances in age on the gastrointestinal microflora of beagle dogs. J Vet Med Sci 1992; 54:703–706.
2. Buddington RK. Postnatal changes in bacterial populations in the gastrointestinal tract of dogs. Am J Vet Res 2003; 64:646–651.
3. Tannock GW, Munro K, Harmsen HJM, Welling GW, Smart J, Gopal PK. Analysis of the fecal microflora of human subjects consuming a probiotic product containing *Lactobacillus rhamnosus* DR20. Appl Environ Microbiol 2000; 66:2578–2588.
4. Delles EK, Willard MD, Simpson RB, et al. Comparison of species and numbers of bacteria in concurrently cultured samples of proximal small intestinal fluid and endoscopically obtained duodenal mucosa in dogs with intestinal bacterial overgrowth. Am J Vet Res 1994; 55:957–964.
5. Davis CP, Cleven D, Balish E, Yale CE. Bacterial association in the gastrointestinal tract of beagle dogs. Appl Environ Microbiol 1977; 34:194–206.
6. Tannock GW. The normal microflora: an introduction. In: Tannock GW, ed. Medical Importance of Normal Microflora. London: Kluwer Academic Publishers, 1999:1–23.
7. Lode H, Von der Hoh N, Ziege S, Borner K, Nord CE. Ecological effects of linezolid versus amoxicillin/clavulanic acid on the normal intestinal microflora. Scand J Infect Dis 2001; 33:899–903.
8. Sullivan Å, Edlund C, Nord CE. Effect of antimicrobial agents on the ecological balance of human microflora. Lancet Infect Dis 2001; 1:101–114.
9. Nord CE, Gajjar DA, Grasela DM. Ecological impact of the des-F(6)-quinolone, BMS-284756, on the normal intestinal microflora. Clin Microbiol Infect 2002; 8:229–239.
10. Johnston KL. Small intestinal bacterial overgrowth. In: Simpson KW, ed. The Veterinary Clinics of North America, Small Animal Practice. Saunders, Philadelphia: Progress in Gastroenterology, 1999:523–550.

11. Greetham HL, Giffard C, Hutson RA, Collins MD, Gibson GR. Bacteriology of the Labrador dog gut: a cultural and genotypic approach. J Appl Microbiol 2002; 93:640–646.

12. Rinkinen ML, Koort JMK, Ouwehand AC, Westermarck E, Björkroth JK. *Streptococcus alactolyticus* is the dominating culturable lactic acid bacterium species in canine jejunum and feces of four fistulated dogs. FEMS Microbiol Lett 2004; 230:35–39.

13. Martineau B. Comparison of four media for the selection of bifidobacteria in dog fecal samples. Anaerobe 1999; 5:123–127.

14. Beynen AC, Baas JC, Hoekemeijer PE, et al. Faecal bacterial profile, nitrogen excretion, and mineral absorption in healthy dogs fed supplemental oligofructose. J Anim Physiol Anim Nutr 2002; 86:298–305.

15. Grieshop CM, Flickinger EA, Fahey GC, Jr. Oral administration of arabinogalactan affects immune status and fecal microbial populations in dogs. J Nutr 2002; 132:478–482.

16. Simpson JM, Martineau B, Jones WE, Ballam JM, Mackie RI. Characterization of fecal bacterial populations in canines: effects of age, breed, and dietary fiber. Microb Ecol 2002; 44:186–197.

17. Swanson KS, Grieshop CM, Flickinger EA, et al. Fructooligosaccharides and *Lactobacillus acidophilus* modify gut microbial populations, total tract nutrient digestibilities, and fecal protein catabolite concentrations in healthy adult dogs. J Nutr 2002; 132:3721–3731.

18. Willard MD, Simpson RB, Cohen ND, Clancy JS. Effects of dietary fructooligosaccharide on selected bacterial populations in feces of dogs. Am J Vet Res 2000; 61:820–825.

19. Johnston KL, Swift NC, Forster-van Hijfte M. Comparison of the bacterial flora of the duodenum in healthy cats and cats with signs of gastrointestinal tract disease. J Am Vet Med Assoc 2001; 218:48–51.

20. Hartemink R, Rombouts FM. Comparison of media for the detection of bifidobacteria, lactobacilli, and total anaerobes from faecal samples. J Microbiol Methods 1999; 36:181–192.

21. Buogo C, Burnens AP, Perrin J, Nicolet J. Presence of *Campylobacter* spp *Clostridium difficile*, *C. perfringens*, and salmonellae in litters of puppies and in adult dogs in a shelter. Schweiz Arch Tierheilkd 1995; 137:165–171.

22. Spain CV, Scarlett JM, Wade SE, McDonough P. Prevalence of enteric zoonotic agents in cats less than 1 year old in central New York State. J Vet Intern Med 2001; 15:33–38.

23. Marks SL, Kather EJ, Kass PH, Melli AC. Genotypic and phenotypic characterization of *Clostridium perfringens* and *Clostridium difficile* in diarrheic and healthy dogs. J Vet Intern Med 2002; 16:533–540.

24. Sanchez S, Lee MD, Harmon BG, Maurer JJ, Doyle MP. Animal issues associated with *Escherichia coli* O157:H7. J Am Vet Med Assoc 2002; 221:1122–1126.

25. Hackett T, Lappin MR. Prevalence of enteric pathogens in dogs of north-central Colorado. J Am Anim Hosp Assoc 2003; 39:52–56.

26. Staats JJ, Chengappa MM, DeBey MC, Fickbohm B, Oberst RD. Detection of *Escherichia coli* Shiga toxin (stx) and enterotoxin (estA and elt) genes in fecal samples from non-diarrheic and diarrheic greyhounds. Vet Microbiol 2003; 94:303–312.

27. Drolet R, Fairbrother JM, Harel J, Helie P. Attaching and effacing and enterotoxigenic *Escherichia coli* associated with enteric colibacillosis in the dog. Can J Vet Res 1994; 58:87–92.

28. Hammermueller J, Kruth S, Prescott J, Gyles C. Detection of toxin genes in *Escherichia coli* isolated from normal dogs and dogs with diarrhea. Can J Vet Res 1995; 59:265–270.

29. Beutin L. *Escherichia coli* as a pathogen in dogs and cats. Vet Res 1999; 30:285–298.

30. Starčič M, Johnson JR, Stell AL, et al. Haemolytic *Escherichia coli* isolated from dogs with diarrhea have characteristics of both uropathogenic and necrotoxigenic strains. Vet Microbiol 2002; 85:361–377.

31. Holland RE, Walker RD, Sriranganathan N, Wilson RA, Ruhl DC. Characterization of *Escherichia coli* isolated from healthy dogs. Vet Microbiol 1999; 70:261–268.

32. Weese JS, Staempfli HR, Prescott JF, Kruth SA, Greenwood SJ, Weese HE. The roles of *Clostridium difficile* and enterotoxigenic *Clostridium perfringens* in diarrhea in dogs. J Vet Intern Med 2001; 15:374–378.

33. Sasaki J, Goryo M, Asahina M, Makara M, Shishido S, Okada K. Hemorrhagic enteritis associated with *Clostridium perfringens* type A in a dog. J Vet Med Sci 1999; 61:175–177.

34. Thiede S, Goethe R, Amtsberg G. Prevalence of (2 toxin gene of Clostridium perfringens type A from diarrhoiec dogs. Vet Rec 2001; 149:273–274.

35. Struble AL, Tang YJ, Kass PH, Gumerlock PH, Madewell BR, Silva J, Jr. Fecal shedding of *Clostridium difficile* in dogs: a period prevalence survey in a veterinary medical teaching hospital. J Vet Diagn Invest 1994; 6:342–347.

36. Cave NJ, Marks SL, Kass PH, Melli AC, Brophy MA. Evaluation of a routine diagnostic fecal panel for dogs with diarrhea. J Am Vet Med Assoc 2002; 221:52–59.

37. van Duijkeren E, Houwers D. *Salmonella enteritis* in dogs, not relevant? Tijdschr Diergeneeskd 2002; 127:716–717.

38. LeJeune JT, Hancock DD. Public health concerns associated with feeding raw meat diets to dogs. J Am Vet Med Assoc 2001; 219:1222–1225.

39. Joffe DJ, Schlesinger DP. Preliminary assessment of the risk of *Salmonella* infection in dogs fed raw chicken diets. Can Vet J 2002; 43:441–442.

40. Hill SL, Cheney JM, Taton-Allen GF, Reif JS, Bruns C, Lappin MR. Prevalence of enteric zoonotic organisms in cats. J Am Vet Med Assoc 2000; 216:687–692.

41. Tauni MA, Österlund A. Outbreak of *Salmonella typhimurium* in cats and humans associated with infection in wild birds. J Small Anim Pract 2000; 41:339–341.

42. Ketley JM. Pathogenesis of enteric infection by Campylobacter. Microbiology 1997; 143:5–21.

43. Steinhauserova I, Fojtikova K, Klimes J. The incidence and PCR detection of Campylobacter upsaliensis in dogs and cats. Lett Appl Microbiol 2000; 31:209–212.

44. Allos BM. *Campylobacter jejuni* infections: update on emerging issues and trends. Clin Infect Dis 2001; 32:1201–1206.

45. Torre E, Tello M. Factors influencing fecal shedding of *Campylobacter jejuni* in dogs without diarrhea. Am J Vet Res 1993; 54:260–262.

46. Fernandez H, Martin R. *Campylobacter* intestinal carriage among stray and pet dogs. Rev Saude Publica 1991; 25:473–475.

47. Moser I, Rieksneuwöhner B, Lentzsch P, Schwerk P, Wieler LH. Genomic heterogeneity and O-antigenic diversity of *Campylobacter upsaliensis* and *Campylobacter helveticus* strains isolated from dogs and cats in Germany. J Clin Microbiol 2001; 39:2548–2557.

48. Engvall EO, Brandstrom B, Andersson L, Baverud V, Trowald-Wigh G, Englund L. Isolation and identification of thermophilic *Campylobacter* species in faecal samples from Swedish dogs. Scand J Infect Dis 2003; 35:713–718.

49. Burnens AP, Angeloz-Wick B, Nicolet J. Comparison of *Campylobacter* carriage rates in diarrheic and healthy pet animals. J Vet Med 1992; 39:175–780.

50. Sandberg M, Bergsjo B, Hofshagen M, Skjerve E, Kruse H. Risk factors for *Campylobacter* infection in Norwegian cats and dogs. Prev Vet Med 2002; 55:241–253.

51. Stanley J, Jones C, Burnens A, Owen RJ. Distinct genotypes of human and canine isolates of *Campylobacter upsaliensis* determined by 16S rRNA gene typing and plasmid profiling. J Clin Microbiol 1994; 32:1788–1794.

52. Shen Z, Feng Y, Dewhirst FE, Fox JG. Coinfection of enteric *Helicobacter* spp. and *Campylobacter* spp. in cats. J Clin Microbiol 2001; 39:2166–2172.

53. Prescott JF, Munroe DL. *Campylobacter* jejuni enteritis in man and domestic animals. J Am Vet Med Assoc 1982; 81:1524–1530.

54. Sandstedt K, Ursing J, Walder M. Thermotolerant *Campylobacter* with no or weak catalase activity isolated from dogs. Curr Microbiol 1983; 8:209–213.

55. Duim B, Vandamme PA, Rigter A, Laevens S, Dijkstra JR, Wagenaar JA. Differentiation of *Campylobacter* species by AFLP fingerprinting. Microbiology 2001; 147:2729–2737.

56. Engvall EO, Brandstrom B, Gunnarsson A, Morner T, Wahlstrom H, Fermer C. Validation of a polymerase chain reaction/restriction enzyme analysis method for species identification of thermophilic campylobacters isolated from domestic and wild animals. J Appl Microbiol 2002; 92:47–54.

57. Patton CM, Shaffer N, Edmonds P, et al. Human disease associated with "*Campylobacter upsaliensis*" (catalase -negative or weakly positive *Campylobacter* species) in the United States. J Clin Microbiol 1989; 27:66–73.

58. Fox JG, Maxwell KO, Taylor NS, Runsick CD, Edmonds P, Brenner DJ. "*Campylobacter upsaliensis*" isolated from cats as identified by DNA relatedness and biochemical features. J Clin Microbiol 1989; 27:2376–2378.

59. Baker J, Barton MD, Lanser J. *Campylobacter* species in cats and dogs in South Australia. Aust Vet J 1999; 77:662–666.

60. Happonen I, Linden J, Saari S, et al. Detection and effects of helicobacters in healthy dogs and dogs with signs of gastritis. J Am Vet Med Assoc 1998; 213:1767–1774.

61. Strauss-Ayali D, Simpson KW. Gastric *Helicobacter* infection in dogs. Vet Clin North Am Small Anim Pract 1999; 29:397–414.

62. Happonen I, Linden J, Westermarck E. Effect of triple therapy on eradication of canine gastric helicobacters and gastric disease. J Small Anim Pract 2000; 41:1–6.

63. Neiger R, Simpson KW. *Helicobacter* infection in dogs and cats: facts and fiction. J Vet Intern Med 2000; 14:125–133.

64. Buczolits S, Hirt R, Rosengarten R, Busse HJ. PCR-based genetic evidence for occurrence of *Helicobacter pylori* and novel *Helicobacter* species in the canine gastric mucosa. Vet Microbiol 2003; 24:259–270.

65. Handt LK, Fox JG, Stalis IH, et al. Characterization of feline *Helicobacter pylori* strains and associated gastritis in a colony of domestic cats. J Clin Microbiol 1995; 33:2280–2289.

66. El-Zaatari FA, Woo JS, Badr A, et al. Failure to isolate *Helicobacter pylori* from stray cats indicates that H. pylori in cats may be an anthroponosis—an animal infection with a human pathogen. J Med Microbiol 1997; 46:372–376.

67. Simpson KW, Strauss-Ayali D, Straubinger RK, et al. *Helicobacter pylori* infection in the cat: evaluation of gastric colonization, inflammation and function. Helicobacter 2001; 6:1–14.

68. Stanley J, Linton D, Burens AP, et al. *Helicobacter canis* sp. nov. a new species from dogs: an integrated study of phenotype and genotype. J Gen Microbiol 1993; 139:2495–2504.

69. Foley JE, Marks S, Munson L, et al. Isolation of *Helicobacter canis* from a colony of Bengal cats with endemic diarrhea. J Clin Microbiol 1999; 37:3271–3275.

70. Fox JG, Shen Z, Xu S, et al. *Helicobacter marmotae* sp. nov. isolated from livers of woodchucks and intestines of cats. J Clin Microbiol 2002; 40:2513–2519.

71. Torrey JC. The regulation of the intestinal flora of dogs through diet. J Med Res 1918; 39:415–477.

72. Hussein HS, Flickinger EA, Fahey GC, Jr. Petfood applications of inulin and oligofructose. J Nutr 1999; 129:1454S–1456S.

73. Swanson KS, Grieshop CM, Flickinger EA, et al. Effects of supplemental fructooligosaccharides plus mannanoligosaccharides on immune function and ileal and fecal microbial populations in adult dogs. Arch Anim Nutr 2002; 56:309–318.

74. Flickinger EA, Schreijen EM, Patil AR, et al. Nutrient digestibilities, microbial populations, and protein catabolites as affected by fructan supplementation of dog diets. J Anim Sci 2003; 81:2008–2018.

75. Sparkes AH, Papasouliotis K, Sunvold G, et al. Bacterial flora in the duodenum of healthy cats, and effect of dietary supplementation with fructo-oligosaccharides. Am J Vet Res 1998; 59:431–435.

76. Sparkes AH, Papasouliotis K, Sunvold G, et al. Effect of dietary supplementation with fructo-oligosaccharides on fecal flora of healthy cats. Am J Vet Res 1998; 59:436–440.

77. Molis C, Flourie B, Ouarne F, et al. Digestion, excretion, and energy value of fructooligosaccharides in healthy humans. Am J Clin Nutr 1996; 64:324–328.

78. Terada A, Hara H, Kato S, et al. Effect of lactosucrose (4G-beta-D-galactosylsucrose) on fecal flora and fecal putrefactive products of cats. J Vet Med Sci 1993; 55:291–295.

79. Saarela M, Mogensen G, Fondén R, Mättö J, Mattila-Sandholm T. Probiotic bacteria: safety, functional and technological properties. J Biotechnol 2000; 84:197–215.

80. Benyacoub J, Czarnecki-Maulden GL, Cavadini C, et al. Supplementation of food with *Enterococcus faecium* (SF68) stimulates immune functions in young dogs. J Nutr 2003; 133:1158–1162.
81. Kanasugi H, Hasegawa T, Goto Y, Ohtsuka H, Makimura S, Yamamoto T. Single administration of enterococcal preparation (FK-23) augments non-specific immune responses in healthy dogs. Int J Immunopharmacol 1997; 19:655–659.
82. Rinkinen M, Jalava K, Westermarck E, Salminen S, Ouwehand AC. Interaction between probiotic lactic acid bacteria and canine enteric pathogens: a risk factor for intestinal *Enterococcus faecium* colonization? Vet Microbiol 2003; 92:111–119.
83. Vahjen W, Männer K. The effect of a probiotic *Enterococcus faecium* product in diets of healthy dogs on bacteriological counts of *Salmonella* spp. *Campylobacter* spp. and *Clostridium* spp. in faeces. Arch Anim Nutr 2003; 57:229–233.
84. Pasupathy K, Sahoo A, Pathak NN. Effect of *Lactobacillus* supplementation on growth and nutrient utilization in mongrel pups. Arch Tierernahr 2001; 55:243–253.

20

The Gastrointestinal Microbiota of Farm Animals

Alojz Bomba, Zuzana Jonecová, Soňa Gancarčíková, and Radomíra Nemcová
Institute of Gnotobiology and Prevention of Diseases in Young,
University of Veterinary Medicine, Kosice, Slovak Republic

INTRODUCTION

The colonization of the digestive tract in animals begins soon after birth or hatching and the normal microbiota changes dramatically during the life of the host. The composition of gastrointestinal microbiota differs between animal species, between individuals within the same species and between the body sites of the host. The gut microbiota is a complex interactive community of organisms and its functions are the result of activities of all microbial components. Together with the host, the microorganisms constitute an ecological system, beneficial for the host, as well as for the microbial species. In principle, the role of gut microbiota in animals is the same as in humans—salvaging energy from the undigested feed components through fermentation, providing the basis for a barrier that prevents pathogenic bacteria from invading the gastrointestinal tract, protective functions together with the gut immune system, a role in metabolism of xenobiotics and contribution to the vitamin and amino acids requirements of the animals (1). Some of these functions are emphasized in farm animals with regard to their environment, character of their feed and the economy of farm animals' rearing. The composition and metabolism of the gastrointestinal microbiota affects the performance of farm animals in many ways, especially in the young, which are subjected to many stressful conditions.

Farm animals can be divided into three main groups according to the degree of development of their gastrointestinal tract and efficacy of feed digestion: (1) omnivorous animals—the feed of plant origin with small content of cellulose and lignin, as well as the feed of animal origin is easily and quickly digested with a help of enzymes produced in the gastrointestinal tract of the animal (pigs), (2) carnivorous animals—under natural conditions they consume mostly feed of animal origin, (3) herbivorous animals—consume feed of plant origin with high content of cellulose and lignin, which the animal is able to digest exclusively through microbial fermentation by its gastrointestinal microbiota (ruminants, horses). Herbivorous animals have some part of their gastrointestinal tract adapted to microbial fermentation. The ruminants are polygastric animals with foregut capacity 150–180L in adult cows. In horses, which are monogastric, the caecum with capacity 100–140L is developed for microbial fermentation of lignin and cellulose.

The greatest differences in the composition of the microbiota of the gastrointestinal ecosystem have been shown to occur between ruminants and monogastric animals. Gradual changes in the composition of the gastrointestinal microbiota that take place within an animal species are related to age (2). At an early age the microbiota of the digestive tract of young animals is very similar. With the exception of poultry, this similarity is related to the intake of maternal milk. During the suckling period, bacteria, which can utilize the components of milk, predominate in the upper tract, and the milk constituents evidently largely determine which microbe can be implanted in the intestines. The forestomachs of ruminants have not yet started functioning and the physiology of the digestive tract compares to that of monogastric animals. After the animals start to consume creep feed and they are finally weaned, an adult type of microbiota begins to develop in the upper and lower intestinal tract. At the same time the main site of bacterial fermentation changes from the stomach to the large intestine or, in ruminants, to the rumen.

Due to progressing of age, changes in the composition of the ingested feed and a different morphological and functional development of the gastrointestinal tract, certain differences gradually occur in the composition of the microbiota in calves, lambs, suckling piglets and chicks that are typical for the given farm animal species. The gut ecosystem of adult animals is stable and changes only due to the effects of external factors of an adequate intensity (long-lasting change of feeds, stress, administration of antibiotics).

MICROBIOTA OF THE GASTROINTESTINAL TRACT IN FARM ANIMALS

The gastrointestinal ecosystem of animals is a complex, open, interactive system involving the animal's environment and diet, the animal itself, and many microbial species. This system regulates the course of the successional events and the population levels and geographic distribution of the climax communities once they are formed. In adult animals the microbial communities occupy many niches in habitats distributed from the center of the lumen to the depths of the crypts, and from the oral cavity to the anus. Depending upon the animal species any or all habitats may be occupied. The microbial communities occupying the habitats are usually composed of autochthonous (indigenous) microbes. A sample from any given habitat may at any given time yield allochthonous (non-indigenous) microbes as well as indigenous ones. The allochthonous microbes derive from what the animal ingested (feed, water, faeces) or from habitats above the one in question.

The gastrointestinal microbiota interact profoundly with their animal host, influencing its early development, quality of life, ageing and resistance to infectious diseases. One of the functions of the microbiota is to degrade dietary components such as fiber in order to provide short-chain fatty acids and other essential nutrients that are absorbed by the host. Animal hosts have incubation chambers such as the rumen (cattle, sheep, goat) or the caecum (horse, chicken) in which bacterial fermentation proceeds under optimal conditions. Those animals that have only small caeca, (pigs), have a microbiota which has adapted to use "fast food" such as simple carbohydrates and proteins that are consumed with the diet and available in the host's secretions such as saliva or mucus (3).

In horses and poultry, so-called hind gut fermenters, the caecum fulfill a function that is similar to that of the rumen in ruminants. The caecum is found in the anterior part of the large intestine and its microbial activity can provide for about 30% of the nutritional requirements of these animals.

In monogastric animals, the enzymes of the host ensure digestion of the feed despite the fact that their digestive tract is rather short. Of the farm animal species pigs are typical representatives of this group of animals. Humans are equipped with a similar type of digestive tract. The large intestinal microbiota of pigs is the most numerous and most varied one. Recent knowledge indicates a pronounced similarity of the ruminal, caecal and large intestinal microbiota in animals.

Regulation of the composition and localization of microbial communities in the gastrointestinal tract is a multi-factorial process in which any or all of these numerous forces may come into play (4). Stability of the microecosystem of the digestive tract is maintained by the interrelations of the microecosystem and the macroorganism as well as by the interactions of the microorganisms in the ecosystem. On the part of the host, both endogenous (age, host immunity, digestive tract motility and length, acidity) and exogenous factors (diet) play an important role (5). On the other hand, the microbiota of the digestive tract greatly affects the development of the host animal, mainly at an early age, and plays a very important role in the animal's resistance to infectious diseases. The interactions between microorganisms are mediated by competition for gut receptors and nutrients as well as by the production of antimicrobial substances (6,7). The mechanisms of bacterial interactions also mediate the barrier effect (8) or competitive exclusion (9), which is the ability of the indigenous microbiota to prevent the implantation of allochthonous microbes in the gastrointestinal tract. Knowledge of the mechanism of bacterial interactions is an inevitable presupposition if optimization of the composition of the gastrointestinal microbiota and stimulation of the beneficial effects of the latter on the host animal are desired (10).

Pigs

The gastrointestinal tract of the piglets at parturition is sterile, but the gut microbiota develops very rapidly. The first bacteria, which become established in the digestive tract of the piglet, originate from the dam or the environment, but they are not the most abundant ones of the ecosystems encountered by the young (11). The newborn possesses very efficient selection systems enabling it to favor certain bacterial species among the bacteria of the different ecosystems. Many factors might be involved in this selection—diet, environmental conditions such as hygienic stage, temperature, the microbial interactions in the digestive tract and the barrier effect of the dominant microbiota against the environmental bacteria.

The indigenous microbiota exerts a profound influence on both the morphological structure and on the digestive and absorptive capabilities of the gastrointestinal tract (12). From the stomach of suckling piglets significant populations of microorganisms have been isolated upto 10^7 viable counts per 1 cm^2 of the tissue (13). The microbial population adhering to the pars esophagea varies little from birth until after weaning and the anaerobic microbiota, particularly lactobacilli, might be important in maintaining the pars esophagea free from colonization by other microorganisms. The stratified squamosus epithelium, of which the pars esophagea is composed, is continuously desquamating releasing cells with attached bacteria into the lumen and may serve as a continuous inoculum of specific lactic acid bacteria into the gastric contents (14).

In the small intestine, a fast transit time and digestive secretions such as bile acids limit bacterial numbers and diversity. The gastrointestinal microbiota of the young piglets is composed of facultatively anaerobic microorganisms in the proximal intestine (duodenum, jejunum) whose number ranges from 10^3 to 10^7 per g content (11). This number increases progressively in the ileum, and in the last parts of the digestive tract strictly anaerobic bacteria are found among the dominant microbiota. In very young piglets, *Escherichia coli* is the dominant microbe of all gut segments, together with species of the genera *Lactobacillus* and *Streptococcus*. The microbiota of the piglet progressively changes with age, the number of *Escherichia coli* decreases in all segments and the lactobacilli and streptococci constitute the dominant microbiota of the proximal intestine. The presence of lactobacilli as a constituent of the normal microbiota of the gastrointestinal tract is considered to be beneficial to the porcine host (15). The strictly anaerobic microbiota becomes more diversified in the distal segments, where *Bacteroides, Eubacterium, Peptostreptococcus* and many *Clostridium* species are found (11).

The change of the gut environment occurs in connection to weaning of the piglets. Weaning and weaning age have significant effects on microbial population and volatile fatty acids concentration (16). During the first week after weaning, pH and the content of dry matter decrease, as well as the count of lactobacilli, while the number of coliform bacteria increases (17). These changes contribute to low weight gains and predisposition to diarrhea. Associated with weaning there are marked changes to the histology and biochemistry of the small intestine, such as villous atrophy and crypt hyperplasia, which caused decreased digestive and absorptive capacity (18) and contribute to post-weaning diarrhea. The major factors implicated in the etiology of these changes are: change in nutrition, stress due to separation from mother and littermates, new environment, the withdrawal of milk-borne growth promoting factors, as well as enteropathogens and their interactions with the gut microbiota. Enterotoxigenic *Escherichia coli* strains are generally considered to be the main cause of diarrhea at weaning and the period immediately thereafter. The colonizing of the small intestine by enterotoxigenic *E. coli* strains may be possible for several reasons (19): (1) the brush border of the intestinal epithelium of newly weaned pigs may be damaged by components in the feed or by viruses allowing *E. coli* to adhere and colonize the damaged epithelium, (2) after weaning the pigs are no longer protected by the milk of the sow, an important factor that prevents *E. coli* colonization during the suckling period, (3) newly weaned pigs have a shortage of digestive enzymes and feed is poorly digested and absorbed.

Concentrations of bacteria in contents of the gastrointestinal tract of pigs are much higher in the caecum and in colon than in more proximal portions of the tract. The microbiota is dominated by strict anaerobes and the most numerous species are members of the genera *Bacteroides, Selenomonas, Butyrivibrio, Lactobacillus, Peptostreptococus* and *Eubacterium* (20). The development of a complex microbiota in the large intestine takes 2–3 weeks after weaning. Starch and some oligosaccharides are mainly digested in the small intestine of monogastric animals by enzymes of the salivary glands, pancreas and intestinal brush border. Cellulose, hemicelluloses, pectins and some oligosaccharides are partly digested by the microbiota of the large intestine. Fiber total digestibility varies considerably and depends on the nature of the fiber and the animal species. It is less than 10% in chickens, whereas pigs seem to digest fibers as well as sheep (21). Dietary fiber may contribute up to 30% of the maintenance energy needs of growing pigs. Higher energy contributions may be obtained from dietary fiber fed to sows, along with some improvements in reproduction, health, and well-being. Swine microbiota constitutes highly active ruminal cellulolytic and hemicellulolytic bacterial species, which include *Fibrobacter succinogenes (intestinalis), Ruminococcus albus, Ruminococcus flavefaciens, Butyrivibrio species, and Prevotella (Bacteroides) ruminicola* (22). Additionally, a new highly active cellulolytic bacterium, *Clostridium herbivorans*, has been isolated from pig large intestine (23). The populations of these microorganisms are known to increase in response to the ingestion of diets high in plant cell wall material. The numbers of cellulolytic bacteria from adult animals are approximately 6 to 7 times greater than those found in growing pigs. None of these highly active cellulolytic bacterial species are found in the human large intestine. Thus, the pig large intestinal fermentation of fiber seems to more closely resemble that of ruminants than that of humans (22).

Poultry

Bacterial colonization of the intestinal tract of poultry occurs after hatching when the young bird starts to receive the feed. The esophagus of gallinaceous poultry creates the crop, which serve as a store of the feed. The ingested feed in the crop is softened by water and by secretion of salivary glands and the glands of esophagus. In water poultry, the esophagus is able to widen throughout its length. The gastric juice produced in the gizzard helps in chemical digestion of the feed. The gut of poultry is short and the caecum is doubled. Soft feed passes through the digestive tract very fast (2 to 4 hours), crude feed takes much longer (up to 20 hours). The poultry should be fed with feed of high nutritive value due to the shortness and fast transit time of the intestinal content.

Lactobacillus microbiota lining the crop of the chicken gastrointestinal tract becomes established within a few days after hatching and the specific adherence of avian associated lactobacilli onto the crop epithelium plays a role in the colonization (24). From the third day of life, large numbers of lactobacilli are present throughout the alimentary tract (25). Recent research showed that freshly isolated lactobacilli from chickens are able to adhere to the epithelium of crop, as well as to the follicle-associated epithelium and the apical surface of mature enterocytes of intestinal villi (26).

Enterobacteriaceae and enterococci are present in large numbers in 3-day-old broilers but they start to decrease with the age. Lactobacilli, however, remain stable during the growth of broilers. The presence of volatile fatty acids is responsible for the reduction of *Enterobacteriaceae* in the broiler chicken. The amounts of acetate, butyrate and propionate increase from undetectable amounts in 1-day-old broilers to high concentrations in 15-day-old broilers (27). Facultative anaerobic microbiota (streptococci, lactobacilli and *E. coli*) comprise the predominant microbiota of the small intestine and Salanitro and coworkers (28) found that the above-mentioned bacteria represent 60–90% of the isolated bacteria. While the number of aerobic and anaerobic bacteria in duodenum and ileum were in their study very similar, they found 10^{11} anaerobic bacteria per g of dry tissue in the caecum and the latter exceeded aerobe plate count by at least a factor 100. The use of anaerobic methods developed for rumen bacteria have shown that the dominant microbiota of the caecum is composed of strict anaerobes and the most frequently isolated genera were *Eubacterium, Clostridium, Fusobacterium, Bacteroides, Bifidobacterium, Peptostreptococcus*, and *Lactobacillus* (28,29). Scanning electron microscopy of the intestinal epithelia of 14-day-old chickens revealed populations of microbes on the duodenal, ileal and caecal mucosa surfaces (28).

The study of intestinal microbiota composition has relied almost exclusively on the quantitative cultivation of microbes from samples. Culture results obtained in these studies compose between 50 and 80% of total microscopic counts (30). Culture-based techniques can be very selective, but never capture the total microbial community of complex anaerobic habitats such as the avian gastrointestinal tract. Apajalahti and coworkers (31) analyzed broiler chickens from eight commercial farms in Southern Finland for the structure of their gastrointestinal microbial community by a non-selective DNA-based method, percent G+C-based profiling and, in addition, a phylogenetic 16S rRNA gene-based study was carried out to aid interpretation of the percent G+C profiles. Most of the

16S rRNA sequences found could not be assigned to any previously known bacterial genus or they represented an unknown species of one of the taxonomically heterogeneous genera such as *Clostridium, Bacteroides* and *Eubacterium*. Bacteria related to ruminococci and streptococci were the most abundant members observed. The source of the feed and feed amendment changed the bacterial profile significantly.

Horses

The intestinal tract of horses and other monogastric herbivores is characterized by a combination of a large caecum and an even larger colon where fermentation and absorption occurs. Bacteriological studies have shown that the equine intestinal ecosystems contain several hundreds of microbial species, of which most are strict anaerobes (32) and metabolic products from this microbiota provide the horse with a significant part of its energy requirements. There is little information about the microbiota of the small intestine in horses. However, like in the other species of animals, the total microbial counts as well as *E. coli* and streptococci rise continuously from duodenum to ileum; lactobacilli predominate in the duodenum (33). The acetate concentration increases along the length of the small intestine and molar proportion of acetate, propionate and butyrate 85:10:3 were found in hindgut (34). Acetate is a common fermentation end product from intestinal anaerobes of the genera *Bacteroides, Bifidobacterium, Eubacterium, Propionibacterium* and *Selenomonas* (35), and it is indicative for a diet that is low in rapidly fermentable sugars or concentrates. From the data given by Colinder and coworkers (36), horses have a lower total concentration of faecal short-chain fatty acids than pigs, rats and man and even lower than the values in cows. The significantly higher proportion of acetate can depend on its correlation to high-fiber diets and reflects a difference in diets between horses and other monogastric species. Reduced faecal excretion of absorbable compounds, as short-chain fatty acids, is probably due to prolonged stay of digesta in the hindgut; four days or more (37). Daly and Shirgazi-Beechey (38) obtained quantitative data on the predominant bacterial populations inhabiting the equine large intestine by using group-specific oligonucletide probes. Results showed the *Spirochetaceae*, the *Cytophaga-Flexibacter-Bacteroides* assemblage, the *Eubacterium rectale-Clostridium coccoides* group and unknown cluster C of *Clostridiaceae* to be the largest populations in the equine gut, each comprising 10–30% of the total microbiota in each horse sampled. Other detected notable populations were the *Bacillus-Lactobacillus-Streptococcus* group, *Fibrobacter* and unknown cluster B, each comprising 1–10% of the total microbial community.

Ruminants

The forestomach of cattle, sheep and goats consists of the reticulum, rumen and omasum that are followed by the abomasum; the latter is an analogy of the stomach of monogastric animals.

In young ruminants after birth, only the fourth stomach (abomasum) is functional and its capacity is about twice that of the other compartments. In the adult ruminants, abomasum represents only 8% of the total capacity. The volume of the rumen represents 80% of the total (39). The difference between ruminants and non-ruminant animals results from the morphological adaptation of their gastrointestinal tract to the consumption and utilization of cellulose as well as their adaptation to utilization of the end products from the rumen fermentation. The rumen provides an ideal environment for fermentation with relative stable temperature and a continuous supply of the nutrients (40). The ruminal pH value in a healthy animal is 6.2–6.8 and it is influenced by food, buffer capacity of the saliva, by products of fermentation and by the animals' ability to absorb the latter through the rumen wall. The microbial ecosystem of the rumen is one of the most complex, with wide variety of interactions between microorganisms, between microorganisms and the host and between microorganisms and the feed (41). The rumen microbial population consists of bacteria, protozoa and fungi. The amount of rumen protozoa depends on the diet, but usually ranges from 10^4 to 10^7 per ml of rumen digesta. Because of their sensitivity to low pH and sufficient amount of nutrients, they can completely disappear from the rumen content. The rumen anaerobic fungi take part in rumen fiber digestion (42).

The population of rumen bacteria is characteristic and indispensable for the ruminal ecosystem. Bacteria in the rumen adhere to the epithelium of the rumen wall, to feed particles, or they move freely in the contents (43). Bacteria adhering to the epithelium of the rumen wall are considered to be the regulating factor of the rumen microbiota (44). At the age of 9 to 13 weeks the ruminal microbiota of the calf is similar to that of an adult animal. The number of rumen bacteria ranges from 10^9 to 10^{11} per ml of rumen digesta and depends on the diet and the time of sampling after feeding (45). The permanent microbiota consists of more than 60 species of bacteria and the concentration of dominant species ranges from 10^8 to 10^{10} per ml of rumen digesta. The most important species are divided in to metabolic groups according to their main substrates which they are able to ferment (46)—cellulolytic *(Bacteroides succinogenes, Ruminococcus albus, Ruminococcus flavefaciens)*, amylo- and dextrinolytic *(Bacteroides amylophylus, Streptococcus bovis, Succinomonas amylolytica, Succinivibrio dextrino- solvens)*, saccharolytic *(Bacteroides ruminicola, Butyrivibrio fibrisolvens, Megasphaera elsdenii, Selenomonas ruminantium)* and hydrogen-utilizing bacteria *(Methanobacter ruminantium, Vibrio succinogenes)*. The most important attributes of the ruminal microbiota are the ability to hydrolyse cellulose, synthesize amino acids, produce volatile fatty acids and vitamins. In the young of ruminants, lactate-utilizing bacteria, among them *Megasphaera elsdenii, Veillonella alcalescens and Selenomonas ruminantium* (47), are of great importance. Comparative Polymerase Chain Reaction (PCR) assays were developed for enumeration of the rumen cellulolytic bacterial species: *Fibrobacter succinogenes, Ruminococcus albus* and *Ruminococcus flevefaciens* (48). Enumeration of the cellulolytic species in the rumen and alimentary tract of sheep found *Fibrobacter succinogenes* dominant; 10^7 per ml of rumen digesta compared to *Ruminococcus* species ($10^{4–6}$ per ml). All three species were detected in the rumen, omasum, caecum, colon and rectum, the numbers at these sites varied within and between animals.

INFLUENCING THE ECOSYSTEM OF THE DIGESTIVE TRACT IN FARM ANIMALS

In farm animals the microbiota of the digestive tract plays an important role both in the process of optimal development and growth of the organism as well as in securing the resistance of animals to diseases. However, due to various adverse impacts, disturbances of optimum growth, production and health state of the animals are rather frequent in animal production.

Abrupt change of feed, weaning, stress, administration of antibiotics at therapeutical dosage and pathogenic microorganisms can all be classified among these adverse factors. All of them disturb the stability and composition of the natural microbiota of the digestive tract, thus disturbing physiological processes and resistance of the organism to diseases;

they slow down growth, decrease the performance or lead to diseases of farm animals. From these facts it is obvious that in order to minimize the negative effects of adverse factors it is essential to give targeted and efficient support to the beneficial microbiota of the digestive tract that plays an important part in the physiological processes and in the resistance of the organism to diseases. In order to ensure optimum growth, production and health of the farm animals the beneficial microbiota of the ecosystem of the digestive tract can be supported by manipulation of the diet and application of probiotic microoraganisms. Growth-promoting antibiotics will be banned in the European Union by 2006 and similar measures may be expected in other countries in the future. From this point of view, it is necessary to search for naturally occurring alternatives to antibiotics. The manipulation of the gastrointestinal microbiota by diet and application of probiotics could represent such safe alternative to antimicrobials.

Manipulation of the Gastrointestinal Microbiota by Diet

Dietetic methods can be used to positively influence the development of the rumen microbiota of young ruminants during the period of milk nutrition and transition from milk to plant feeding; in monogastric animals, mainly pigs, these methods can be used for the same purpose mainly at the time of weaning.

The influence of the feed amount and quality upon the ecosystem of the digestive tract is of extraordinary importance (49). If the diet is changed from roughage to grain, the rumen microbiota and microfauna and the final products of these elements undergo changes as well (50). Dietetic stimulation of the rumen microbiota of ruminants comprises several ways of manipulating the feeds offered to the animals, among them changing the composition of the feeds, the form of the feeds as well as the time of starting feeding dry feeds to milk-fed animals. Adverse factors such as regulation of milk feeding or feeding frequency may also be used to influence the development of the rumen microbiota or rumen digestion. A gradual decrease of the amount of milk forces the animals to supplement the missing nutrients by taking in dry grain and later forage feeds, which accelerates the functional and morphological development of the rumen (51–53). Cruywagen and Horn (54) point at the possibility of influencing dry fodder intake by the composition of the liquid diet. According to these authors a factor is present in the bovine colostrum that stimulates the intake of dry concentrate feed. Bush and Nicholson (55) also stated that it would be possible to increase the intake of dry feeds during the period of milk nutrition and thus to affect changes in the microbiota of the digestive tract of calves by the addition of formic acid. In these animals feeding a pre-starter mixture and weaning at an early age have a very positive effect on the functional development of the rumen (56).

Feed composition is of decisive importance for the stimulation of rumen digestion in ruminant young in the period of predominant milk nutrition. The amount of dry feeds is only of secondary importance. Easily fermentable grains are vital for the development of the amylolytic microbiota while roughage, silage, hayage and hay are decisive for the cellulolytic one. With respect to the development of the functions of the forestomachs intake of high-quality hay and grain is of vital importance (57,58). Since calves do not consume great amounts of hay in the first eight weeks of life, the level of rumen metabolism during the period of milk nutrition can be positively affected mainly by a suitable composition of the starter mixture (59). With progressing age and maturation of the rumen, cellulolytic microbiota gradually develops and increased amounts of hay, hayage and silage can be offered to the calves. The cellulolytic activity of rumen bacteria is stimulated by isoacids that develop during the catabolism of certain amino acids. Isoacid levels in the rumen can be increased by a diet that is rich in concentrate and proteins (60).

Dietetic methods can also be used to influence the microecosystem of the intestinal tract in piglets during weaning. At this period important morphological and functional changes occur in the digestive tract of piglets that are also accompanied by changes in the composition of the gut microbiota (17,61). In the first days after weaning *Lactobacillus* populations decrease considerably whereas the numbers of coliforms increase. In piglets the brush border of the intestinal epithelium can be damaged by feed components (62) or viruses (63); such damage enables enterotoxigenic *E. coli* to colonize the injured epithelium. Important factors that the piglets had been receiving by maternal milk and that prevented *E. coli* from colonizing the gut (64) are no more at the animals' disposal. All these changes support the tendency to low weight gain and predispose to the occurrence of the diarrheic syndrome. Several researchers tried to influence the morphological and functional development of pigs during the weaning period in order to optimize digestion and to minimize the danger of the post-weaning diarrheic syndrome. Adjustment of the form of feeds seems to positively influence morphological development of the intestinal epithelium in weaned piglets. On days 8 and 11 after weaning Deprez and coworkers (65) observed the intestinal villi to be higher in the piglets fed pulpy feeds than in those receiving the same composition in pellets. The higher villi observed in the piglets receiving pulpy feeds may reflect an increased level of energy intake. This assumption has been confirmed by the findings of Partridge and coworkers (66) who stated weanlings receiving dry feed in the form of a pulp consume more feed and grow more rapidly than piglets receiving the same feed as pellets. Beers-Schreurs and coworkers (67) concluded a decreased energy intake during the post-weaning period to be the main cause of villar atrophy. If it is our aim to influence the development of the digestive tract during the weaning period, then the finding of Kelly and coworkers (12) according to whom continuous presence of feeds in the lumen plays an important part in the integrity of intestinal morphology and function is of extreme importance. McCracken (68) stated that a low intake of feed after weaning might cause morphological and functional changes in the intestinal tract. Pluske (69) pointed out that if nutritional stress caused by discontinuation of feed intake at weaning could be overcome, transition from maternal milk to solid feeds would be less traumatic to the piglets. Milk intake after weaning seems to have pronounced stimulating effects upon growth and functioning of the mucosa; it promotes the integrity of the small intestine and supports the growth of piglets by increasing or maintaining the digestive and absorption capacity. Pluske and Williams (70) demonstrated that the height of villi and depth of crypts in weanlings can be maintained by feeding fresh cow's milk at two-hour intervals immediately after weaning.

It is important to stress that current modern methods of rearing frequently employ early and abrupt weaning, which increases the predisposition to diseases of the digestive tract. The most pronounced changes in the morphology of the intestine, its enzyme capacity, in the physiology of digestion and the microbiota of the digestive tract occur in the period after weaning. For this reason the composition of feeds during the period of transition from milk to plant-based nutrition should take into account the morphological changes of the digestive tract and the level of its functional development.

Manipulation of the Gastrointestinal Microbiota by Application of Probiotic Microorganisms

Administration of preparations based on autochthonous microorganisms is a very effective method of affecting the microbiota of the gastrointestinal tract in farm animals. In this way development of the microbiota of the young at an early age and around weaning can be influenced.

Development of the rumen microbiota in calves and lambs can be supported by microbial preparations mainly at the start of dry feeding. Effective use of microbial preparations in the young depends also on the level of knowledge of the so-called environmental factors in the rumen which determine the age at which a given microorganism may colonize the rumen and enable the development of cellulolytic microbiota (71). The specificity of using probiotics in calves, lambs and goatlings consists in the possibility of influencing the formation of the ruminal ecosystem; application of selected strains of rumen microorganisms lays the foundation of a future population showing a high fermentation activity. Colonization with selected cultures of living microorganisms should enable an earlier and more stable onset of the ruminal type of digestion. Controlled action on the rumen microbiota in the young during milk nutrition is mainly related to the effect upon development of the microbiota adhering to the epithelium of the rumen wall. The effects of stimulation can be expected to be most pronounced at the period of the most rapid development of the adherent microbiota, at 2 to 3 weeks of age. Autochthonous species colonizing the rumen immediately after birth are of decisive importance. This microbiota, though simple at the beginning, enables the development of a cellulolytic population and that of ruminal digestion. Strains of *Streptococcus bovis* may be used to stabilize rumen fermentation. During a 4-week administration of a colonizing preparation containing *S. bovis* AO 24/85 to lambs the numbers of *S. bovis* germs adhering to the rumen epithelium were significantly increased ($p < 0.001$) and so was their alpha-amylase activity (72). In order to promote the development of the ruminal microbiota Kopečný and Šimunek (73) used a mixture of rumen bacteria that contained amylolytic, cellulolytic, hemicellulolytic, saccharolytic, proteolytic and lactate-utilizing strains.

It is of great importance to influence the intestinal microbiota of calves, piglets and poultry at an early age since this is the period when the danger of diarrhea-accompanied diseases of the digestive tract reaches its maximum. Due to their high morbidity and mortality rates such diseases present an extraordinarily serious health and economic issue. Preventive application of probiotics at an early age helps to optimize the composition of the gut microbiota and has an inhibitory effect upon the pathogens of the digestive tract in the young of farm animals. Preventive application of *Lactobacillus casei* at a dose of 1.10^8 germs decreased the counts of enterotoxigenic *E. coli* O101:K99 adhering to the small intestinal mucosa of gnotobiotic lambs by 99.1% and 76.0% on day 2 and 4 after inoculation, respectively (74). Perdigon and coworkers (75) found the preventive effect of *L. casei* and yoghurt against *Salmonella typhimurium* infections in mice to depend on the duration of administration. The short-term preventive application of *Lactobacillus paracasei* (76) induced slight decrease in number of *E. coli* adhered to jejunal mucosa of gnotobiotic piglets, while continuous application led to significant ($p < 0.05$) decrease (Fig. 1). Thomke and Elwinger (77) and Mead (78) suggested that it seems possible to lower enteropathogens (*E. coli* and *Salmonella*) but not to control them by administering *Lactobacillus acidophilus*. Increased lactic acid production in the small intestine of pigs fed lactobacilli and yeast caused a decrease in intestinal pH and the presence of *E. coli* within in intestinal content (79).

Potentiation of the probiotic effect of microorganisms seems to be possible by combining them with synergically acting components of natural origin. As such, prebiotics (mainly oligosaccharides), substrates and metabolites of microorganisms and phyto-components are taken into consideration. Bomba and coworkers (80) showed that the administration of *L. paracasei* alone had almost no inhibitory effect on the adhesion of *E. coli* to the jejunal mucosa of gnotobiotic and conventional piglets while *L. paracasei* administered together with maltodextrin decreased the number of *E. coli* colonizing the

log 10.cm^{-2}

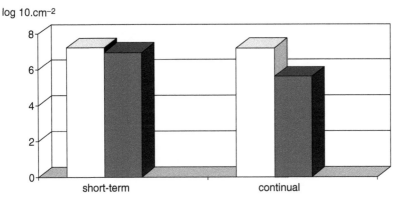

Figure 1   Colonization of the jejunal mucosa of gnotobiotic piglets by *Escherichia coli* 08: K88 at short-term and continual preventive application of *Lactobacillus paracasei*. (□) Control group E; (■) experimental group L-E. *Source*: From Ref. 76.

jejunal mucosa of conventional piglets by 2.7 logarithm (4.75 log 10/cm^2) in comparison to the control group (7.42 log 10/cm^2, $p < 0.05$).

Findings reported by Nemcová and coworkers (81) pointed at the fact that the probiotic effect of microorganisms could be potentiated by combining them with prebiotics. The application of *L. paracasei* combined with fructooligosaccharides to piglets for the first 10 days of life and 10 days after weaning revealed an effect upon bacterial counts in the faeces that was significantly more positive than that of lactobacilli only. With this combination significantly increased counts of *Lactobacillus* species, *Bifidobacterium* species, total anaerobes and aerobes as well as significantly decreased counts of enterococci were stated in the faeces when compared to the control as well as the *Lactobacillus* only group. Comparison with the controls revealed the combination of lactobacilli and fructooligosaccharides to result in a significant decrease of *Clostridium* and *Enterobacteriaceae* and an insignificant decrease of coliform counts in the faeces of piglets. These results prove a synergically positive effect of *L. paracasei* and fructooligosaccharides in the faecal microbiota of piglets (Table 1).

Our results showed that the application of *L. paracasei* combined with fructooligosaccharides and maltodextrin decreased the preweaning mortality of piglets (Fig. 2). The field trial lasted eight months and comprised 4000 heads of 1–35 days old piglets and the results were compared with the same period of the previous year in which antibiotic feed additivies were used.

Competition for receptors on the intestinal wall is one of the mechanisms that mediates the inhibitory effect of probiotic microorganisms on the adhesion of pathogens to the intestinal mucosa. Based on this fact it can be hypothesized that an increase in the number of probiotic microorganisms colonizing the intestinal epithelium may potentiate their probiotic effect. From this point of view the findings of Ringø and coworkers (82) about the effects of lipids containing feeds on the gastrointestinal microbiota and especially on the population of lactobacilli are of great interest. According to Kankaanpää and coworkers (83) higher concentrations of polyunsaturated fatty acids inhibited the growth and mucus adhesion of selected lactobacilli whilst growth and mucus adhesion of *Lactobacillus casei* Shirota was promoted by low concentrations of γ-linolenic acid and arachidonic acid. In gnotobiotic piglets oral administration of oil that contained polyunsaturated fatty acids significantly increased the numbers of *Lactobacillus paracasei*

Table 1 Composition of Fecal Microbiota in Weanling Pigs Receiving *Lactobacillus paracasei* and Mixture of *Lactobacillus paracasei* and Fructooligosaccharides

Organisms	Group 1	Group 2	Group 3
Total anaerobes	9.8 ± 0.2	9.8 ± 0.3	10.2 ± 0.2 a[*], b[*]
Total aerobes	8.0 ± 0.5	8.2 ± 0.2	9.3 ± 0.7 a[*], b[*]
Bifidobacterium	7.5 ± 0.3	7.1 ± 0.7	8.3 ± 0.3 a[*], b[*]
Lactobacillus	9.9 ± 0.1	9.9 ± 0.3	10.3 ± 0.1 a[**], b[*]
Enterococcus	9.3 ± 0.1	9.3 ± 0.3	8.2 ± 0.2 a[***], b[***]
Clostridium	8.1 ± 0.1	7.4 ± 0.4 a[*]	7.7 ± 0.3 a[*]
Enterobacteriaceae	7.9 ± 0.4	6.5 ± 0.9 a[*]	5.9 ± 0.9 a[**]
Coliforms	6.8 ± 0.7	6.3 ± 0.7	5.8 ± 0.8

Values are mean $\pm$ SEM of log bacteria counts per gram of wet feces (n = 7). Group 1—control.
Group 2—Lactobacillus paracasei. Group 3—Lactobacillus paracasei and FOS.
(a) Significantly different from control group.
(b) Significantly different from *Lactobacillus paracasei* group.
*p < 0.05; **p < 0.01; ***p < 0.001.
Source: From Ref. 81.

adhering to the jejunal mucosa as compared to the control group (84). It is suggested that polyunsaturated fatty acids could modify the adhesion sites for gastrointestinal microorganisms by changing the fatty acid composition of the membranes of the intestinal epithelial cells (82). The ability of probiotics to adhere to mucosal surfaces is a presupposition of their health-promoting effects. The stimulatory effect of polyunsaturated fatty acids upon the adhesion of lactobacilli could be used to enhance the effectiveness of probiotics in inhibiting the pathogens of the digestive tract.

Early colonization of the gut by an autochthonous microbiota protects chickens from *Salmonella* infection. The direct competition for the site of attachment is suggested to be the prime mechanism for the competitive exclusion (85) and development of a biofilm of protective microbiota was observed using scanning electron microscopy. The method of competitive exclusion constitutes an additional prophylactic method that may be applied directly in the animal to enhance its resistance towards *Salmonella* infection (86). It is also considered a possible application in preventing colonization of poultry with *E. coli* O157

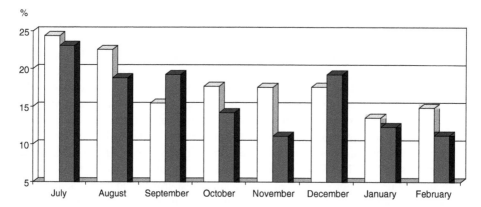

Figure 2 Total preweaning mortality of the piglets during control period July 2000–February 2001 and during experimental period July 2001–February 2002. (☐) 2000, 2001 (■) 2001, 2002.

and *Campylobacter jejuni* (78). Optimal protection against *S. typhimurium* was observed when broiler chicks were treated with a culture of caecal microbiota in combination with dietary lactose (87). The same results were described in turkey poultry (88) and layer chicks (89). In poultry, lactose can also be considered a prebiotic because of absence of the endogenous lactase. The lactose is converted into lactic acid by fermentation of hindgut microbiota. The decrease of intestinal pH results in reduction of the *S. typhimurium* concentration.

THE USE OF GNOTOBIOTIC ANIMALS IN STUDIES OF THE GASTROINTESTINAL MICROBIOTA IN FARM ANIMALS

Gnotobiotic animals proved to be a very useful model for studying the physiology of the digestive tract. They mainly enable observation of the role of microorganisms in the process of the functional and morphological development of the digestive tract and the investigation of bacterial interactions and their influence on the macroorganism. A key experimental strategy for defining the conversations that occur between microorganisms and their hosts is to first define cellular function in the absence of bacteria (under germ-free conditions) and then to evaluate the effects of adding a single or defined population of microbes. The power of germ-technology lies in the ability to control the composition of the environment in which a multicellular organism develops and functions. The combined use of genetically manipulatable model organisms and gnotobiotic has the potential to provide new and important information about how bacteria affect normal development, establishment and maintenance of the mucosa-associated immune system, and epithelial cell functions. Gnotobiology can help to provide new insights into the aetiology of infectious diseases. The combination of gnotobiotics and molecular genetics should provide a deeper understanding of how pathogens arise, how they gain control of their habitat, and what contributions are made by the "normal" gut inhabitants to the pathogenesis of diseases. Such understanding, in turn, could lead to the development of novel chemicals and microbes for use in prebiotic and probiotic strategies in order to prevent or cure infectious diseases and perhaps also immune disorders. For a more extensive review on research with germ-free and gnotobiotic animals, see the chapter by Norin and Midtvedt.

Gnotobiotic Ruminants in Studies into the Microbiota of the Gastrointestinal Tract

Gnotobiotic ruminants can be used to observe the development of the rumen ecosystem as well as to study the relations between rumen and its microbiota. The rumen microbiota directly affects the development of the rumen epithelium and the level of intermediary metabolism by the action of rumen fermentation and its final metabolites. Fonty and coworkers (90), using meroxenic lambs demonstrated that the functions of the rumen and the stability of the ecosystem depended on the complexity and diversity of the microbiota. In the light of the present knowledge it is not possible to precisely determine the composition of the minimum microbiota enabling rumen development and function. Fonty and coworkers (91) also studied the role of rumen microbiota in the development of the rumen ecosystem and functional development of the rumen at an early age. Their results suggest that the rumen microbiota of the very young lamb plays an essential role in the establishment of the rumen ecosystem and in the onset of the digestive functions. Those bacterial species that colonize the rumen immediately after birth when this organ is not yet active, contribute to a biotope favoring the establishment of cellulolytic strains and the set-up of digestive processes that affect both degradation of the lignocellulose-rich feeds and fermentation of the resulting soluble compounds. Ecological factors controlling the establishment of cellulotytic bacteria and ciliate protozoa in the lamb rumen were studied in meroxenic lambs (92). The results obtained in this study suggested that the establishment of cellulolytic bacteria and protozoa required an abundant and complex microbiota and was favored by an early inoculation of the animals. All above-mentioned results point at the extremely important role of the microbiota in the development of the rumen. There is a good relationship between the development of rumen function and the complexity of its microbiota. The presence of a simple microbiota cannot assure the digestive function as properly as a complex microbiota can. Bomba and coworkers (93) used the gnotobiotic approach to observe the development of rumen fermentation in lambs from birth up to 7 weeks of age in association to the complexity of the digestive tract ecosystem. The results obtained indicated that complexity of rumen microbiota significantly affected the development of rumen fermentation both from the quantitative and the qualitative viewpoint.

The fact that early inoculation of the animals is a factor favoring fermentation and digestive activities in the rumen is probably related to the action of bacteria on the development of papillae, the rumen mucosa and the digestive tract (94). A complex microbiota presents an inevitable presupposition of optimal development of the alimentary tract in ruminants.

Colonization of the individual gut segments by lactobacilli and the inhibitory effect of Lactobacillus casei upon the adhesion of enterotoxigenic *E. coli* K 99 to the intestinal wall were also studied in gnotobiotic lambs (74). Soares and coworkers (95) and Lysons and coworkers (96) compared several parameters of the morphological and functional development in germ-free, gnotobiotic and conventional lambs.

Monogastric Gnotobiotic Animals in Studies of the Gastrointestinal Microbiota

Monogastric gnotobiotic animals were also used to study the functional and morphological development of the digestive tract. Nemcová and coworkers (97) studied the colonization ability of selected strains of lactobacilli in the small intestine of gnotobiotic piglets. Studies were also aimed at the effects of lactobacilli on the intestinal metabolism during the first 3 weeks of life (98). The numbers of lactobacilli adhering to the jejunal and ileal mucosa and found in the jejunal and ileal contents were comparable to the data reported by other authors (99,100) in conventional and gnotobiotic piglets. Bomba and coworkers (101) investigated the effect of the inoculation of three *Lactobacillus* strains upon organic acid levels in the mucosal film and intestinal contents of gnotobiotic pigs. In the jejunum of inoculated animals, the mucosal film revealed significantly increased levels of lactic, propionic and acetic acids when compared to the contents. In the ileum of gnotobiotic pigs propionic acid levels in the mucosal film were significantly higher than those in the contents. The above results suggest that significantly increased levels of the lactobacilli-produced organic acids in the intestinal mucosal film may present an efficient barrier to inhibit the adherence of digestive tract pathogens to the intestinal mucosa.

Gnotobiotic animals present a very good model to determine bacterial interactions in the digestive tract. The interactions of lactobacilli and enterotoxigenic *E. coli* in the intestinal tract of gnotobiotic piglets were observed by Bomba and coworkers (80). In experiments carried out in gnotobiotic animals the interest focused on the effects of the microbiota upon morphology, motility, secretion and absorption in the digestive tract (102,103). The use of germ-free, gnotobiotic and conventional animals facilitated considerable progress in the knowledge of the complex ecological system of the gastrointestinal tract in birds (104).

CONCLUSION

The gastrointestinal microbiota plays a very important role in the physiology of farm animals. Despite substantial knowledge of this ecosystem, it is necessary to obtain additional information on the mechanisms mediating their interactions. Such knowledge will facilitate the optimization of the development and function of gastrointestinal microbiota of, especially young, farm animals. It can be expected, that new biotechnological and natural methods for manipulation of gastrointestinal microbiota will be developed. These methods will enable to replace prophylactic antibiotic use in farm animals' diet and will contribute to the production of healthy and safe foods while at the same time benefiting the environment. Several useful in vitro methods are used to study gastrointestinal microbiota. It seems that germ-free and gnotobiotic animals could represent, in conjunction with in vitro methods, a helpful base for the complex study of gastrointestinal ecosystem in farm animals.

REFERENCES

1. Salminen S, Boulez C, Boutron-Ruault M-C, et al. Functional food science and gastrointestinal physiology and function. Brit J Nutr 1998; 1:147–171.
2. Smith HW. The development of the flora of the alimentary tract in young animals. J Path Bact 1965; 90:495–513.
3. Carman RJ, Roger L, van TMS, Tracy DW. The normal intestinal microflora: ecology variability and stability. Vet Human Toxicol 1993; 35:11–14.
4. Savage DC. Microbial ecology of the gastrointestinal tract. Ann Rev Microbiol 1977; 31:107–133.
5. Tannock GW. Effects of dietary of environmental stress on the gastrointestinal microbiota. In: Hentges DJ, ed. Human Intestinal Microbiota in Health and Disease. London: Academic Press, 1983:517–539.
6. Piard JC, Desmazeaud M. Inhibiting factors produced by lactic acid bacteria. 1. Oxygen metabolites and catabolism and products. Lait 1991; 71:525–541.
7. Freter R. Factors affecting the microecology of the gut. In: Fuller R, ed. Probiotics: The Scientific Basis. London: Chapman and Hall, 1992:111–114.
8. Ducluzeau R, Bellier M, Raibaud P. Transit digestif de divers inoculums bactériens into duits "per os" chez des souris axéniques ou "holoxéniques" (conventionelles) effect antagoniste de la microteore dutractus gastrointestinal. Zbl Bacteriol I Abt Orig 1970; 213:533–548.
9. Lloyd AB, Cumming RB, Kent RD. Prevention of *Salmonella typhimurium* infection in poultry by pretreatment of chickens and poults with intestinal extracts. Austral Vet J 1977; 53:82–87.
10. Stavric S, Kornegay ET. Microbial probiotic for pigs and poultry. In: Wallace RJ, Chesson A, eds. Biotechnology in Animal Feeds and Animal Feeding. Weinheim, Germany: VCH Verlagsgesellschaft mbH, 1995:205–231.
11. Ducluzeau R. Implantation and development of the gut flora in the newborn piglet. Pigs News Inf 1985; 4:415–418.
12. Kelly D, Begbie R, King TP. Postnatal intestinal development. In: Valey MA, Williams PEV, Lawrence TLJ, eds. Neonatal Survival and Growth. Edinburgh, UK: Occasional Publication, British Society of Animal Production, 1992; 15:63–79.

13. McGillvery DJ. Anaerobic microflora associated with the pars oesophagea of the pig. Res Vet Sci 1992; 53:110–115.

14. Fuller R, Barrow PA, Brooker BE. Bacteria associated with the gastric epithelium of neonatal pigs. Appl Environ Microb 1978; 35:582–591.

15. Tannock GV. The microecology of lactobacilli inhabiting the gastroinetstinal tract. Adv Microb Ecol 1990; 11:147–171.

16. Franklin MA, Mathew AG, Vickers JR, Clift RA. Characterization of microbial populations and volatile fatty acid concentrations in the jejunum, ileum, and cecum of pigs weaned at 17 versus 24 days of age. J Anim Sci 2002; 80:2904–2910.

17. Jensen BB. The impact of feed additions on the microbial ecology of the gut in young pigs. J Anim Feed Sci 1998; 7:45–64.

18. Pluske JR, Hampson DJ, Williams IH. Factors influencing the structure and function of the small intestine in the weaned pig: a review. Liv Prod Sci 1997; 51:215–236.

19. Nabuurs MJA. Microbiological, structural and functional changes of the small intestine of pigs at weaning. Pig News Inf 1995; 16:93–97.

20. Robinson IM, Allison MJ, Bucklin JA. Characterization of the cecal bacteria of normal pigs. Appl Environ Microbiol 1981; 41:950–955.

21. Champ M. Carbohydrate digestion in monogastric animals. Reprod Nutr Dev 1985; 25:819–842.

22. Varel VH, Yen JT. Microbial perspective on fiber utilization by swine. J Anim Sci 1997; 75:15–22.

23. Varel VH, Tanner RS, Woese CR. *Clostridium herbivorans* sp.nov. a cellulolytic anaerobe from the pig intestine. Int J Syst Bacteriol 1995; 45:490–494.

24. Fuller R. Ecological studies of the *Lactobacillus* flora associated with the crop epithelium of the fowl. J Appl Bacteriol 1973; 36:131–139.

25. Barnes EM, Impey CS, Cooper DM. Manipulation of the crop and intestinal flora of the newly hatchet chicks. Am J Clin Nutr 1980; 33:2426–2433.

26. Edelman S, Westerlund-Wikstrom B, Leskela S, et al. In vitro adhesion specificity of indigenous lactobacilli, within the avian intestinal tract. Appl Environ Microb 2002; 68:5155–5159.

27. van der Wielen PW, Biersterveld S, Notermans S, Hofstra H, Urlings BA, van Knapen F. Role of volatile fatty acids in development of the cecal microflora in broiler chickens during growth. Appl Environ Microb 2000; 66:2536–2540.

28. Salanitro JP, Blake IG, Muirhead PA, Maglio M, Goodman JR. Bacteria isolated from duodenum, ileum and cecum of young chicks. Appl Environ Microb 1978; 35:782–790.

29. Adami A, Cavazzoni V. An experimental standard pattern of assessing the cecal microflora of chicken. Ann Microb Enzimol 1993; 43:329–336.

30. Tannock GW. Molecular assesment of intestinal microflora. Am J Clin Nutr 2001; 73:410–414.

31. Apajalahti JH, Kettunen A, Bedford MR, Holben WE. Percent G+C profiling accurately reveals diet-related differences in the gastrointestinal microbial community of broiler chickens. Appl Environ Microb 2001; 67:5656–5667.

32. Julliand V. Microbiology of the equine hindgut. 1. European conference on the nutrition of the horse. Pferdeheilkunde: Sonderausgabe, 1992:42–47.

33. Frey K, Sasse HHL. Zur darmflora des pferdes-eine literaturstudie. Pferdeheilkunde 1996; 6:855–863.

34. Mackie RI, Wilkins CA. Enumeration of anaerobic bacterial microflora of the equine gastrointestinal tract. Appl Environ Microb 1988; 54:2155–2160.

35. Holdeman LV, Cato EP, Moore WEC. 4th ed. Anaerobe laboratory manual. Blacksburg: Virginia Polytechnic Institute and State University, 1977.

36. Collinder E, Lindholm A, Midtvedt T, Norin E. Six intestinal microflora-associated characteristics in sport horses. Equine Vet J 2000; 32:222–227.

37. Cuddeford D, Pearson RA, Archibald RF, Miurhead RH. Digestibility and gastro-intestinal transit time of diets containing different proportions of alfalfa and oat straw given to Thoroughbreds, shetland ponies, highland ponies and donkeys. Anim Sci 1995; 61:407–417.

38. Daly K, Shirazi-Beechey SP. Design and evaluation of group-specific oligonucleotide probes for quantitative analysis of intestinal ecosystems: their application to assessment of equine colonic microflora. FEMS Microb Ecol 2003; 44:243–252.

39. Roy JHB. The calf. London: Butterwors, 1980.

40. Russell JB, Hespell RB. Microbial rumen fermentation. J Dairy Sci 1981; 64:1153–1169.

41. Gall LS. Significance of microbial interaction in control of microbial ecosystem. Biotech Bioeng 1970; 12:333–340.

42. Bauchop T. The anaerobic fungi in rumen fiber digestion. Agri Environ 1981; 6:339–348.

43. Cheng KJ, McCowan RP, Costerton JW. Adherent epithelial bacteria in ruminants and their roles in digestive tract function. Amer J Clin Nutr 1979; 32:139–148.

44. Cheng KJ, Stewart CS, Dinsdale D, Costerton JW. Electron microscopy of bacteria involved in the digestion of plant cell walls. Anim Feed Sci Tech 1984; 10:93–120.

45. Bartoš S. Microbiology and biochemistry of digestion in rumen of ruminants. Praha: Academia, 1987.

46. Baldwin RL, Allison MJ. Rumen metabolism. J Anim Sci 1983; 2:461–477.

47. Kmeť V, Baran M, Kalačnjuk GI. Manipulation of rumen ecosystem in calves and lambs by microbiological preparations. Bratislava: Veda, 1990.

48. Koike S, Kobayashi Y. Development and use of competitive PCR assays for the rumen cellulolytic bacteria: *Fibrobacter succinogenes*, *Ruminococcus albus* and *Ruminococcus flavefaciens*. FEMS Microb Lett 2001; 204:361–366.

49. Smith HW. The development of the bacterial flora of the faeces of animals and man: the changes that occur during ageing. J Appl Bact 1961; 24:235–241.

50. Berger LL. Effects of diet composition on rumen fermentation. ISI Atl Sci Anim Plant Sci 1988; 1:178–182.

51. Lynch GP, Bond J. Relationship of lactose to nitrogen matabolism of artificially reased beef calves. J Dairy Sci 1983; 66:2544–2550.

52. Stewart GD, Schingoethe DJ. Evaluation of high starch and high fat rations for dairy calves. J Dairy Sci 1984; 3:598–605.

53. Kertz AF, Reutzel LF, Mahoney JH. *Ad libitum* water intake by neonatal calves and its relationship to calf starter intake, weight gain, feces score and season. J Dairy Sci 1984; 67:2964–2969.

54. Cruywagen CW, Horn JG. Pre-weaning growth and feed intake of dairy calves receiving different combinations of soya bean flavour, whey powder and colostrum. S Afr J Anim Sci 1985; 15:11–14.

55. Bush RS, Nicholson JWG. Effect of two acids and formalin in calves milk on faed consumption and performance. Can J Anim Sci 1987; 67:1129–1131.

56. Klein RD, Kincaid RL, Hodgson AS, Harrison JH, Hillers JK, Cronrath JD. Dietary fiber and early weaning on growth and rumen development of calves. J Dairy Sci 1987; 70:2095–2104.

57. Laksesvela B, Slagsvold P, Krogh A, Ommundsen A, Landsverk T. Indigestion in young calves. II. The influence of ground barley, coarse and fine hay. Acta Veter Scand 1977; 18:416–425.

58. Jagoš P, Dvorák R, Skrivánek M. The influence of early intake of concentrates and roughage feeds on development of digestive processes of calves in high concentration calves house. Vet Med (Praha) 1986; 31:257–264.

59. Bomba A, Lauková A, Reiffová K. The effect of starter diet formula on rumen metabolism in calves. Živ Výr 1993; 38:1003–1114.

60. Flachowsky G, Matthey M, Ochrimenko WI, Schneider M. Profile of isoacids in rumen fluid and influence of added isoacids on *in sacco* dry matter disappearance of untreated and ammonia treated wheat straw. Arch Anim Nutr Berlin 1988; 38:431–439.

61. Risley CR, Kornegay ET, Lindermann HD, Wood CM, Eigel WN. Effects of feeding organic acids on selected intestinal content measurements at varying times postweaning in pigs. J Anim Sci 1992; 70:196–206.

62. Kik MJ, Huisman J, van der Poel AF, Mouwen MJ. Pathologic changes of the small intestine mucosa of piglets after feeding of *Phaseolus vulgaris* beans. Vet Pat 1990; 27:329–334.

63. Lecce JG, Balsbaugh RK, Clare DA, King MW. Rotavirus and hemolytic enterophatogenic *Escherichia coli* in weanling diarrhoea of pigs. J Clin Microbiol 1982; 16:715–723.

64. Deprez P, Van den Hende C, Muylle E, Oyaert W. The influence of the administration of sows milk on the post weaning excretion of the hemolytic, *E. coli* in the pig. Vet Res Com 1986; 10:469–478.

65. Deprez P, Deroose P, van den Hende C, Muylle E, Oyaert W. Liquid verces dry feeding in weaned piglets: The influence on small intestine morphology. J Vet Med 1987; B34:254–259.

66. Partridge GG, Fischer J, Gregory H, Prior SG. Automated wet feeding of weaner pigs versus conventional dry feeding effects on growth rate and food consumption. Anim Prod 1992; 54:484 (abstract).

67. Beers-Schreurs van, HMG, Nabuurs MJA, Vellenga L, Breukink HJ. The effect of weaning and diets on vilous height and crypt depth in the small intestines of piglets. In: Proceedings of the IX-th International conference on Production Diseases in farm animals. Berlin: Germany, 1995; 103.

68. McCracken KJ. Effect of diet compositions on digestive development of early weaned pigs. Proc Nutr Soc 1984; 43:109.

69. Pluske JR. Psychological and nutritional stress in pigs at weaning: Production parameters, the stress response and histology and biochemistry of the small intestine. Ph.D. Thesis, The University of Western Australia, 1993.

70. Pluske JR, Williams IH. Reducing stress in piglets as a means of increasing production after weaning: administration ton of amperozide or co-mingling of piglets during lactation? Anim Sci 1996; 62:121–130.

71. Fonty G, Gouet P, Jouany J, Senaud J. Ecological factors determing establishment of cellulolytic bacteria and protozoa in the rumens of meroxenic lambs. J Gen Microb 1983; 129:213–223.

72. Kmeť V, Bomba A. Control of rumen microflora development. Ann Rep Inst Anim Phys Slov Acad Sci 1987; 7:44–52.

73. Kopečný J, Šimúnek J. Probiotics in calves based on rumen microorganisms. In: Proceeding on new knowledge in using of additives in calves nutrition. Uhriněves: ČSVTS, 1989:82–88.

74. Bomba A, Kravjanský I, Kaštel R, et al. Inhibitory effects of *Lactobacillus casei* upon the adhesion of enterotoxigenic *Escherichia coli* K99 to the intestinal mucosa in gnotobiotic lambs. Small Ruminant Res 1996; 23:199–206.

75. Perdigon G, Alvarez S, Rachid M, Agnero G. Immune system stimulation by probiotics. J Dairy Sci 1995; 78:1597–1606.

76. Bomba A, Gancarčíková S, Kaštel R, Nemcová R, Herich R. Concentration of organic acids in the digestive tract of gnotobiotic piglets after preventive application of lactobacilli. Slov Vet Čas 1998; 23:321–326.

77. Thomke S, Elwinger K. Growth promotants in feeding pigs and poultry. III. Alternatives to antibiotic growth promotants. Ann Zootech 1998; 47:245–271.

78. Mead GC. Prospects for "competitive exclusion" treatment to control salmonellas and other feedborne pathogens in poultry. Vet J 2000; 159:111–123.

79. Kovacs- Zomborszky M, Kreinzinger F, Gombo S, Zomborszky Z. Data on the effects of the probiotic "Lacto-Sacc". Acta Vet Hung 1994; 42:3.

80. Bomba A, Nemcová R, Gancarčíková S, Herich R, Kaštel R. Potentiation of the effectiveness of *Lactobacillus casei* in prevention of *E. coli* induced diarrhea in conventional and gnotobiotic pigs. In: Paul PS, Francis DH, eds. Mechanisms in the Pathogenesis of Enteric Diseases 2. New York: Kluwer Academic/Plenum Publishers, 1999:185–190.

81. Nemcová R, Bomba A, Gancarčíková S, Herich R, Guba P. Study of the effect of *Lactobacillus paracasei* and FOS on the faecal microflora in weanling piglets. Berl Munch Tieraztl Wschr 1999; 112:225–228.

82. Ringø E, Bendiksen HR, Gausen SJ, Sundsfjord A, Olsen RE. The effect of dietary fatty acids on lactic acid bacteria associated with the epithelial mucosa and from faecalia of Artic charr, *Salvelinus alpinus* (L.). J Appl Microb 1998; 85:855–864.

83. Kankaanpää PE, Salminen SJ, Isolauri E, Lee YK. The influence of polyunsaturated fatty acids on probiotic growth and adhesion. FEMS Microb Lett 2001; 194:149–153.

84. Bomba A, Nemcová R, Gancarčíková S, et al. The influence of ω- 3 polyunsaturated fatty acids (ω-3 pufa) on lactobacilli adhesion to the intestinal mucosa and on immunity in gnotobiotic piglets. Berl Munch Tieraztl Wschr 2003; 116:312–316.

85. Soerjadi AS, Rufner R, Snoeyenbos GH, Weinack OM. Adherence of Salmonellae and native gut microflora to the gastroinestinal mucosa of chicks. Avian Dis 1982; 26:576–584.

86. Mether U. Administration of autochthonus intestinal microflora—a method to prevent Salmonella infections in poultry. Dtsch Tierarztl Wochenschr 2000; 107:402–408.

87. Nisbet DJ, Corrier DE, DeLoach JR. Effect of mixed cecal microflora maintained in continuous culture and of dietary lactose on Salmonella typhimurium colonization in broiler chicks. Avian Dis 1993; 37:528–535.

88. Corrier DE, Hinton A, Jr., Kubena LF, Ziprin RL, DeLoach JR. Decreased Salmonella colonization in turkey poults inoculated with anaerobic cecal microflora and provided dietary lactose. Poult Sci 1991; 70:1345–1350.

89. Corrier DE, Hargis B, Hinton A, Jr., et al. Effect of anaerobic cecal microflora and dietary lactose on colonization resistance of layer chicks to invasive Salmonella enteritidis. Avian Dis 1991; 35:337–343.

90. Fonty G, Gouet P, Ratefiarivelo H, Jouany JP. Establishment of Bacteroides succinogenes and measurement of the main digestive parameters in the rumen of gnotoxenic lambs. Can J Microbiol 1988; 34:39–46.

91. Fonty G, Jouany JP, Chavarot M, Bonnemoy F, Gouet P. Development of the rumen digestive functions in lambs placed in a sterile isolator a few days after birth. Reprod Nutr Develop 1991; 31:521–528.

92. Fonty G, Gouet P, Jouany JP, Senaud J. Ecological factors determining the estabilishment of cellulolytic bacteria and protozoa in the rumen of meroxenic lambs. J Gen Microbiol 1983; 129:213.

93. Bomba A, Žitňan R, Koniarová I, et al. Rumen fermentation and metabolic profile in conventional and gnotobiotic lambs. Arch Anim Nutr 1995; 48:231–243.

94. Lysons RJ, Alexander TJL, Wellstead D, Hobson PN, Mann SO, Stewart CS. Defined bacterial populations in the rumens of gnotobiotic lambs. J Gen Microbiol 1976; 94:257–269.

95. Soares JH, Leffel EC, Larsen RK. Neonatal lambs in a gnotobiotic environment. J Anim Sci 1970; 31:733–740.

96. Lysons RJ, Alexander TJL, Wellstead D, Jennings W. Observations on the alimentary tract of gnotobiotic lambs. Res Vet Sci 1976; 20:70–76.

97. Nemcová R, Bomba A, Herich R, Gancarčíková S. Colonization capability of orally administered Lactobacillus strains in the gut of gnotobiotic piglets. Dtsch Tierärztl Wschr 1997; 105:199–200.

98. Bomba A, Gancarčíková S, Nemcová R, et al. The effect of lactic acid bacteria on intestine metabolism and metabolic profile of gnotobiotic pigs. Dtsch Tieraztl Wschr 1998; 105:384–389.

99. Sarra PG, Cantalupo R, Massa S, Trovatelli LD. Colonization of the gastrointestinal tracts of conventional piglets by Lactobacillus strains. J Gen Appl Microb 1991; 37:219–223.

100. Tortuero F, Rioperex J, Fernandez E, Rodriguez ML. Response of piglets to oral administration of lactic acid bacteria. J Food Protect 1995; 58:1369–1374.

101. Bomba A, Kaštel R, Gancarčíková S, Nemcová R, Herich R, Čížek M. The effect of lactobacilli inoculation on organic acid levels in the mucosal film and the small intestine contents in gnotobiotic pigs. Berl Munch Tieraztl Wschr 1996; 109:428–430.

102. Gordon HA, Pesti L. The gnotobiotic animal as a tool in the study of host parasite relationships. Bacteriol Rev 1971; 35:390–429.

103. Yokota H, Coates ME. The uptake of nutrients from the small intestine of gnotobiotic and conventional chicks. Br J Nutr 1982; 47:349–356.

104. Vanbelle M, Teller E, Focant M. Probiotics in animal nutrition: a review. Arch Anim Nutr 1990; 40:543–567.

Index

9 780367 390747